QUANTUM FLUCTUATIONS OF SPACETIME

World Scientific Series in Contemporary Chemical Physics

Editor-in-Chief: M. W. Evans (AIAS, Institute of Physics, Budapest, Hungary)

Associate Editors: S. Jeffers (York University, Toronto)
D. Leporini (University of Pisa, Italy)
J. Moscicki (Jagellonian University, Poland)
L. Pozhar (The Ukrainian Academy of Sciences)
S. Roy (The Indian Statistical Institute)

Vol. 7   Dynamic Kerr Effect: The Use and Limits of the Smoluchowski Equation and Nonlinear Inertial Responses
by J.-L. Dejardin

Vol. 8   Dielectric Relaxation and Dynamics of Polar Molecules
by V. I. Gaiduk

Vol. 9   Water in Biology, Chemistry and Physics: Experimental Overviews and Computational Methodologies
by G. W. Robinson, S. B. Zhu, S. Singh and M. W. Evans

Vol. 10 The Langevin Equation: With Applications in Physics, Chemistry and Electrical Engineering
by W. T. Coffey, Yu P. Kalmykov and J. T. Waldron

Vol. 11 Structure and Properties in Organised Polymeric Materials
eds. E. Chiellini, M. Giordano and D. Leporini

Vol. 12 Proceedings of the Euroconference on Non-Equilibrium Phenomena in Supercooled Fluids, Glasses and Amorphous Materials
eds. M. Giordano, D. Leporini and M. P. Tosi

Vol. 13 Electronic Structure and Chemical Bonding
by J.-R. Lalanne

Vol. 14 The Langevin Equation: With Applications to Stochastic Problems in Physics, Chemistry and Electrical Engineering, 2nd Edition
by W. T. Coffey, Yu. P. Kalmykov and J. T. Waldron

Vol. 15 Phase in Optics
by V. Perinova, A. Luks and J. Perina

Vol. 16 Extended Electromagnetic Theory: Space Charge in Vacuo and the Rest Mass of the Photon
by S. Roy and B. Lehnert

Vol. 17 Optical Spectroscopies of Electronic Absorption
by J.-R. Lalanne, F. Carmona and L. Servant

Vol. 18 Classical and Quantum Electrodynamics and the B(3) Field
by M. W. Evans and L. B. Crowell

Vol. 19 Modified Maxwell Equations in Quantum Electrodynamics
by H. F. Harmuth, T. W. Barrett and B. Meffert

Vol. 20 Towards a Nonlinear Quantum Physics
by J. R. Croca

Vol. 21 Advanced Electromagnetism and Vacuum Physics
by P. Cornille

Vol. 22 Energy and Geometry: An Introduction to Deformed Special Relativity
by F. Cardone and R. Mignani

Vol. 23 Liquid Crystals, Laptops and Life
by M. R. Fisch

Vol. 24 Dynamics of Particles and the Electromagnetic Field
by S. D. Bosanac

World Scientific Series in Contemporary Chemical Physics – Vol. 25

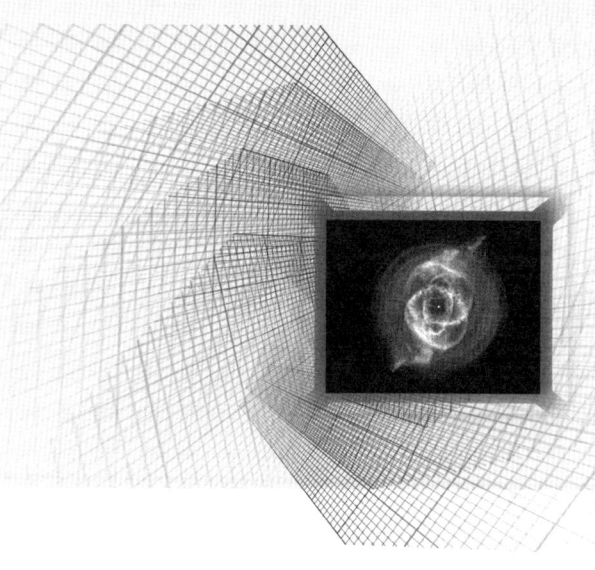

QUANTUM FLUCTUATIONS OF SPACETIME

Lawrence B Crowell

Alpha Institute of Advanced Study, Hungary

NEW JERSEY · LONDON · SINGAPORE · BEIJING · SHANGHAI · HONG KONG · TAIPEI · CHENNAI

Published by

World Scientific Publishing Co. Pte. Ltd.
5 Toh Tuck Link, Singapore 596224
USA office: 27 Warren Street, Suite 401-402, Hackensack, NJ 07601
UK office: 57 Shelton Street, Covent Garden, London WC2H 9HE

British Library Cataloguing-in-Publication Data
A catalogue record for this book is available from the British Library.

First published 2005
Reprinted 2006

QUANTUM FLUCTUATIONS OF SPACETIME

Copyright © 2005 by World Scientific Publishing Co. Pte. Ltd.

All rights reserved. This book, or parts thereof, may not be reproduced in any form or by any means, electronic or mechanical, including photocopying, recording or any information storage and retrieval system now known or to be invented, without written permission from the Publisher.

For photocopying of material in this volume, please pay a copying fee through the Copyright Clearance Center, Inc., 222 Rosewood Drive, Danvers, MA 01923, USA. In this case permission to photocopy is not required from the publisher.

ISBN 981-256-515-9

Printed in Singapore by B & JO Enterprise

To Junko

Preface

It is possible that physics in the early 21st century is near the possibility of experimental tests of quantum gravity. This is an expository on the subject, where phenomenology and experimental prospects play a crucial part. With the exception of the final three chapters of this book, much of the discussion here may become empirically relevant within a couple of decades. By this it is hoped that the topics will become at best "old physics" in the light of subsequent developments within a few decades into the 21st century.

This book is written to become a part of a process that leads to an empirical understanding of quantum gravity, as well as the development of a possible final theory of physics. This final theory may at best be a theory that describes the universe up to the limits of what is understandable. Here there are proposed quantum gravity experiments which could be performed in the next one or two decades. These may help to initiate the study of quantum gravity as a really empirical science.

Despite the incredible difference in scale between the Planck scale and the scale of current experimental physics, it has to be recognized that there are similar precedents in the past. The ancient world had no means to examine the atomic structure, but there were considerable speculations on the nature of atoms stemming from Democritus. The Epicurean cosmos framed the cosmos on three fundamental principles: materialism, mechanism, and atomism. Epicurus thought the universe as an infinitude of space and events were played out according to matter within a void. Epicurus upholds Democritus theory that all matter is composed of indestructible atoms, with eternal properties that can neither be created or destroyed. However, whereas Democritus thought the number of atomic sizes and shapes as infinite, Epicurus argued that their number, while large, is nevertheless finite. Lucretius wrote that atoms could be any size, where he argued that the slow motion of oil through a sieve indicated the largeness of the atoms that constituted oil. Democritus claimed atoms move in straight lines in all directions and always in accordance with the iron laws of "necessity." In the first century B.C.E. Lucretius further argued in his *The Nature of the Universe* that atomic motion required that:

"If one supposes that heavier atoms on a straight course through empty space should outstrip lighter ones and fall on top of them from above, thus causing impacts that might give rise to generative motions, he is going far astray from the path of truth. The reason why objects falling through water or thin air vary in speed according to their weight is simply that the matter composing water or air cannot obstruct all object equally, but is forced to give way more speedily to heavier ones. But empty space can offer no resistance to any object in any quarter at any time, so as not yield free passage as its own nature demands. Therefore,

through undisturbed vacuum all bodies must travel at equal speed though impelled by unequal weights."

Physical events were governed by natural laws, where motion occurs through the effective operation of immutable and eternal physical laws. These concepts are remarkably in line with the reasoning of Galileo and Newton. There are other examples of remarkable insights into the nature of physics. Nicolous Copernicus in his *Revolutions of Heavenly Spheres*, section five wrote:

For every apparent change in place occurs on account of the movement either of the thing seen or of the spectator, or on account of the necessarily unequal movement of both. For no movement is perceptible relatively to things moved equally in the same direction; I mean relatively to the thing seen and the spectator.

This is a remarkable insight that was most completely codified by Einstein in 1905.

This suggests that insights of today can potentially lead to unexpected empirical results. Given the current age of rapid technological development, barring any sort of upheaval similar to the end of the ancient world and the so called "dark ages," the lag time between insight and measurement should be much shorter.

The intention is to connect various theories and ideas about quantum gravity and quantum fluctuations. Many of the current theoretical constructs of this physics are discussed in connections with the main theoretical point advanced here: quantum gravity requires a quantum uncertainty principle more general than obtained in current quantum physics. Quantum gravity is also discussed in the light of possible experimental tests that might be performed in the near future. This connection between various theories and possible experimental tests is meant to initiate additional research and possible experimentation.

As Einstein stated, "Subtle is the Lord, but malicious He is not." In many ways that is a statement of optimism advanced here. It almost seems impossible that the universe could give rise to intelligent observers, but where the ultimate basis of physics are forever unobservable by these observers. This is not in reference to the objectivism matter raised by hidden variable theorists, but with respect to the unification of gravity and quantum mechanics. While the scale of unification appears extreme by our high energy physics capabilities, there are likely to be large scale aspects of quantum gravity that may be tested by techniques accessible by modern technology and experimentation. It is likely the case that we may never have the ability to explore quantum gravity to the fullness that we probe subjects such as photons and atoms, but some elements of empirical tests are certainly possible.

The two main current theories of quantum gravity are superstring theory and loop space theory. While superstrings generally have the intellectual higher ground neither of these are completely satisfactory. In the development of spacetime fluctuations it is shown that there is a gauge-like underlying structure that leads itself of a possible fundamental underpinning to both of these approaches to quantum gravity. This underlying structure is nonassociative quantum field theory, with an error code content. There are also indications that this is a route towards M-theory. The more fundamental approach to quantum gravity operates to a scale of $\sim$ 1.2 times the Planck length $L_p = \sqrt{G\hbar/c^3}$, which contrast to the string length of $L = \sqrt{8\pi}L_p$. This scale is provided by the polyhedral structure in the 8 dimensional space of the octonions. If this is the case then this more fundamental approach may provide a unification of string theory and loop space variables.

Acknowledgments

Some of the contents in this book were worked by Robert Betts, Department of Mathematics, University of Massachusetts Boston, Ma. His contributions are noted within.

Lawrence B. Crowell
July 2005

Contents

Preface	vi

Chapter 1 Why Quantum Fluctuations of Spacetime?   1
     1.1   Quantum Fluctuations and Spacetime   9
     1.2   Detecting Quantum Gravity Fluctuations   14
     1.3   Connections to Strings and Membranes   18
     1.4   Beyond Standard Constructions   19

Chapter 2 Quantum Fluctuations and Spacetime   21
     2.1   Introduction   21
     2.2   Fluctuations in Quantum Mechanics   22
     2.3   ADM Relativity, the Wheeler-DeWitt Equation and Torsional Flucutations   34
     2.4   Thermalized Torsional Fluctuations   41
     2.5   Squeezed States, Wormholes and Nascent Cosmologies   49
     2.6   Spacetime Fluctuations and Wormholes   53
     2.7   The Geometry of Quantum Mechanics   65
         2.7.1   Projective Structure in Spacetime and Quantum Mechanics   66
         2.7.2   Bell's Inequality, Metrics and Expectation   69
         2.7.3   The Hibert Space CP^2 and Lorentz Group $PSO(5,1)$ 73
         2.7.4   Quantum Geometry and Extended Field Theories   75
     2.8   Quantum Computing Backwards in Time and $P \subset NP$ 76
     2.9   Summary on Spacetime Fluctuations   81

Chapter 3 Detection of Quantum Gravity Fluctuations   85
     3.1   Can Quantum Gravity Be Experimentally Tested?   85
     3.2   Atomic Detection of Quantum Spacetime Fluctuations   87
     3.3   Detecting Virtual Black Holes   99
     3.4   Quantum Optical Detection of Quantum Gravity Induced Decoherence   102
     3.5   Ring Gyro Test of the Holographic Principle   106
     3.6   Measurement of Unruh-Hawking Physics with Bose-Einstein Condensates   110
     3.7   Testing Gravitational Squeezing of the Vacuum State   114

3.8	Cosmological Gravity Waves	118
3.9	The Universe as a $\sum(all) = 0$	125

Chapter 4 Quantum Gravity Fluctuations, Strings and Two Dimensional D-Branes   139

4.1	Quantum Gravity Fluctuations in Modern Physics	139
4.2	Finite Temperature Theory of Rotating Strings	141
4.3	Nonlinear Spacetime Transformations, Quantum Theory and Gravity	149
4.4	Overview of Basic Topological Field Theory	154
4.5	Supersymmetry and Cohomology of the Horizon Algebra	160
4.6	Moduli Space Constructions of Horizons	167
4.7	Brane Dynamics and Black Holes	181
4.8	Supersymmetric Moduli Space	182
4.9	Moduli Spaces With Non-Hausdorff Zariski Topology	184
4.10	Modularity and Fractal Aspects of Moduli with Zariski Topology	188
4.11	Julia Sets and NonHausdorff Measures over Moduli Spaces	190

Chapter 5 Topology, Extra Large Dimensions and the Higgs Field   193

5.1	Extra Dimensions and the Horizon $D2$-Brane and $D5$-Brane	193
5.2	Time Symmetry and Violations of Topological Quantum Numbers	200
5.3	TEV Physics From D-Branes and the Higgs Field	206
5.4	Quantum Group Structure With Ricci Flows and Three-Manifold Surgery	218

Chapter 6 Zeta Functions, Topological Quantum Numbers and M-Theory   228

6.1	Wrapped Spaces and Internal Symmetries	232
6.2	Instantons Between String Types, Topology, and Vacuum Ambiguity	233
6.3	String Types, M-Theory and Octonions	237

Chapter 7 The Generalized Uncertainty Principle   243

7.1	The Vacuum Problem	243
7.2	Differential Structure of Fields in Spacetime	245
7.3	Nonlocalizability of Energy and Noncommutative Geometry	251

	7.4	Quantization of Fields in Spacetime and Noncommutative Geometry	254
	7.5	Connections to One-Loop Quantum Gravity	259
	7.6	The Noncommutative Algebra	262
	7.7	General Relativity and Quantum Mechanics as Subsets of a Single Theory	265
	7.8	A Proposed General Mathematics for the B^* Algebra	268
	7.9	Concluding Words	274

Chapter 8 Octonionic Quantum Gravity — 276

8.1	Algebra and Basis Elements	276
8.2	Triality and Right/Left Eigenvectors Over $\mathcal{O}$	281
8.3	Quantum Mechanical Interpretation	282
8.4	Eigenvalues, Clifford Algebras and Quaternions	283
8.5	Holographic Principle from the Quantum Theory of Spin	285
8.6	Associative and Co-Associative SubAlgebra and Spacetime	292
8.7	Polyhedra and Jordan Algebraic Representations of Octonions	295
8.8	Nonassociative Fields and Topological Supersymmetry	299
8.9	Bohr's Gravitization of the Quantum	304
8.10	Quantum Black Holes as Strings and Error Correction Algorithms	310
8.11	Quantum States for Hamming and Golay Error Correction Codes	316
8.12	Phase Structure of Black Holes	322
8.13	Higher Polyhedral Representations	328

Chapter 9 Physical Law from No Law — 333

9.1	Universal Computation In Quantum Cosmology and the Decidability of Cosmic Censorship	334
9.2	PreGeometry and Mathematical Undecidability	339
9.3	Recovering Geometry From PreGeometry	348
9.4	Gödel's Theorem In Quantum Mechanics	355
9.5	The Anthropic Cosmological Principle? — Maybe	359

Index 365

1: Why Quantum Fluctuations of Spacetime?

The primary function of physics is to understand the universe according to a single set of principles that can be stated and formally manipulated in a consistent manner. The history of physics illustrates grand syntheses in theoretical physics from Copernicus to Einstein. On the theoretical front there are cases of individuals who performed grand intellectual feats, such as Newton, Maxwell and Einstein, as well as other instances where the unification of physical reality is more of a collaborative effort, such as with quantum mechanics. It is also important to keep in mind those crucial experimental results which played a vital role as well, from Galileo' s observation of Jovian moons to the Michelson-Morely experiment. The most important experiments are those that counter the conventional thinking of the day.

For quantum gravity to become serious physics, as opposed to abstract mathematics and theory, experimental tests must be performed. These measurements may involve gravity waves or neutrinos detected from the early universe, subtle effects in particle physics or sensitive tests of quantum gravity fluctuations. If the future presents the physics community with no experimental tests of theories, or with the futility in attempting to make such observations, then the topics of quantum gravity and the very early cosmology are not properly topics of physics, but are arenas of mathematics with interpretations that border on theology. Only with experimentation will quantum gravity be developed as an arena of physics in manner similar to the history of the founding of quantum mechanics.

This problem lies with the nature of the Planck length. It is sometimes illustrative to go back and see where this fundamental scale comes from. For a black hole with a mass M its Schwarzshild radius is known to be $r = 2GM/c^2$. We ignore the issue of rotating black holes. An obvious question is what is a scale where a black hole has a deBroglie wave length $\lambda = h/p$, p the momentum of the black hole, equal to its diameter defined by the Schwarschild radius. This is a spacetime momentum that is $p = E/c = Mc$. Further, the mass of the black hole is equated to $M = h/c\lambda$. The analyst who puts this all together then finds that the deBroglie wavelength of this black hole is,

$$\lambda_P = \sqrt{\frac{Gh}{c^3}} = 4.051 \times 10^{-33} cm. \tag{1.1}$$

This is generalized for a two dimensional wave on the event horizon to obtain the Planck length $L_p = \sqrt{G\hbar/c^3} = .1616 \times 10^{-33}$cm. If one converts this length into energy units, since energy is based on $E = hc/\lambda$, the energy associated with a wave quanta on this scale is 10^{19}GeV. As a comparison the Fermi length for a nucleon is 10^{-13}cm, so the energies required to probe these lengths are 20 orders of magnitude larger that what is used in nuclear physics.

From an experimental physics viewpoint this is terrible news. The current accelerators push particles to $10^3 GeV$. This means that based on current technology with accelerators that have diameters of several kilometers, such machines must be scaled up by a factor of 10^{16}. A light year distance is approximately $10^{13} km$, which means that an accelerator would have to have the diameter of some 10^3 light years. Such an accelerator would encompass a significant portion of our Milky Way galaxy. So this is indeed terrible news. One might switch to cosmic ray physics, yet the highest energy recorded so far is some is 10^9 GeV. The author has addressed the question of whether an implosive shock wave from a nuclear explosion might produce Planck scale energy at its core. This appears more realistic by some measures. However, estimates are not encouraging. Under highly idealized conditions it would require a nuclear explosive of some 10^5 megatons, more than the current nuclear explosive capacity currently arrayed in the world. Rayleigh-Taylor instabilities of shock wave fronts bring about serious questions about the feasibility of this. This means such a device would likely have explosive power several magnitudes beyond this to achieve a shock wave front that could implode material into a quantum black hole. Based on the accrued costs of nuclear weapons research $\simeq$ $\$10^{13}$ by both the United States and Russia, it is unlikely that a "Planck bomb" will be built, where such would likely be politically unpopular. It is left to the reader familiar with nuclear explosives to calculate an estimated energy required to produce a quantum black hole whose energy output is greater than the explosive energy necessary to produce it.

So the question must be asked on how quantum gravity can experimentally explored. At the same time, one might ponder how well quantum gravity is theoretically understood. Einstein stated that nature was subtle, but not malicious. It might well be that our ideas about quantization are not appropriate for the subject of quantum gravity. The above argument for the Planck scale is based upon rather elementary concepts of quantum waves and their relationship with the area of a black hole event horizon. The viewpoint rests upon the idea that gravitation is quantized field. It may be best to assume the point of view that gravitation is a field that is unified with quantum mechanics into a larger field theory. If this is so quantum mechanics and gravitation are low energy results from a more general symmetry at very high energy.

Canonical spacetime operators may be derived which satisfy a finite temperature theory. Here the temperature is determined by the gravity associated with a quantum fluctuation. Further these fluctuations exhibit gravitational self-squeezing. A self-parametrically down shifted or squeezed vacuum state may then determine the tunneling probability vacuum modes may tunnel into a nascent cosmology. This is then developed further to illustrate how this fits into a conformal field theory, where this may be derived by coupled fluctuations of a string with a two dimensional membrane on a black hole membrane.

This may demonstrate how the Planck scale is a fundamental cut off in measurable physics. Any wavelength shorter than the Planck scale will be concealed behind the event horizon of a Planck mass black hole. So physics with energy greater than the Planck energy is not measurable or observable. This argument rests upon the assumption the gravitation coupling parameter G, Newton's gravitational constant, is truly constant. It also assumes that the Planck unit of action is truly constant. The Planck and gravitational constants are amongst the smallest of physical constants. They also point to a curious difference between quantum gravity and other quantized fields. Gauge theories fields are quantized with their coupling parameters held constant, where renormalization of that constant is a derived result. With quantum gravity there exist two limits: one where $\hbar \rightarrow 0$ and the other where $G \rightarrow 0$. These two limits recover classical gravitation and quantum mechanics respectively. Quantum gravity may well be a domain of physics where both of these constants may be renormalized to larger values for interaction energies less than the Planck energy scale. If both of them are renormalized to larger values as the interaction energy increases this would mean that the Planck length is in reality larger than what is currently thought.

Andre Sakharov suggested a foundation of gravitation as a metric elasticity [1]. This metric elasticity was based upon a microscopic structure analogous to molecular structure behind material elasticity. Wheeler proposed pregeometry as more fundamental to physics than geometry. In this viewpoint the gravitational coupling parameter is seen to emerge from a grand sum over many pregeometric entities which have oscillation modes. A renormalization of physics view may then be invoked so very high frequency oscillations are removed from this contribution, at least for physics at low energy. This suggests the foundations of gravitation lay in the quantum physics of the vacuum.

It is then reasonable to propose a Lagrangian that describes the zero-point energy of the geometrodynamic vacuum. This indicates a Lagrangian as a power series in spacetime curvature

$$\alpha \hbar \int k^3 dk + \beta \hbar (\int k dk) R^{(4)} + \hbar \int \frac{dk}{k} (\gamma R^{(4)2} + \delta R^{\mu\nu} R_{\mu\nu}) + O(R^{(4)3}). \quad (1.2)$$

Here α, β, γ, δ are coupling coefficients of order unity and k are the modes of the pregeometric entities. The corrections to Einstein's general relativity advanced by Sakharov are a reflection of departures at high energy due to the breakdown of spacetime. The first term is expected to be removed by renormalization arguments. From the second term it is seen that $c^3/16\pi G = -\hbar\beta \int k dk$. This leaves open the question of whether the speed of light or the gravitational constant is determined by this factor. Since the speed of light is an aspect of the metric signature of spacetime $(-1, 1, 1, 1)$ this might be assumed to be the ultimate invariant in nature.

This introduction is written as the question, "why quantum fluctuations of spacetime?" This question is ultimately at the heart of problems that needs to be addressed in the future. From a theoretical perspective the physical vacuum of the standard model is $\sim 246 \text{GeV}$, while the cosmological constant suggests that it should be no more than about $10^{-44} GeV^4$. This is an incredible observational conflict of some 120 orders

4 *Quantum Fluctuations of Spacetime*

of magnitude. The standard model of electroweak interactions has a good track record of experimental tests and astronomy has produced decent constraints on the cosmological constant. So these two observational results lead to a theoretical contradiction. The second reason for the question is that quantum fluctuations may well be experimentally testable with sufficiently sensitive measurements. It is possible to detect metric fluctuation to verify their existence, where later the nature of quantum gravity fluctuations may be probed to obtain data that may help resolve this issue by supporting or refuting theories of quantum gravity.

The approach of this monograph is to consider the nature of quantum gravity, more specifically quantum fluctuations in spacetime, and in a context that holds some prospect for experimental tests or astronomical observations. This is also written with the expectation that our views on the foundations of the universe are likely to change radically over the first two decades of the 21st century. This may come from the LIGO observation of the gravity wave universe, as well as possible experimental detection of quantum gravity fluctuations. This will further mean that many of our ideas about quantum gravity, strings and unification will be radically changed. As such this monograph is directed significantly towards the theory and phenomenology of spacetime fluctuations and possible experimental measurements. This is done in order to provide a practical direction for research in the future, research that holds the prospect of undoing much of what is canonized as theoretically tentative.

John Wheeler stated there are three levels of gravitational collapse: black holes, cosmological big bang and quantum gravity fluctuations. In each of these cases there exist event horizons. Event horizons are themselves not detectable, nor is it a surface that matter impacts, but the physics near them is. In the case of black holes the gravitational field removed from the event horizon is such that its influence on incoming matter is detected. black holes has been largely detected, where they further appear to be rather astrophysical aspect of the universe. black hole event horizons also have the curious property of freezing matter just above them. Another event horizon is very easy to "see," but still difficult to recognize. This is the night sky, which outside the light of stars, exhibits a blackness of a horizon. This is a result of the cosmological horizon due to the big bang. The red shifting of light emitted by galaxies and quasars and radiation released during the transition to the current matter dominated period, as seen with the microwave background, accumulate near this horizon. The quantum gravity fields that decoupled from other fields around $t > 10^{-40}$sec into the big bang should exist as gravity waves. These have not yet been detected.

Event horizons are two dimensional achronal surfaces. The achronal nature of event horizons means that observable fields are time dilated and length contracted as they accumulate near them. Recognition of this has lead to the holographic principle. The curvature of spacetime in the environment of an event horizon and this accumulation of fields near them leads to the striking prospect that all fields may be described according to surfaces of two dimensions that accumulate near event horizons. This means that the data for fields and gravity need only to be specified in two dimensions rather than within three. From a computational viewpoint, in particular if a problem is being run on a computer, this is a great saving in the number of calculations required. Further, it suggests a new paradigm in spacetime physics where the most fundamental object of dynamics is the surface within a Planck unit of distance from an event horizon. The field theoretic content of this membrane is purely gravitational, as all other fields with wave lengths larger than the Planck length have been red shifted to near infinity. As such this surface in two dimensions is regarded as a quantum membrane. It is a D-brane under a spacetime target map and interacts with strings according to $U(n)$ Chan-Paton factors.

This monograph is in part concerned with D-branes of two dimensions and other D-branes of other dimensions correlated with D2-branes by duality principles. Higher dimensional D-branes will be discussed, but not as extensively developed. The initial thrust here will be quantum gravity in a weak coupling limit, where the stronger coupling limit is considered later. If aspects of quantum gravity are ever to be experimentally explored it will certainly be this domain that is tested first. In the case of the universe future observations of gravity waves produced in the big bang will reveal structure very close to the initial event. The gravity waves will be due to the tunneling of the universe out of the vacuum or from the decoupling of gauge fields from quantum gravity. If Light Interferometric Gravitational Observatories (LIGO) are able to probe this aspect of the universe then measurements of quantum gravity, at least in the weak limit, are possible. In this case future astronomers may make observations of the affect of a D-brane in two dimensions.

Before engaging in a discussion of spacetime quantum fluctuations it is best to consider quantum fluctuations in nonrelativistic quantum mechanics. This will lay the foundation for discussions of spacetime fluctuations. The approach that is taken is to use the Bohm approach to quantum mechanics to derive stochastic fluctuations. This is done not to promote any "interpretation" of quantum mechanics or to advocate for hidden variables, but since this formalism is a convenient approach to initiate a discussion of quantum fluctuations. In this presentation it will also be indicated that this viewpoint on quantum mechanics is fundamentally no different from the standard approach. The derivation of a path integral is given as a demonstration that this "particle-pilot wave" approach to quantum mechanics is not fundamentally different from the more standard Hilbert space and operator approach.

A fluctuation of the spatial three dimensional manifold $\Sigma^{(3)}$ will accelerate a test mass. If the particle horizon is a length L^2/L_p from the test mass this acceleration will be on the order of $c^2 L_p/L^2$, which can become quite large. It is then apparent that this mass under this acceleration will have a particle horizon within $c^2/a \simeq L^2/L_p$ of the particle. This accelerated particle will couple to this fluctuation, with a corresponding event horizon, which has a finite temperature. From a physical viewpoint it makes sense that a particle under any acceleration, constant or not, will interact with a thermal vacuum. The vacuum such a particle interacts with will have a finite temperature. Whether the zero point energy for quantum gravity is finite temperature up to the Planck length is a question to be addressed later. Yet apparently the quantum gravity vacuum for scales that approach the Planck length will be finite temperature. This temperature will then be associated with the event horizon.

The Unruh effect demonstrates now quantum theory does not have a one to one correlation with a state in Hilbert space and a particle. The existence of a particle is not completely correlated with a Hamiltonian of the form $H = \hbar\omega a^\dagger a$. A vacuum according to an accelerated observer contains particles in a thermal distribution. Such a vacuum will have a Hamiltonian with terms $a^\dagger a^\dagger$ and aa, where this additional term to the Hamiltonian is essentially a squeezed state operator.

These properties are known with quantum fields in curved spacetime. An open question is whether these properties of the quantum gravity vacuum apply to gravitation itself. Intuitively it makes sense that it should, at least in the weak limit. The one major difference is that the vacuum in this case interacts with itself. This presents some problems, but some preliminary results are developed here. This self-interaction results in a generalized uncertainty principle for quantum gravity. A most interesting find from this generalized uncertainty principle is that the quantum gravity vacuum acts as its own parametric amplifier. The vacuum is then parametrically down shifted and that the Planck length is shifted to a larger value. The Planck energy is down shifted to energies approaching $\sim 10^{16}$GeV. This puts quantum gravity at the same energy as GUT unification energy. This also raises the experimental prospect that particles accelerated by parametrically amplified coherent states of photons may interact by entanglement with the quantum gravity vacuum squeezed to high values. The experimental prospect then exists to detect departures from standard quantum uncertainty principle by effectively renormalizing $L_p = 1.6 \times 10^{-33}$cm to larger values. Further, this self-squeezing of the quantum gravity vacuum is also a mechanism that may permit virtual cosmologies to tunnel out of the vacuum state.

For spacetime quantum fluctuations associated with event horizons the relevant physical fields involved with this fluctuation are those tied to the event horizon. These fields then determine a virtual two dimensional membrane. These membranes are then identified with virtual black holes, Cauchy horizons and Lanzcos junctions with wormholes. In the case of a virtual cosmology they are identified with cosmological horizons. These are the three levels of collapse indicated by Wheeler. As these two dimensional D-branes are due to the holographic principle and they capture the three levels of collapse, it is then advanced that they codify the dynamics of quantum gravity in the weak limit.

The quantum gravity fluctuation is developed with teleparallel connections and Finsler geometry. Finsler geometry, with its Lagrangian-like structure, is a natural way to develop this theory of fluctuations. Teleparallel connections define paths on a manifold that are not geodesic in nature, such as the lines of latitude on a globe. Physically these describe accelerated motion, which provides a framework for the description of quantum gravity fluctuations. From here the physical basis for quantum gravity fluctuations is developed, where various implications for the physical world are discussed. In particular possible implications for GUT theories and quark physics are discussed.

If quantum fluctuations of spacetime exist they should be measurable. The dimension of an atom is such that quantum gravity fluctuations should be measurable with sensitive enough technology. For a region with length scale L the curvature induced by quantum fluctuations is [2]

$$\Delta R \sim \frac{L_p}{L^3}, \qquad (1.3)$$

where $L_p = \sqrt{G\hbar/c^3} \sim 1.6 \times 10^{-33} cm$. For the region contained in an atom, $L \simeq 10^{-8} cm$, these curvature fluctuations will be on the order of $\Delta R \simeq 10^{-9} cm^{-2}$. This appears small, but when compared to the curvature near the Earth $G\rho/c^2 \simeq 4 \times 10^{-28} cm^{-2}$ it is in fact quite large. An atomic physics experiment should detect spacetime fluctuations. The comparison of the gravity force due to metric fluctuations with electrostatic forces in an atom is quite small with a ratio of $F_{fluc}/F_e \simeq 10^{-21}$. So the effect is subtle, but potentially within the bounds of a possible measurement.

Of all the fields that exist in the universe the best understood, which has spawned much technology as well, is the electromagnetic field. Further, the interaction of photons and atoms is one of the most highly developed arenas of quantum physics. It is then reasonable to consider this as a potential probe of quantum gravity fluctuations. The physics required to detect quantum spacetime fluctuations is an atom-photon interaction laser. A theoretical basis for this is developed in the theory of coherent states. The matters of stochastic processes from quantum and thermal noise are addressed. It is found that the technology is quite possible and that this experiment could be performed.

With very high powered lasers an electron may be accelerated up to $a = 10^{22} cm/sec^2$ or beyond. This acceleration may then lead to a parametric down shifting of the quantum state of the electron. This may be further "primed" into action by the employment of squeezed laser states, which enter into an entanglement with the electron. It is possible a greater uncertainty in state measurement may be detected. If this is the case with higher accelerations this process may be extended to measure these departures further. In effect the Planck length and equivalently the Planck constant are renormalized so that $\hbar \rightarrow \hbar^*$ with $\hbar^* > \hbar$. The pursuit of such experiments and long term improvements in high powered lasers hold the prospect that the Planck length could be renormalized to $10^{-13} cm$. This would in the long run result in the ultimate source of energy where 100% of matter can be converted to energy. In effect a virtual black hole could be renormalized to the scale of a nucleus and nuclei could be converted into positrons and photons. If this energy technology does emerge in the future it will provide both abundant energy, as well as a way of completely disposing of nuclear wastes from the nuclear fission age of the 21st century.

The nature of quantum fluctuations and virtual black holes is examined in the light of the horizon, or a surface one Planck length from the horizon, as a D2-brane under a target map. This construction is tied to the horizon algebra for the black hole and its supersymmetric extension. From here the topology of the virtual black holes is examined. The moduli space for the horizon algebra is found where the singularities in the moduli space correspond to the horizon. In a pseudoEuclidean metric the nature of the cones in projective space is addressed where a Virasoro sector is derived with weighted projective spaces. Here the weighting of projective lines $z_i \to z_i^n$ results in a periodic structure from which a Virasoro algebra emerges. These singularities suggest structure beyond the string scale, where an internal structure to the string emerges from the blow up of a point. A gauge theory with a quaternionic Polyakov action is seen to emerge from this construction. This leads to the prospect of a theory beyond string theory at high energy and at the Hagedorn temperature.

This squeezed state renormalization suggests high energy physics at the TeV scale might be impacted by the nature of quantum fluctuations and virtual black holes. The compactification of space on the horizon leads to the existence of gauge connections threaded within wrapped dimensions pinned to the horizon. This leads to the possibility the Higgs field may be related to quantum black holes. The Higgs field may have a quantum entanglement with a quantum black hole, where there is a small probability that the Higgs field may act as a black hole that violates baryon numbers. Under the renormalization of the Planck scale at high energy the possibility exists that black holes could be produced in the laboratory. The major departure with such similar schemes is that ordinarily the black hole exists on a much higher energy scale than the Higgs, but under circumstances where the experiment can be conducted with the parametric amplification of states the Planck scale renormalization can bring the black hole down the TeV scale. This brings forth large extra dimensions at experimentally testable scales.

This then is the start of a more formal investigation into the nature of the quantum gravity vacuum. A new uncertainty principle, partly indicated prior to this point, is found. This vacuum is described by a multilinear bracket structure outlined by Peano. This vacuum is described by an orthomodular lattice that puts relativity and quantum theory on the same footing. Gravitation and quantum mechanics are then aspects of the same theory. With this uncertainty principle it is argued that the divergences in quantum gravity and the strong coupling limit in string theory may be absorbed into the uncertainty. Such divergences are the energy cut-off in emergent one-loop quantum gravity, where in the vacuum these divergences may be absorbed.

The octonions form the most accessible multilinear bracket structure as a nonassociative algebra. This extension of field theory as a model for quantum gravity is argued for on physical grounds. The mathematical structure of this theory also indicates that nature may be fundamentally unitary. This will preserve quantum information in black holes. The octonions have connections to E_8 lattice codes where quantum information is ultimately processed by the quantum gravity "computer" in a way the preserves quantum information. This further has connections to string theory and the adS/CFT correspondence. Octonionic field theory when topologically gauged also naturally leads to a supersymmetric form of gauge fields. The fundamental discrete structure for this theory are the Mathieu groups, the first 5 elementary sproadic groups. It is possible that the other sporadic groups, the Janko group and the Conway group for instance might provide the permutation structure for higher multilinear structures of Peano that lay beyond the octonions. This may extend up to the monster group.

Beyond this a more philosophical question is addressed whether the ultimate fundamental laws of physics are knowable. Quantum theories involve the propagation of a field within a space. Even string theories invoke a background structure. Yet this approach approach to physics may be terribly limited. In the case of the quantum cosmology in its first few Planck units of time one clearly has quantum gravity modes propagating on themselves. There is the possibility that quantum gravity may exhibit axiomatic incompleteness due to Gödel' s theorem. In effect the problem of fields propagating on themselves is similar to a quantum computer that uses its own programming structure as its data input. If this is the case structures that approach the Planck scale may indeed explode on us and these structures may emerge for no particular reason at all. As such we may never come up with a final theory of quantum gravity. In fact maybe the "final theory" of quantum gravity is that there is no complete axiomatic based final theory.

In the past 25 year considerable progress has been made theoretically towards an understanding of quantum gravity. From Hawking black hole radiation, strings, loop space variables and now with membranes a formalistic framework has been developed whereby quantum gravity can be discussed. Of course as yet nothing empirical has been found, so much of this machinery is today just formalistic possibilities. Yet physics may be near where theoretical work could begin to gel into a more consistent whole, where initial experimental and observational tests are possible. It this does occur within the next 25 years, then quantum gravity may emerge as a realistic physical theory that may begin to rightfully call itself science. It is the hope that the work presented here will at least in some small way contribute to this long term effort.

1.1 Quantum Fluctuations and Spacetime

Now consider the elementary notion of quantum fluctuations. Elementary courses in modern physics illustrate how the uncertainty principle of Heisenberg $\Delta E \Delta t \simeq \hbar$ indicates a small fluctuation in the energy of a particle associated with a probability that particle may pass over a potential energy barrier. This quantum effect is well documented in physics such as radioactive decay. This suggests a quantum particle in motion exhibits stochastic deviations from its classical motion. This is seen in the Ehrenfest theorem

$$\frac{d}{dt}\langle\,|\mathbf{r}|\,\rangle \;=\; \frac{1}{m}\langle\,|\mathbf{p}|\,\rangle, \tag{1.4}$$

where both sides of the equation have a sum over all states. The velocity of any particle is an expectation from a sum over velocities, where most deviate from a classical result. The inference is that if one makes a momentum measurement of the particle there is a probability that the particle is under a momentum fluctuation. A measurement has a probability of finding the momentum of the particle different from the classical result.

Quantum mechanics indicates quantum wave functions are nonlocal, where different states or probable outcomes are entangled in a nonlocal fashion. The Einstein-Podolsky-Rosen (EPR) and Bell's theorem illustrate this is an inescapable aspect of quantum mechanics. Early in the 20th century this was troubling to many, including Einstein. David Bohm presented the Schrödinger equation in a real part that is a modified Hamilton-Jacobi equation and an imaginary part that is a continuity equation. The modified Hamilton-Jacobi equation contains a "quantum potential" interpreted as due to a pilot wave. However, this formalism did not eliminate the strangeness of nonlocality, for under an experiment the quantum potential must nonlocally readjust itself to fit the outcome of any experiment.

There are some of situations where Bohm's approach to quantum mechanics may be useful. One of those is with quantum fluctuations. With a wave function in polar form $\psi = \rho e^{iS/\hbar}$ the application of the momentum operator $\hat{p} = -i\hbar \nabla$ gives the result

$$\hat{p}\psi = (\nabla S - i\hbar \nabla \rho)\psi, \qquad (1.5)$$

which is a classical momentum plus an imaginary term interpreted as a fluctuation in the momentum. This imaginary term will be illustrated how this term is interpreted as a fluctuation through the continuity equation. The expectation of the momentum is

$$\langle\, |\hat{p}|\, \rangle = \langle\, |\nabla S|\, \rangle + \langle\, |\delta p|\, \rangle, \qquad (1.6)$$

where the expectation of the momentum is $\langle\, |\hat{p}|\, \rangle = \langle\, |\nabla S|\, \rangle$ and so $\langle\, |\delta p|\, \rangle = 0$. The expectation value of the square of fluctuations is nonzero and is delta function correlated

$$\langle\, |\delta p(t)\delta(t')|\, \rangle = (\Delta p(t))^2 \delta(t - t'). \qquad (1.7)$$

The term $i\hbar \nabla \rho$ is demonstrated to obey the same Markovian result. This development is examined further, where fluctuations are independent of the expectations $\langle\, |\nabla S|\, \rangle$ and canonical transformations on the expectations and fluctuations are also independent. This leads to the fact that a canonical transformation of the fluctuation momentum can change the quantum potential in the modified Hamilton-Jacobi equation. As such the modified Hamilton-Jacobi equation only describes one path in the entire "sum over histories" in a path integral. Hence a path integral emerges from Bohm's quantum theory.

The purpose of this exercise is to introduce the notion of the quantum fluctuations and to illustrate that this Bohmian starting point does not infer any "quirky" notions of hidden variables or some underlying "subquantum" causality to quantum fluctuations.

The exercise simply uses this approach to quantum mechanics as a tool to develop quantum spacetime fluctuations, not to promote any sort of hidden variable agenda.

The Unruh effect is the thermalization of quantum vacua as measured on an accelerated frame. This is due to the inability to specify fields on the other side of the split horizon $\mathcal{H}_1 \oplus \mathcal{H}_2$. The instantaneous planes of simultaneity, Σ_n, the accelerated observer passes through intersect one another on the other side of $\mathcal{H}_1 \oplus \mathcal{H}_2$. As such fields the observer measures are thermally distributed to prevent future information from being observed in the past. Given time increments s, not all fields on Σ_{-3s} are causally tied to fields on Σ_{3s} as detected by the accelerated observer. Figure 1.1 illustrates the intersection of instantaneous planes of simultaneity on the plane of simultaneity of an inertial observer Σ_0, where the accelerated frame comes to an instantaneous stop. This means there are nonlocal entanglements between states on different planes of simultaneity which are hidden by the particle horizon. These hidden entanglements result in a decoherence of the particle states.

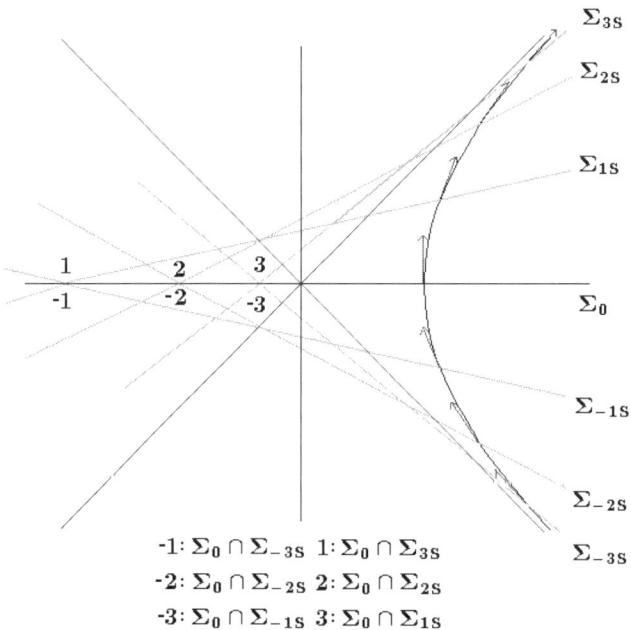

-1: $\Sigma_0 \cap \Sigma_{-3s}$ 1: $\Sigma_0 \cap \Sigma_{3s}$
-2: $\Sigma_0 \cap \Sigma_{-2s}$ 2: $\Sigma_0 \cap \Sigma_{2s}$
-3: $\Sigma_0 \cap \Sigma_{-1s}$ 3: $\Sigma_0 \cap \Sigma_{1s}$

Geometry of the accelerated reference frame.
Figure 1.1

As indicated above quantum gravity fluctuations are likely to be associated with a horizon analogous to the Unruh effect. It is argued that this is in general the case. The first situation examined is a rotating open string, where the quantum vacuum is examined.

12 *Quantum Fluctuations of Spacetime*

The thermal vacuum for a rotating string is equivalent to a string that interacts with a black holes with a Hawking temperature. This situation is argued to obtain for general accelerations. The time period for this fluctuation T and the frequency of modes ν must satisfy $T \sim .1/\nu$. The situation where $T \sim 1/\nu$ is likely to occur near the strong coupling domain of physics, where this approach to quantum fluctuations of spacetime is likely to break down. This domain is discussed in chapters 7 through 9.

Gravitation has an uncertainty in the energy-momentum content of a 3-d volume of space. A spatial surface evolves according to the diffeomorphisms of spacetime physics. This makes it difficult to assign a region that contains a certain mass-energy content. Then one has the uncertainty principle of quantum mechanics, which gives a commutator structure for the uncertainty between conjugate observables. The "fusion" of these uncertainties illustrates how the vacuum dynamics has an uncertainty in $a^\dagger a$ or particle number and in topological numbers. This leads to a generalized uncertainty in the commutation of variables [3]. Under sufficiently large gravity fields the uncertainty in two conjugate variables will both increase. The raising and lowering operators on either side of an acceleration induced horizon satisfy

$$\alpha_i^b = \hat{a}_j^b + a_i^b e^{-\pi \mathcal{H}/g}, \quad {\alpha^b_i}^* = \hat{a}_j^{b\dagger} + a_i^{b\dagger} e^{-\pi \mathcal{H}/g}, \qquad (1.8)$$

where $\hat{a}_j^b$ and $\hat{a}_j^{b\dagger}$ are operators for fields on the observer's side of the horizon with $[\hat{a}_j^b, a_j^{b\dagger}] = [\hat{a}_j^{b\dagger}, a_j^b] = 0$. This generalized uncertainty principle is given by the commutation between raising and lower operators according to

$$[\alpha_j^b, \alpha_k^{c*}] = \left(1 + e^{-2\pi\mathcal{H}/g}\right)\delta_{jk}\eta^{bc} - \frac{\pi}{g}e^{\pi\mathcal{H}/g}\left(\alpha_j^{0b*}\alpha_k^{0c} + \alpha_j^{0b}\alpha_k^{0c*}\right)e^{-\pi\mathcal{H}/g}, \qquad (1.9)$$

where $H = \hbar\omega\sum_j a_j^\dagger a_j$.

The gravitationally induced squeezing of the vacuum state may permit a virtual cosmology to tunnel out of the vacuum state. The uncertainty in the particle number, which results from the vacuum not being strictly a dynamical eigenstate of a diagonal Hamiltonian, permits the creation of a cosmology out of "nothing". Freeman Dyson suggested, in a critique of Feynman's quantum electrodynamic theory, the following artificial "toy theory". Suppose the electric charge transform as $e \rightarrow ie, i = \sqrt{-1}$. Virtual quanta by $e^2 \rightarrow -e^2$ will not recombine, but rather the virtual charges repel. Consequently the vacuum would be terribly unstable and spew electron-positron pairs or radiation. This is a quantum electrodynamic analogue of a process where the vacuum is unstable so that fields and particles may be generated.

Figure 1.2 illustrates two wormhole openings. At $r = r_0$ is the Lanzcos junction defined by the Reissnor-Norstrom metric. The horizon is identified from the exterior region $r > r_0$ and the interior region $r < r_0$. This interior region is topologically identical to sewing two three dimensional balls together, through the identification of points on their respective two dimensional spherical boundaries, which defines a 3-ball. Across a Lanzcos junction on the 2-spherical boundaries is a jump in G^{00}. This junction is similar to a horizon, or a shock wave front, that changes $t \to -t$ and $r \to -r$. This is analogous to the situation Dyson illustrated. In the old "ict" way of doing relativity this also is equivalent to an effective $m \to im$, for $m =$ mass, where if this virtual cosmology is flat enough this run away situation similar to Dyson's will occur. The fluctuation, which is the wormhole itself, leads to a flattening of this virtual cosmology by gravitational self-squeezing of the vacuum state. This situation is unstable and the uncertainty in particle quantum number, eg. the Witten topological index or Atiyah-Singer index, leads to the generation of a spacetime cosmology literally out of nothing. Further, there is no "prime mover" or extra causal agent: the process is completely due to stochastic fluctuations and is completely spontaneous.

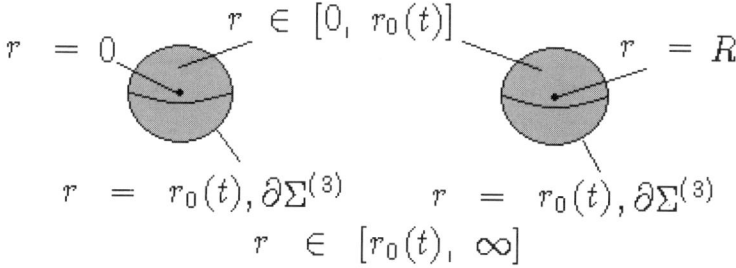

Two worm hole openings with a virtual
spacetime cosmology in the interior region.
Figure 1.2

Thomas Aquinas wrote in his Summa contra Gentiles III, "Materia artificialium est a natura, naturalium vero per creationem a Deo". Yet it appears as if nothingness or the quantum gravitational vacuum permits the creation of something from nothing "Creati Ex Nihilio". This will doubtless have a profound impact on physics and cosmology, as well as on the cultural and social context of the world at large. This has the quality of Creati Ex Nihilio, where the nothingness is the quantum gravity vacuum. This vacuum is defined by virtual excitons of the gravity field, black holes, wormholes, cosmologies and gravity wave fluctuations. The quantum uncertainty principle with the inability to locate mass-energy means that fluctuations of energy can tunnel out of the vacuum and spontaneously give rise to cosmologies.

This implies the existence of other cosmologies, where quantum spacetime may give rise to other nascent cosmologies. However, this junction at $r = r_0$ is a causal boundary and we are prevented from ever looking into the new cosmology. The vacuum may be unstable in a way analogous to Dyson's illustration with QED, but the creation of other cosmologies is forever hidden from our view. These other cosmologies are sealed off from us by horizons or achronal boundaries. As such information about them is inaccessible. This is different from standard quantum mechanics where one can prepare identical systems and obtain an empirical table of amplitudes. All that is possible is to study the nature of quantum gravity fluctuations and infer certain signatures that a nascent cosmology was generated, but observers are forbidden from ever determining anything about these cosmologies or to tap their mass-energy.

This is then concluded with a discussion on a potential H-theorem for quantum gravity and spacetime fluctuations with a finite temperature. A potential approach to this is explored according to knot theory and connections to loop space variables.

1.2: Detecting Quantum Gravity Fluctuations

An electron wave function adjusted by a parameter associated with a loop in space will experience a phase shift. The curvature two-form $\mathcal{R}^\mu{}_\nu = \mathcal{R}^\mu{}_{\nu\rho\sigma} dx^\rho \wedge dx^\sigma$ from quantum gravity fluctuations will produce a phase proportional to

$$\int_\mathcal{A} \mathcal{R}^\mu{}_\nu \mathbf{e}_\mu \otimes dx^\nu, \qquad (1.10)$$

where $\mathcal{A}$ is the area in space determined by the cyclic evolution of a parameter.

Consider atoms periodically injected through a cavity at a frequency equal to the Rabi frequency ν_R of the atom interacting with photons. Each atom in a cavity during a time $\frac{1}{2}\nu_R$ it will contribute a photon by stimulated emission. The entangled atom and photon states will exhibit a phase shift due to the zero point energy of the quantum gravity vacuum.

Consider a post-Newtonian form of gravity for brevity. The post-Newtonian gravito-electric and gravito-magnetic fields are similar to Maxwell fields and are written as a linear operator expansion without the difficulties of gravitational nonlinearity. The metric in post-Newtonian gravity is expanded around a flat spacetime:

$$g_{\mu\nu} = \eta_{\mu\nu} + g^{(2)}_{\mu\nu} + g^{(4)}_{\mu\nu} + \dots \quad \{\mu = \nu = 0, \text{ or } \mu, \nu = i, j\},$$

$$g_{i0} = g^{(3)}_{i0} + g^{(5)}_{i0} + \dots . \qquad (1.11)$$

$\eta_{\mu\nu}$ is the flat spacetime metric. Since the spacetime is source free $g^{(2)}_{\mu\nu} = -(2GM/r)\delta_{\mu\nu} = 0$. The next highest terms within the harmonic condition are $g^{(3)}_{0i} = \mathcal{A}_i$, where $\mathcal{A}$ is a vector potential and $g^{(4)}_{00} = -2\phi$ a scalar field. The connections to lowest order are [4],

$$\Gamma^{(4)i}{}_{00} = \frac{\partial \phi}{\partial x^i} + \frac{\partial \mathcal{A}_i}{\partial t}, \quad \Gamma^{(3)i}{}_{0j} = \frac{1}{2}\left(\frac{\partial \mathcal{A}_i}{\partial x^j} - \frac{\partial \mathcal{A}_j}{\partial x^i}\right). \qquad (1.12)$$

Now define $\mathcal{E}_i = -\Gamma^{(4)i}{}_{00}$ and $\mathcal{B}_i = \epsilon_{ijk}\Gamma^{(3)j}{}_{0k}$ so these fields are:

$$\mathcal{E}_i = -\frac{\partial \mathcal{A}_i}{\partial t} - \nabla_i \phi, \quad \mathcal{B}_i = \frac{1}{2}(\nabla \times \mathcal{A})_i. \tag{1.13}$$

In the harmonic condition $\nabla \cdot \mathcal{A} = 0$ it is easy to find the Maxwell type of equations

$$\nabla \times \mathcal{E} - \frac{2}{c}\frac{\partial \mathcal{B}}{\partial t} = 0 \tag{1.14a}$$

$$\nabla \times \mathcal{B} + \frac{1}{2c}\frac{\partial \mathcal{E}}{\partial t} = 0. \tag{1.14b}$$

Now expand the gravito-electromagnetic field into normal modes with the operators b and $b^\dagger$ analogous to QED. The linearized gravity vector potential expanded as spatial eigenmodes quantized in a box is,

$$\mathcal{A}_i = i\epsilon_i \sum_k \sqrt{\frac{\hbar}{2\omega\epsilon_g V}}\left(b^\dagger e^{-ikx} - b e^{ikx}\right). \tag{1.15}$$

Here ϵ_g is a vacuum permittivity for gravity, analogous to the permittivity in electromagnetism. The electric and magnetic analogues of the gravity field are quantum operators for linearized gravitons.

Consider a "toy model" of this gravito-electromagnetic field coupling to an atom. Let two atomic eigenstates be given by σ_z and the transition operator between these states be the operators $\sigma_\pm = \frac{1}{2}(\sigma_x \pm i\sigma_y)$. The coupling between gravitons and the atom is a quadrupole interaction due to the conservation of momentum and the spin 2 property of the graviton. An approximate quadrupole interaction is $H_2 \simeq g' \int yz\mathcal{E}d^3r$, where $g' = g/V$. The quadrupole term written with $\sigma_\pm$, σ_z gives a Hamiltonian in a rotating wave analogous approximation as

$$H = \frac{\hbar\omega}{2}\sigma_z - \hbar g(b\sigma_+ - b^\dagger\sigma_-)\sigma_z, \tag{1.16}$$

where $g \sim (L_P/L^2)c \simeq 10^{-7} Hz$. The gravity fluctuations are assumed constant through the volume. With only virtual gravitons $b^\dagger b|\rangle = 0$ and $b^\dagger b$ is dropped from the Hamiltonian. The occurrence of a linear graviton coupled to this system is associated with a $\Delta j = 2$ in the atom, so there is an implicit spin change in the electron. If an electromagnetic field is quantized in the cavity with operators a and $a^\dagger$. The photons are tuned to interact with the atomic state, so the total Hamiltonian is:

$$H = \frac{\hbar\omega}{2}(\sigma_z + 2a^\dagger a) + \hbar\kappa(a\sigma_+ + a^\dagger\sigma_-) - \hbar g(b\sigma_+ - b^\dagger\sigma_-)\sigma_z = H_0 + H_1 + H_2. \tag{1.17}$$

Physically the atom executes a Rabi oscillation from its interaction with the electromagnetic field.

Within the interaction picture the evolution of H_1 gives

$$\frac{\partial \sigma_\pm}{\partial t} = \mp i\omega \sigma_\pm, \quad \sigma_\pm = \sigma_\pm(0) e^{\mp i\omega t}. \tag{1.18}$$

The parameters which define the quadrupole moment of the linear gravitational interaction then evolve. This then gives the term:

$$\frac{\nabla_{\sigma_+} H_2 \times \nabla_{\sigma_-} H_2}{(\Delta E)^2} = -(\hbar g)^2 \frac{[b, b^\dagger]\sigma_z^2}{2(\Delta E)^2} = -(\hbar g)^2 \frac{1_{2\times 2}}{2(\Delta E)^2}. \tag{1.19}$$

The expectation of his term evaluated on the area defined by the evolution of the polarization vectors defines a Berry phase. This phase is given by,

$$\phi = -(\hbar g)^2 \int\int_{area} \frac{d^2 x}{2(\Delta E)^2}. \tag{1.20}$$

Evidently the electron needs to have a high radial quantum number to get an enhanced quadrupole moment from an elliptical electron orbit and to minimize the ΔE between the atomic states. This will induce a nondynamical phase on the atomic wave function that is entirely due to modes of the zero point energy of the quantum gravity field that couple to the atom.

For a dipole coupling constant $\sim .1\omega$ and transitions between high Rydberg states $n > 300$ the energy gap may be as small as $\Delta E \simeq 10^{-8} ev$, a quick estimate illustrates that this Berry phase is $\phi \simeq 10^{-29}$ per Rabi oscillation. The total phase shift per second is $\simeq 10^{-18}$, where if the experiment is conducted for a year in a one atom laser the phase would be $\simeq 10^{-11} - 10^{-10}$.

The quadrupole moment interaction with quantum gravity fluctuations is required for their observation. This suggests that the orbit of the electron should be a Bohr-Sommerfeld type of orbit with a large eccentricity. This is best achieved if the atom is a high Rydberg atom with an electron wave packet in an elliptical orbit. There is considerable work yet to be done with real atomic wave functions that will give precise theoretical expectations for the phase. The tolerances on the apparatus will have to be comparable to tolerances required of the Light Interferometric Gravitywave Observatory (LIGO).

Additional quantum optical experiments of this nature are proposed to test the holographic principle. The larger scale effects of holographic induced fluctuations should be more amenable to such experimental techniques. The phase terms computed are significantly larger. Through the remainder of this text the holographic principle is explored as a rich foundation for deeper foundations of physics.

Experimental tests of quantum fields in spacetime are an emerging reality. Recently it was announced that gravitational Bohr quantization was measured with cold neutrons [5].

This result is analogous to the Bohr atom derived early in the 20th century. Observing spacetime fluctuations is then likely the next step for measurements on the quantum effects of gravitation.

The second approach towards the detection of quantum gravity is the measurement of the departure in the uncertainty between complementary variables in the presence of strong gravity fields. By the Einstein equivalency principle the acceleration of a particle by a force in flat spacetime is equivalent to any force that deviates the geodesic motion of a particle in curved spacetime. With high powered lasers an electron may be accelerated to values as large as $g = 10^{21} cm/s^2$ and beyond in the not too distant future. This electron will interact with a thermal vacuum with a temperature of $T = (g/10^{21} cm/s^2) \times 1K$. The Unruh effects may be observable within the next few decades, where experiments involving Bose-Einstein condensates are proposed. Assume that electrons are accelerated to huge values by a high powered laser beam, or realistically a pulse, with a good degree of coherence to the beam. This laser beam may further require some degree of parametric amplification as a "primer" required to induce squeezed states associated with the accelerated electron. If a beam of electrons is accelerated then the synchrotron radiation emitted by these electrons will induce a stimulated emission of synchrotron radiation. This uses a laser to generate a laser, analogous to the wiggler laser. The employment of a homodyne detector will measure the degree of squeezing of this light, where further if there are departures in the uncertainty in observables.

Obviously this experiment will present some difficulties. This involves quantum optical techniques in the X-ray region of the electromagnetic spectrum. The technology for these types of measurements will doubtlessly have to be considerably improved in the not too distant future.

Another test of quantum gravity may be with the detection of gravity waves. The earliest phase of the universe was one where gravity was quantized and unified with the other gauge fields. The inflationary process set up the decoupling of the forces of nature may well have also involved the decoupling of gravity from the other forces of nature. With this decoupling gravity assumed a classical nature, but where quantum gravity fluctuations may have been imprinted on classical spacetime. With the rapid expansion of spacetime quantum fluctuations in spacetime may have been also expanded rapidly and frozen into the classical manifold. This would then manifest itself as a cosmic background of gravity waves. Planck scale fluctuations may have been frozen out as gravity waves, which may now exist on a scale of $\sim 1 - 100m$. This spectrum of gravity waves should be due to quantum spacetime fluctuations when the observable universe was 10^{-14}cm to 10^{-16}cm in radius. This background spectrum of gravity wave radiation is analogous to the microwave background of electromagnetic radiation due to the end of the radiation dominated period of the universe.

With Light Interferometric Gravity Observatories (LIGO) it may be possible to detect this background of gravitational radiation. This would extend our observation of the universe as close to the cosmological event horizon as is likely possible. It would also be an astronomical indication of the existence of quantum gravity fluctuations in the early universe. The spectrum of this radiation will contain information about the nature of the extremely early universe. This is one the reasons why much of this monograph discusses quantum gravity from a low energy and potential empirical viewpoint. If this remnant from the quantum gravity period of the universe is detected it will doubtless lead to an incredibly different view of the universe, where much of the highly developed theories devised today will be either revised or abandoned.

1.3: Connections to Strings and Membranes

The holographic principle indicates that quantum fields approach an event horizon, pile up near the horizon and are redshifted out of view. This implies that event horizons are associated with a Planck length thick membrane of fields, where only their quantum gravitational modes are observed. These fields have wavelengths at the Planck scale, where all other field amplitudes on a larger scale are redshifted away to "infinity". In the case of a black hole this membrane consists of all available quantum information that composes the black hole. This membrane is then considered as a two dimensional membrane, or D2-brane, under a target map. For a virtual black holes this membrane may have a negative energy fluctuation. In this case the virtual black hole is a virtual wormhole.

The construction of fluctuations up to this point involves Lanzcos junctions that are discontinuities in G^{00} with analogues to the discontinuity in the electric displacement vector at the boundary of a dielectric. Further, if this discontinuity is zero then the negative energy deficit required to sustain this is removed and the 2-d sphere of the Lanczos junction is the event horizon of a black hole. As a quantum fluctuation this will involve the fluctuation in a Planck mass δm_p plus a fluctuation in a negative energy component $-\delta\epsilon$, where on average $|\delta m_p| > |\delta\epsilon|$ according to the quantum interest conjecture. So in general quantum gravity fluctuations larger than $\sim 10 \times L_p$ will involve a preponderance of virtual black holes, where the occurrence of wormholes is a measure of the squeezing of the vacuum state.

The connection to membranes is discussed in the light of the 't Hooft horizon algebra [6]. This algebra is defined by the commutation of momentum and position variables for fields entering and leaving a black hole. For $x_t = x_t(\sigma)$ the transverse coordinates to the horizon, where σ is an arbitrary coordinate, then

$$[\partial_\sigma x^\mu(\sigma),\ \partial_\sigma x^\nu(\sigma')] = i\epsilon^{\mu\nu}{}_\rho \partial_\sigma x^\rho(\sigma')\delta^2(\sigma - \sigma'). \qquad (1.21)$$

These commutators define an $SO(2,1)$ algebra, where this conveniently reduces the dimension of the Lorentz algebra. A basis of connection forms and curvature forms is then derived from this algebra. This system is then extended into a supersymmetric form. With the application of ghost fields and Brecci-Rouet-Stora-Tyupin (BRST) quantization this theory may be expressed according to topological quantum numbers. Here the membrane of fields a Planck unit away from the event horizon of a black hole is characterized by topological indices in an elementary manner.

The horizon algebra contains a moduli space of solutions. The principle bundle that defines this moduli space is split at the event horizon $\mathbf{P} \to \mathbf{P}_1 \oplus \mathbf{P}_2$ with an associated singularity in the moduli space. This dimension of this moduli space is given by the Donaldson theorem, which has curious implications for manifolds in four dimensions. An examination of the moduli space near the singularity leads to a surprising result. A physical ansatz is invoked to define the analog for the elliptic complex for a pseudoEuclidean metric. The moduli space near the singularity assumes the form of a cone in the weighted projective space CP_w^2. A weighted projective space is where the coordinates in CP^2 are mapped into $z \to z^n$. This weighted projective space naturally has a Virasoro structure with an associated Kac-Moody algebra. This structure has the form of a quantized gauge theory. This gauge theory is later to be examined with respect to Chan-Paton factors for strings. This leads to the suggestion that strings at high energy are in fact made of constituents or have an underlying structure identified with the blow up of a point in the moduli space. This gauge theory, if it is psuedoEuclidean, will in turn have a moduli space with a similar singularity structure. This leads to a fractal-like situation with respect to the foundations of physics.

These structures lead to various approaches with respect to topological quantum numbers and indices. These are subsequently discussed as possible outcomes of this view of the universe.

1.4 Beyond Standard Constructions

Beyond this point it is possible that the universe is structured according to multilinear brackets more general than the standard commutators of quantum theory or the parallel transport construction of general relativity. These lead to a general structure labelled as the B^* algebra based on Peano's multilinear structure of geometry. A possible realization of this according to the octonions is also developed. In this view the foundations of quantum gravity is posited to be due to the 24-cell and its error correction code basis. In this model Planck units of volume in a 24-cell are distinguishable and their relationship is given by a Hamming or Golay code. This construction appears to underlie the 26 bosonic string, where these is an E_8 lattice construction as well. The underlying structure for this is the Mathieu group $C24$, which is the fifth of the 26 sporadic groups. The current superstring theories may then have this sort of underlying basis according to such constructions.

On a scale smaller than the 24-cell it is likely that this approach fails. This might mean the 24 cell is defined for overlapping Planck units of volume, where the distinguishability of states fails. If the 24-cell is not compressed this domain then may involve structures beyond the octonions. However, this implies physics has an algebraic-geometric structure that no longer is a normed division algebra. It might be possible to order more generalized multilinear brackets for the B^* algebra around the Janko or Conway sporadic groups. This path would lead to the monster group. The octonions or some extended B^* structure may be the end of all possible geometry. This breakdown might reflect how quantum gravity and cosmology ultimately involves quantum logical propositions that are no longer decidable. The physical implication is that physical law may ultimately be random. Physical law may then emerge for reasons that are ultimately random. The emergent physical laws may then be due to the undecidability of the truth of quantum logical propositions. This is an even more speculative look at the foundations of quantum gravity, but one that is shown to have a basis in standard quantum mechanics and Chaitan's halting probability

References

[1] A. D. Sakharov. *Doklady Akad. Nauk. SSR*, **177**, 70-71 (1967), (english) *Soviet Phys. Doklady*, **12** 1040-1041 (1967).

[2] C. W. Misner, K. S. Thorne, J. A. Wheeler, *Gravitation*, Freeman Press (1973).

[3] L. Crowell, *Found. Phys. Lett.*, **12**, 6, 585 (1999).

[4] S. Weinberg (1972), *Gravitation and Cosmology*, p. 215, John Wiley and Sons, New York.

[5] V. Nesvizhevsky, H. G. Börner, A. K. Petukhov, H. Abele, S. Bae (17 Jan 2002), *Nature* **415**, 297 - 299.

[6] G. 't Hooft, *Class. Quant. Grav.* **16** (1999) 395-405.

[7] S. K. Donaldson, P. B. Kronheimer, *The Geometry of Four Manifolds*, Clarendon Press, Oxford (1990).

2: Quantum Fluctuations and Spacetime

2.1: Introduction

Teleparallel connections and torsion enter into general relativity with nongeodesic curves in spacetime. Within ADM general relativity torsion may be used as a model of quantum fluctuations of the metric, where such a fluctuation deflects the motion of a particle away from a classically determined geodesic. The torsional connection on the spacial surface of evolution is an extrinsic curvature tensor that defines the gravity g on the particle horizon induced by the fluctuation. This analysis leads to a form of quantum gravity operators as finite temperature operators which obey a more generalized uncertainty principle and exhibit gravitational self-squeezing of the vacuum state. This means the quantum gravity vacuum has an intrinsic instability and permits the tunnelling of nascent cosmologies out of the vacuum.

Physically a quantum gravity fluctuation of a spacelike manifold of evolution will accelerate a test mass away from its geodesic motion on the classical spacetime. The Unruh effect demonstrates how accelerations that deviate a particle away from a geodesic causes this particle to interact with a thermal vacuum[1]. As a result quantum gravity fluctuations involve connections that are teleparallel or non-holonomic, where these fluctuations are associated with a thermal set of states. This then leads to a torsion model for thermal quantum gravity fluctuations. Further, this temperature can be determined by the torsional connection, for it defines a Killing tensor that determines g on the particle horizon induced by the fluctuation.

The development of quantum gravity fluctuations is initiated with the application of the Bohm approach to quantum mechanics. The concept of quantum fluctuations is then developed with this machinery. Quantum fluctuations are an aspect of the Schrödinger as this acts on a polar wave function to give forward and backward Fokker-Planck equations. This is then developed to illustrate their basic physical properties, such as Markovian statistics. Bohm's approach to quantum mechanics is fundamentally no different from the standard approach to quantum mechanics. This approach to the theory of fluctuations is employed to examine spacetime fluctuations. There are no hidden variables or any "inner causality" process beyond standard quantum mechanics.

The ADM formalism of general relativity is according to Gauss' fundamental forms, where its quantum analogue is the Wheeler-DeWitt equation. The Wheeler-DeWitt equation assumes the form of a Schrödinger equation if an auxiliary field is imposed in spacetime. Forward and backward Fokker-Planck equations are derived. This approach to quantum fluctuations is extended to the Wheeler-DeWitt equation with an auxiliary field. Torsional connections are introduced as a model of quantum fluctuations in gravitation. The resulting torsional connection on a spacial surface is an extrinsic curvature tensor that is a Killing tensor. One special form of the torsional extrinsic tensor is the product of a

Killing vector and a spin tensor. This is extended to a more general form of the extrinsic torsional tensor. This derives a gauge-like structure of gauge fluctuations and operators for low energy quantum gravity. These operators are demonstrated to have a more general form than those in standard quantum mechanics. These operators then have connections to Finsler geometry of bundles with horizontal and vertical portions of the bundle.

Quantum mechanics and general relativity appear to be two completely theories. Further quantum mechanics is found to be completely independent of spacetime language. Quantum states exist independent of spacetime, where a wave function is simply a representation of a quantum state vector in coordinates. In what is developed further quantum mechanics and general relativity are found to be categorically equivalent $[GR] \sim [QM]$ as well as independent. From this the two exist as equivalent pairs of subsets in quantum gravity $[QM] \cup [GR] \subset QG$

This chapter, as well as chapter 4, involves several ideas about the nature of quantum gravity and fluctuations. Not all these ideas may be consistent with one another, but they are presented as possible alternatives. Since quantum gravity involves great empirical ignorance it is likely that unexpected structures may exist. Potentially many structures may exist currently unknown. For this reason various ideas and possibilities about the nature of quantum gravity and fluctuations are addressed. The intention here is that some of them might be explored theoretically and maybe experimentally in the future.

2.2: Fluctuations in Quantum Mechanics

Since quantum mechanics was discovered to be the proper description of atomic physics there has been an uncomfortable sense the description of subatomic particles involves an epistemological viewpoint, while the measuring apparatus is treated as classical and ontologically real. This is the Copenhagen interpretation of quantum mechanics. The epistemological implications of quantum mechanics stem from the fact that the actual state of a system cannot be ascertained until a measurement. A classical measurement apparatus is described by real states through its evolution. The apparatus has an ontological status absent from the quantum system it measures. A quantum wave function gives no such definite being to a particle, but defines amplitudes for the probable occurrence of a particle. Hence the process of measurement is somehow different from the wave mechanics of the quantum system measured. This is in contrast to classical mechanics, where the system observed and the measurement apparatus obey the same set of physical laws.

This leads to some further conundrums in how to interpret quantum mechanics. Einstein, Podolsky and Rosen [2] demonstrated that measurement inferred nonlocality. Their case involves two spin 1/2 particles resulting from a decay process of a spin 0 boson. This system is described by a single wave function as an entanglement of the states for the two resultant particles. The spin directions these two particles are in a superposition of spin up and down. For a measurement of the two spins performed at two spacetime points with a spacial interval that is noncausal, conservation of angular momentum demands that the two spins are measured to be in opposite directions. This gives the appearance the spins are tied to each other by noncausal means.

There are two positions one can take: either quantum mechanics has a classical underpinning, or the measuring apparatus is ultimately quantum mechanical so the measurement process obeys the laws of quantum mechanics. The latter case is the Many Worlds View of quantum mechanics of Everett and DeWitt, where there exists a superposition of universes and a measurement just involves the observation of one particular eigen-branching of the world [3][4]. Whether these two interpretations are related to each other is beyond the scope of this discussion. How this ties into the mind of the observer is problematic, for one does not see a superposed universe in a measurement with superposed observers and brains. Consequently the issue of consciousness is safely removed from the problem. The first approach is the topic of this paper, but where the presumed classical underpinning to quantum mechanics is shown to be illusory. David Bohm demonstrated the Schrödinger equation may be written according to real and imaginary parts, where the real part is a Hamilton-Jacobi equation modified by a quantum potential and the imaginary part is a continuity equation, similar to the Navier-Stokes equation, for a pilot wave [5]. This is seen if one considers a polar wave function $\psi = Re^{iS/\hbar}$, where S is the action. The Schrödinger equation is then split into two parts

$$\frac{\partial S}{\partial t} + \frac{1}{2m}(\nabla S)^2 + V - \frac{\hbar^2}{2m}\frac{\nabla^2 R}{R} = 0 \qquad (2.1a)$$

$$\frac{\partial R^2}{\partial t} + \frac{1}{m}\nabla \cdot (R^2 \nabla S) = 0. \qquad (2.1b)$$

The last term in equation 2.1a is the quantum potential. With $p = \nabla S$ equation 2.1a is the modified Hamilton-Jacobi equation and equation 2.1b is a continuity equation for the pilot wave. This program was developed to illustrate how hidden variables may underlie the quantum world. The modified Hamilton-Jacobi equation is an energy constraint for the motion of a physical particle, often called a beable. The application of a gradient operator on this energy constraint, equation 2.1a, gives a Newton's law equation of motion with a quantum force, where the continuity equation in principle provides information for the calculation of this quantum force.

This physical picture appears to offer a "hidden variable" view of quantum theory [5]. By this argument Bohm proposed quantum theory has the same ontological status as classical mechanics. There is an apparent real trajectory for a particle, termed an active channel by Bohm and Hiley [6]. From a path integral point of view all the other paths in the "sum over histories" are regarded as inactive channels. This then offers a way around the dichotomy between a quantum system and the classical apparatus involved with a measurement. What has been referred to as the "wave function collapse" may now be thought of as virtually identical to a classical collapse, such as what occurs with a roulette wheel or dice. Further, hidden variables are presumed to recover locality in quantum mechanics.

In spite of this there has yet to be any experimental evidence or additional theoretical demonstrations of hidden variables or a beable with an ontological interpretation of a wave function beyond quantum theory. After some 75 years of quantum physics, experimental evidence has substantiated the view point that there is no inner particle associated with a wave function. The theorems of Bell and of Kochen-Specker [7] [8] illustrate that the assumption of a hidden variable does not change the physical interpretations of quantum probability, or impose any sort of locality to quantum mechanics. It appears that hidden variables, or ontological beables, are excess metaphysical baggage that can be eliminated from physics with no observable consequence. Despite this there continues to be a small industry devoted to an ontological interpretation of quantum mechanics that may recover locality.

Along with laying the foundations for quantum fluctuations, it would then appear reasonable to ask whether Bohm's quantum formalism does demand a particle ontology in quantum mechanics. Bohm's theory appears to stand alone in the world of quantum study, where it is advanced that ontology and locality are an aspect of quantum physics.

Consider Bohm's theory according to symplectic variables. Commonly Bohm's theory is written according to configuration variables q, $\dot{q}$, yet there is nothing that restricts us from transforming Bohm's interpretation of quantum mechanics to symplectic variables q, p. This in many ways is more natural for this explicitly involves the classical-like Hamiltonian seen in equation 2.1a. Quantum fluctuations then involve these two canonically conjugate variables in a manner so the quantum potential is entirely an aspect of fluctuations of conjugate variables.

Start with the polar wave function

$$\psi = Re^{iS/\hbar}. \tag{2.2}$$

The action of momentum operator on the wave function gives

$$\hat{p}\psi = p\psi = \frac{\hbar}{i}\nabla(Re^{iS/\hbar}) = \left(\nabla S + \frac{\hbar}{i}\frac{\nabla R}{R}\right)\psi. \tag{2.3}$$

The momentum is demonstrated to be

$$p = \langle p \rangle + \delta p. \tag{2.4}$$

$\langle p \rangle$ and δp are respectively the greatest expectation of the momentum and a fluctuation around the classical extremum.

The identification of $\frac{\hbar}{i}\frac{\nabla R}{R}$ as a fluctuation is seen by considering the continuity equation for $\rho = \psi^*\psi$

$$\frac{\partial \rho}{\partial t} = (i\hbar/2m)(\psi^*\nabla^2\psi - \nabla^2(\psi^*)\psi), \tag{2.5}$$

which may be also written as

$$= (i\hbar/2m)\nabla \cdot (\psi^*\nabla\psi - \nabla(\psi^*)\psi). \tag{2.6}$$

For ψ in polar form $\psi = Re^{iS/\hbar}$, $\rho = R^2$ equation 2.4 becomes

$$\frac{\partial \rho}{\partial t} = -\frac{1}{2m}\nabla \cdot (\rho \nabla S) + \frac{i\hbar}{4m}\left(\frac{\nabla^2 \rho}{\rho} - \frac{1}{2}\frac{(\nabla \rho)^2}{\rho^2}\right)\rho$$

$$-\frac{1}{2m}\nabla \cdot (\rho \nabla S) - \frac{i\hbar}{4m}\left(\frac{\nabla^2 \rho}{\rho} - \frac{1}{2}\frac{(\nabla \rho)^2}{\rho^2}\right)\rho. \tag{2.7}$$

The imaginary part of the Schrödinger equation are recovered when the opposite $\nabla^2 \rho$ and $(\nabla \rho)^2/\rho$ terms are subtracted

$$\frac{\partial \rho}{\partial t} + \frac{1}{m}\nabla \cdot (\rho \nabla S) = 0. \tag{2.8}$$

However, before performing the subtraction a curious thing is apparent. Consider the density as composed of two parts

$$\frac{\partial \rho}{\partial t} = \psi^* \frac{\partial \psi}{\partial t} + \left(\frac{\partial \psi^*}{\partial t}\right)\psi$$

$$= \rho^{1/2}e^{-iS/\hbar}\frac{\partial}{\partial t}\rho^{1/2}e^{iS/\hbar} + \left(\frac{\partial}{\partial t}\rho^{-1/2}e^{-iS/\hbar}\right)\rho^{1/2}e^{iS/\hbar}, \tag{2.9}$$

which gives the two Fokker-Planck equations

$$\left.\frac{\partial \rho}{\partial t}\right|_+ = -\frac{1}{2m}\left(\nabla \cdot (\rho \nabla S)\right) + \frac{i\hbar}{4m}\left(\nabla^2 \rho^{1/2} - \frac{1}{2}\frac{(\nabla \rho)^2}{\rho}\right) \tag{2.10a}$$

$$\left.\frac{\partial \rho}{\partial t}\right|_- = -\frac{1}{2m}\nabla \cdot (\rho \nabla S) - \frac{i\hbar}{4m}\left(\nabla^2 \rho - \frac{1}{2}\frac{(\nabla \rho)^2}{\rho}\right). \tag{2.10b}$$

These two coupled inhomogeneous Fokker-Planck equations sum together to give the continuity equation. The first Fokker-Planck equation is a forword Fokker-Planck equation, as it comes from differentials on $\psi \sim e^{-i\omega t}$, while the second is the backward Fokker-Planck equation from $\psi^* \sim e^{i\omega t}$. That the two equations sum into a continuity equation is an illustration that the diffusions associated with these equations conserve probability.

The continuity equation can be expressed according to a current $J = (1/2m)\rho \nabla S$

$$\partial \rho / \partial t + \nabla \cdot J = 0. \tag{2.11}$$

The forward Fokker-Planck equation is now expressed as

$$\left.\frac{\partial \rho}{\partial t}\right|_+ + \nabla \cdot \left(J - \frac{i\hbar}{4m}\nabla \rho\right) = 0. \tag{2.12a}$$

For the differential equation

$$\left.\frac{\partial \rho}{\partial t}\right|_+ + \nabla \cdot \left(J - \frac{i\hbar}{4m}\nabla \rho\right) = 0 \tag{2.12b}$$

the current may be generalized to

$$\mathcal{J} = J - \frac{i\hbar}{4m}\frac{\nabla R}{R}, \qquad (2.13)$$

where $\mathcal{J} + \mathcal{J}^* = 2J$. The momentum of a particle by the Fokker-Planck equation is of the form

$$p = \nabla S - \frac{i\hbar}{m}\frac{\nabla R}{R}. \qquad (2.14)$$

The analysis is the same with the backward Fokker-Planck equation.

By the association of the Fokker-Planck equation with a Langevin equation the diffusion term is identified as the second order term. This then identifies $(\hbar/2i)\nabla\rho/\rho$ as the fluctuation. It is then apparent that the fluctuation in the momentum has no expectation value:

$$\langle p \rangle = \langle \langle p \rangle \rangle + \langle \delta p \rangle = \langle p \rangle + \langle \delta p \rangle, \qquad (2.15)$$

whereby $\langle \delta p \rangle = 0 = -i\hbar \langle \nabla R/R \rangle$. It is only the square of the momentum fluctuation that has a nonvanishing expectation.

For ease of calculation write $R = e^r$ so the momentum operator acts upon the wave function as

$$\hat{p}\psi = \frac{\hbar}{i}\nabla e^{r + iS/\hbar} = (\nabla S - i\hbar\nabla r)\psi, \qquad (2.16)$$

with $\delta p = -i\hbar\nabla r$. It is then apparent that the momentum has a form analogous to the momentum of an electrically charged particle in a magnetic field $P = p + ieA$. The analogy extends to the observation that $\langle \delta p \rangle$ and $\langle A \rangle$ both vanish and are not directly observable. Quantum fluctuations are similar to gauge potentials, but where this vector potential-like object is purely stochastic.

Consider the quantum potential written according to $r = ln(R)$

$$\mathcal{Q} = -\frac{\hbar^2}{2m}(\nabla^2 r + (\nabla r)^2) = \frac{1}{2m}\left(\frac{\hbar}{i}\nabla \cdot \delta p + (\delta p)^2\right). \qquad (2.17)$$

The quantum potential involves a "bare potential," that depends on the divergence of a fluctuation, plus a term that is the product of fluctuations. Given two fluctuations at t and t' these fluctuations are delta function correlated as

$$\langle \delta p(t)\delta p(t') \rangle = (\Delta p(t))^2 \delta(t - t'), \quad \Delta p(t) = \sqrt{\langle p(t)^2 \rangle - \langle p(t) \rangle^2}. \qquad (2.18)$$

The quantum potential contributes a kinetic energy term to the Hamiltonian delta function correlated in time. These quantum fluctuations in the momentum are related to quantum fluctuations in the position. The commutator of the momentum and position operators is well known. This commutator as it acts upon the wave function is expanded out so the commutator is identified with the Poisson bracket of the fluctuations

$$[\hat{p}, \hat{q}]\psi = \frac{\hbar}{i}\{\delta p, \delta q\}_{PB}\psi = \frac{\hbar}{i}\psi. \qquad (2.19)$$

It is then apparent that the quantum potential is not the same type of potential function, $V = V(q)$ often found in the Schrödinger equation. The quantum potential is a function of momentum fluctuations, while the standard conservative potential is a function of position:

$$V(q) = V(\langle q \rangle + \delta q) \simeq V(\langle q \rangle) + \nabla_q V(\langle q \rangle)\delta q. \tag{2.20a}$$

This potential is dependent on the coordinate variable and its fluctuation. The quantum potential is dependent on the momentum and its fluctuation. Hence the quantum potential and the standard potential are not in involution with each other:

$$[\mathcal{Q}, V(q)] = -\frac{i\hbar}{m}\delta p \nabla_q V(q). \tag{2.20b}$$

This involutory nature of the two potentials indicates they are not physically equivalent.

Fluctuations are similar to gauge potentials, with transformation properties which maintain the same type of invariance principle in gauge theories. Fluctuations also determine a quantum potential with fundamentally different properties than more ordinary potentials. To start an examination of this difference and its implications consider a canonical transformation on the fluctuations. A new Hamiltonian $H' = H'(t)$ is related to the Hamiltonian $H = H(t)$ by the generating function F as

$$H' = H + \frac{\partial F}{\partial t}. \tag{2.21}$$

Consider a canonical transformation restricted to the fluctuations $\delta p \to \delta P$ and $\delta q \to \delta Q$. It is permissible to consider only canonical transformations of the fluctuations since they are completely independent of the classical or extremal variable. The only requirement is the transformed fluctuation has the same physical units. A transformation of the variables δq and δQ according to the different Hamiltonians is

$$\delta Q = \frac{\partial H'}{\partial(\delta P)}, \quad \delta q = \frac{\partial H}{\partial(\delta p)}. \tag{2.22}$$

By direct substitution this results in

$$\delta Q = \frac{\partial \mathcal{Q}'(p, \delta P)}{\partial(\delta P)}, \quad \delta q = \frac{\partial \mathcal{Q}(p, \delta p)}{\partial(\delta p)}. \tag{2.23}$$

It is then of interest to find the correlations between the two fluctuations δq and δQ in $\langle \delta q \delta Q \rangle$ Quantum mechanically the fluctuation is $\delta q = \hat{q} - \langle q \rangle$, with the same for δQ, so that this is

$$\langle \delta q \delta Q \rangle = \langle |\hat{q}|\hat{Q}| \rangle - \langle q \rangle \langle Q \rangle. \tag{2.24}$$

The insertion of completeness terms $\mathbf{1} = \int dq |q\rangle\langle q| = \int dQ |Q\rangle\langle Qq|$ this this correlation as

$$\langle \delta q \delta Q \rangle = \int dq' \int dQ' \langle |\hat{q}|q'\rangle\langle q'|Q'\rangle\langle Q'|\hat{Q}| \rangle - \langle q \rangle\langle Q \rangle. \tag{2.25}$$

The sum over the dummy variables q' and Q' will give the expectation for these two representations of the position. The term $\langle q'|Q\rangle$ is then a unitary operator that will transform the $|\ \rangle$ into a basis appropriate for the operator $\hat{Q}$,

$$\langle \delta q \delta Q \rangle = \int dq' \int dQ' \langle \ |\hat{q}|q'\rangle\langle Q'|\hat{Q}U|\ \rangle - \langle q\rangle\langle Q\rangle. \tag{2.26}$$

which over the sum gives $\langle q\rangle\langle Q\rangle$ for the first right hand side term. Therefore in general these fluctuations are independent by

$$\langle \delta q \delta Q \rangle = 0 = \langle \partial_{\delta p}\mathcal{Q}(p, \delta p)\partial_{\delta P}\mathcal{Q}'(p, \delta P)\rangle. \tag{2.27}$$

Thus the two quantum potentials under the different representations of quantum fluctuations are utterly independent. Under this convolution these potentials involve quantum forces, due to fluctuations that deflect a particle away from a classical path, with orthogonal modes. This permits a description of the quantum system according to the two Hamilton-Jacobi equations

$$-\frac{\partial S}{\partial t} = H \simeq \frac{1}{2m}\langle p\rangle^2 + V(\langle q\rangle) + \nabla_q V(\langle q\rangle) + \mathcal{Q}(p, \delta p)$$

$$-\frac{\partial S'}{\partial t} = H' \simeq \frac{1}{2m}\langle p\rangle^2 + V(\langle q\rangle) + \nabla_q V(\langle q\rangle) + \mathcal{Q}'(p, \delta P). \tag{2.28}$$

This is an interesting result, for it clearly shows that the Bohm trajectory is simply a manifestation of the fluctuation variables chosen by any canonical transformation. For a Schrödinger equation written according to other fluctuating variables in general the quantum potentials will not be the same.

This calls into question the ontological interpretation of a particle in Bohm's program for quantum mechanics. It now is difficult for anyone to put their finger on a path and state conclusively that the particle travelled along this path and not any other. It appears impossible to distinguish an active channel from an inactive one. For a particular problem transformed into various fluctuation coordinates the various analysts who examine Bohm trajectories arrive at different Newton-like dynamical equations for the particle. Yet with the philosophy of Bohmian ontology each of these analysts will insist that the particle (beable) follows their chosen path. It appears there is no way to determine which analyst is right. It is then best to abandon ontology in favor of democracy and conclude that all the analysts are right. Each analyst has an interpretation that defines one path superposed with all the other paths. In fact it appears that even within one set of classical coordinates the problem involves many possible paths determined by the set of all possible canonical transformations of the fluctuation variable. This gives the prospect that Bohm's program for quantum mechanics defines a type of path integral as a grand sum over all possible paths which deviate from a classical path according to quantum fluctuations.

This indicates that the quantum potential is a nonlocal object. The quantum potential must change its configuration in response to a measurement along spacelike intervals.

This indicates the quantum potential is then determined by nonlocal quantum effects. Quantum fluctuations reflect the probability amplitude for the occurrence of a measured variable that deviates from the classical or most probable value. The quantum potential is dependent upon nonlocal quantum fluctuations which may be arbitrarily represented by canonical transformations. Thus a particular "beable path" is simply a choice of fluctuation coordinates.

For a quantum system the density matrix is defined by the outer product of the state vectors at a given time

$$\rho_t = |\psi_t\rangle\langle\psi_t|. \tag{2.29}$$

It is a standard result of quantum mechanics that

$$Tr\rho_t = \int dq_t \langle q_t|\psi_t\rangle\langle\psi_t|q_t\rangle = 1. \tag{2.30}$$

The density matrix is now redefined for an infinitesimal increment in time as

$$\rho_t = \lim_{\delta t \to 0} |\psi_t\rangle\langle\psi_{t-\delta t}|. \tag{2.31}$$

This definition of the density matrix may be used to examine the overlap of a wave function at different times. These density operators applied in an infinite product may be used to construct the overlap according to

$$\langle\psi_{t'}|\psi_t\rangle = \lim_{\delta t \to 0} \lim_{N \to \infty} \langle\psi_{t'}|\rho_{t'-\delta t}\rho_{t'-2\delta t}\ldots N \; times \ldots \rho_{t+2\delta t}\rho_{t+\delta t}|\psi_t\rangle,$$

$$N\delta t = constant. \tag{2.32}$$

This is justified by using completeness relations, defined as

$$1 = \lim_{\delta t \to 0} \int dq_t |q_t\rangle\langle q_{t-\delta t}| \tag{2.33}$$

and the trace of the density matrix in equation 2.30. This allows the wave overlap to be written according to

$$\langle\psi_{t'}|\psi_t\rangle = \lim_{\delta t \to 0} \lim_{N \to \infty} \prod_{n=0}^{N} \int dq_{t+n\delta t} \langle\psi_{t+n\delta t}|q_{t+n\delta t}\rangle\langle q_{t+(n-1)\delta t}|\psi_{t+(n-1)\delta t}\rangle. \tag{2.34}$$

This description of the overlap is then according to "snapshots" determined by projectors, where in the limit the time increment vanish they recover the density matrix.

We now focus on a product defined on one particular time slice. Each infinitesimal overlap is related to the polar wave function by

$$\int dq_t \langle\psi_t|q_t\rangle\langle q_{t-\delta t}|\psi_{t-\delta t}\rangle = \int dq_t \psi^*(q,t)\psi(q,t-\delta t) =$$

$$\int dq_t e^r e^{iS(t)} e^r e^{iS(t-\delta t)}. \tag{2.35}$$

Since the polar form of the wave function is for a system in the position representation, this integrand as an infinitesimal overlap is written as

$$\psi^*(q,t)\psi(q,t-\delta t) = \psi^*(t)\Big(\psi(q,t) + \delta t \frac{d\psi}{dt}(q,t) + \frac{1}{2}\delta t^2 \frac{d^2\psi}{dt^2}(q,t) + O(\delta t^3)\Big). \tag{2.36}$$

The term to $O(\delta t)$ is easily seen to be

$$\delta t \frac{d\psi}{dt}(q,t) = \delta t \Big(\frac{\partial \psi}{\partial t}(q,t) + \frac{dq}{dt}\nabla_q \psi(q,t)\Big) =$$

$$\frac{i\delta t}{\hbar}\Big(\frac{dq}{dt}p - \big(\frac{1}{2m}\langle p\rangle^2 + V(q) + \mathcal{Q}(p,\delta p)\big)\Big)\psi(t). \tag{2.37a}$$

The terms dr/dt reproduce the continuity equation which is zero. The final equality contains $p\dot{q} - H$, where the Hamiltonian contains the quantum potential. The first term on the right hand side is the invariant phase space volume, where the system is considered to be conservative. The rest is the Hamiltonian with quantum potential. The term to $O(\delta t^2)$ is

$$\delta t^2 \frac{d^2\psi}{dt^2}(q,t) = -\delta t^2 \Big(\frac{1}{\hbar^2}(ip\dot{q} - H(p,q))^2 + -\frac{i}{\hbar}\frac{dH(p,q)}{dt}\Big)\psi(q,t), \tag{2.37b}$$

where the Hamiltonian is the classical-like observable

$$H(p,q) = \frac{1}{2m}\langle p\rangle^2 + V(q) + \mathcal{Q}(q,\delta p). \tag{2.38}$$

This second order term is a power term which results from the explicit time dependence of the Hamiltonian due to fluctuations. This is a fluctuation in the total energy during the infinitesimal time slice. The $O(\delta t^2)$ term is a partial time derivative of the Hamiltonian plus a term that involves the square of the Hamiltonian. During this small time interval the Hamiltonian will fluctuate from H to H'. The total time derivative of H is then determined by a Poisson bracket between H and H', where $H' = H(t + \delta t)$ is the Hamiltonian at the end of the interval. The integrand of the infinitesimal overlap is

$$\psi^*(q,t)\psi(q,t-\delta t) = e^{\left[\frac{i\delta t}{\hbar}\left(\dot{q}p - \frac{1}{2m}\langle p\rangle^2 - V(q) - \mathcal{Q}(p,\delta p)\right)\right]}\psi^*(t)\psi(t) \tag{2.39}$$

$$+ \frac{i\delta t^2}{2\hbar}\{H', H\}_{PB}\psi^*(t)\psi(t) + \text{higher bracket terms}.$$

In quantum mechanics this Poisson bracket may be replaced by a commutator

$$\{H', H\}_{PB} \to \frac{i}{\hbar}[H', H]. \tag{2.40}$$

This commutator determines to the minimum uncertainty between these two variables. This determines the uncertainty relationship between the two Hamiltonians as $\Delta H \Delta H' = i/2[H, H']$. Since the two Hamiltonians are very close to each other for small enough time intervals, the uncertainties in the two modified Hamiltonians are regarded as equal. However, the term $\frac{\delta t^2}{\hbar^2}\Delta H^2$ may be ignored as $\delta t \simeq \Delta t$, where $\Delta t \Delta E \sim \hbar$ so in the sequence of time slices this contributes an overall phase e^i. So the quantum overlap may be expressed with the integrand

$$\psi^*(q,t)\psi(q,t-\delta t) = e^{\left[\frac{i\delta t}{\hbar}\left(\dot{q}p - \frac{1}{2m}\langle p\rangle^2 - V(q) - \mathcal{Q}(p,\delta p)\right)\right]}\psi^*(q,t)\psi(q,t) \quad (2.41)$$

Physically a particle that travels on a path, quantum trajectory or active channel is perturbed into other channels by quantum fluctuations. As these fluctuations are arbitrary it is impossible to specify deterministically how the particle "hops" from one channel to another. It then appears best to consider a sum over all possible fluctuations, where the particle then has probability amplitudes for existing in each of these channels.

To construct a the path integral take this result and insert it into equation 2.34 and replace the overlap with $Z[q]$

$$Z[q] = \lim_{\delta t \to 0} \lim_{N \to \infty} \prod_{n=0}^{N} \int dq_{t+n\delta t}\rho(q_{t+n\delta t}) \times$$

$$e^{\left[\frac{i\delta t}{\hbar}\left(\dot{q}p - \left(\frac{1}{2m}\langle p\rangle^2 + V(q) + \mathcal{Q}(p,\delta p)\right)\right)\right]}\bigg|_{t+n\delta t}, \quad (2.42)$$

where $\rho(q,t) = \psi^*(q,t)\psi(q,t)$. The Hamiltonian involves a purely classical Hamiltonian plus the quantum potential, so the invariant phase space volume involves a fully quantum (classical plus fluctuation) variable. Now write the quantum potential in the Hamiltonian as

$$\mathcal{Q}(p,\delta p) = -\frac{1}{2m}(p \cdot \delta p + \delta p^2). \quad (2.43)$$

This gives the modified Lagrangian with $\mathcal{Q} \simeq -\frac{m}{2}\dot{q}\cdot\delta\dot{q}$, which for $\langle p\rangle^2 = m(\dot{q}^2 - 2\dot{q}\delta\dot{q}$ gives

$$\mathcal{L} \to \mathcal{L} + \tilde{\mathcal{Q}} = \frac{m}{2}\dot{q}^2 - V(q) - \frac{m}{2}\dot{q}\cdot\delta\dot{q}, \quad (2.44)$$

where $\dot{q} = \langle\dot{q}\rangle + \delta\dot{q}$. The wave function overlap assumes the form

$$Z[q] = \lim_{\delta t \to 0} \lim_{N \to \infty} \prod_{n=0}^{N} \int dq_{t+n\delta t}\rho(q_{t+n\delta t})e^{-i\frac{m\delta t}{2\hbar}\dot{q}_{t+n\delta t}\delta\dot{q}_{t+n\delta t}}e^{\frac{i}{\hbar}\int_{t_0}^{T}dt\mathcal{L}}. \quad (2.45)$$

The measure of the integral is a product of integrals that are defined at particular coordinate positions. There are canonical transformations which induce the time evolution from $\dot{q}(t)$ and $\delta\dot{q}(t)$ to $\dot{q}(t+\delta t)$ and $\delta\dot{q}(t+\delta t)$ during each time interval. Now compress this information by setting $\lim_{\delta t \to 0} \lim_{N \to \infty} \prod_{n=0}^{N} dq_{t+n\delta t} \rho(q_{t+n\delta t}) = \mathcal{D}[q]$. The integration measure is then $\mathcal{D}[q]exp((imt/2\hbar)\dot{q}(t)\delta\dot{q}(t))$ and contains additional information due to the quantum potential. The path integral is then

$$Z[q] = \int \mathcal{D}[q(t)] e^{-i\frac{mt}{2\hbar}\dot{q}(t)\delta\dot{q}(t)} e^{\frac{i}{\hbar}\int_{t_0}^{T} dt\mathcal{L}}. \tag{2.46}$$

A full analysis of the path integration measure $\mathcal{D}[q]exp(-(imt/2\hbar)\dot{q}(t)\delta\dot{q}(t))$ is likely to be difficult. This is beyond the scope of this study currently. However, estimates at this stage are obtained. To start approximate the terms with the quantum potential according to,

$$e^{i\frac{mt}{2\hbar}\mathcal{Q}(q,\,\delta q)} \simeq 1 - \frac{imt}{2\hbar}\dot{q}(t) \cdot \delta\dot{q}(t) -$$

$$\frac{m^2 t^2}{8\hbar^2}\left(\dot{q}(t) \cdot \delta\dot{q}(t)\right)^2. \tag{2.47}$$

We then let the variables in this expansion be replaced by their average by the substitution $q^n \delta q^n \to \langle q^n \delta q^n \rangle$, where $q^n = (\langle q \rangle + \delta q)^n$. For Markovian fluctuations $\langle \delta p \rangle = 0$, so the path integral may be written as

$$Z[q] \simeq \int \mathcal{D}[q]\left(1 + \frac{t^2}{8m^2\hbar^2}(\Delta p(t))^2\right) e^{\frac{i}{\hbar}\int_{t_0}^{T} dt\mathcal{L}}, \tag{2.48}$$

where the uncertainty relationship is derived in the usual manner. This illustrates the measure of the path integral involves the uncertainty of a wave function with time. The square of the momentum spread is the uncertainty in the kinetic energy of the particle and may be approximated as the uncertainty of one half the total energy

$$Z[q] \simeq \int \mathcal{D}[q] \cos\left(\frac{t\Delta E}{2\hbar}\right) e^{\frac{i}{\hbar}\int_{t_0}^{T} dt\mathcal{L}}. \tag{2.49}$$

This term is a manifestation of the Fubini-Study metric for the projective Hilbert space, where this term is a holonomy due to the parallel transport on the Hilbert space with a fibration over the projective Hilbert space [9]. This term contributes a nondynamical phase that is related to the Berry phase[10].

There are two extreme cases that may exist. The first and easiest is where the uncertainty in the energy is constant through the sum over histories. This would correspond to a process where the uncertainty is constant and reflects in a coarse grained manner the measurement time, where the uncertainty outcome is contained in this phase term. The other case is where measurement or parametric amplification is a process that changes the uncertainty of the energy through the evolution. Here the uncertainty will change with each δt of the measurement. In this case this phase term will involve the squeezing of states.

It is clear Bohm's formalism of quantum mechanics has no inherent structure beyond the standard formalism of quantum mechanics. In order to derive a theory of quantum mechanics different from the standard formalism one must impose different structures on quantum fluctuations. These fluctuations are Markovian, where this can be seen in the delta function correlations in equations 2.15 and 2.18. Fluctuations at one time and position in space contain no information about fluctuations at other times and positions in space. A modification of quantum mechanics with hidden variables would then have to have a nonMarkovian structure.

One may consider the overlap of the wave functions as

$$\langle\psi(t)|\psi(t')\rangle = \prod_n \langle\psi(t)|e^{i(t\ +\ n\delta t)\mathcal{L}(t\ +\ n\delta t)}|\psi(t)\rangle \simeq \langle e^{i\sum_n \delta t \mathcal{L}(t)}\rangle, \qquad (2.50)$$

where the analog of this in statistical mechanics is a partition function. In a case with Markovian fluctuations the analog in statistical mechanics is equilibrium. If one assumes that the fluctuations are superMarkovian the quantum mechanical analog would be a system described by a master equation, where various decoherent processes, such as spontaneous emission and quantum noise, are taken into account. SuperMarkovian fluctuations are those that bury information with time, where fluctuations at one position time erase quantum information at later times. This corresponds to the breakdown in coherence in a quantum system. Such a system is a finite quantum system coupled to an open environment. The evolution of such a system will experience decoherence as quantum information is lost to the environment, where the phase space volume of the finite quantum system changes due to its coupling to a thermal or quantum noisy environment. This change in the phase space volume is such that quantum information contained on the energy surface is buried.

The difficulty with any hidden variable, or a preferred trajectory that contains the beable, is this infers fluctuations must be subMarkovian. Here fluctuations must act in consort or synchronicity so that quantum information, rather than being lost to an environment, is spontaneously assembled from the environment. This is a situation where fluctuations assemble information that is contained in the classical-like trajectories so that only a few, or potentially only one, quantum trajectory becomes the active channel. It is physically apparent that the statistical mechanics analog of such a system is one where thermal fluctuations are able to decrease the entropy of a system. Alternatively, one might consider this as a thermodynamic analog of situations where temperatures are negative. Such a quantum system with subMarkovian fluctuations is one that must be held to serious skepticism. To explore this question additional research must be completed. Currently it appears that a system with subMarkovian fluctuations that generate an "inner ontology" to quantum mechanics should be regarded with suspicion.

Bohm's decomposition of the Schrödinger equation does not lead to any physics that is significantly different. A path integral may be derived with an integration measure that depends upon the quantum potential. The integration measure is then seen as influenced by the "smearing out" of observables, where this is seen in the approximation in equation 2.49. The quantum potential correlates a given quantum trajectory with all others and gives a probability for that path. That Bohm's formalism of quantum mechanics constructs a path

2.3: ADM Relativity, the Wheeler-DeWitt Equation and Torsional Flucutations

The Wheeler-DeWitt equation is a quantum form of the Hamiltonian constraint equation in the ADM space plus time formalism of general relativity. Spacetime is foliated by spacial surfaces that are "pushed" forward by lapse functions. This treats the spacetime manifold $\mathcal{M}$ as decomposed into $\Sigma^{(3)} \times \mathbf{R}$, where $\Sigma_t^{(3)}$ is a spacelike region and $\mathbf{R}$ is a real interval [11]. On $\mathcal{M}$ there is the metric covariantly constant $g_{\mu\nu}$ with respect to the operator ∇_μ. The spacetime $\mathcal{M}$ is foliated into spacelike slices $\Sigma_t^{(3)}$, where the time parameter is a defined along $\mathbf{R}$ and defines the vector field $t_\mu = \partial_\mu t$. Thus for some $\Sigma_t^{(3)}$ there exists a future directed normal vector n_μ and the Gauss fundamental forms

$$h_{\mu\nu} = g_{\mu\nu} + n_\mu n_\nu, \quad K_{\mu\nu} = -\frac{1}{2}\mathcal{L}_n h_{\mu\nu}, \tag{2.51}$$

where $\mathcal{L}_n$ denotes the Lie derivative. Given a projector operator $P_i{}^\mu$ its action on $h_{\mu\nu}$ is to define $h_{i\nu}$, which projects $h_{\mu\nu}$ from $\mathcal{M}$ to $\Sigma_t^{(3)}$ according to $P_i{}^\mu P_j{}^\nu h_{\mu\nu} = h_{ij}$. This defines the Gauss fundamental forms on $\Sigma_t^{(3)}$. The lapse and shift functions for the foliation are then $N = -t^\mu n_\mu$ and $N^i = h^i{}_\mu t^\mu$ with $t^\mu = Nn^\mu + N^\mu$.

The Ricci scalar $R = R_{\mu\nu} g^{\mu\nu}$ defines the Lagrangian for the Hilbert-Palatini action

$$S = \int d^4x (-g)^{1/2} R. \tag{2.52}$$

This is fraught with some interpretive problems. In classical mechanics the notion of time exists because there is a clock that exists outside the system. With general relativity there is no such external clock to a cosmology. Time is then a measure of the diffeomorphism of the spacial surface and this can be reparameterized arbitrarily. In this manner time can be thought to "disappear" from cosmology. The Hamiltonian constraint $NH = 0$ then defines the arbitrary reparameterizations of the diffeomorphisms of general relativity. Similarly with quantum gravity the Wheeler-DeWitt equations $\mathcal{H}\Psi = 0$ defines the reparameterizations of all trajectories in superspace.

However, there is a way out of this "no time" problem if the topology remains unchanged. It is possible to appeal to DeSitter spacetime construction with an added dimension and the group of diffeomorphisms of general relativity embeds into this space $SO(3,1) \subset SO(3,2)$[12]. This additional dimension defines a time or a "pseudotime" that permits the Wheeler-DeWitt equations to be integrated in the same manner as the Schrödinger equation. The spacelike manifold of evolution is described by the Wheeler-DeWitt equation, which is canonically quantized according to conjugate observables π_{ij}, g^{ij}, the spacelike momentum and metric respectively. The momentum operator is written as $\hat{\pi}_{ij} = -i\hbar \delta/\delta g^{ij}$. Classically the momentum is defined by the extrinsic curvature K_{ij} according to $\pi_{ij} = g^{1/2}(g_{ij} TrK - K_{ij})$. The Hamiltonian constraint in the ADM formalism of general relativity carries over to an operator H that acts on the wave functional $\Psi[g]$ as $\mathcal{H}\Psi[g] = 0$, often called the wave functional of the universe. Introduce a scalar field ϕ onto the spacetime that extends the wave functional to $\Psi[g, \phi]$. The Wheeler-DeWitt equation takes the form [13]

$$-\left(G^{ijkl}\frac{\delta^2}{\delta g_{ij}\delta g_{kl}} + R^{(3)} - \frac{\partial^2}{\partial\phi^2} + \phi^2\right)\Psi[g,\phi] = 0. \tag{2.53}$$

Here G^{ijkl} is the supermetric for the minisuperspace of 6 dimensions.

This added dimension in the de Sitter cosmology is treated in the Wheeler-DeWitt equation as an additional scalar field with harmonic oscillator modes. The wave function may then be expanded into a set of modes of the field ϕ as $\Psi[g, \phi] = \sum_n \Psi_n[g] a_n(\phi)$, so that

$$-\frac{1}{2}\left(\frac{d^2}{d\phi^2} - \phi^2\right)a_n(\phi) = (n + 1/2)a_n(\phi). \tag{2.54}$$

Here the scalar wave has been rescaled as $\phi \rightarrow \phi/(\sqrt{2\pi}\sigma R)$, where R is the radius of the cosmology and σ is a normalization factor. The eigenvalues may then be interpreted as due to the application of a time derivative on a basis of elements $\psi_n(\phi) = \psi_{n0} e^{-i(n+1/2-E_0)t}$,

$$\mathcal{H}\Psi_n[g] = i\frac{\partial}{\partial t}\psi_n[g] = (n + 1/2 - E_0)\psi_n[g], \tag{2.55}$$

where the time or pseudotime derivative measures the numbers of Planck units. This scalar field then acts to define a time parameter so that the Wheeler-DeWitt equation becomes a Schrödinger-like equation. This is a curious element to the Wheeler-DeWitt equation, for without some auxiliary field it is simply a constraint equation with no sense of time or parameterization of fields. It is the introduction of an auxiliary field which permits one to consider the Wheeler-DeWitt equation as an equation that describes the evolution of a cosmological wave function. For the metric of the DeSitter cosmology the solution to the Wheeler-DeWitt equation is numerically integrated and the solution to the wave function appears in figure 2.1. Within a quantum mechanical setting the metric momentum variable is replaced with an operator. In a metric representation of the theory this means that the conjugate momentum becomes the operator

$$\hat{\pi}_{ij} = \frac{\hbar}{i}\frac{\delta}{\delta g^{ij}}. \tag{2.56}$$

The Poisson bracket for classical ADM gravity is replaced by a commutator in the canonical quantization picture of quantum gravity. Assume that the cosmological wave functional is in polar form $\Psi[g] = R[g]exp(iS/\hbar)$, where $R[g]$ is an amplitude that in a semiclassical WKB limit is slowly varying. This wave function is a form used in the Bohm formalism of quantum mechanics, which is useful here in illustrating metric fluctuations. The classical momentum is

$$\langle \pi_{ij} \rangle = \frac{\delta S}{\delta g^{ij}}, \tag{2.57}$$

36 Quantum Fluctuations of Spacetime

which means that the momentum operator acts on the wave function to give

$$\hat{\pi}_{ij}\Psi[g] = \langle \pi_{ij}\rangle \Psi[g] + \frac{\hbar}{i}\frac{\delta R}{\delta g^{ij}}R^{-1}\Psi[g]. \tag{2.58}$$

This is interpreted as the expected conjugate momentum metric plus quantum corrections that arise due to uncertainty fluctuations.

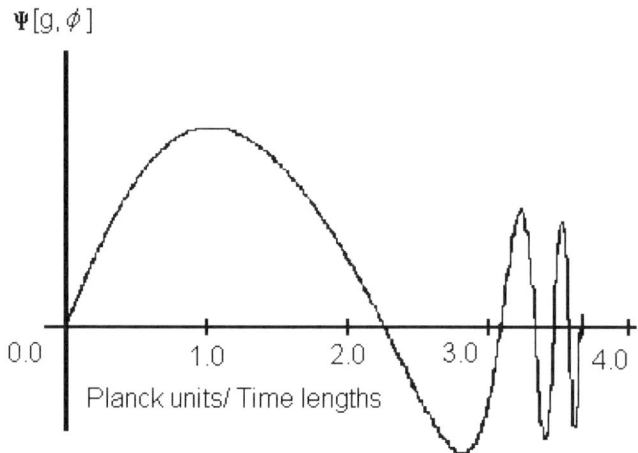

Solution to the Wheeler-DeWitt Equation.
Figure 2.1

Now consider accelerations $\mathbf{a} = d\mathbf{u}/d\tau$, with $\mathbf{u}^2 = -1$. This means that $\mathbf{a} \cdot \mathbf{u} = 0$. Assume an observer in a rest frame observes this accelerated system so at that instant $\mathbf{u} = \mathbf{n}$. The accelerated frame can be seen according to the covariant constancy of the first fundamental form $h_{\mu\nu}$

$$0 = \frac{dn_\mu}{d\tau}n_\nu, \tag{2.59}$$

where the second fundamental form may be expressed with the Lie derivative as,

$$\mathcal{L}h_{\mu\nu} = n_{\mu;\nu} + n_{\nu;\mu} + n_\mu a_\nu + n_\nu a_\mu. \tag{2.60}$$

The acceleration is then $a^\mu = n^\nu \nabla_\nu n^\mu$, with ∇_ν being the covariant derivative. This then means

$$n_{\mu;\nu} = \frac{1}{2}n_{(\mu;\nu)} + \frac{1}{2}n_{[\mu;\nu]} = -K_{\mu\nu} - n_\mu a_\nu + \frac{1}{2}n_{[\mu;\nu]}. \tag{2.61}$$

In the description of a particle in motion with an acceleration the normal n^μ pertains to a classical spacial manifold. Since $t_\mu = \partial_\mu t$ it is obvious that $n_{[\mu;\nu]} = 0$.

Consider a normal n_μ which pertains to a manifold $\Sigma^{(3)}$ within a foliation of spacial manifolds that define a quantum superposition of spacial manifolds. This foliation is defined by a path integral that defines the transition from g_{ij} to g'_{ij}

$$\langle g_{ij}|g'_{ij}\rangle = \int \mathcal{D}[g_{ij}]e^{iS/\hbar}. \tag{2.62}$$

This foliation in the spacial metric indicates the existence of quantum fluctuations in the spacial manifold that have their classical analogues in variations. The variation in the action is due to the volume between one $\Sigma^{(3)}$ and $\Sigma'^{(3)}$. This variation is a bubble in spacetime $\delta \mathbf{n} \cdot \mathbf{\Sigma}^{(3)}$, where $\mathbf{\Sigma}^{(3)}$ is a vector dual to the volume form defined by $\Sigma^{(3)}$. The standard action

$$S = (-g)^{-1/2}\int d^4x R = (-g)^{-1/2}\int d^4x\big((TrK)^2 - Tr(K^2) + R^{(3)}\big) \tag{2.63}$$

is seen to be derived from

$$S = (-g)^{-1/2}\int d^4x\big(n^\mu{}_{[;\nu}n^\nu{}_{;\mu]} + R^{(3)}\big), \tag{2.64}$$

where $n^\mu{}_{[;\nu}n^\nu{}_{;\mu]}$ is the kinetic energy component of the Lagrangian. The variation in the action results in

$$\delta^\mu{}_\sigma(g^{\rho\mu}n^\mu{}_{;\nu\rho} - \delta^\rho{}_\mu n^\nu{}_{;\nu\rho}) + \frac{1}{2}R^{(3)}{}_{;\sigma} = 0. \tag{2.65}$$

This means the covariant derivative of the extrinsic curvature is

$$K_{\mu\nu;\rho} - g_{\mu\nu}K_\sigma{}^\sigma{}_{;\rho} + (n_\mu a_\nu)_{;\rho} - \frac{1}{2}n_{[\mu;\nu];\rho} = 0. \tag{2.66}$$

The term $n_{[\mu;\nu]}$ is set to zero, but with quantum fluctuation is not generally zero. Consider a fluctuation in the scalar $t = t_0 + \delta t$, where δt may be complex valued in general. The time vector $t^\mu = t_0{}^\mu + \delta t^\mu$, where δt^μ is a fluctuation in the vector field, is physically motivated by above discussions of quantum fluctuations. As $t^\mu = Nn^\mu + N^\mu$ then $\delta n^\mu = \frac{1}{N}\delta t^\mu$. Then write $\delta n^\mu = n^\mu{}_{;\nu}\delta x^\nu$. Clearly $dt = t_\mu dx^\mu$, but with fluctuations $t_\mu dx^\mu = (t_\mu + \delta t_\mu)dx^\mu$. δt_μ is not determined by a gradient or differential action on a function on $\sigma^{(3)}$. Since the fluctuation δt_μ is not the result of the application of a differential operator on a 0-chain it is a cocycle. By the definition of n^μ the same applies. A cocyle exists for the fluctuations in t^μ and n^μ. Consider this fluctuation as a loop that takes t_0 to $t = t_0 + \delta t$ and back, as parameterized according to an additional variable such as the scalar field ϕ. A loop integral $\oint dt$ gives by Stokes' law $\int t_{\mu;\nu}dx^\mu \wedge dx^\nu$. This then evaluates the antisymmetric tensor $n_{[\mu;\nu]}$ on a two-chain. Since quantum fluctuations vanish under expectations $\langle \delta t\rangle = 0$, expectations of this antisymmetric term then in turn vanishes. However, quadratic terms in the fluctuation and this antisymmetric tensor are delta function correlated and reflect Markovian statistics.

Define the fluctuation connection term $\rho^\alpha{}_{\mu\nu}$ by $n_{[\mu;\nu]} = \rho^\alpha{}_{\mu\nu}n_\alpha$. This connection term is antisymmetric in the $\mu\nu$ indices. This is then a torsional term [14] that describes a teleparallel connection associated with quantum fluctuations. The Ricci identity on $\Sigma^{(3)}$ is then

$$P^\alpha{}_\mu P^\beta{}_\nu R_{\alpha\sigma\rho\beta} n^\sigma n^\rho = n^\sigma [\nabla_\mu, \nabla_\sigma] n_\nu = \quad (2.67)$$

$$\mathcal{L}_\mathbf{n} K_{\mu\nu} + K_{\mu\eta} K^\eta{}_\nu + \mathcal{L}_\mathbf{n}(\rho^\alpha{}_{\mu\nu}) n_\alpha + \rho^{\alpha\sigma}{}_\mu \rho^\beta{}_{\sigma\nu} n_\alpha n_\beta + (\rho^\alpha{}_{\sigma\mu} K^\sigma{}_\nu + \rho^\alpha{}_{\sigma\nu} K^\sigma{}_\mu) n_\alpha$$

Here the projector operators have an antisymmetric portion in addition to $\delta^\mu{}_\nu + n^\mu n_\nu$. With quantum fluctuations a reasonable choice for the projector is $-i\sigma^\mu{}_\nu + \delta^\mu{}_\nu + n^\mu n_\nu$, where the spin tensor is $\sigma^\mu{}_\nu = (i/2)[\gamma^\mu, \gamma_\nu]$. The first two terms in equation 2.67 are the standard Ricci tensor and the next two define the curvature due to torsion and the last term is a cross term. This ADM curvature is the same as the torsional curvature presented by Vargas and Torr in equation 58 of [15].

From these the Lagrangian density $16\pi\mathcal{L} = (-g^{(4)})^{1/2} R$ is then

$$(-g^{(4)})^{1/2} R = (-g^{(4)})^{1/2} \big(R^{(3)} + (Tr\ K)^2 - Tr(K^2) + \rho^\alpha{}_{\mu\nu} \rho_\beta{}^{\mu\nu} n_\alpha n^\beta \big), \quad (2.68)$$

where as a result the action may be written as $S = S_0 + \rho^\alpha{}_{\mu\nu} \rho_\beta{}^{\mu\nu} n_\alpha n^\beta$, where last term in the action is due to torsional curvature from quantum fluctuations. From here the path integral may be expressed as

$$\langle g_{ij} | g'_{ij} \rangle = \int \mathcal{D}[g_{ij}] e^{iS_0} e^{i/16\pi \int d^4 x (-g^{(4)})^{1/2} \big(\rho^\alpha{}_{\mu\nu} \rho_\beta{}^{\mu\nu} n_\alpha n^\beta \big)}. \quad (2.69)$$

The dynamics induced by the action $\rho^\alpha{}_{\mu\nu} \rho_\beta{}^{\mu\nu} n_\alpha n^\beta$ is determined by the differential equation

$$\rho^\alpha{}_{\mu\nu;\sigma} n_\alpha + \rho^\alpha{}_{\mu\nu} n_{\alpha;\sigma} = 0, \quad (2.70)$$

which is equivalent to $\rho^\alpha{}_{\mu\nu} n_\alpha = constant$. The torsion then projected onto the velocity is a constant. This means that the torsional connection defines Killing fields $\rho_\beta{}^{\mu\nu} \rho^\alpha{}_{\mu\nu} n_\alpha = \xi_\beta \xi^\alpha n_\alpha$, for ξ^α the Killing vector for an accelerated reference frame. This means contractions over the $\mu\ \nu$ indices leads to a constant. So as an ansatz let $\rho^\alpha{}_{\mu\nu} = \xi^\alpha \sigma_{\mu\nu}$. This is the definition of surface gravity normal to the horizon induced by the fluctuation [16]

$$\big(\rho^\alpha{}_{\mu\nu} \rho_\alpha{}^{\mu\nu} \big)_{;\beta} = -2g\xi_\beta, \quad (2.71)$$

where g is the acceleration associated with the fluctuation. Since $\rho^\alpha{}_{\mu\nu} \rho_\alpha{}^{\mu\nu} = \xi^\alpha \xi_\alpha$ this is the square of the gravitational redshift factor due to torsional acceleration. The gravity on a test mass induced by a fluctuation is

$$g^2 = -\rho^\alpha{}_{\mu\nu}{}^{;\beta} \rho_\alpha{}^{\mu\nu}{}_{;\beta}. \quad (2.72)$$

This gravity is induced by a quantum fluctuation as it pushes a test mass away from its classical geodesic.

Torsion connections do not result in the violation of conservation of momentum. In the ADM formalism with torsion the second fundamental form is related to the standard one $K_{\mu\nu}$ by

$$k_{\mu\nu} = K_{\mu\nu} + \rho^\alpha{}_{\mu\nu} n_\alpha, \qquad (2.73)$$

where $\rho^\alpha{}_{\mu\nu} = -\rho^\alpha{}_{\nu\mu}$. The $G_{\mu\nu}$ component of the Einstein field equation is

$$G_{\mu\nu} = G_{\mu\nu}|_{\rho=0} + \frac{1}{2} g_{\mu\nu}(\rho^{\alpha\gamma\lambda}\rho^\beta{}_{\gamma\lambda}n_\alpha n_\beta) + \rho^\alpha{}_{\mu\nu;n} n_\alpha, \qquad (2.74)$$

where $,n$ reflects differentiation along the direction normal to $\Sigma^{(3)}$ and $G_{\mu\nu}|_{\rho=0}$ is the standard result of classical gravity without torsion. The field due to the torsional acceleration is then $\frac{1}{2} g_{\mu\nu}(\rho^{\alpha\gamma\lambda}\rho^\beta{}_{\gamma\lambda}n_\alpha n_\beta)$ and $\rho^\alpha{}_{\mu\nu,n}$. Further, from equation 2.74

$$\rho^\alpha{}_{\mu\nu;n}n_\alpha = \rho^\alpha{}_{\mu\nu;\sigma}n^\sigma n_\alpha = -2\rho^\alpha{}_{\mu\nu}n_\sigma n_\alpha{}^{;\sigma} = 0, \qquad (2.75)$$

where the annulment of this is seen by the symmetry of the α and σ indices and that $n^\alpha n_\alpha{}^{;\sigma} = 0$. This infers that the antisymmetric term in the stress-energy is zero. This torsion reflects the momentary violation of momentum due to quantum fluctuations according to the uncertainty principle of quantum mechanics. There is an expectation

$$\langle \rho^\alpha{}_{\mu\nu}(x) \rho_\beta{}^{\mu\nu}(x') \rangle = g^2 f(\phi) \delta^\alpha{}_\beta \delta^3(x-x'), \qquad (2.76)$$

where f is a scaling function of a scalar field ϕ and κ is a constant near unity. This term identified as $Tr\rho^2$ in the Bohm description is from the quantum potential. A reasonable choice for $f(\phi)$ is $\kappa(\nabla\phi)^2/\phi^2$ and κ is a constant.

The torsional Lagrangian

$$\mathcal{L} = \frac{1}{16\pi} g_{\mu\nu}(\rho^{\alpha\mu}{}_\gamma \rho_\beta{}^{\nu\gamma} n_\alpha n^\beta) = \frac{1}{16\pi} \rho^\alpha{}_{\mu\nu} \rho_\beta{}^{\mu\nu} n_\alpha n^\beta, \qquad (2.77a)$$

where the stress-energy is $T^{\mu\nu} = 2\frac{\partial \mathcal{L}}{\partial g_{\mu\nu}} - g^{\mu\nu}\mathcal{L}$,

$$T^{\mu\nu} = \frac{1}{4\pi}\left(\rho^{\alpha\mu}{}_\gamma \rho_\beta{}^{\nu\gamma} n_\alpha n^\beta - \frac{1}{4} g^{\mu\nu} \rho^\alpha{}_{\gamma\sigma} \rho_\beta{}^{\gamma\sigma} n^\alpha n^\beta\right). \qquad (2.77b)$$

For the torsional tensor determined by a potential term

$$\rho^\alpha{}_{\gamma\lambda} n_\alpha = i\sqrt{2\pi}(\mathcal{D}_\lambda \omega_\gamma - \mathcal{D}_\gamma \omega_\lambda), \qquad (2.78)$$

where $\mathcal{D}_\mu = \partial_\mu + ig\omega_\mu$, the stress-energy assumes the form of a Yang-Mills theory. Thus torsion is a gauge field with the form $\mathcal{L} = -\frac{1}{4} F^{\mu\nu} F_{\mu\nu}$. By equation 2.77 it is evident that $Tr\rho^2 \geq 0$.

This theory is similar to Yilmaz theory of gravity, which is proposed as an alternative to classical gravitation as described by general relativity. The principle difference is that the above additional stress-energy term has a Yang-Mills structure, whereas Yilmaz advances a scalar field that gives a contribution to the stress-energy. Yilmaz theory advances a scalar field to recover a Newtonian limit, which appear not to be correct. Evidently the quantum fluctuations of the spacetime metric are fields which obey Yang-Mills equations. It is then tempting to consider that these fluctuations are tied to local gauge transformations of a real Yang-Mills field theory, where these gauge transformations are due to quantum fluctuations of the Yang-Mills field. It is reasonable to consider this gauge field theory is the unified field theory, either $SU(5)$ or $SO(10)$, of the electromagnetic, weak and nuclear interactions. This is similar in its thrust to the derivation of equation 2.53, where an auxiliary field was used to find a time evolution equation from the Wheeler-DeWitt equation. This suggests fluctuations in the metric of space are tied to fluctuations of the gauge fields on that manifold.

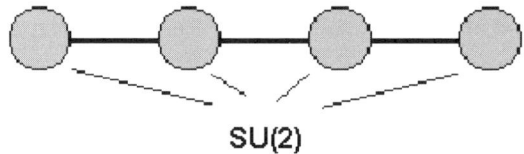

SU(5) Dynkin diagram.
Figure 2.2

This then introduces something unique. Assume that the fluctuations of spacetime obey the SU(5) algebra of the simplest GUT model. With SU(5) there is the Dynkin diagram in figure 2.2. For each of these $SU(2)'s$ there exist the Cartan elements H_i, $i = 1, ..., 4$. For $U(5)$ there are the additional H_5 formed by combinations of the other Cartan elements H_i, $i = 1, ..., 4$. H_5 defines an $U(1)$ that makes $U(5)$ non-semisimple. Hence within $U(5)$ are embedded two copies of $SU(5)$ with a $U(1)$ rotation between them. So assume that gravity has a "virtual gauge field" associated with the $SU(5)$ GUT gauge field. The total GUT theory as $U(5)$ infers the existence of an additional gluon to the theory corresponding to the $U(1)$. This has profound implications, for as energy approaches the Planck energy the charges of quarks are 0 and ±1 as given in the Han Nambu model. The fractional charges of quarks are due to a fractional quantum Hall effect in QCD at lower energy. The notion of $U(5)$ GUT and the additional gluon with integral quark charges was suggested by P. Rowlands [17]. Of course this can be embedded into larger gauge theories that concern strings. Greene, Schwarz and Witten demonstrate how $U(n)$ groups will embed into a heterotic string theory.

The emergence of this mirror gauge field from torsional connections occurs on the level of curvature tensors. This is different from the Kaluza-Klein theory where gauge fields emerge on the level of connections. These gauge fields are derived from basis forms ω_μ^I, where the upper index is the additional spacial dimensions $I > 4$. With the cyclicity condition, no derivatives of spacetime forms with respect to the additional dimensions, a gauge field emerges of the form $d\omega_\mu^I = F_{\mu\nu}^I \wedge dx^\nu$. Gauge theory emerges from a

Kaluza-Klein extension of relativity. Since the group product for gauge fields emerges from torsional connections and from Kaluza-Klein theory, with the total gauge theory $U(n) = SU(n) \times U(1)$ this is then a gauge field that spans the torsional curvature and nontorsional connection levels.

This leads to considerations of the cosmological constant. With the identification $\langle Tr\rho^2 \rangle \sim \kappa(\nabla\phi)^2/\phi^2$ this leads to a Lagrangian that is similar to the Armandariz-Mukhanov-Steinhardt Lagrangian that describes the acceleration of dust in a cosmology [18]. If this term is modified on a large scale according to

$$\langle Tr\rho^2 \rangle = p_k(\nabla\phi)^2/\phi^2, \qquad (2.79)$$

where p_k is the pressure due to dust, then this recovers the AMS Lagrangian. This Lagrangian leads to an equation of state with a negative pressure. This still leads to a positive mass-energy density, but also results in a negative momentum density associated with this negative pressure. This induces a continuous acceleration of particles in the universe. The similarity between the form of $\langle Tr\rho^2 \rangle$ and the AMS Lagrangian infers that this inflationary "quintessence" may be ultimately induced by quantum fluctuations of spacetime.

If the vacuum energy density of a cosmology is constant the expansion of that universe will change the net vacuum energy. Thus something must absorb work from the cosmology to conserve energy. It then appears that the vacuum may be structured in such to provide the required negative pressure required.

In section 2.4 gravitational squeezing of the quantum gravity vacuum is considered. This property of the vacuum to act as its own parametric amplifier has the affect of renormalizing $\hbar \rightarrow \hbar'$. The evaluation of $Tr\rho^2$ for squeezed states depends upon the magnitude of the squeezing parameter, $|z|^2$. Equation 2.79 will then involve gravitational self-squeezing of the vacuum. This then suggests that the Planck length scale will on average be renormalized upward, where this may then permit the coincidence of gauge coupling parameters at $\sim 10^{15}$ GeV and give a potential solution to the gauge hierarchy problem. This is further the value of the Planck energy suggested in [17] with the additional 25th generator in the $U(5)$ group. Further, this suggests that the problem of the cosmological constant is strongly related to the gauge hierarchy problem.

2.4: Thermalized Torsional Fluctuations

Quantum fluctuations of spacetime give a gauge theory that mirrors physical gauge theories. This may be good news, for it could mean that the states for quantum gravity could be considered with gauge theoretic calculations. Gravity could then be treated as approximately classical, or at a one-loop Feynman level. At this point the thermal nature of quantum gravity has not been discussed. Finsler connections are used explicitly to derive the thermal nature of the quantum gravity vacuum.

Consider the 3-velocity as $n^\alpha = U^\alpha$ and the spacelike momentum $\pi_{ij} = g^{1/2}(g_{ij}TrK - K_{ij})$. The generalized spacelike momenta that includes quantum fluctuations is

42   Quantum Fluctuations of Spacetime

$$\Pi_{ij} = \pi_{ij} + \rho^\alpha{}_{ij} U_\alpha. \tag{2.80}$$

This is given quantum mechanically by the operator $\hat{\pi}_{ij} = i\delta/\delta g^{ij}$, which acts on the metric variable, where since the torsion tensor describes an acceleration $\rho^\alpha{}_{ij}U_\alpha$ acts on the tangent bundle to the metric. This then describes an operator with a Finsler connection. An overview of Finsler geometry is given below. $D = c^2/g$ is the distance from an accelerated test mass and the particle horizon. Let $\hat{\sigma}_{ij} = i(\gamma_i\gamma_j + \frac{\delta}{\delta v^{ij}})$, where v^{ij} are components of the tangent metric space to $\delta\Sigma^{(3)}$ due to the change in the spacial surface by quantum fluctuations. This then requires an extrinsic tensor field on the spacial manifold $\Sigma^{(3)}$ be written as

$$\phi_{ij}(g,\ v) = \int \frac{d^3 p}{2p_0} a_{ij} e^{-i\pi_{kl}g^{kl} - D\rho^\alpha{}_{kl}v^{kl}U_\alpha}, \tag{2.81}$$

where a_{ij} is a lowering operator. Its role is similar to those in Sen connections. These analyses will be considered in spinor form below. The momentum operator is generalized as

$$\hat{\Pi}_{ij} = \frac{1}{i}\left(\frac{\delta}{\delta g^{ij}} + D^{-1}\frac{\delta}{\delta v^{ij}}\right) \tag{2.82}$$

The form of the phase term is motivated by the action of the operator $\hat{\Pi}_{ij}$ on the phase term in equation 2.81 produces the result

$$\tilde{g}_{ij} = g_{ij} - iDv_{ij}. \tag{2.83}$$

$\tilde{g}_{ij}$ is a general metric, where the first term on the right hand side is the standard metric for $\Sigma^{(3)}$ and the last term is the for a tangent bundle metric to $\Sigma^{(3)}$. In quantum field theory it is common to consider such fields as Hermitian conjugates. The above field is the positive mode $\phi^+{}_{ij}$. The total field is then a sum of a positive and negative field $\phi_{ij} = \phi^+{}_{ij} + \phi^-{}_{ij}$. The fields are expanded as

$$\phi^-{}_{ij}(g,\ v) = \int d\omega a^\dagger{}_{ij}(\omega) e^{i\pi_{kl}g^{kl} - Dn^\alpha \pi^{kl} v_{kl} U_\alpha}, \tag{2.84}$$

where $\pi_{ij} = \pi_{ij}(\omega)$. Now construct the propagator with the commutation of these two fields

$$[\phi^-{}_{ij}(g,\ v),\ \phi^+{}_{kl}(g',\ v')] = i\Delta_{ij}(g,\ g',\ v,\ v')\delta_{ik}\delta_{jl}. \tag{2.85}$$

As pointed out by Brandt [19], who evaluated scalar fields with Finsler connections in Minkowski space, this type of propagator is nonzero for spacelike separations if the acceleration is sufficiently large. This then changes in principle the nature of causal domains.

We now consider the real valued term in the phase $Dn^\alpha \rho^\alpha{}_{ij} n_\alpha v^{ij} U_\alpha$. Under the equivalence of the Lagrangian and Hamiltonian pictures of mechanics $\rho^\alpha{}_{ij}n_\alpha \propto v_{ij}$. As an ansatz consider this phase term to be a kinetic energy part of the Hamiltonian multiplied by a constant. As a result it is possible to replace $(tr\pi)^2$ with just the diagonal entries in the form $a^\dagger a$, under normal ordering. In the case of a constant acceleration the Killing vector defines the Killing parameter v by $\xi^\alpha \nabla_\alpha v = 1$. This is in coordinates $U = t - x$ and $V = t + x$, where $v = (1/a) ln|V|$. The term $\xi^\alpha U_\alpha$ is then equal to $U^\alpha \nabla_\alpha v$. By using $ds^2 = dU dV$ it is possible to perform the following replacement

$$D n^\alpha \pi^{ij} v_{ij} U_\alpha \rightarrow \frac{1}{2}\tau \int dw a^\dagger(w) a(w) = \tau H \qquad (2.86)$$

where $\tau = (1/g) ln(1/2(U+V))$ is seen below to be the Killing time. This has been done for a constant acceleration. However, an elementary fluctuation with $\nu \sim 10\nu_p$ should be considered as one where the acceleration increases from 0 to some maximum and then back to 0. The fluctuation may be described according to small intervals of constant Killing time, each with a unique acceleration g.

The kinetic energy Hamiltonian for the Wheeler DeWitt equation is obviously modified by

$$\hat{H} = -G^{ijkl}\frac{\delta^2}{\delta g^{ij}\delta g^{kl}} + iD^{-1}\frac{\delta^2}{\delta v^{ij}\delta v_{kl}}. \qquad (2.87)$$

The occurrence of D^{-1} instead of D^{-2} is due to an implicit tangent space metric term with a term D required to cancel the implicit D^{-1} in the tangent space momentum variable $\rho^\alpha{}_{ij}$. An $i = \sqrt{-1}$ rotation of time defines a Euclidean metric. Hence the occurrence of this term in the action gives an imaginary valued action that enters the phase as

$$exp\left(-c^2 \rho^\alpha{}_{ij} \rho_\alpha{}^{ij}/g\right) = e^{-\epsilon/g}. \qquad (2.88)$$

This then has the form of a thermal partition function for $g \propto T$, where ϵ is the energy of the vacuum state.

Consider momentum metric fluctuations with the ADM Hamiltonian constraint and a wave functional in polar form. The Hamiltonian operator in the Wheeler-DeWitt equation,

$$\hat{H}\Psi[g] = -\left(G_{ijkl}\frac{\delta^2}{\delta y^{ij}\delta g^{jl}} - g^{1/2} R^3(g)\right)\Psi[g] = 0. \qquad (2.89)$$

The momentum operator $\hat{\pi}_{ij} = \frac{\hbar}{i}\frac{\delta}{\delta g^{ij}}$, which enters this Hamiltonian, acts on $\Psi[g] = A[g]e^{iS/\hbar}$ so the fluctuations in the momentum are

$$\pi_{ij} = \langle\pi_{ij}\rangle + \delta\pi_{ij}, \qquad (2.90)$$

where the momentum metric fluctuation is

$$\delta\pi_{ij} = \frac{\hbar}{i} A^{-1}[g]\frac{\delta A[g]}{\delta g^{ij}}. \qquad (2.91)$$

$A[g]$ is the amplitude of the cosmological wave functional $\Psi[g]$. For low energy, or large scale physics, where $\sqrt{G\hbar/c^3}\omega \ll c$ these fluctuations are small in comparison to the expectation values and semiclassical relativity is recovered. Consider scales where $\sqrt{G\hbar/c^3}\omega < 10c$ in general. From here on set $c = 1$ where $\sqrt{G\hbar}$ is the Planck length. These momentum metric fluctuations are related to the metric fluctuations according to

$$[\hat{g}_{ij}(r), \hat{\pi}^{kl}(r')]\Psi[g] = \frac{\hbar}{i}\{\delta g_{ij}(r), \delta\pi^{jl}(r')\}_{PB}\Psi[g] = \frac{\hbar}{i}\delta_i{}^k\delta_j{}^l\delta^3(r-r')\Psi[g]. \quad (2.92)$$

Given there is an "arrow of time" $t^\mu = Nn^\mu + N^\mu$ the momentum operator may be expressed as

$$\pi_{ij} = \langle\pi_{ij}\rangle - \frac{\partial\pi_{ij}}{\partial N}\delta N. \quad (2.93)$$

This is computed with the Poisson bracket with the classical ADM Hamiltonian as

$$\pi_{ij} = \langle\pi_{ij}\rangle - \{H, \pi_{ij}\}_{PB}\delta N. \quad (2.94)$$

The corresponding term that involves fluctuations in the shift parameter are easily demonstrated to vanish. The fluctuation is evaluated to be

$$\pi_{ij} = \langle\pi_{ij}\rangle + g^{1/2}\frac{\partial R^{(3)}}{\partial g^{ij}}\delta N. \quad (2.95)$$

The Killing vector is a vector that is constant along the momentum of a particle. Cryptically this is written as

$$\mathbf{p}\cdot\xi_\nu = const. = \frac{d}{d\lambda}\xi_\nu = p^\mu\nabla_\mu\xi_\nu. \quad (2.96)$$

The fluctuation in momentum of a particle, as it is carried along with a fluctuation in the metric momentum, is then in components

$$\hat{p}_\mu = (\hat{p}_0, \hat{\pi}_{ij}n^j), \quad (2.97)$$

which means that the fluctuation component induced by the momentum operator is seen to be

$$\delta p_\mu = -\left(\delta p_0, -g^{-1/2}\frac{\delta R^{(3)}(g)}{\delta g^{ij}}n^j\right)\delta N. \quad (2.98)$$

When this momentum fluctuation is projected onto the Killing field this results in the constant vector

$$\delta p^\mu\nabla_\mu\xi^\nu = -\left(\delta p_0\partial_t\xi_0, -g^{-1/2}\frac{\delta R^{(3)}(g)}{\delta g^{ij}}n^j\nabla_i\xi_k\right)\delta N. \quad (2.99)$$

This projection further satisfies

$$\delta p^\mu\nabla_\mu\xi^\nu = \delta(p^\mu\nabla_\mu\xi_\nu) - p^\mu\nabla_\mu(\delta\xi_\nu), \quad (2.100)$$

where the first right hand side term vanishes. Now let

$$g^{-1/2}\frac{\delta R^{(3)}(g)}{\delta g^{ij}}n^j\delta N = \sqrt{G\hbar}\omega^2\delta N\left(\frac{\partial}{\partial x}\right)_i. \tag{2.101}$$

Since the fluctuation is massless and propagates along a null direction, then by equation 2.100 the Killing vector is

$$\chi_\mu = \frac{\delta\xi_\mu}{\delta N} = \sqrt{G\hbar}\omega^2\left(x\left(\frac{\partial}{\partial t}\right)_\mu - t\left(\frac{\partial}{\partial x}\right)_\mu\right), \tag{2.102}$$

where the Heisenberg uncertainty principle is invoked and the Planck constant explicitly shown. The acceleration is $g = \sqrt{G\hbar}\omega^2$. Let the fluctuation occur along a single spacial direction. The Killing vector is expressed according to (U, V) coordinates as

$$\chi_\mu = \sqrt{G\hbar}\omega^2\left(V\left(\frac{\partial}{\partial V}\right)_\mu - U\left(\frac{\partial}{\partial U}\right)_\mu\right). \tag{2.103}$$

V is the inertial time defined for the classical spacetime. The quantum fluctuation determines a Killing time v defined on the horizon h_+ as

$$\xi^\mu\nabla_\mu v = 1. \tag{2.104}$$

From the Killing vector this equals

$$\sqrt{G\hbar}\omega^2 V\frac{\partial v}{\partial V} = 1 \tag{2.105a}$$

or

$$v = \frac{1}{\sqrt{G\hbar\omega^2}}ln|V|. \tag{2.105b}$$

The field expanded into normal modes is

$$\phi(V, y, z) = \int_0^\infty d\omega\left(f^+(V, y, z)a(\omega) + f^-(V, y, z)a^\dagger(\omega)\right). \tag{2.106}$$

The Killing time is given by fluctuations around the horizon. As such the phases are restricted to the horizon according to

$$f^\pm = \begin{cases} g(y, z)e^{\pm i\omega v}, & V > 0 \\ 0, & V < 0 \end{cases}. \tag{2.107}$$

A Fourier transform of this phase is

$$\mathcal{F}_\omega(\Omega, y, z) = \frac{1}{\sqrt{2\pi}}\int_{-\infty}^\infty e^{i\Omega V}f_\omega(V, y, z)dV. \tag{2.108}$$

46 *Quantum Fluctuations of Spacetime*

For $V > 0$ the phase function has the generator V. The analytical extension into the complex plane with $V = i\theta$ is

$$\ln V = \ln(i\theta) = i\frac{\pi}{2} + \ln\theta. \qquad (2.109)$$

The Fourier transform then assumes the form

$$\mathcal{F}_\omega(\Omega, y, z) = \frac{i}{\sqrt{2\pi}} exp\left(-\frac{\pi a^\dagger a}{\sqrt{G\hbar\omega^2}}\right) \int_{-\infty}^\infty e^{-\omega\theta} exp\left(\frac{-ia^\dagger a}{\sqrt{G\hbar\omega^2}}\ln\theta\right) d\theta. \qquad (2.110)$$

It is then possible to examine the same Fourier integral for $\Omega \to -\Omega$, which corresponds to $V \to -V$, which gives

$$\mathcal{F}_\omega(-\Omega, y, z) = -exp\left(-\frac{\pi a^\dagger a}{\sqrt{G\hbar\omega^2}}\right)\mathcal{F}_\omega(\Omega, y, z). \qquad (2.111)$$

As this corresponds for $V < 0$ this is an extension of the wave function across the horizon. It is apparent that,

$$\phi_i = \phi_{iI} + exp\left(-\frac{\pi a^\dagger a}{\sqrt{G\hbar\omega^2}}\right)\phi_{iII}. \qquad (2.112)$$

This field is then represented by fields that are on both sides of the horizon, ϕ_{iI} on the "observer's" side of the horizon and ϕ_{iII} on the other side.

The metric momentum variable is written according to the second fundamental form restricted to $\Sigma^{(3)}$ as $\pi^{ij} = g^{1/2}(g^{ij}Tr\,K - K^{ij})$. This momentum metric and the spacial metric g_{kl} satisfy the Poisson bracket $\{g^{ij}(x), \pi_{kl}(x')\} = \delta^i_k\delta^j_l\delta^3(x-x')$. Further, the field operators are assumed to arise from spacetime variables through the Ashtekar variables [20] $g_{ij} = \sigma^b{}_i e^b_j$ and $\pi_{ij} = \sigma^b{}_i p^b_j$ and with [21]

$$e^b_i = \int_0^\infty \frac{d\omega}{\omega}(a^{b\dagger}{}_i(\omega)\mathcal{F}^* + a^b_i(\omega)\mathcal{F}) \qquad (2.113a)$$

$$p^b_i = i\int_0^\infty d\omega(a^{b\dagger}{}_i(\omega)\mathcal{F}^* - a^b_i(\omega)\mathcal{F}). \qquad (2.113b)$$

The index b is an internal spin index. The term $\mathcal{F}$ is the oscillatory function expanded into some basis of orthogonal functions. The Einstein field G^{00} is then determined by

$$G^{00} = -\frac{1}{2}R^{(3)} + g^{-1}(Tr\pi^2 - \frac{1}{2}(Tr\pi)^2), \qquad (2.114)$$

where the kinetic energy portion when expanded in normal ordered harmonic oscillator modes is

$$g^{-1}(Tr\pi^2 - \frac{1}{2}(Tr\pi)^2) = \int_0^\infty d\omega \big(a^{b\dagger}_i(\omega)a^b_i(\omega)\mathcal{F}^*(\omega)\mathcal{F}(\omega)\big) +$$

$$\int_0^\infty d\omega \big(a_i^{b\dagger}(\omega)a_i^{b\dagger}(\omega)\mathcal{F}^*(\omega)\mathcal{F}^*(\omega) + a_i^b(\omega)a_i^b(\omega)\mathcal{F}(\omega)\mathcal{F}(\omega)\big). \tag{2.115}$$

The last term has zero expectation under $\int d\omega \langle n(\omega)\mathcal{O}n(\omega)\rangle$, where $\mathcal{O}$ is the last operator in equation 2.115.

Equations 2.113 a and b infer the operators a_i^b and $a_i^{b\dagger}$ written in affine coordinates are in Killing coordinates

$$\alpha_i^b = \hat{a}_i^b + a_i^b e^{-\pi\mathcal{H}/g}, \quad \alpha_i^{b*} = \hat{a}_j^{b\dagger} + a_i^{b\dagger} e^{-\pi\mathcal{H}/g}, \tag{2.116}$$

where $\hat{a}_j^b$ and $\hat{a}_j^{b\dagger}$ are operators for fields on the observer's side of the horizon with $[\hat{a}_j^b, a_j^{b\dagger}] = [\hat{a}_j^{b\dagger}, a_j^b] = 0$. These commutators vanish as any delta function condition on the spacial position of these fields must be zero by definition. These operators of quantum gravity are then generalized to account for accelerations induced by quantum fluctuations.

These modified operators lead to a generalized uncertainty principle. A commutation of these two operators is [22]

$$[\alpha_j^b, \alpha_k^{c*}] = \big(1 + e^{-2\pi\mathcal{H}/g}\big)\delta_{jk}\eta^{bc} - \frac{\pi}{g}e^{\pi\mathcal{H}/g}\big(\alpha_j^{0b*}\alpha_k^{0c} + \alpha_j^{0b}\alpha_k^{0c*}\big)e^{-\pi\mathcal{H}/g}, \tag{2.117}$$

where $\alpha_j^b = \hat{a}_j^b + \alpha_j^{0b}$, $\alpha_j^{0b} = a_j^b e^{-\pi\mathcal{H}/g}$. Further, η^{bc} is the metric for the internal $SO(4)$ group. It is a canonical exercise to demonstrate that the Heisenberg uncertainty relationship between the momentum and position variables in quantum mechanics is

$$(\Delta p_j^b)^2(\Delta e_k^c)^2 \geq \frac{1}{4}\big(i\langle[p_j^b, e_j^c]\rangle\big)^2. \tag{2.118}$$

Within the context of these modified operators the position and momentum operators are $\alpha_j^b = 1/\sqrt{2}(e_j^b + ip_j^b)$ and $\alpha_j^{b*} = 1/\sqrt{2}(e_j^b - ip_j^b)$, which leads to the fact,

$$(\Delta p_j^b)^2(\Delta e_k^c)^2 \geq \frac{1}{4}\big(\langle[\alpha_j^b, \alpha_k^{c*}]\rangle\big)^2. \tag{2.119}$$

This gives the uncertainty relationship expressed according to thermalized creation and annihilation operators with the form

$$(\Delta p_j^b)^2(\Delta e_k^c)^2 \geq \frac{1}{4}\bigg(\big(1 + \langle e^{-2\pi\mathcal{H}/g}\rangle\big)\delta_{jk}\eta^{bc} -$$
$$\frac{\pi}{g}\langle e^{\pi\mathcal{H}/g}\big(\alpha_j^{0b*}\alpha_k^{0b} + \alpha_j^{0b}\alpha_k^{0b*}\big)e^{-\pi\mathcal{H}/g}\rangle\bigg)^2. \tag{2.120}$$

For a vacuum spacetime the Einstein field equation reduces to $R_{\mu\nu} = \frac{1}{2}Rg_{\mu\nu}$. It is then apparent that the above commutator for vacuum spacetime gives a correction to the Ricci curvature for $\Sigma^{(3)}$ as

$$\mathcal{R}_{jk} = \frac{1}{2}\Big(R\big(1 + e^{-2\pi\mathcal{H}/g}\big)g_{jk} -$$

$$\int_0^\infty d\omega \frac{\pi}{g} e^{\pi \mathcal{H}/g} \big(\alpha_j^{0b}(\omega)^* \alpha_k^{0b}(\omega) + \alpha_j^{0b}(\omega)\alpha_k^{0b*}(\omega)\big) e^{-\pi \mathcal{H}/g}\Big), \qquad (2.121)$$

which is a form similar to that suggested by Sakharov and Ruzmaikin [23][24].

In the above uncertainty relationships the quadrature phases of the field, e_j^b and p_j^b exhibit a minimum uncertainty. This leads to the conclusion that conjugate variables do not exist in quadrature phase. Consider the squeeze state operator

$$S(z) = exp\big(\frac{z}{2} a_j^{b\dagger 2} - \frac{z^*}{2} a_j^{b\, 2}\big). \qquad (2.122)$$

where the squeezing parameter z is $z = re^{i\theta}$. This operator when applied to these generalized operators $\beta_j^b = S(z) \alpha_j^b S^\dagger(z)$ gives

$$\beta_j^b = S(z)\alpha_j^b S^\dagger(z) =$$

$$e^{2\pi\omega_j(z a_j^{\dagger 2} - z^* a_j^2)/g}\big(\alpha_j^{0b} cosh(|z|) - \alpha_j^{0c} \frac{z}{|z|} sinh(|z|)\big). \qquad (2.123)$$

This is a Bogoliubov transformed annihilation operator. Hence

$$S(z) = e^{2\pi\omega(z a_j^{\dagger 2} - z^* a_j^2)/g} \qquad (2.124)$$

is the gravitation induced squeezed state operator. This defines the commutator between the squeezed creation and annihilation operators

$$[\beta_j^b, \beta_k^{b*}] = \big(1 + \mathcal{S}^2(z) e^{-2\pi \mathcal{H}/g}\big)\delta_{jk}$$

$$- \frac{\pi\omega_j}{g} \mathcal{S}^2(z) e^{\pi \mathcal{H}/g}\big((\alpha_j^{0b}\alpha_k^{0b*} + \alpha_j^{0b*}\alpha_k^{0b})(cosh^2(|z|) - \frac{z^2}{|z|^2}sinh^2(|z|)) \qquad (2.125)$$

$$- \frac{2z}{|z|}(\alpha_j^{0b}\alpha_k^{0b} + \alpha_j^{0b*}\alpha_k^{0b*}) cosh(|z|) sinh(|z|)\big) e^{-\pi \mathcal{H}/g}.$$

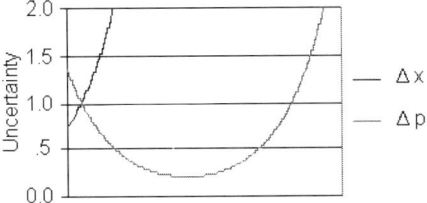

Increasing squeezing parameter,
units are arbitrarily defined.
Figure 2.3

Given a squeezing of quantum gravity states, or any departure from quadrature phase, quantum fluctuations will enhance this squeezing. When applied to an uncertainty relationship the uncertainty in both conjugate variable increases for a sufficiently large squeezing parameter.

2.5: Squeezed States, Wormholes and Nascent Cosmologies

Physically the occurrence of gravitational squeezing involves the dynamics of quantum wormholes. The Lanzos junction between two exterior regions of a wormhole is a membrane or D2-brane. Then the topology of the space is one where photons will wrap endlessly around the wormhole. In the case where the two multiply connected junctions are separated by a lightlike interval in the exterior region there exists a Cauchy horizon. Fields wrap around the wormhole and converge from both a past and future region. In the region future to the Cauchy horizon the wormhole is effectively a time machine. As photons in the past region, with frequency ν, wrap through the wormhole and converge to the Cauchy horizon their occurrence rate increases and surpasses their periodicity $\sim 1/\nu$. At this point the existence of such a photon on a path becomes highly uncertain. An observer who measures the recurrence of a mode wrapping through the wormhole will do so by performing measurements of that mode's frequency. As the recurrence rate of the photon near the Cauchy horizon, as measured by an observer far from the junctions, approaches half that of the periodicity of the mode the recurrence rate of the photon becomes uncertain. Here the observer uses the frequency of the mode as a sampling rate, which is the Nyquist frequency required to sample the rate a mode wraps around the wormhole. Once the frequency by which modes warp around the wormhole is greater than twice the Nyquist frequency $\nu_m \leq \frac{1}{2}\nu_{rec}$ the observer loses complete certainty on the occurrence of these modes.

An observer who approaches the Cauchy horizon will continue to measure the frequency of modes, but where time measurements of their occurrence are utterly lost. This appears to infer that the quantum fields that wrap through the wormhole are becoming increasingly uncertain in time, which if they remain coherent would mean that energy or frequency measurements should become more certain. This argument points to the fact that the vacuum structure is being squeezed by the wormhole. This further avoids the problem just illustrated, for the squeezing of light is accomplished through parametric down conversion, where a photon of frequency ν is converted into two photons of frequency $\frac{1}{2}\nu$. The two photons are in an entangled state and behave as a single photon with frequency ν. This would then appear to avoid the Planck scale problem. This infers that quanta and vacuum modes are continuously parametric down converted from surpassing the Planck frequency. Near the Cauchy horizon it becomes uncertain whether a particular quantum state pertains to an unique measurable particle.

The quantum fluctuation may then be represented as the junction condition in the energy across the mouth of the wormhole. Start with the Reissner-Nordstrom metric [25]

$$ds^2 = -F(r)dt^2 + \frac{1}{F(r)}dr^2 + d\Omega^2, \tag{2.126}$$

that describes many spacetimes, for $F(r) = 1 - 2M/r + Q/r^2 + \Lambda r^2$. Here M is the mass, Q is the charge and Λ is the cosmological constant. For $Q = \Lambda = 0$ this recovers the Schwarzschild black hole. For the wormhole two such geometries that are surgically connected at a radius $r = r_0(t)$ that in general is time dependent. This gives a multiple connecting bridge between these regions [25]. The first is where two external regions are glued at $r \in [r_0(t), \infty]$ and the other are the two internal regions $r \in [0, r_0(t)]$, that together define a 3-sphere. In both cases one has a geometry without discontinuity, but with a junction kink or discontinuity in its derivative as $r = r_0(t)$. In the first case this is a wormhole, where two manifolds are glued together at $r = r_0(t)$, are joined again at $r = \infty$. In the second interior case this is a spherical cosmology. The junction conditions at the boundary $r_0(t)$ determine the topology of the solution. The evolution of the vacuum is governed by the discontinuity in the Einstein field at $r = r_0$. This discontinuity is then given by

$$\lim_{\epsilon \to 0} \int_{-\epsilon}^{+\epsilon} G^{00} dn = \rho = \frac{1}{2\pi r_0}\sqrt{F(r_0) - (\dot{r}_0 + U)^2}, \qquad (2.127)$$

where $U = \sqrt{U^\mu U_\mu}$ is the speed of the one of the wormhole faces. This junction condition has its analogue in electromagnetism as a change in the electric displacement vector at the boundary of a dielectric. If the change in the dielectric results in $\delta\mathbf{D} = (\epsilon_0 - \epsilon_1)\mathbf{E}$ this change in the electric displacement vector can be a model of as a quantum fluctuation, where $[\delta\mathbf{D}, \delta\mathbf{H}] = \hbar$. The fluctuation in the electromagnetic field as changes in the permativity of space due to virtual quanta. In an analogous manner the membrane in two dimensions that separate two regions of spacetime can be thought as due to a fluctuation in the spacetime metric, with a "kick" that results in changing the uncertainty in $\mathbf{D}$. The conjugate observables g_{ij} and π^{ij} are analogous to $\mathbf{D}$ and $\mathbf{H}$. For a virtual wormhole M and Λ vanish as time averaged quantities, where Q is either zero or where one has $\pm Q$ attached separately to the two wormhole openings.

For an average squeezing parameter $|z| \sim 5.5$ the renormalization of the Planck length scale will give a Planck energy in the range of 10^{15} GeV. An estimate is possible of what is the most probable value for the absolute value of the squeezing parameter. With parametric down conversion then $k' = \frac{1}{2}k_{planck}$. Assume the wave functional is $\psi \sim e^{p^2/\sigma}e^{ipx}$, then the value of the wave functional has been changed by $\psi \to e^{-1/4}\psi \simeq .3863\psi$. Physically the wave functional under this reparameterization must obey conservation of probability, where the average value of $|z|$ is such this occurs. This then means that the average value for the magnitude of the squeezing parameter will be ~ 5.18. This leads to an average renormalization of the Planck energy scale to $\sim 1.2 \times 10^{15}$GeV. This approximate average renormalized value for the Planck energy is just the value required for quantum gravity to unify with gauge theories at the same scale.

It has been illustrated how squeezed states of the quantum gravity vacuum may lead to the tunnelling of a nascent cosmology out of the vacuum [26]. This shows that quantum gravity fluctuations will spontaneously lead to the occurrence of nascent cosmologies through the "self-squeezing" of the quantum gravity vacuum. The squeezing parameter for this tunnelling is

$$|z| = ln2 + \frac{1}{3\Lambda}, \qquad (2.128)$$

which gives the value $|z| \simeq 10^{60}$. This means that a single quantum fluctuation has a probability of $\sim 10^{-120}$ of causing the tunnelling of a nascent cosmology out of the vacuum. For a squeezing parameter z this leads to a cosmology with an entropy

$$S = ln\ cos(2|z|). \tag{2.129}$$

This entropy results from a coarse graining of a state, with a resulting temperature. These results indicate how the spacetime cosmology can tunnel out of the vacuum, or *Creati Ex Nihilio*.

Torsion is the result of accelerations that appear as an anholonomy in the parallel translation of vectors around a closed loop. Such nongeodesic motion was examined by Elie Cartan as a way to extend the general theory of relativity [27]. These efforts were further pursued by Einstein as a way of unifying the fundamental forces of nature to include gravitation. Cartan's efforts were more for purely mathematical pursuits. The overall spirit here has been more of a physical nature, but some mention should be given to how this theory connects with mathematics of manifolds and bundles.

With the above construction of the Gaussian second fundamental form in equation 2.61, the connection coefficients in the theory are

$$\omega^\mu = K^\mu{}_\nu dx^\nu - \rho^\mu{}_{\sigma\nu} n^\sigma dx^\nu. \tag{2.130}$$

These one-forms are analogous to the forms $\omega^0 = \gamma(dt - u^0 dx)$ and $\omega^i = \gamma(dx^i - u^i dt)$, for acceleration in flat spacetime. These define lifting conditions when they vanish on the manifold of solution. $K^\mu{}_\nu dx^\nu$ defines the horizontal portion of the bundle and $\rho^\mu{}_{\sigma\nu} n^\sigma dx^\nu$ defines the vertical portion (or bridge) of the bundle.

The constructions in Finsler geometry as similar to the standard construction of fibre bundles in Riemannian geometry. Given an atlas of charts $\{\mathcal{U}_i\}$ on a manifold $\mathcal{M}$. Define a fibre at a point x in $\{\mathcal{U}\}$ as $\pi^{-1}(\{\mathcal{U}\})$, where this defines the space $E = \{\pi(x),\ \forall\ x\ \in\ \mathcal{M}\}$. The fibre bundle is further mapped onto the space $\phi_i : \pi^{-1}(\{\mathcal{U}_i\}) \to \{\mathcal{U}_i\} \times R^n$, so that the diagram below is commutative.

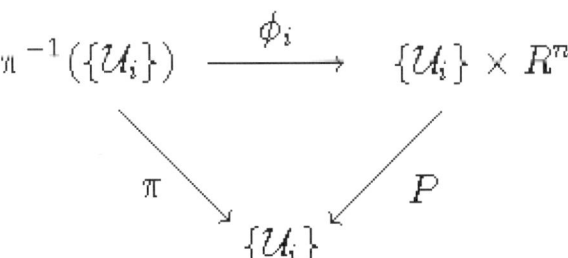

Fibre construction in Finsler geometry.
Figure 2.4

The diffeomorphisms ϕ_i are transition functions and on the overlap between charts $\mathcal{U}_i \cap \mathcal{U}_j$ they define connection coefficients.

From here tangent spaces may be constructed. Define the map $\pi': T\mathcal{M} \to \mathcal{M}$ and also define $\pi'': TT\mathcal{M} \to T\mathcal{M}$. The kernal of this map is then

$$ker\pi''_z = x \in T\mathcal{M}, \; \pi''(x) = 0. \tag{2.131}$$

for all $z \in T\mathcal{M}$. This defines the vertical tangent space, which is the basis for Finsler geometry. The vertical bundle is then $VTM = \cup_i ker\pi''_i$. The horizontal bundle defines a gauge field or gravitation. The vertical tangent bundle then give the velocity component in a basis of one forms $\omega^\mu = \Lambda^\mu{}_\nu(dx^\nu - u^\nu dt)$. Where this form vanishes the Euler-Lagrange equation is $0 = (\frac{d}{dt}(\partial L/\partial x_\mu) - \partial L/\partial x_\mu)dt$. So the vertical tangent bundle defines a Lagrangian space. For quantum fluctuations this Euler-Lagrange equation is one with a stochastic force term.

The use of Finsler geometry can lead to a complex valued theory, where the complex nature of time fluctuations tie into the Kähler-Dirac theory. The subject of teleparallel theory and the Kähler-Dirac theory is illustrated in [28]. Further, Finsler geometry with the application of the vertical bundle permits a description of supermanifold theory. These results are illustrated in Bejancu [29].

The definition of the curvature with a Finsler connection is

$$R^\alpha{}_{\beta\mu\nu} = \partial_\nu\Gamma^\alpha{}_{\beta\mu} - \partial_\mu\Gamma^\alpha{}_{\beta\nu} + \Gamma^\alpha{}_{\mu\gamma}\Gamma^\gamma{}_{\beta\nu} - \Gamma^\alpha{}_{\nu\gamma}\Gamma^\gamma{}_{\beta\mu} + \rho^\gamma{}_{\mu\nu}C^\alpha{}_{\beta\gamma}, \tag{2.132}$$

where $[X_\beta, X_\gamma] = C^\alpha{}_{\beta\gamma}X_\alpha$ and $\{X_\alpha\}$ define the frames on the bundle. Clearly in our definition of the torsional connection in equations 2.67 - 2.70 that $C^\alpha{}_{\beta\gamma}X_\alpha = \rho^\alpha{}_{\beta\gamma}X_\alpha$. This reflects that fact that the bundle in gravitation is defined by connection forms intrinsic to the manifold.

The cocycle associated with the time fluctuation with the definition of $n_{[\mu;\nu]}$ and the fibration above defined by π' and π'' suggests that quantum gravity fluctuations are defined by a cohomology

$$ker\pi''/im\pi'. \tag{2.133}$$

If the maps π' and π'' are given by the same operator Q, with $Q^2 = 0$ then this is the BRST quantization condition [30]. This is in fact suggested by the cohomology associated with $n_{\mu;\nu}$. This would make the construction of quantum gravity a finite temperature theory commensurate with superstring theories.

A quantum fluctuation with a connection $\rho^\mu{}_{\nu\sigma}$ defines the G^{00} in equation 2.127 according to equation 2.74. The junction energy is then a manifestation of a quantum vacuum that is a cohomological field.

It is reasonable to state that the universe is primarily stochastic and quantum fluctuations are stochastic. It has been argued that torsional connections may be used to model fluctuations in quantum gravity. These fluctuations then have surprising properties, such as gravitational self-squeezing. This then illustrates that the quantum gravity vacuum has a fundamental instability to it that leads to the spontaneous creation of spacetime cosmologies. This implies that fluctuations of the spacetime manifold in our universe is generating nascent cosmologies at a huge rate. An estimate indicates that quantum gravity fluctuations on a scale $\sim 10L_p$ have a probability of $\sim 10^{-120}$ per Planck unit of time of generating a nascent cosmology. This then implies that spacetime within our cosmological horizon is associated with as many as $\sim 10^{103}$ new cosmologies every second on the Hubble frame! Further, as these cosmologies also have quantum fluctuations they in turn generate nascent cosmologies.

It should further be noted that the above work relies on the existence of additional parameters and that spacetime is ultimately embedded into a manifold of higher dimension. This suggests that further work can be performed where this theory involves supergravity and that these parameters at low energy are compactified into Calabi-Yau spaces. Further, these structures must tie into the Yang-Mills structure for quantum fluctuations. It also appears that there is a deep relationship between gauge hierarchy and the cosmological constant. In the quintessence model it is proposed that the value of the cosmological constant reaches an optimal value, much as a classical system that reaches a basin of attraction. Perhaps this proposed dynamics has the gravitational self-squeezing of the vacuum as an underlying mechanism.

The argument above shows that the Cauchy Horizon of a wormhole acts as a parametric amplifier of the vacuum state. This mechanism is likely tied into the creation of a spacetime cosmology. This mechanism is explored in some greater detail below. This discussion will also indicate that this process is far from understood. This leads to the question of to what degree black holes dominate over the existence of wormholes in the quantum gravity vacuum. In what follows it will be argued that on a scale significantly larger than the Planck scale that black holes dominate. This suggests that wormholes are not stable objects in spacetime if they involve any more than a few quantum numbers or Planck lengths. Quantum gravity in the weak string coupling limit or at the one-loop Feynman graph level is expected to have a vacuum where virtual black holes dominate.

2.6: Spacetime Fluctuations and Wormholes

A dynamical wormhole may give rise to a nascent cosmology. This starts with a static wormhole as a necessary starting point. Start with the Reissner-Nordstrom metric

$$ds^2 = -F(r)dt^2 + \frac{1}{F(r)}dr^2 + d\Omega^2, \quad (2.134)$$

that describes many spacetimes. Consider a sphere with radius $r_0(t)$ that is a junction or thin shell between two regions. The normal vectors to this sphere are

$$n^\mu = \pm\left(\frac{\dot{r}_0}{F(r_0)}, \sqrt{F(r_0) - \dot{r}_0^2}, 0, 0\right), n_\mu = \pm\left(\dot{r}_0, \frac{\sqrt{F(r_0) - \dot{r}_0^2}}{F(r_0)}, 0, 0\right). \quad (2.135)$$

The sign is an indication of the direction of the normal on the sphere. The extrinsic curvature is computed as $K_{\mu\nu} = \frac{1}{2}n^\sigma \partial_\sigma g_{\mu\nu}$ and the components are

$$K_{\theta\theta} = \pm\frac{1}{r_0}\sqrt{F(r_0) - \dot{r}_0^2}$$

$$K_{tt} = \mp \frac{1}{2} \frac{1}{\sqrt{F(r_0) - \dot{r}_0^2}} \left(\frac{\partial F(r_0)}{\partial r_0} - \ddot{r}_0 \right). \tag{2.136}$$

The ADM curvature may then be computed, where these extrinsic curvatures contribute to $(TrK)^2 - Tr(K^2)$ so that

$$-G^{00} = \frac{1}{2} R^{(3)} - \frac{1}{2a} \left(\frac{\partial F(r_0)}{\partial r_0} - \ddot{r}_0 \right), \tag{2.137}$$

where $R^{(3)}$ is the curvature of the spacial manifold.

In the thin shell compute the tension and energy density of the membrane. These quantities are related to the discontinuity in the extrinsic curvature on the membrane. The above G^{00} is integrated on a pillbox configuration to find the jump in the surface energy on the membrane, which is the jump in extrinsic curvature at $r = r_0(t)$. For a static membrane the tension equals the energy density so there is an overall conservation of the stress-energy on the membrane. The energy density is

$$\rho = -\frac{1}{4\pi} K_{\theta\theta} = \frac{1}{2\pi r_0} \sqrt{F(r_0) - \dot{r}_0^2} \tag{2.138}$$

and the tension is

$$\tau = -\frac{1}{4\pi} (K_{\theta\theta} - K_{tt}). \tag{2.139}$$

The static membrane then infers the following equation

$$\frac{F(r_0)}{r_0^2} - \left(\frac{\dot{r}_0}{r_0} \right)^2 = 4\pi^2 \tau^2. \tag{2.140}$$

The conservation of energy $\partial \rho / \partial t$ applied to this constraint equation results in the evolution equation

$$\ddot{r}_0 + \frac{1}{r_0} \left(F(r_0) - \dot{r}_0^2 \right) - \frac{\partial F(r_0)}{\partial r_0} = 0. \tag{2.141}$$

Now consider the Reissnor-Nordstrom metric with

$$F(r) = 1 - \frac{2M}{r} + \frac{Q^2}{r^2} - \frac{\Lambda}{3} r^2. \tag{2.142}$$

The terms are M for mass, Q for charge and Λ is the cosmological constant. With this form of the Reissnor-Nordstron metric a numerical solution to this differential equation may be found. For simplicity consider a solution where $M = Q = \Lambda$. For a particle outside the well the path escapes to infinity with a constant "velocity." Conversely if the "particle" is trapped in the well it will exhibit a periodic motion. This solution is represented by figure 2.5. Let the wormhole connect two regions of a single space so there is a finite distance between the two openings. This means there is the region $r \in [0, r_0(t)]$ identified with $r \in [0, r_1(t)]$, where r_1 is obtained from r_0 by a Poincaré transformation. If the internal region is identified by $\mathcal{S}$ the wormhole is defined by $r \in [r_0(t), \Sigma^{(3)} - \mathcal{S}]$

and $r \in [r_1(t), \Sigma^{(3)} - \mathcal{S}]$. For two openings of the wormhole initially on a plane of simultaneity define their relative displacements of points on the two openings of the wormhole is

$$r_1{}^\mu = R^\mu + r_0{}^\mu. \tag{2.143}$$

Here $r_0{}^\mu$ and $r_1{}^\mu$ are the coordinates of each point on the membrane as determined by with a coordinate basis determined by an observer in the external region. Here the velocity of points on the r_1 face of the wormhole relative to those of the r_0 face is

$$\frac{dr_1{}^\mu}{ds} = \frac{dr_0{}^\mu}{ds} + \frac{dR^\mu}{ds} = \dot{r}_0^\mu + U^\mu \tag{2.144}$$

and the acceleration is

$$\ddot{r}_1^\mu = \ddot{r}_0^\mu + \frac{d^2 R^\mu}{ds^2} = \ddot{r}_0^\mu + a^\mu. \tag{2.145}$$

Now consider the r_1 face as accelerated outward and then back towards the r_0 face. This means at some point the two faces of the wormhole will have a light-like separation which defines the Cauchy horizon. This then involves two coupled differential equations by the term d^2R/ds^2.

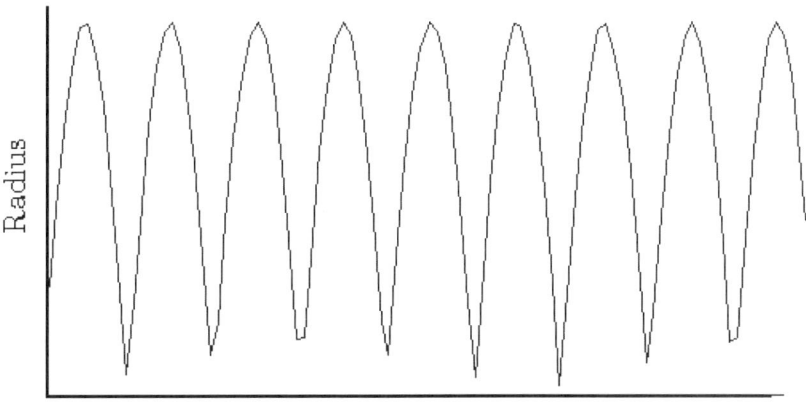

Time

Worm hole dynamics.

Figure 2.5

Now consider modes that wind through the wormhole, which reduces the problem of a field in one dimension. A scalar field is expanded according to

$$\phi(x) = \int_0^\infty d\omega (a(\omega)e^{ik_\mu x^\mu} - a^\dagger(\omega)e^{-ik_\mu x^\mu}). \tag{2.146}$$

Given a field on a winding loop through the wormhole, the momentum representation is a loop integral through the wormhole is

$$\phi(k) = \frac{1}{2\pi} \oint dx \phi(x) e^{ikx}. \tag{2.147}$$

This field is then an element of the $H^1(M, R)$ cohomology and reflects the multiply connected topology of the space. Further, viewed from an observer on a timelike geodesic midway between the faces of the wormhole the occurrence of these fields will becomes increasingly rapid. Let λ_i identify the ith null geodesic through the wormhole, then the proper time separation between their occurrence is a Cauchy sequence

$$\tau(\lambda_i, \lambda_{i+1}) < \tau(\lambda_{i-1}, \lambda_i) \,\forall i, \qquad (2.148)$$

that converses to a final null geodesic λ_∞ with finite τ. This is dependent on the acceleration a^μ changing sign, or being an oscillatory function. The correlation of the field $\phi(x^\mu)$ at various two proper intervals as measured by the observer is

$$K(\Omega) = \frac{1}{2\pi} e^{i\Omega\delta\tau} \int_{\lambda(t)}^{\lambda(t+\delta\tau)} d\lambda \langle 0'|\phi(\lambda_{t+\delta\tau})\phi(\lambda_t)|0\rangle, \qquad (2.149)$$

where Ω is the frequency of the occurrence of the field winding through the wormhole. Here the vacuum state $|0'\rangle$ illustrates that the vacuum zero point energy is modified by the accumulation of modes that wrap though the wormhole. For $\Omega << \omega$ this represents only a phase shift in the field as $\Omega\tau \sim r = const$, where the measurement of the two occurrences of the field are not temporally distinguishable. This makes the correlation function assume the form

$$K(r/\delta\tau) \simeq \frac{1}{2\pi} e^{ir} \int d\omega \left(\frac{1}{2} + \langle 0'|a^\dagger(\omega)a^\dagger(\omega)e^{-2i\omega\delta t} - a(\omega)a(\omega)e^{2i\omega\delta t})|0\rangle\right). \qquad (2.150)$$

This correlation function then contains two parts to it. The first is an infinite sum over frequencies and the second is the expectation value of an off diagonal operator that is nonvanishing if the $\langle 0'|$ has been acted on by a squeezed state operator, where is the parametric down shift of $\langle\, 0|$.

Evidently the correlation function will vanish if the vacuum $\langle 0'|$ is the Hilbert space dual of $|0\rangle$. Yet there is a squeeze state operator that determines the evolution of the vacuum state. Further, the vacuum zero point energy increases near the Cauchy horizon. The evolution of the vacuum is governed by the discontinuity in the Einstein field at $r = r_0$. This discontinuity is then given by

$$\lim_{\epsilon \to 0} \int_{-\epsilon}^{+\epsilon} G^{00} dn = \rho = \frac{1}{2\pi r_0}\sqrt{F(r_0) - (\dot{r}_0 + U)^2}, \qquad (2.151)$$

where $U = \sqrt{U^\mu U_\mu}$ is the speed of the r_1 face. A vacuum mode is then changed with each winding as $|0\rangle \to e^{i\rho\delta t}|0\rangle$, where δt is the time the mode takes to cross the thin shell. The energy density further contains off diagonal terms a^2 and $(a^\dagger)^2$. Further, as the membrane evolves to sweep out a world tube the windings increase rapidly near the Cauchy horizon $n(t) \to \infty$ as $\lambda \to \lambda_\infty$. This means the number of these iterated evolutions of the vacuum rapidly increase to infinity. So the total evolution of the vacuum is generically determined by

$$|0'\rangle = \lim_{n(t) \to \infty} \lim_{\delta t \to 0} e^{in(t)\rho\delta t}|0\rangle \sim e^{i\kappa\rho}|0\rangle, \qquad (2.152)$$

where κ is an undetermined constant.

This result infers several things. The first is there is a piling up of vacuum modes around the Cauchy horizon. The wormhole is an attractor of vacuum modes, which indicates that the vacuum energy density diverges around the Cauchy horizon. Secondly, the existence of a squeeze state operator in the auto correlation function for fields that wrap around the wormhole shows the vacuum structure is in a squeezed state. The membrane on the wormhole face has negative energy, but there is an accumulation of positive energy from vacuum modes that pass through the wormhole mouth. This raises the question of whether the wormhole can continue to exist if the stress-energy associated with the wormhole becomes positive as these vacuum modes wind through it. The wormhole may evolve into two black holes before the openings reach the Cauchy horizon.

A quantum or virtual wormhole may be described according to the Wheeler-DeWitt equation. The momentum operator is

$$\hat{\pi}^{ij} = -i\frac{\delta}{\delta g_{ij}}, \quad \hat{\pi}^{tt} = -i\left(\frac{\partial F(r)}{\partial r}\right)^{-1}\frac{\partial}{\partial r}, \quad \hat{\pi}^{\theta\theta} = -\frac{i}{2r}\frac{\partial}{\partial r}. \quad (2.153)$$

The kinetic energy operator is then $\hat{\pi}^{tt}\hat{\pi}^{\theta\theta} + \hat{\pi}^{\theta\theta}\hat{\pi}^{tt}$ in the Wheeler-DeWitt equation of the form

$$-\left(\frac{\partial F(r)}{\partial r}\right)^{-1}\frac{\partial^2 \Psi(r)}{\partial r^2} + \frac{1}{2}\left(\left(\frac{\partial F(r)}{\partial r}\right)^{-2}\frac{\partial^2 F(r)}{\partial r^2} + \frac{1}{r}\left(\frac{\partial F(r)}{\partial r}\right)^{-1}\right)\frac{\partial \Psi(r)}{\partial r} +$$

$$\left(\frac{\partial F(r)}{\partial r}\phi - \ddot{\phi}r + rR^{(3)}\right)\Psi(r) = 0. \quad (2.154)$$

This introduction of time or pseudotime into the Wheeler-DeWitt equation with the field ϕ permits the Wheeler DeWitt equation to assume the form $H\Psi[g] = i\partial\Psi[g]/\partial t$, where for a polar wave functional this wave equation has real and imaginary parts. The auxiliary field is then considered to be a source of the gravity field, which defines a time scale so Wheeler DeWitt equation may be integrated. It is then possible to solve the Wheeler-DeWitt equation numerically in an elementary manner. To start begin with the static brane where the wormhole openings are infinitely separated, or connected by two separate manifolds. Physically this may correspond to a quantum fluctuation that connects two spacetime cosmologies. Numerically, the solution appears in figure 2.6.

This numerical solution (figure 2.6) introduces a small amplitude at $r = \epsilon < 0$, where for $r \in (.1, \sim 10)$ Planck lengths this is this wave function. Further, the occurrence and general form of this wave form is independent of the initial amplitude introduced. This solution is then apparently a quantum fluctuation or a small quantum number exciton that occurs between .1 and 10 Planck lengths. Further a frequency spectrum of this solution has harmonic oscillator states $E = n\hbar\omega$ that are integral numbers of tenths of Planck frequencies. Physically the wormhole openings between the two spacetime open and close, where the time for this opening is determined by the energy fluctuation of the wormhole.

58 Quantum Fluctuations of Spacetime

This fluctuation is then an uncertainty in the energy density given by equation 2.138. The time interval for the fluctuation is then on the order of 10 Planck units of time.

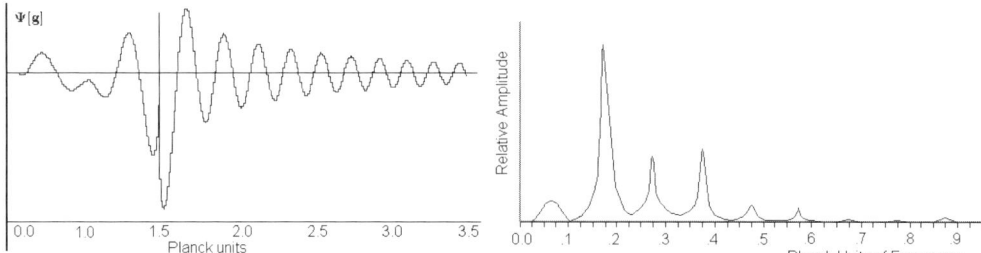

Dynamics of a wormhole with openings on two disconnected manifolds.
Figure 2.6

The lack of low frequency modes tend to further indicate that the solution is on the order of at most 20 Planck lengths. Now consider the case where the two openings of the wormhole are separated by a finite distance. This is the case where the wormhole connects two regions in the same spacetime cosmology. Further, let the two openings of the wormhole move apart from each other and then back again. A wormhole that connects two regions within the same spacetime would have its openings separate and return to the same region. A virtual wormhole will involve this type of motion. Numerically this motion is introduced "by hand" by requiring this motion to occur on a time $T \sim \hbar/\Delta E$. This then approximates the wave function for the vacuum state with virtual production of wormholes. This motion classically is known to result in a Cauchy horizon and to a time machine. Quantum mechanically it is expected that the achronal nature of virtual wormholes will be removed on a large scale. The attenuation of the wave function at larger numbers of Planck scale tends to imply that the vacuum contains virtual wormholes that are dominant only within a few Planck lengths. This would suggest that any quantum with a wave length much larger than the Planck length does not interact with the tiny achronal regions associated with the quantum gravity vacuum and virtual wormholes. The numerical solution in metric and momentum metric coefficients exhibits a large low frequency peak reflects the motion of the two wormhole openings, as seen in figure 2.7. The spectrum exhibits a suppression of high frequencies near the Planck scale. This lack of $1/f$ type of behavior shows that this solution has no large scale $>> 1.6 \times 10^{-33} cm$ behavior. The appearance of cusps in the metric solution to the Wheeler-DeWitt equation is dependent on the cosmological constant. If the cosmological constant is reduced to .3 times the value the solution illustrated above indicates that the cusp is shifted inward. Further if it is reduced to .1 the solution is seen in figure 2.8. If the cosmological constant is increased to .3 the solution appears in figure 2.9.

Quantum Fluctuations and Spacetime 59

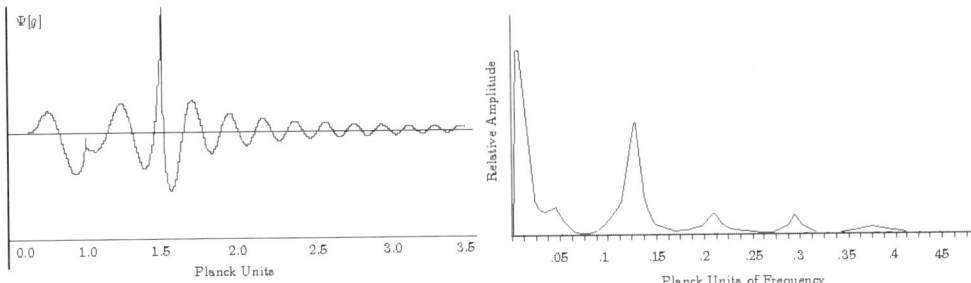

Quantum wormhole connecting two manifolds where one opening is Lorentz boosted.
Figure 2.7

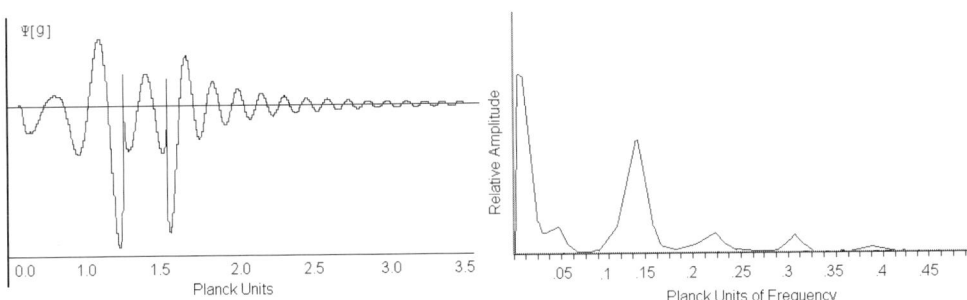

Quantum wormhole with a relative boost of one opening in the same spacetime and $\Lambda = 1$.
Figure 2.8

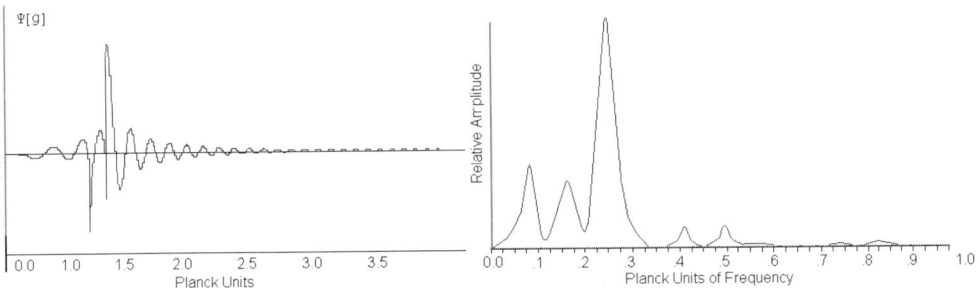

Quantum wormhole with a relative boost of one opening in a single spacetime and $\Lambda = 3$.
Figure 2.9

As the cosmological constant is reduced, $\Lambda \rightarrow 0$ the expansion rate of space, where in turn the repulsive acceleration between the two wormhole faces are reduced.

The exact solution to the Wheeler-DeWitt equation is difficult to arrive at. However, it is possible to find a generic or approximate form of the wave function to illustrate the physics of gravitational squeezing of the vacuum state. The Schrödinger equation form of the Wheeler-DeWitt equation is

$$-\left(\frac{\partial F(r)}{\partial r}\right)^{-1}\frac{\partial^2 \Psi(r)}{\partial r^2} + \frac{1}{2}\left(\left(\frac{\partial F(r)}{\partial r}\right)^{-2}\frac{\partial^2 F(r)}{\partial r^2} + \frac{1}{r}\left(\frac{\partial F}{\partial r}\right)^{-1}\right)\frac{\partial \Psi(r)}{\partial r} +$$
$$rR^{(3)}\Psi(r) = i\frac{\partial \Psi(r)}{\partial t}, \qquad (2.155a)$$

with

$$i\frac{\partial \Psi(r)}{\partial t} = \left(\frac{\partial F(r)}{\partial r}\dot\phi - r\ddot\phi\right)\Psi(r), \qquad (2.155b)$$

and where t is the pseudo-time parameter. The term $\frac{\partial F(r)}{\partial r}$ for $r > 1.0$ Planck units is numerically seen to have nearly linear behavior. Physically the region larger than the Planck length is relevant. Regions smaller than the Planck scale are not observable, so the wave function within this region is not as crucial to the physical interpretation of the problem. Further, it appears that the generic form of the wave function is similar for regions on the order of 2.0 Planck lengths. A generic form of the wave function is used to describe the problem according to creation and annihilation operators and to describe gravitational self-squeezing of the vacuum state.

Now write the differential equation as:

$$-\frac{1}{\eta r}\frac{\partial^2 \Psi(r)}{\partial r^2} + \frac{1}{\eta r^2}\frac{\partial \Psi(r)}{\partial r} + rR^{(3)}\Psi(r) = (r\eta\dot\phi - r\ddot\phi)\Psi(r), \qquad (2.156)$$

where η is determined by the linearity of $\frac{\partial F(r)}{\partial r} \simeq \eta r$.

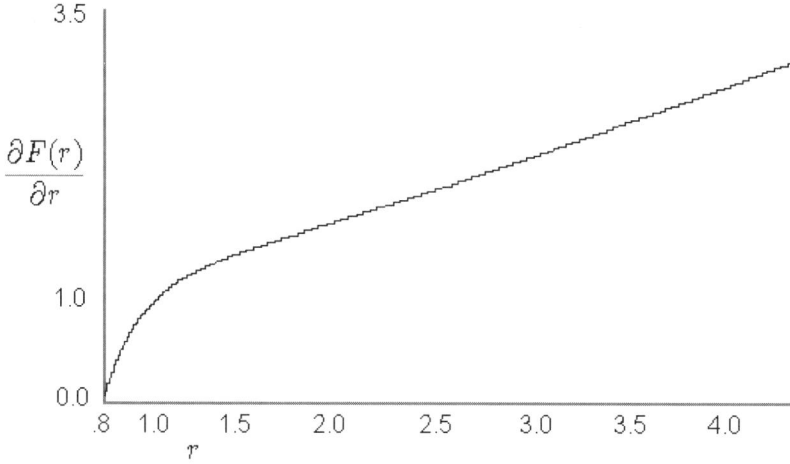

The first derivative of the Reissner-Nordstrom metric terms with respect to the radius.
Figure 2.10

The solution to this differential equation can be found with Laplace transforms $L[f(r)] = f(s) = \int e^{-sr}f(r)$. For $\phi \simeq cos(\omega r)$ and $R^{(3)} = const$ this differential equation then transforms into

$$s^2\psi'(s) + s\psi(s) = -\eta^2\left(\frac{-24s^4}{(s^2+\omega^2)^4} + \frac{36s^2}{(s^2+\omega^2)^3} - \frac{-6s}{(s^2+\omega^2)^2}\right) -$$

$$\eta\omega^2\left(\frac{4s^3}{(s^2+\omega^2)^3} - \frac{6s}{(s^2+\omega^2)^2}\right). \tag{2.157}$$

We assume that ω has a real and a smaller negative imaginary part to give a damping of the oscillation for large r. By simple methods of integrating factors the solution to this differential equation is

$$\psi(s) = \eta^2\left(\frac{6}{s\omega^5}tan^{-1}(s/\omega) - \frac{4}{(s^2+\omega^2)^3} - \frac{15}{\omega^4(s^2+\omega^2)},\right.$$

$$\left. -\frac{6}{\omega^2 s^2} - \frac{8}{\omega^2(s^2+\omega^2)^2}\right) - \tag{2.158}$$

$$\eta\omega^2\left(\frac{4}{s(s^2+\omega^2)} - \frac{3}{2\omega^2 s(s^2+\omega^2)} - \frac{3}{\omega^4 s}ln(s^2/(s^2+\omega^2))\right).$$

The inverse Laplace transform gives the approximate solution to the differential equation as

$$\Psi(r) \simeq -A\left\{\eta^2\left(4\sqrt{\pi}\frac{t}{2\omega^5}J_1(\omega r) + \frac{4}{\omega^5}(sin(\omega r)+(\omega r)cos(\omega r)) + \frac{15}{\omega^5}cos(\omega r) - \frac{6}{\omega^5}Si(\omega r)\right)\right.$$

$$\left. + A\eta\omega^2\left(\frac{3}{2\omega^4}(1 - cos(\omega r)) - 4e^{-\omega r} + \frac{6}{\omega^4}Ci(\omega r)\right)\right\}, \tag{2.159}$$

where $J_1(\beta r)$ is the Bessel function of the first kind, the functions Si and Ci are the sine and cosine integrals $\int_{-\infty}^{x}(sinx')dx'/x'$ and $\int_{-\infty}^{x}(cosx')dx'/x'$ and A is the amplitude. If there is a Fourier sum over frequencies then this general solution will have the maximal amplitudes found numerically in figures 2.6 through 2.7 near the Planck length scale. Hence this solution is within the assumptions imposed reasonable for the wave functional. This solution is approximate and assumes that $\partial^2 F/\partial r^2$ is a constant and rather artificial conditions are imposed on ϕ. An exact description based on the differential equation 2.155 is far more difficult. This difficulty is further compounded by the fact that this model does not involve the creation and destruction of the virtual wormhole. This appears to require a solution to the classical problem of two black holes, which appears to involve an infinite number of symmetries. The problem lacks a complete solution, either exactly or numerically.

Now consider the autocorrelation function with this wave function envelope written as $\mathcal{F}$. Further, the field operators are assumed to arise from spacetime variables through the Ashtekar variables $g_{ij} = \sigma^b{}_i e^b_j$ and $\pi_{ij} = \sigma^b{}_i p^b_j$ and with

$$e^b_i = \int_0^\infty \frac{d\omega}{\omega}(a^{b\dagger}{}_i(\omega)\mathcal{F}^*(\omega) + a^b_i(\omega)\mathcal{F}(\omega)) \tag{2.160}$$

$$p_i^b = i\int_0^\infty d\omega(a^{b\dagger}{}_i(\omega)\mathcal{F}^*(\omega) - a_i^b(\omega)\mathcal{F}(\omega)). \tag{2.161}$$

$\mathcal{F}(\omega)$ is approximately related to the Fourier transform of equation 2.159. The index b refers to an internal space with a spinor basis. Initially these variables are expanded according to fields in the exterior region or wormhole. For simplicity consider these operators written according to ladder operators on both sides of the junction. The Einstein field G^{00} is then determined by

$$G^{00} = -\frac{1}{2}R^{(3)} + g^{-1}\left(Tr\pi^2 - \frac{1}{2}(Tr\pi)^2\right), \tag{2.162}$$

where the kinetic energy portion expanded in harmonic oscillator modes is

$$g^{-1}\left(Tr\pi^2 - \frac{1}{2}(Tr\pi)^2\right) = \int_0^\infty \int_0^\infty d\omega d\omega' (a_i^{b\dagger}(\omega)a_i^b(\omega')\mathcal{F}^*(\omega)\mathcal{F}(\omega')) +$$

$$\int_0^\infty \int_0^\infty d\omega d\omega' (a_i^{b\dagger}(\omega)a_i^{b\dagger}(\omega')\mathcal{F}^*(\omega)\mathcal{F}^*(\omega') + a_i^b(\omega)a_i^b(\omega')\mathcal{F}(\omega)\mathcal{F}(\omega')). \tag{2.163}$$

Consider the thin shell as possessing a delta function in the time variable so the energy function is

$$\rho = \lim_{\epsilon \to 0}\int_{-\epsilon}^{+\epsilon} G^{00} dn =$$

$$\int_0^\infty d\omega\left(a_i{}^\dagger(\omega)a_i(\omega)|\mathcal{F}(\omega)|^2\right) + \tag{2.164}$$

$$\int_0^\infty d\omega\left(a_i{}^\dagger(\omega)a_i{}^\dagger(\omega)(\mathcal{F}^*(\omega))^2 + a_i(\omega)a_i(\omega))(\mathcal{F}(\omega))^2\right).$$

The terms quadratic in a_i and $a_i{}^\dagger$ are only nonzero if this operator acts on states on the right and left that are not equal. In particular the vacuum $\langle 0'|$ is not the dual of $|0\rangle$. It is a requirement that

$$\langle 0'|\left(a_i{}^\dagger(\omega)a_i{}^\dagger(\omega)(\mathcal{F}^*(\omega))^2 + a_i(\omega)a_i(\omega))(\mathcal{F}(\omega))^2\right)|0\rangle \tag{2.165}$$

be real valued. If it is assumed that

$$\langle 0'| = \langle 0|\left(a_i{}^\dagger(\omega)a_i{}^\dagger(\omega)(\mathcal{F}(\omega))^2 - a_i(\omega)a_i(\omega))(\mathcal{F}^*(\omega))^2\right), \tag{2.166}$$

then the above matrix element will be real valued.

If the term with $a^\dagger a$ is just a phase the evolution of the quantum vacuum is determined by the operator

$$|0'\rangle = exp\left(i\kappa \int_0^\infty d\omega\left(a_i{}^\dagger(\omega)a_i{}^\dagger(\omega)(\mathcal{F}(\omega))^2 - a_i(\omega)a_i(\omega))(\mathcal{F}^*(\omega))^2\right)\right)|0\rangle. \tag{2.167}$$

Consider the time (pseudotime) variable as complex valued $t = \tau + is$. This gives a squeezed state operator which depends upon the imaginary component of the pseudotime variable. Define the squeeze state operator as $\mathcal{S}(r, \mathcal{F})$ with an imaginary component to κ. Now consider the action of this operator on $a^a{}_i(\omega)$ with the result

$$\mathcal{S}(r,\mathcal{F})a_i(\omega)\mathcal{S}^\dagger(r,\mathcal{F}) = a_i(\omega)cosh(r\mathcal{F}^*) - a_i{}^\dagger(\omega)\mathcal{F}^{*-2}sinh(r\mathcal{F}^*). \qquad (2.168)$$

This modifies the Einstein field term G^{00} to

$$G^{00} = -\frac{1}{2}R^{(3)} + g^{-1}(Tr\pi^2 - \frac{1}{2}(Tr\pi)^2) + iTr\pi, \qquad (2.169)$$

where $Tr\pi = e_\mu p^\mu + p^\mu e_\mu$. This term defines the squeeze state operator. For large quantum numbers $\langle\,|e_\mu p^\mu + p^\mu e_\mu|\,\rangle = 0$, which shows this does not contribute to the classical field equation.

So far field amplitudes have been treated without consideration of their occurrence in the spacetime geometry. To an observer that approaches the Cauchy horizon it will appear that field modes are being generated, while in the future of the Cauchy horizon it will appear that field modes are being absorbed. Further, within the region $r < r_0$ there exist another set of modes. The exterior of the spacetime is separated into the regions to the future and the past of the Cauchy horizon $\mathcal{C}$, designated as $\mathcal{J}^+(\mathcal{C})$ and $\mathcal{J}^-(\mathcal{C})$. Consider the modes $a^\dagger$, $a \in [0, r)$ and $b^\dagger$, $b \in [r, \infty)$ within the region $\mathcal{J}^-(\mathcal{C})$ and the modes $c^\dagger$, c and $d^\dagger$, d within the region $\mathcal{J}^+(\mathcal{C})$. The solution to the Wheeler-DeWitt equation on the two sides of the Cauchy horizon is then

$$\psi_-(t) = \sum_i (a_i \mathcal{F}_i + b^\dagger{}_i \mathcal{F}^*{}_i) \in \mathcal{J}^-(\mathcal{C})$$

$$\psi_+(t) = \sum_i (c_i \mathcal{F}_i + d^\dagger{}_i \mathcal{F}^*{}_i) \in \mathcal{J}^+(\mathcal{C}). \qquad (2.170)$$

The indices i, j represent a summation over modes. The spacetime indices have been suppressed. This description of the cosmological wave function is forced on to us since the junction $r = r_0$ separates fields in these regions. These operators satisfy a Bogoliubov algebra. The evolution of the wave function $\psi_-(t)$ to $\psi_+(t)$ is governed in part by the squeeze state operator $S(r) = exp(r(cd\mathcal{F}_- - c^\dagger d^\dagger \mathcal{F}_-))$ with $\psi_+(t) = U(t)S(r)\psi_-(t)$. Here $U(t) = exp(1/2(c^\dagger c + d^\dagger d))$ is the unitary evolution operator. The unitary evolution operator only acts on the modes so that

$$a_i = c_i\, cosh(r\mathcal{F}_i^*) - \mathcal{F}^{*-2}d_i{}^\dagger sinh(r\mathcal{F}_i^*) \qquad (2.171a)$$

$$b_i{}^\dagger = d_i{}^\dagger cosh(r\mathcal{F}_i^*) - \mathcal{F}^{*-2}c_i\, sinh(r\mathcal{F}_i^*). \qquad (2.171b)$$

The transition probability for the creation of modes in the internal $r < r_0$ region of the solution is

$$T^2 = \frac{c^\dagger c}{a^\dagger a} = \frac{1}{cosh^2(r\mathcal{F}^*)} - \frac{d^\dagger d}{a^\dagger a} tanh^2(r\mathcal{F}^*). \qquad (2.172)$$

The exterior modes which are squeezed enter from the future direction in the region $\mathcal{J}^-(\mathcal{C})$ and from the past direction in $\mathcal{J}^+(\mathcal{C})$. The reflection coefficient is from $T^2 + R^2 = 1$ equal to

$$R^2 = \left(\frac{d^\dagger d}{a^\dagger a} - 1\right) \tanh(r\mathcal{F}), \tag{2.173}$$

which requires for probabilities $\in [0,1]$ that $\langle d^\dagger d\rangle \geq \langle a^\dagger a\rangle$. The total vacuum energy for the quantum fluctuation is then

$$\langle E\rangle = \frac{1}{2}\langle a^\dagger a\rangle + \frac{1}{2}\langle d^\dagger d\rangle = 0, \tag{2.174}$$

where $\langle a^\dagger a\rangle$ corresponds to a DeSitter cosmology with a nonnegative energy. This means that $\langle d^\dagger d\rangle \geq 0$ corresponds to the "vacuum response" of the observed cosmology.

This interesting piece of information has some profound implications. An observer in a spacetime cosmology finds the quantum vacuum is lower in energy than what would be predicted without gravitational self-squeezing of the vacuum. The stress-energy computed for this contribution to the vacuum is

$$T_{\mu\nu} = \sum_i \omega_i \left(d^\dagger_{\mu i} d_{\nu i} - \frac{1}{2}g^{\mu\nu} d^{\rho\dagger}{}_i d_{\rho i}\right). \tag{2.175}$$

We see that the T^{00} component is

$$T_{00} = \sum_i \omega_i \left(d^\dagger_{0i} d^0{}_i + \frac{1}{2}F(r) d^{\rho\dagger}{}_i d_{\rho i}\right) \tag{2.176}$$

and spacial components are

$$T_{rr} = \sum_i \omega_i \left(d^\dagger_{ri} d^r{}_i - \frac{1}{2F(r)} d^{\rho\dagger}{}_i d_{\rho i}\right)$$

$$T^{\theta\theta} = \sum_i \omega_i \left(d^\dagger_{\theta i} d_{\theta i} - \frac{1}{2}r^2 d^{\rho\dagger}{}_i d_{\rho i}\right) \tag{2.177}$$

$$T^{\phi\phi} = \sum_i \omega_i \left(d^\dagger_{\phi i} d_{\phi i} - \frac{1}{2}r^2 \sin(\theta) d^{\rho\dagger}{}_i d_{\rho i}\right).$$

Apparently for the region $r > r_0$ that $F(r) > 0$ and it is then evident that the weak energy condition $T^{00} \geq 0$ is satisfied. This infers a spatial stress-energy that are negative. This prevents problems with the quantum interest conjecture, which is posited to prevent various causality violations. $\sum_i \omega_i d^{0\dagger}{}_i d^0{}_i = \rho$ is the energy density and $\frac{1}{2}d^{j\dagger}{}_i d^j{}_i = P$ is a current projected onto a spatial surface as a pressure of the field. It is then apparent that the pressure term satisfies $3P \sim -\rho$ so the cosmology will exhibit inflationary expansion. The energy term is positive due to Cauchy horizon virtual modes which wrap pile up and introduce a positive energy into the wormhole. The wormhole will be dominated by positive energy as the two opening approach the Cauchy horizon. It is then reasonable on physical grounds the wormhole becomes a pair of black holes. This implies a virtual wormhole will transform into two virtual black holes before it becomes a virtual time machine. A cosmos with negative pressure is one that will be accelerating outward, but one where the energy principle is not violated. This currently appears to be the case with the observable universe. This pressure term is then what will contribute to equation 2.79.

2.7: The Geometry of Quantum Mechanics

This section involved considerable input from Robert Betts.

In this section the geometry of quantum mechanics is presented in order to illustrate its independence from spacetime geometry. This is done to first indicate that the nonlocality of quantum mechanics is an essential feature. It further gives imputus for quantum gravity as a pairing of quaternions, one for spinorial gravity and the other for quantum field theory, so this pair defines octonionic quantum gravity.

The geometry and logic of quantum mechanics exhibit nonlocality. Mathematical proofs from Von Neumann [31], Gleason [32] and Kochen-Specker [33], demonstrate the impossibility for local hidden variables that derive quantum mechanics in a noncontextual manner. Bell's Theorem asserts that reality is nonlocal by an inequality for rotated Stern Gerlach apparatuses. Bell's theorem may be derived without context to spacetime, but instead with projective Hilbert state space CP^1 for two measured states and the normalizer bundle over these states defined on CP^1. This manifold contains its own metric and is isomorphic to the heavenly sphere for the null cone $S^2 \simeq SO(3,1)/x : x^2 = 0$. A map between CP^1 and the heavenly sphere S^2, a measurement projection, is completely random. The random projection map $CP^1 \rightarrow S^2$ is examined for higher dimensional Lorentz algebras.

Among the basic axioms of Quantum Mechanics are the following [34]:

[A] The state of a physical quantum system at a time t is defined by a state vector $|\psi(t)\rangle$ which is an element of the state space.

[B] A measurement of some observable quantity on the physical system in the state $|\psi(t)\rangle$, yields a definite result which is the eigenvalue. The state vector is projected into the eigenspace to obtain the appropriate eigenvector in the state space.

[C] A quantum state space is determined by the diagonalized linear Hermitian operator, whose eigenvalues determine the eigenvectors and eigenstates for the given measurement. A finite dimensional complex valued quantum state space has a $U(1)$ fibration with the projective Hilbert space CP^n. Here the positive integer n depends on the number of phase entangled states in the system [35] and CP^n has a Fubini-Study metric.

A distance between two quantum states, or overlap, on the projective Hilbert space manifold CP^n is determined by the Fubini-Study metric

$$ds^2 = 4(1 - |\langle\psi(t)|\psi(t + dt)\rangle|^2) = 4(1 - |cos(\alpha/2)|^2). \tag{2.178}$$

This is an angle measure independent from a Minkowski metric. The complex quantum projective Hilbert state space $\mathcal{C}P^n$ has an orthonormal basis with a metric derived from [35]

$$h(X, Y) = Tr(XY), \qquad (2.179)$$

where X and Y are Hermitian operators representing observables. This is the distance between two rays in an abstract Hilbert space [35], where each ray is considered to be a single point in $\mathcal{C}P^n$. All points on a ray in Hilbert space differ by a phase factor belonging to the same point in $\mathcal{C}P^n$.

In reference [36] are comparisons between the metric for the projective quantum Hilbert state space $\mathcal{C}P^n$ and the metric for the Minkowski $(3+1)$ manifold for special relativity

$$R^3 \times R, \qquad (2.180)$$

which in the case of Lorentz boosts has Lorentz signature $(-1, 1, 1, 1)$. From reference [36] there is a significant difference between the metric defined for the hyperbolic Minkowski manifold with a Lorentz signature and with Lorentz group $SO(3,1)$, and the metric defined for a complex quantum projective Hilbert state space. The two metrics are different from each other, with a connection on null cones in the Lorentz space. Hidden variable constructions with the exchange information faster than light are untenable, since quantum states have no spacetime context.

2.7.1: Projective Structure in Spacetime and Quantum Mechanics

There are four natural Lorentz spaces: $2+1, 3+1, 5+1$ and $9+1$. The proper Lorentz groups for these metric spaces are [37]

$$SO(2, 1), \; SO(3, 1), \; SO(5, 1), \; SO(9, 1). \qquad (2.181)$$

The proper Lorentz group $SO(3, 1)$ specifies special relativity as $3+1$ Minkowski spacetime with own world lines and Lorentz boosts. There are unitary representations of the projective Poincaré group, $PSL(2, \mathcal{C})$ on a Hilbert space [38]. Spinors are then defined at the points of a flat Minkowski spacetime, which leads to the group identifications [37]

$$SL(2, \mathbf{R}) \simeq SO(2, 1), \; SL(2, \mathcal{C}) \simeq SO(3, 1),$$

$$SL(2, \mathcal{H}) \simeq SO(5, 1), \; SL(2, \mathcal{O}) \simeq SO(9, 1), \qquad (2.182)$$

where $\mathbf{R}$, $\mathcal{C}$, $\mathcal{H}$ and $\mathcal{O}$ are respectively the real, the complex field, the quaternionic and octonionic division algebra vector spaces.

In [35] the universal bundle

$$\mathcal{C}P^n = U(n+1)/\big(U(n) \times U(1)\big), \; n = 1, 2, 3, \ldots \qquad (2.183)$$

is identified with n entangled particles. $\mathcal{C}P^1$ defines the Fubini-Study metric for a single spin 1/2 particle, which might be undergoing precession in a homogenous magnetic field,

$calCP^2$ is for two entangled spin 1/2 particles, where if one particle precesses in a magnetic field the other does as well even if in a magnetic field free region. In reference [36] this has an identification with the "heavenly sphere," or the Riemann sphere S^2, which is the space of all null rays on the light cone. Hence any measurement of spin on a single particle by two observers, there is a mathematical relationship between the Fubini-Study metric and a null vector $X = x + y + z + l$ and $X \cdot X = 0$. The relationship between S^2 and CP^1 is a central subject of this paper.

Using methods outlined in [37], the following Theorem will be proved:

Theorem 1: The quantum Hilbert state space for two measured states, CP^1 is, up to isomorphism, equivalent to the heavenly sphere (Riemann sphere S^2) for the 3 + 1 dimensional null cone.

There exists a stereographic projection for a continuous bijection between unique points on the unit sphere and complex plane $\mathcal{C}$. The projective Lorentz group defines the heavenly sphere as $SO(3, 1)/(x: x^2 = 0) \subset PSO(3, 1)$. These are just the set of null projective rays. In the language of the Jordan algebra $(u, a) \odot (v, b) = \langle u, v \rangle - ab$, where for a null ray $ab = 1$ with vectors of the form $(u, 1)$ for $u^2 = 1$. The identification with a point at infinity for the complex plane with an antipodal point on the unit sphere defines elements from the Möbius group

$$\{(w + ix)/(y + iz) \in \mathcal{C}, wz - yx \neq 0\} \qquad (2.184)$$

and this group is $Aut(\mathcal{C}P^1)$. The Möbius transformation is nonsingular, where in $\mathcal{C}$ identify the point $\{\infty\}$ so that $\mathcal{C} \cup \{\infty\} \simeq CP^1 \simeq S^2$. Thus on this one point blow up identify the lightcone at ∞ relative to any finite point. Since the Möbius group is isomorphic to $PSL(2, \mathcal{C})$ (and is identified also with $PSO(3, 1)$), the heavenly sphere S^2 is a subset of the Möbius group.

First some definitions found in [37] are required:

Definition 1: For either a real, inner product vector space V of dimension n, or a normed division algebra K which is a vector space of dimension n, a spin factor

$$J(V) \text{ (or } J(K)), \qquad (2.185a)$$

is a Jordan algebra

$$h_2(V) \text{ (or } h_2(K)) \qquad (2.185b)$$

generated by V modulo $v^2 = ||v||^2$, $v \in V$ (or $K/(r^2 = ||r||^2)$, $r \in K$).

Definition 2: Let X be an element from a Jordan algebra of spin factors. The lightcone is defined as being a subset of the Jordan algebra as spin factors for which

$$X = (x, y, z, l), \ X \cdot X = 0, \ X \in J(V) \text{ (or } X \in J(K)). \qquad (2.186)$$

The heavenly sphere, which is the Riemann sphere, denoted in [37] by $S(K \oplus R)$, has group actions from $J(K \oplus R)$ and the projective line KP^1 has group actions from $h_2(K)$. Also the following isomorphism holds:

$$\phi : h_2(K) \rightarrow J(K \oplus R) \qquad (2.187)$$

It is also clear from [37] that $J(K \oplus R) \simeq S(K \oplus R)$.

Since the Riemann sphere is also homeomorphic to the projective line. Theorem 1 is restated as:

Theorem 1: The projective Hilbert state space, $\mathcal{C}P^1$, is isomorphic to the heavenly sphere (Riemann sphere S^2) for the 3 + 1 dimensional null cone.

Proof: Let $K = \mathcal{C}$, $K \oplus \mathbf{R} = \mathcal{C} \oplus \mathbf{R}$. Then the heavenly sphere $S(\mathcal{C} \oplus \mathbf{R})$ associated with $\mathcal{C} \oplus \mathbf{R}$ is equivalent to the projective line $\mathcal{C}P^1$ associated with $h_2(\mathcal{C})$ and $\mathcal{C}P^1 \simeq S(\mathcal{C} \oplus \mathbf{R})$, since $h_2(\mathcal{C}) \simeq J(\mathcal{C} \oplus \mathbf{R})$ [37].

It follows from Theorem 1 that for the measurement of any two quantum states in $\mathcal{C}P^1$, the Fubini-Study metric is identified with a null geodesic on the light cone. This implies that information communicated about measurements is constrained on the lightcone, or a time-like region. There is no spacelike communication of information between two observers at the moment of measurement. On the null cone the Lorentz metric reduces to

$$s^2 = (x_1)^2 + (x_2)^2 + (x_3)^2 - (x_4)^2 = 0. \qquad (2.188)$$

The heavenly sphere S^2 that connects the two measurements is that point in spacetime where light-like communications from A and B intersect. The information communicated to this intersection point is equivalent up to homeomorphism to the projective Hilbert state space manifold $\mathcal{C}P^1$.

The metric for the Hilbert state space is the Fubini-Study metric

$$ds^2 = 4(1 - |\langle \psi(t)|\psi(t+dt)\rangle|^2) = 4(1 - |cos(\alpha/2)|^2). \qquad (2.189)$$

Measurement information is communicated on the light cone, which defines the set of projective rays for a two-sphere. Thus there is a map from the Fubini-Study metric on $\mathcal{C}P^1$ to the projective rays defined by light cones that connect the two points of measurement by two observers. The two dimensional space of nonlocality is then $\mathcal{C}P^1$ defined by the normal bundle, where there is a $U(1)$ phase involved with the embedding of this space. This space modulo this phase $\mathcal{C}P^1 = \mathcal{C} \cup \{\infty\}$ is the two sphere plus lightcone of the other observer.

The two sphere involved with nonlocality is then associated with the real space of measurements as defined on a light cone. Although there is an "inducement" from this metric to a metric on the Riemann sphere [35] (or, equivalently, the heavenly sphere), the Lorentz and Fubini-Study metrics are not equivalent. On the Lorentz spacetime manifold distance is defined as the distance between two different events in Minkowski spacetime, while the Fubini-Study metric gives the distance between two states in quantum state space. In quantum mechanics distance is an angular measure identified with the Riemann sphere with unit radius. The angles cannot have a distance greater than some finite multiple of 2π. Angular measure on a Lorentzian manifold is different, where vectors subtended by angles may have inner products greater than 1. In fact it is possible for the dot product between two four-vectors to approach infinity [39].

2.7.2: Bell's Inequality, Metrics and Expectation

Terminology is first introduced for this section, which will be used to model closely Bell's argument [41]. Assume the observers Alice and Bob execute measurements on a two particle quantum system in a singlet spin state, with the total spin equal to zero. Each experimenter makes three consecutive measurements along three axes, one axis which is assumed to exist for the hidden variable spin vector. In this section let $\theta(r, t)$ denote the "hidden variable" expectation value determined by Alice and Bob after their first spin measurement. Alice measures the spin state along axis r and Bob measures the spin state along axis t. $\theta(r, s)$ is the hidden variable expectation value they both determine for the second pair of measurements. Alice takes the second measurement along axis r and Bob makes a measurement along the hidden variable axis axis s and $\theta(t, s)$ is the hidden variable expectation value they compute after their third and final pair of measurements. On the third measurement Alice measures her spin value along axis t (Bob's axis) and Bob measuring his spin value along the hidden variable axis s (see Figure 1). The hidden variable axis s along which the "hidden variable spin vector" lies, which makes an angle of θ with Alice's axis. The angle between axis r and axis t is ϕ. From [41], the quantum mechanical expectation prediction will be taken to be $\theta_{QM}(r, t) = -cos(\theta)$, while $\theta(r, t) = 2\phi/\pi - 1$.

Assume Bob and Alice are separated by a great distance, where at the midpoint between them a spin 0 particle decays into a pair of phase entangled electrons in a singlet spin state. The decay imparts a momentum to each electron so the wave function is an expanding S-wave. The electrons interact with a Stern-Gerlach analyzer at the observation points. The pair of daughter product electrons are in the state vector

$$|\psi\rangle = \frac{1}{\sqrt{2}}(|\uparrow\rangle_1|\downarrow\rangle_2 + |\downarrow\rangle_1|\uparrow\rangle_2). \tag{2.190}$$

This wave vector measured by both Alice and Bob for each of their three measurements are reduced to $|\uparrow\rangle_1|\downarrow\rangle_2$ or $|\downarrow\rangle_1|\uparrow\rangle_2$ [40], where the arrows indicate spin up or down. However the two states which are actually measured are the states "Alice and Bob measure the up and down spins" and "Alice and Bob measure down and up spins."[41].

The proof to Bell's Theorem posits a locally real hidden variable angle, θ which determines the outcomes for the three measurements, then proceeds to show that, although the hidden variable expectation values observe this inequality the quantum mechanical expectation variable $\theta_{QM}(r, t)$ does not, which implies that a theory of reality with locally real hidden variables is impossible. The machinery for Bell's proof depends on the following:

I: Alice points her measuring device along axis r, Bob does the same along t, then they compute $\theta(r,\ t)$.

II: Bob rotates his measuring device along hidden variable axis s, while Alice makes no adjustment from r. They then compute $\theta(r,\ s)$.

III: Alice now rotates her device along t, while Bob keeps his along s. They then compute expectation value $\theta(t,\ s)$.

The measuring devices Bob and Alice use detect the sign + or − for the projection of the (hidden variable) spin vector along the axes. Thus the possible states turn out to be [39]:
$$A = sgn(cos(\theta)),$$
$$B = sgn(-cos(\theta - \phi)),\ AB = sgn(-cos(\theta)cos(\theta - \phi)), \quad (2.191)$$
where the hidden variable θ is in any one of the intervals
$$[-\pi/2,\ -\pi/2 + \phi],\ [-\pi/2 + \phi,\ \pi/2],\ [\pi/2,\ \pi/2 + \phi],\ [\pi/2 + \phi,\ 3\pi/2]. \quad (2.192)$$
Suppose all three axes are parallel and pointing in the same direction. Bell's inequality states
$$|\theta(r,\ t) - \theta(r,\ s)| \leq 1 + \theta(t,\ s). \quad (2.193)$$
A detailed derivation of the Bell Inequality can be found in Chapter seven, Section 7.7, in [42]. Numerous experiments in [44] and [45] and by others have shown that the quantum expectation value
$$\theta_{QM}(r,\ t) = -cos(\phi) \quad (2.194)$$
only agrees with the hidden variable expectation value $\theta(r,\ t)$ when ϕ is equal only to 0, $\pi/2$ and π. For example in the proof to Bell's Theorem the inequality
$$|\theta_{QM}(r,\ t) - \theta_{QM}(r,\ s)| \leq 1 + \theta_{QM}(t,\ s) \quad (2.195)$$
fails to hold when $\phi = \pi/3$. The conclusions of Bell's theorem are [44]:

(1) Reality is nonlocal
(2) Inductive system of logical inference must be abandoned
(3) Einstein separability violated and spacelike causality exists.

To demonstrate that (3) is a false assumption when applied to "quantum signals" the following is proven.

Theorem 2: Let the frames for Alice and Bob be such such that the two frames are mutually at rest. Let the frame midway between their two locations, the site of particle decay, also be

Quantum Fluctuations and Spacetime 71

at rest. The distance metric between any quantum spin states that Alice and Bob measure for two phase entangled particles in an isotropic singlet state is completely independent of any timelike or spacelike Lorentz distance interval s between their laboratories when $s \neq 0.\odot$

Proof: The two measured spin states lie on the time-orientable, future (or past) part of the light cone where $s^2 = 0$ and on the light cone $s = 0$. It is along these null intervals that information about the experiment must be communicated, where this information contains nothing that involves spacelike or timelike intervals. Hence irrespective of the Lorentz interval between their two laboratories information that is communicated involves only angle measurements. This information is encoded by a Fubini-Study distance between the unknown states on $\mathcal{CP}^1$ before measurement and is communicated along null rays on the null cone after the measurements.

The measurement of an angle in $\mathcal{CP}^1$ is communicated to that angle in S^2 by the isomorphism of the two spaces, where further this measurement is obtained in space is due to the projector $P = |r\rangle\langle r|$ on the state vector. This gives

$$|\psi(r)|^2 = \langle\psi|r\rangle\langle r|\psi\rangle, \tag{2.196}$$

so a probability measure over some region is then

$$\int_{vol} d^3r |\psi(r)|^2 = \int_{vol} d^3r \langle\psi|r\rangle\langle r|\psi\rangle. \tag{2.197}$$

Since $1 = \int_{R^3} d^3r |r\rangle\langle r|$ then

$$\int_{R^3} d^3r |\psi(r)|^2 = \int_{R^3} d^3r \langle\psi|r\rangle\langle r|\psi\rangle = \langle\psi|\left(\int_{R^3} d^3r |r\rangle\langle r|\right)|\psi\rangle = 1. \tag{2.198}$$

A measurement with the outcome $|\phi\rangle$ at location r from the state $|\psi\rangle$ is modelled with the projectors $P_r = |r\rangle\langle r|$ and $P = |\phi\rangle\langle\phi|$,

$$|\psi\rangle \rightarrow |r\rangle\langle r|\phi\rangle\langle\phi|\psi\rangle, \tag{2.199a}$$

so that

$$\langle r'|\psi\rangle \rightarrow \psi(r')_{meas} = \langle r'|r\rangle\phi(r)\langle\phi|\psi\rangle, \tag{2.199b}$$

with

$$|\langle\psi|\psi_{meas}\rangle|^2 = |\langle\phi|\psi\rangle|^2 = \cos^2(\alpha/2). \tag{2.200}$$

For a density matrix $|\psi\rangle\langle\psi| = \rho$

$$Tr\rho = 1 = Tr\rho_{meas}. \tag{2.201}$$

For the trace defined as $Tr(\rho\rho_{meas}) = \langle\psi\rho\rho_{meas}\psi\rangle$ then $Tr(\rho\rho_{meas}) = Tr\rho Tr\rho_{meas}$, $= |\langle\psi|\psi_{meas}\rangle|^2$ which then means $\cos^2(\alpha/2) = 1$. Then $\alpha/2 = 2n\pi$, $n = 0, 1, 2, \ldots$.

Here α is the angular distance between the two rays on which the two state lie. The above distance exists on a space $\mathcal{C}P^1$ that is mapped to the heavenly sphere by the projector $P_r = |r\rangle\langle r|$. The projector defines a the map $\mathcal{C}P^1 \to S^2$ which is the geometry appropriate for the nonlocality of quantum mechanics. The distance defined by the projector is also independent of the distance in Minkowski space.

To conclude this section a connection between the Fubini-Study metric and the Bell inequality is made explicit. Here the hidden variable parameter employed in the Bell theorem is demonstrated to be the distance measure on the projective Hilbert space

Theorem 3: For the hidden expectation values discussed in this paper and for a Bell probability measure $d\lambda$, where λ is a hidden variable, the inequality

$$|\theta(r,\ t)\ -\ \theta(r,\ s)|\ \leq\ \Lambda\ +\ \int_0^\Lambda \theta(t,\ \lambda)\theta(s,\ \lambda)d\lambda \quad (2.202)$$

is dependent on a quantum state space probability measure $d\lambda$ equivalent to

$$d\lambda\ =\ \sqrt{2(1\ -\ |\langle\psi|\psi'\rangle|^2)}\ =\ \sqrt{2(1\ -\ |cos(\alpha/2)|^2)}, \quad (2.203)$$

where $cos(\alpha/2) = \langle\psi|\psi'\rangle$ and $\alpha/2 = \epsilon$, or $\pi - \epsilon$.

Proof: For $r \to r + \delta r$

$$|\theta(r + \delta r,\ t) - \theta(r + \delta r,\ s)| \leq |\theta(r,\ t) - \theta(r,\ s)| + |\theta(\delta r,\ t) - \theta(\delta r,\ s))|. \quad (2.204)$$

With $\theta(r,\ t) = \langle\psi|\hat{\theta}(r,\ t)|\psi\rangle$, the variation of the operator by the Schrödinger equation in the Heisenberg representation gives

$$\theta(r + \delta r,\ t) - \theta(r,\ t) = -iH\theta(r,\ t)(\partial T/\partial r)\delta r - \frac{1}{2}H^2\theta(r,\ t)(\partial T/\partial r)^2\delta r^2, \quad (2.205)$$

where T is time. This leads to the conclusion that

$$\theta(r + \delta r,\ t) = \langle\psi|\hat{\theta}(r,\ t)|\psi\rangle - i\langle\psi|H\hat{\theta}(r,\ t)|\psi\rangle\frac{\partial T}{\partial r}\delta r - \frac{1}{2}\langle\psi|H^2\hat{\theta}(r,\ t)|\psi\rangle\left(\frac{\partial T}{\partial r}\right)^2\delta r^2, \quad (2.206)$$

with the result that

$$|\theta(r + \delta r,\ t) - \theta(r + \delta r,\ s)| - |\theta(r,\ t) - \theta(r,\ s)| =$$

$$|\theta(r,\ t) - \theta(r,\ s)|\sqrt{(1 - (\partial T/\partial r)^2\delta r^2(\langle H^2\rangle - \langle H\rangle^2)}, \quad (2.207)$$

where the term under the radical is the distance measure for the Fubini-Study metric and is $d\lambda$. Now apply the Bell theorem and set $\theta(t,\ s) \leq \theta(\lambda, t)\theta(\lambda,\ s)$, to track the dependency on the integration variable, so that

$$|\theta(r+\delta r,\ t) - \theta(r + \delta r,\ s)| - |\theta(r,\ t) - \theta(r,\ s)| \leq (1 + |\theta(\lambda,\ t)\theta(\lambda,\ s))d\lambda. \quad (2.208)$$

An integration of $d|\theta(r, t) - \theta(r, s)|$ then gives

$$|\theta(r, t) - \theta(r, s)| \leq \Lambda + \int_0^\Lambda |\theta(\lambda, t)\theta(\lambda, s)|d\lambda. \tag{2.209}$$

An integration over the probability measure λ from $[0, 1]$ then leads to the Bell theorem.

2.7.3: The Hilbert Space $\mathcal{C}P^2$ and Lorentz Group $PSO(5, 1)$

The complex, quantum mechanical Hilbert state space for four phase entangled particles $\mathcal{C}P^2$ has a topology differing from the topology for $\mathcal{C}P^1$. Although it has dimension four, it is not strictly true that it is homeomorphic to S^4. In fact there is a map

$$\pi : S^5 \to \mathcal{C}P^2, \ with \ fibre \ \pi^{-1}(S^1), \tag{2.210}$$

where there is a homeomorphism

$$S^5/S^1 \to \mathcal{C}P^2, \tag{2.211}$$

which is not strictly the same as S^4. Yet $\mathcal{C}P^2$ has a complex nonoriented disk at the two dimensional line at infinity identified with an antipodal point on S^4 an unoriented disk, or cross cap, there will be a subspace of $\mathcal{C}P^2$ homeomorphic to S^4. In fact it can be shown as an exercise that a subset $U = \{[x_0 : y_0 : z_0] \in \mathcal{C}P^2 | x_0 \neq 0\} \cup \infty$ of $\mathcal{C}P^2$ is homeomorphic to $\mathcal{C}^2$, where there exists a submanifold U' of $\mathcal{C}P^2$ homeomorphic to $\mathcal{C}^2 \cup \infty$.

Theorem 4: There exists a submanifold U' of $\mathcal{C}P^2$ homeomorphic to $\mathcal{H}P^1$ and to the heavenly sphere for $SO(5, 1)$, where $\mathcal{H}$ is the normed division algebra for the quaternions.

Proof: Given that U is homeomorphic to $\mathcal{C}^2$, then one can construct a submanifold U' of $\mathcal{C}P^2$ such that

$$U' = U \cup \infty \simeq \mathcal{C}^2 \cup \infty \tag{2.212}$$

It can be shown that $\mathcal{C}^2 \cup \infty$ is homeomorphic to the unit 4-sphere S^4, with the 4-sphere's antipodal point corresponding to the "point at infinity." Hence

$$\mathcal{C}^2 \cup \infty \simeq S^4. \tag{2.213}$$

However it is also known that $\mathcal{H}P^1 \simeq S^4$ and therefore $U' \simeq \mathcal{C}^2 \cup \infty \simeq \mathcal{H}P^1 \simeq S^4$.

The general universal bundle theorem for a Grassmannian manifold on $\mathcal{C}^n$ is

$$G_k(\mathcal{C}^n) = \frac{U(n)}{U(n-k) \times U(k)}, \tag{2.214a}$$

where the projective space for 4 entangled particles is then

$$G_1(\mathcal{C}^4) = \mathcal{C}P^4 = \frac{U(5)}{U(4) \times U(1)}. \tag{2.214b}$$

Since $O(10) \rightarrow U(5) \times U(1)$ and $O(10) \simeq O(4) \times U(4)$ the trivialization of the line bundle on $\mathcal{CP}^4$ is then $O(4)$, which is the bundle that acts on the projective space. The normalizer or normal subgroup is N so that given any g in the group $g^{-1}Ng = N$. There then exists a normalizer $N = N_{\mathcal{CP}^4/\mathcal{CP}^2}$ for a map

$$\gamma : \frac{U(5)}{U(4) \times U(1)} \rightarrow \left[\frac{U(5)}{U(4) \times U(1)}\right]/N \tag{2.215}$$

For a one-form $\omega \in H^1(\mathcal{CP}^2)$ there exists a form σ such that

$$\int_{\mathcal{CP}^2} \omega = \int_{\mathcal{CP}^4} \omega \wedge \sigma \tag{2.216}$$

Define the coordinates in $\mathcal{CP}^4$ as $\{z_1, z_2, z_3, z_4, C.C\}$. Then let the coordinates of $\mathcal{CP}^2$ be $\{x, \bar{x}\}$ with $x = z_1 z_2 / z_3 z_4$. Then for $\omega = dx$,

$$\omega = \frac{z_1 z_2}{z_3 z_4} \left[\frac{dz_1}{z_1} + \frac{dz_2}{z_2} - \frac{dz_3}{z_3} - \frac{dz_4}{z_4}\right] + H.C. \tag{2.217}$$

A one-form σ is then similarly defined as $\sigma = -dz_1 - dz_2 + dz_3 + dz_4$ so that

$$\omega \wedge \sigma = \frac{z_1 z_2}{z_3 z_4}\left((z_2 - z_1)\frac{dz_1}{z_1} \wedge \frac{dz_2}{z_2} + (z_1 - z_3)\frac{dz_1}{z_1} \wedge \frac{dz_3}{z_3} + (z_1 - z_4)\frac{dz_1}{z_1} \wedge \frac{dz_4}{z_4}\right.$$

$$\left. + (z_2 - z_3)\frac{dz_2}{z_2} \wedge \frac{dz_3}{z_3} + (z_2 - z_4)\frac{dz_2}{z_2} \wedge \frac{dz_4}{z_4} + (z_4 - z_3)\frac{dz_3}{z_3} \wedge \frac{dz_4}{z_4}\right) + H.C. \tag{2.218a}$$

$$= x \sum_{i \leq j = 1}^{4} C_{ij}(z_i - z_j)\frac{dz_i}{z_i} \wedge \frac{dz_j}{z_j} + H.C., \tag{2.218b}$$

where $C_{12} = C_{34} = 1$ and $C_{ij} = -1$ otherwise.

Equation 218b may be written as

$$\omega \wedge \sigma = x \sum_{i \leq j = 1}^{4} C_{ij}(d\,log(z_i) \wedge dz_j + d\,log(z_j) \wedge dz_i) + H.C.. \tag{2.219}$$

Let g_{ik} be a transition function from U_i to U_k as $z_i = g_{ij}z_j$ with

$$\frac{dz_i}{z_i} = \frac{1}{2\pi}\rho_k d\phi_{ik} = -\frac{1}{2\pi i}\rho_k d\,log(g_{ik}) \tag{2.220}$$

and ρ_k a partition of unity on these coordinate charts. The two-form may then be written as

$$\omega \wedge \sigma = \frac{x}{2\pi i}\sum_{i \leq j = 1}^{4} \rho_k d\,log(g_{ik}) \wedge dz_j, \tag{2.221}$$

which determines the Euler characteristic $dim(H^2(\mathcal{C}P^2\otimes\mathcal{C}^2, R))$ This then means that this form is a subbundle of $\mathcal{C}P^2 \otimes \mathcal{C}^2$ and is $Z \times Z$, which is the topology of two three spheres. Hence the heavenly sphere is $SU(2) \times SU(2)$ or $SO(4) \in SO(5, 1)$. The normalizer is then $SO(1, 1)$, that in a Euclideanized metric is homeomorphic to $U(1)$. There is then the decomposition of this group by $SO(4) \rightarrow SU(2) \times U(1)$. A decoherence will reduce this to four copies of $U(1)$ so that four observers of the particles will detect the heavenly sphere S^2 as indicated above.

2.7.4: Quantum Geometry and Extended Field Theories

The conclusion emphasized is that for the moment Alice and Bob make their measurements on two simultaneous frames, Bell's Theorem does not indicate spacelike communication of quantum information is possible. Hence reality is not spacelike due to any superluminal flow of quantum information, although on the quantum state space manifold $\mathcal{C}P^1$ dependent outcomes through correlated quantum states is possible.

One can speculate whether or not the identification through homeomorphism of quantum Hilbert state space with the time-orientable past light cone can offer any insight into J. A. Wheeler's thought experiment known as the Extreme Delayed Choice experiment, concerning a photon traveling from a gravitationally lensed quasar several million light years away. The experimenter is now an astronomer who must decide whether or not to apply a second beam splitter to his apparatus, resulting in two possibilities for observations, one for which he observes two photons, the other for which he observes one. The suggestion is then made that the astronomer somehow altered millions of years of history. This effect might be easily explained if $\mathcal{C}P^1$ is identified with the past time-orientable light cone at the moment the astronomer makes his observation.

The first small conclusion of this section is that quantum mechanics has a geometry and a logic that is fundamentally different from the macroscopic world. The second is that the logic of quantum mechanics has a connection to the structure of relativity on the light cone. It is this connection that indicates a commonality between quantum mechanics and relativity.

This further leads to some additional issues. The number of entangled particles increases the size of $\mathcal{C}P^n$. This further appears to lead to a succession of $SO(n - 1, 1)$ and in doubling appears to emulate the Cayley numbers. However, beyond $N = 8$ there is no algebra. However, the entanglements here do not climb this hierarchy. The size of the projective geometry increases and one can invoke higher dimensional Lorentz algebras, but this does not have the effect of changing the basic structure of quantum mechanics. However, this might change if the energy of a quantum scattering process or measurement approaches the Planck scale. Here this determines a term in the Schwarzschild metric with $M = G\hbar/Rc^2$ for the isolation of an event in a region R. This will result in a loss of locality with a measurement with $[x, x'] \neq 0$. The breakdown of commutivity in quantum mechanics opens the question whether nonassociative structures are required for quantum gravity. This then leads to an octonion theory of quantum gravity that supports the holographic principle [46].

2.8: Quantum Computing Backwards in Time and $P \subset NP$

The occurrence of measured spins in a Stern-Gerlach experiment by two observers obtains via EPR such that the space of quantum states and spacetime are utterly independent. This analysis usually focuses upon nonlocal measurement outcomes that have a spacial separation, as is with most EPR and Bell theorem work. Since the quantum space is completely independent of spacetime this obtains for timelike separations. This appears to conform to the usual causal picture of things. Wheeler's delayed choice experiment implies Bell's theorem is just as valid for observables specified at different points temporally separated as with variables spatially separated. The teleportation of states permits measurable observables to be nonlocally correlated from one region to another with a spacelike separation. This is performed through the entanglement of a quantum system with another, which nonlocally correlates observables to a certain region of space. Since measurement involves the entanglement of a system with a complex environment, this too is a coarse grained analog of teleportation. It is then argued here that a temporal teleportation of states is also possible. The classical state selection process in the Wheeler delayed choice experiment is replaced by its quantum q-bits are teleported backwards in time. More properly it is that states and their observables are entangled with prior states. That this is possible has implications for factorizing algorithms and NP-complete problems. It is demonstrated that the proper outcome of a factorizing algorithm obtains not by an exhaustive loop search, but rather because the the wave function is forced into a configuration corresponding to the outcome. Finally since *classical mechanics* $\subset$ *quantum mechanics* a physical argument is then advanced for the proper inclusion $P \subset NP$.

Consider a photon sent towards a double slit. This photon passes through double slit unobserved, where logically it passes through one slit, through the other or through both. To obtain an interference pattern the wave function must pass through both slits. In order to obtain a particle distribution, the photon must pass through one or the other. The photon's quantum properties are manifested as it passes through the slits. After passing through the slits, the photon is in transit towards the back screen. At the back screen there is available two separate methods of detecting the photon. The choice of detection is made after the photon is known to have passed the slits [47].

Let there be a detection system, eg. a large photoplate, which can measure the horizontal placement of a photon hit, but is not able to distinguish which slit the photon passed through. There are also directed photometers that determine which slit the photon passed through. Either device may be removed quickly after the photon has passed the double slits but before the photon reaches the plane of the screen. This is the experimenter's choice, which is delayed until after the photon has passed the slits.

If the photoplate is removed two sensors are revealed at the screen. The sensors are tightly focused on the narrow space of one slit only. The left photometer watches the left slit, and the right photometer watches the right slit. These will detect a flash of light if the photon went either wholly or in part through the slit on which it is focused. Therefore, the experimenter will obtain information about which path the photon took. Conversely the experimenter may choose to use the photoplate and not the photometers. If so there is then a loss of information about which slit the photon traversed. This means that an interference wave pattern falls on the photoplate, the wave function is reduced on the plate and no information is communicated about which slit the photon traversed.

This is the delayed choice, for a choice of an experiment at a time t determines the quantum configuration of the system at a time $t' < t$. This means that nonlocality exists not just in space but time as well. A choice of state selection in the future can influence the state evolution at an earlier time [47]. Further, the Bell inequality then must obtain for experiments performed on a single wave function at different times.

Given the spin meters A and B there are the outcomes a, a' and b, b' for these meters. For the nth spin particle pair detected there is the following expression

$$Q_n = a_n b_n + a'_n b_n + a_n b'_n - a'_n b'_n \tag{2.222}$$

that has the value between -4 and 4. This may be written as

$$Q_n = a_n(b_n + b'_n) + a'_n(b_n - b'_n). \tag{2.223}$$

So the bounds on this are seen to be further restricted to -2 and 2. There are the following limiting terms,

$$c(a, b) = \lim_{N \to \infty} \frac{1}{N} \sum_{n=1}^{N} (a_n b_n)$$

$$c(a', b) = \lim_{N \to \infty} \frac{1}{N} \sum_{n=1}^{N} (a'_n b_n)$$

$$c(b, a') = \lim_{N \to \infty} \frac{1}{N} \sum_{n=1}^{N} (b_n a'_n)$$

$$c(b', a') = \lim_{N \to \infty} \frac{1}{N} \sum_{n=1}^{N} (b'_n a'_n). \tag{2.224}$$

These are then bounded by

$$|c(a, b) + c(a', b) + c(a, b') - c(a', b')| \leq 2, \tag{2.225}$$

but quantum expectations illustrate that

$$c(a', b) = -\cos(\theta_{a'b}), \quad c(a, b') = -\cos(\theta_{ab'}), \quad c(a', b') = -\cos(\theta_{a'b'}). \tag{2.226}$$

For a parallel to b then $\theta_{ab'} = \theta_{a'b} = \phi$, and $\theta_{a'b'} = 2\phi$ and the inequality is

$$|1 + 2\cos(\phi) - \cos(2\phi)| \leq 2, \tag{2.227}$$

and this is violated for $\phi \in (0, \pi/2)$.

Ordinarily these are considered for different regions of space. This may also be considered for different times, where A and B measure events at different times. The time evolution of the outcomes are

$$a \rightarrow e^{iHt}ae^{-iHt}, \quad b \rightarrow e^{iHT}be^{-iHT}, \tag{2.228}$$

which are given by unitary operators and send the values of B into the future of A if $T > t$. There then exist two values for Q for detectors at different periods of time. Further, since the time transformations are unitary this will have no influence on the basic outcome of the Bell inequality. This is a Bell inequality version of the Wheeler delayed choice experiment.

This may be seen more completely if α and β are operators corresponding to the eigenvalues measured by detectors A and B. For the pair of particles these operators exist in the Hilbert space $\mathcal{H} = \mathcal{C}^2 \otimes \mathcal{C}^2$. Each copy of $\mathcal{C}^2$ corresponds to the eigenvalues for A and B respectively. The operators α and β project the state of each particle according to the azimuthal angles θ, ϕ and correspond to the σ_3 Pauli matrix with

$$\alpha_\phi = |+_1\rangle\langle +_1| - |-_1\rangle\langle -_1|, \quad \beta_\phi = |+_2\rangle\langle +_2| - |-_2\rangle\langle -_2|. \tag{2.229}$$

Further the operator that corresponds to Q in the Bell's inequality for measurements directed along ϕ and χ is

$$\hat{Q} = \alpha_\phi \otimes \beta_\chi + \alpha_\phi \otimes \beta_{\chi'} + \alpha_{\phi'} \otimes \beta_\chi - \alpha_{\phi'} \otimes \beta_{\chi'}. \tag{2.230}$$

The operators measured by the two apparatus at different times along the development of the wave function are then time evolved to these two different time frames. The subsequent time development after measurement is regarded as irrelevant due to wave function decoherence or "collapse." There then exist the time development operators for the two sets of states

$$U_\alpha(\tau) = e^{iE_+\tau}|+_1\rangle\langle +_1| - e^{iE_-\tau}|-_1\rangle\langle -_1|$$
$$U_\beta(\tau) = e^{iE_+\tau}|+_2\rangle\langle +_2| - e^{iE_-\tau}|-_2\rangle\langle -_2|. \tag{2.231}$$

With the above description of the partition time evolution operators it is then possible to describe the operator $\hat{Q}$ according to separate time evolutions for A and B. Now compute the evolution of the state vector $|\psi\rangle = \frac{1}{\sqrt{2}}(|\alpha_+\rangle|\beta_-\rangle - |\alpha_-\rangle|\beta_+\rangle)$. With different times assigned to the α and β states.

$$U_\alpha(\tau)U_\beta(t) = \frac{1}{\sqrt{2}}\left(e^{iE_+\tau}e^{iE_-t}|\alpha_+\rangle|\beta_-\rangle - e^{iE_-\tau}e^{iE_+t}|\alpha_-\rangle|\beta_+\rangle\right). \tag{2.232}$$

This is then used to compute the total expected value for the operator $\hat{Q}$ which evaluates as

$$\langle\psi|\hat{Q}|\psi\rangle = \frac{1}{2}(|\langle\alpha'_+|\alpha_+\rangle|^2 - |\langle\alpha'_-|\alpha_+\rangle|^2 - 1))(1 - |\langle\beta'_+|\beta_-\rangle|^2 + |\langle\beta'_-|\beta_-\rangle|^2) +$$

$$\frac{1}{2}(|\langle\alpha'_+|\alpha_-\rangle|^2 - |\langle\alpha'_-|\alpha_-\rangle|^2 - 1))(1 - |\langle\beta'_+|\beta_+\rangle|^2 + |\langle\beta'_-|\beta_+\rangle|^2) -$$
$$\frac{1}{2}\Big(e^{i(\phi_1 - \phi_2)}(\langle\alpha'_+|\alpha_+\rangle\langle\alpha_-|\alpha'_+\rangle - \langle\alpha'_-|\alpha_+\rangle\langle\alpha_-|\alpha'_-\rangle) -$$
$$e^{-i(\phi_1 - \phi_2)}(\langle\alpha'_+|\alpha_-\rangle\langle\alpha_+|\alpha'_+\rangle) - \langle\alpha'_-|\alpha_+\rangle\langle\alpha_-|\alpha'_-\rangle)\Big) \times$$
$$(\langle\beta'_+|\beta_-\rangle\langle\beta_+|\beta'_+\rangle - \langle\beta'_-|\beta_-\rangle\langle\beta_+|\beta'_-\rangle), \tag{2.233}$$

where $\phi_1 = E_+\tau + E_-t$ and $\phi_2 = E_-\tau + E_+t$.

For the evolution of this wave function ordinarily $t = \tau$ so that $\phi_1 - \phi_2 = 0$. However, suppose that this is not the case. For the wave function at $\psi(t)$ with temporal entanglement, the time deficit $t - \tau$ must exist in some form, since the final time condition for the wave function obtains at a time t at B, yet the observer operator for the apparatus A occurs at the time τ. Thus the phase of the wave function at the time t must contain this time information in the phase associated with the angles α' and β'. The change in the angle $\alpha' - \alpha$ is associated with an operator

$$U_{\Delta\alpha_\pm} = e^{\pm i\Delta\alpha}(|\alpha_\pm\rangle\langle\alpha_\pm| + |\alpha_\mp\rangle\langle\alpha_\mp|) = exp(i\Delta\alpha|\alpha_\pm\rangle\langle\alpha_\pm|), \tag{2.234}$$

with $\Delta\alpha = \alpha - \alpha''$ and α'' some transition angle between α' and α. The angle α'' is what exists at A at the time τ, and equivalently α' obtains at A at the time t. The unitary change in the α' eigenstates is then

$$|\alpha_\pm\rangle = U_{\Delta\alpha_\pm}|\alpha''_\pm\rangle. \tag{2.235}$$

The time evolved part of $\langle\psi|\hat{Q}|\psi\rangle = \langle\hat{Q}\rangle_T$ is then

$$\langle\hat{Q}\rangle_T = \frac{1}{2}\Big(e^{i(\phi_1 - \phi_2)}e^{2i\Delta\alpha}(\langle\alpha'_+|\alpha''_+\rangle\langle\alpha''_-|\alpha'_+\rangle - \langle\alpha'_-|\alpha''_+\rangle\langle\alpha''_-|\alpha'_-\rangle) -$$
$$e^{-i(\phi_1-\phi_2)}e^{-2i\Delta\alpha}(\langle\alpha'_+|\alpha''_-\rangle\langle\alpha''_+|\alpha'_+\rangle) - \langle\alpha'_-|\alpha''_+\rangle\langle\alpha''_-|\alpha'_-\rangle)\Big) \times$$
$$(\langle\beta'_+|\beta_-\rangle\langle\beta_+|\beta'_+\rangle - \langle\beta'_-|\beta_-\rangle\langle\beta_+|\beta'_-\rangle). \tag{2.236}$$

This gives the result $E_+(\tau - t) + E_-(t - \tau) + 2\Delta\alpha = 0$. Hence the additional phase corresponds to a nonlocal teleportation of amplitudes from B backwards in time to A.

This may be further explored by an examination of $\hat{\alpha}' \otimes \hat{\beta}$ and $\hat{\alpha}' \otimes \hat{\beta}'$. The expectations for these operators are

$$\langle\psi(t,\tau)|\hat{\alpha}' \otimes \hat{\beta}|\psi(t,\tau)\rangle = \frac{1}{2}(|\langle\alpha_+|\alpha'_-\rangle|^2 + |\langle\alpha'_+|\alpha_-\rangle|^2 - |\langle\alpha'_+|\alpha_+\rangle|^2 - |\langle\alpha'_-|\alpha_-\rangle|^2) \tag{2.237a}$$

and

$$\langle\psi(t,\tau)|\hat{\alpha}' \otimes \hat{\beta}'|\psi(t,\tau)\rangle = \frac{1}{2}\Big((|\langle\alpha'_+|\alpha_+\rangle|^2 - |\langle\alpha_+|\alpha'_-\rangle|^2)(|\langle\beta_-|\beta'_+\rangle|^2 - |\langle\beta_-|\beta'_-\rangle|^2)$$

$$+ (|\langle\alpha'_-|\alpha_+\rangle|^2 - |\langle\alpha_-|\alpha'_-\rangle|^2)(|\langle\beta_+|\beta'_+\rangle|^2 - |\langle\beta_+|\beta'_-\rangle|^2)\Big)$$

$$-\cos(\phi_1 - \phi_2)(\langle\alpha_-|\alpha'_+\rangle\langle\alpha'_+|\alpha_+ - \langle\alpha_-|\alpha'_-\rangle\langle'_-|\alpha_+\rangle)(\langle\beta'_+|\beta_-\rangle\langle\beta_+|\beta'_+\rangle - \langle\beta'_-|\beta_-\rangle\langle\beta_+|\beta'_+\rangle). \tag{2.237b}$$

Now suppose that A and B are quantum computers, or computing elements nonlocally correlated through time. Further, since A and B are nonlocally correlated at different times we may then access quantum states at a time of our choosing. Recall that A and B are now themselves quantum elements and their classical outputs are accessed in the future of both A and B. Now let p and p' be prime numbers. It is well known that there exists an $N = pp'$. Let N be the input, and we have a command $settime(t, T)$ that teleports states to a time $t < T$, where T is set to the computer clock $T = clock()$. Now set the nth prime number p_n to the number $t_n = n\delta t$ as an input to the register for the time. Consider the following algorithm written in a "pseudo-C" format

input(N)
$T = clock()$
$n = timeregister$
$if(n > 1)$ and $(N \bmod n = 0) goto\ TIME$
$n = 1$
$for\ loop\ n \to 1\ to\ int(sqrt(n)) + 1$
{
    $if(n \bmod N = 0)\ goto\ OUTPUT$
}
$TIME:\ settime(t_n, T)$
$OUTPUT:\ output(n)$
END

Interestingly quantum computing with time entangled states is equivalent to computing with closed timelike curves in general relativity [48]. Outside of trivial problems the inner nested recursive loop is never executed! The existence of this loop is equivalent to how the machine sets the delayed choice quantum computation, but it is never in fact executed. This is similar to the classic science fiction time travel dilemma. For the output obtains at a time T greater than t_n since it is sent by the Wheeler delay choice to t_n. It is as if one were a physicist who invented time travel by receiving a message from the future on how to do it. Then upon building the time machine you send the solution back in time to yourself. However, one should not confuse this with backward time communication, for this concerns nonlocal correlations between an output at one time with quantum states at an earlier time.

A nondeterministic Turing machine has a finite number of states and an infinite tape, where on a given symbol input from the tape there are a finite number of possible states the machine may enter into and a finite choice of symbols in which to replace the read symbol. Nondeterministic Turing machines are capable of efficiently solving problems that do not have a polynomial time solution. Such problems are the traveling salesman problem or the lack of an efficient algorithm that can recognize all graphs with Hamiltonian circuits. A nondeterministic approach with graphs is a "hunt and find" approach, guess edgelinks and verify by hand they are satisfied. The nondeterministic approach then has this "freedom of choice" concerning the procedure. Presumably the brain consists of many processors that act in consort and finds the answer by some maximal entropy approach or by entering some basin of attraction. Of interest is whether NP problems can in general be reduced to P type problems [49]. If the factorizing loop is buried in closed timelike loops the time for this processing is reduced to a very short time period. Analogously with the future state selection in a quantum computer the nonpolynomial time is eliminated. The loop is never actually computed, but the future state selection must conform to the output of the loop. Thus a particular NP type problem has been reduced to a P type problem by the utility of quantum mechanics and temporal teleportation. If time travel is not possible in classical general relativity, the above argument has no analog with a classical time machine computer. A reduction of an NP complete problem to an P complete problem can be done quantum mechanically, but this does not appear as likely classically. This reduction requires violations of Bell's inequalities, and this is not possible for a classical computer that runs an NP complete algorithm. Thus classically $P \subset NP$ and such a reduction requires nonlocal access to q-bits.

This does not constitute a proof of $P \subset NP$. A complete proof for or against this with quantum computing requires a deeper understanding of the classical-quantum correspondence. If $P = NP$ can be demonstrated in the limit that $\hbar \rightarrow 0$ then classical computers could, for sufficiently clever algorithms, solve NP complete problems as P complete.

2.9: Summary on Spacetime Fluctuations

Here the basic theory of quantum spacetime fluctuations has been developed. This theory is less a theory of quantum gravity than it is a phenomenology of fluctuations in geometry. As a theory it suffers from several weaknesses. The first is that it is applicable only for a range of around 10 or more Planck lengths. This is not an exact theory of quantum gravity that approaches the Planck length. It is further apparent that quantum gravity may be inherently finite temperature, where this finite temperature can only be overcome if quantum gravity is embedded into a larger theory. The finite temperature nature of the theory reflects the existence of quantum states that are inherently unavailable to an observer measuring physics at a scale at all significantly larger than the Planck scale. This scrambling of quantum information then determines a temperature associated with quantum gravity.

These issues will be addressed in chapter 4. An examination of D5-brane and string interactions are examined, as well as the possibility that gravitation is a field that emerges at scales beyond the string length. This will hinge on the derivation of a Polyakov action based upon weighted projective spaces. This indicates that gravitation may be a property of the universe that emerges with the tunnelling of the universe out of the vacuum state.

References

[1] W. G. Unruh, R. M. Wald, *Phys. Rev.* **D**29, 1047 (1984).

[2] A. Einstein, B. Podolski, N. Rosen, *Phys. Rev.* **47**, 777.

[3] H. Everett, *Rev. Mod. Phys.*, **29**, 454-462 (1957).

[4] B. S, DeWitt, N. Graham, "The Many Worlds Interpretation of Quantum Mechanics," Princeton Univeristy Press, Princeton, 155-165, 1973.

[5] D. Bohm, *Phys. Rev.* **85**, 166-193 (1952).

[6] D. Bohm, B. Hiley, "The Undivided Universe," Routledge (1993).

[7] J. Bell, *Rev. Mod. Phys.*, **38**, 448 (1966).

[8] S. Kochen, E. Specker, *J. Math. and Mech.*, **17**, 59 (1967).

[9] D. A. Page, *Phys. Rev. A*, **36**, 3479, (1987).

[10] M. V. Berry, *Proc. R. Soc. London*, **A**430, 405 (1985).

[11] C. Miser, K. Thorne, J. Wheeler, "Gravitation," Freeman Press, San Francisco (1973).

[12] W. DeSitter, *Proc. Roy. Acad. Sci.*, (Amsterdam), **19**, 1217, (1917).

[13] J. B. Hartle, S. W. Hawking, *Phys. Rev.* **D 28**, (1983) 2960-75.

[14] J. G. Vargas, D. G Torr, *Gen. Rel. Grav.*, **28**, 4, 451 (1996).

[15] J. G. Vargas, D. G Torr, it Found. Phys., **29**, 2, 145 (1999).

[16] B. S. Kay, R. M. Wald, *Phys. Rep.* **207**49 (1991).

[17] P. Rowlands, J. P. Cullerne, *Proceedings, Vigier 2000*.

[18] C. Armendariz-Picon, V. Mukhanov, Paul J. Steinhardt, *Phys. Rev. Lett.* **85**, 4438-4441 (2000).

[19] H. E. Brandt, *Found. Phys. Lett.* **13**, 4 (2000).

[20] A. Ashtekar, *Phys. Rev. Lett.* **57**, 2244 (1986).

[21] C. Rovelli, *Class. Quant. Grav.*, **8**, 1613, (1991).

[22] L. Crowell, *Found. Phys. Lett.*, **12**, 6, 585 (1999).

[23] A. D. Sakharov. *Doklady Akad. Nauk. SSR*, **177**, 70-71 (1967), (english) *Soviet Phys. Doklady*, **12** 1040-1041 (1967).

[24] R. Ruzmaikin, *Soviet Physics JETP*, **30** 2, 372 (1970).

[25] C. Barcelo, M. Visser, *Nucl. Phys.* **B**584, (2000) 415-435.

[26] A. Feinstein, M. A. P. Sebastián, *Found. Phys. Lett.*, **13**, 2, 133, (2000).

[27] E. Cartan and A. Einstein, *Letters on Absolute Parallelism, 1929-1932*, R. Debever, ed. (Princeton University Press, Princeton, New Jersey, 1979).

[28] J. G. Vargas, D. G. Torr, *Found. Phys.*, **28**, 6, (1998).

[29] A. Bejancu, "Finsler Geometry and Applications," (Ellis Harwood Ltd, West Essex England), (1990).

[30] C. Becchi, A Rouet, R. Stora, *Ann. Phys.* **98**, 287 (1976).

[31] J. von Neumann, "Mathematishce Grundlangen der Quanten-mechanik," Berlin (1932).

[32] A. M. Gleason, "Measures on the Closed Subspaces of a Hilbert Space." *J. math. and mech.*, **6** 885-93 (1957).

[33] S. Kochen, E. P. Specker, "Logical Structures Arizing in Quantum Theory," *Theory of Models* (ed J. Addison, L. Henkin and Q. Tarski), North Holland Pub. Amsterdam (1965).

[34] P. J. E. Peebles, "Quantum Mechanics," Princeton University Press, New Jersey, (1992).

[35] J. Anandan, Y. Aharonov, "Geometry of Quantum Evolution," *Phys. Rev. Lett.*, **65**, 14, (1990).

[36] A. Pandya, A. Nagawat, "A Generalized Definition of the Metric of Quantum States," online article, Department of Physics, University of Rajasthan, Jaipur 302004, India.

[37] J. Baez, "The Octonions," *Bul. AMS*, **39**, 2, (2002).

[38] R. M. Wald, "General Relativity," University of Chicago Press, Chicago, Ill. (1984)

[39] A. Das, "The Special Theory of Relativity: A Mathematical Exposition," Springer-Verlag, (1993).

[40] D. Bohm, "Quantum Theory," Dover, New York, (1989).

[41] G. Greenstein, A. Zajonc, "The Quantum Challenge: Modern Research on the Foundations of Quantum Mechanics," Jones and Bartlett, Massachusetts, (1997).

[42] M. Jammer, "The Philosophy of Quantum Mechanics", John Wiley, 1974.

[43] B. d'Espagnat, "The Quantum Theory and Reality," *Scientific American,* p 158, (1976).

[44] A. Aspect, P. Grangier, G. Roger, "Experimental Test of Realistic Local Theories via Bell's Theorem," *Phys. Rev. Lett.* **47** 460-7 (1981).

[45] J. F. Clauser, M. A. Horne, "Experimental Consequences of Objective Local Theories," *Phys, Rev.* **D10**, 526-35 (1974).

[46] L. Crowell, R. Betts, "Spacetime Holography and the Hopf Fibration," *Found. Phys. Lett.*, **18**, 2, April 2005 (183).

[47] C. F. Boyle, R. L. Schafir, "A delayed-choice thought-experiment with later-time entanglement,"

http://xxx.lanl.gov/ftp/quant-ph/papers/0107/0107098.pdf

[47] T. A. Brun, "Computers with closed timelike curves can solve hard problems," *Found. Phys. Lett.* **16** (2003) 245-253.

[48] S. Cook, "The P versus NP Problem,".
http://www.claymath.org./millennium/P_vs_NP/Official_Problem_Description.pdf

3: Detection of Quantum Gravity Fluctuations and Cosmology

3.1 Can Quantum Gravity Be Experimentally Tested?

At a DPF conference Howard Georgi stated, "I might say something like 'Planck scale physics is an oxymoron because you can't do physics at the Planck scale.'" Currently this appears to be the state of affairs. Young students at both the undergraduate and entering graduate level may start with an interest in the subject of quantum gravity. Yet in time they are confronted with the horrid reality that this sort of physics involves energy so large and scales so small that the prospect for ever doing experimental tests of this physics are simply impossible. As a result many switch their academic and professional interests to other areas of physics. A high energy approach to Planck scale physics appears beyond the technical capabilities of accelerator technology. The LHC will achieve multi-TeV energies, where since the Planck energy is 10^{16}TeV a Planck energy accelerator would be a ring about 10,000 light years in radius.

It is common to consider physics as existing on various independent scales. In modern physics there are atomic physics at 10^{-8}cm, nuclear physics at 10^{-13}cm and elementary particle physics as given by the standard model and QCD at 10^{-17}cm. These areas of physics have been put to considerable experimental test. The two outstanding regions are the GUT region of 10^{-29}cm and the domain of quantum gravity at 10^{-33}cm. The $SU(5)$ GUT has been apparently falsified by the proton lifetime experiments. So at least one test in the 10^{-29}cm domain has been conducted. With quantum gravity no test has been performed.

These domains of physics are nested into one another by length scales. It is generally assumed that the physics at these various scales are not affected by the physics on other scales. Yet this assumption is not entirely correct. The nucleus of an atom has a charge distribution and this contributes to the nature of the nuclear magnetic moment, which factor into hyperfine splitting. This would also be the case if a hydrogen atom is used, for the proton is ultimately made of quarks and gluons, which will also have subtle effects on the nuclear magnetic moment. So the various scaling levels into which physics is compartmented are not entirely sealed off from one another. So if well known physics is perturbed by physics on a much smaller scale, then sensitive enough measurements may be able to register physics at much smaller scales.

This is essentially proposed with the detection of quantum gravity with atomic physics. The curvature of spacetime due to quantum gravity fluctuations in an atom should be on the order of $R \sim 10^{-9} cm^{-2}$, which is considerably larger than the curvature of spacetime due to Earth's gravity $\sim 10^{-28} cm^{-2}$. While the "force" exerted on an electron by quantum spacetime fluctuations is 10^{-21} that of the electrostatic force between the electron and the nucleus, the perturbation may still be observable with sensitive enough measurements.

High energy physics is predicated on the deBroglie theory that a particle with a large momentum will have a short wave length. So very high energy interactions, where large transverse momenta are exchanged, will then probe smaller regions of the world. The LHC will by this wave mechanics probe physics at the 10^{-17}cm scale. It is also possible that the LHC will be the last such accelerator ever built. Budgetary constraints and the growing expense of high energy physics could well spell the end of high energy physics as we know it. Unless there is some dramatic breakthrough in accelerator technology other means must be found to probe quantum gravity.

Gravity is not a force, but the force is what is required to deviate the motion of a particle away from a geodesic. One way to probe the nature of gravity is to deviate the motion of particles in spacetime and measure this force. Similarly electrons that are highly accelerated are physically equivalent to a particle forced to remain at rest near a black hole event horizon through the exertion of a force. In this way the Unruh effect is an analogue to the thermal vacuum detected by an observer hovering over a black hole. Currently devices such as the National Ignition Facility are capable of accelerating electrons to nearly $10^{21} cm/s^2$ [1]. There have also been considerable developments in small high powered lasers [2][3]. It is then possible that in the future the route towards the study of physics involved with the quantum fields in spacetime and quantum gravity will not involve reaching the Planck energy, but by replicating the situation of physics near a black hole with huge accelerations.

The technology for extremely accelerated particles will be improved in the future. In a nonlinear medium a high powered laser beam could give rise to an additional longitudinal electric field. This field could enhance the acceleration of the electron enormously. Further the beam, or pulse, quality could in time be achieved with a high degree of coherence in the photons. This would push the electron closer to its particle horizon situated c^2/a from the electron. This would then permit future experiments to test the gravitational squeezing of the vacuum.

A quantum gravity fluctuation at the Planck scale will induce a small phase on the electronic wave function of an atom. This is the first proposed experiment, where the tiny phase shift may be within an experimentally testable domain in the near future. The holographic principle indicates that the fluctuations of space in a volume are determined by the number of planck units of area on the bounding area of that volume. A number of possible sensitive tests of how holographic induced fluctuations perturb entangled atom-photon systems are presented. Possible quantum optical experimental designs involve single atom lasers, cavity QED designs and a laser ring gyro. These various designs are presented since one may prove more applicable than the others. However, they all involve a sensitive test of a phase shift induced from holographic fluctuations of spacetime.

The Unruh effect plays a central role in the discussion of torsional spacetime fluctuations in chapter 2. A potential test for the Unruh effect is advanced. A rotating Bose-Einstein condensate will exhibt a slight change in the transition temperature to the condensate. On the frame of the condensate there is a slight change in the temperature due to the thermal vacuum on the accelerated frame. A sensitive measurement of the spin dependence of the transition temperature is advanced as a method for detecting the Unruh effect.

Higher energy experiments are proposed to examine the renormalization of the Planck length. High powered lasers which propel a particle, such as an electron or proton, to $\sim 10^{21} cm/sec^2$ [3] will induce a squeezing of the particle state through its interaction with the thermal vacuum state. It should be possible to detect the squeezing of the vacuum state through a homodyne detector.

Finally, in the different scales of the universe the largest is the observable universe. The cosmological horizon observed as the black sky is ultimately an achronal boundary with information that asymptotically piles up near it. Ultimately as one observes closer to the cosmological horizon information concerning the decoupling of quantum gravity from a unified gauge field should appear. These data should appear as a spectrum of gravity waves in analogue to the microwave background due to the end of the radiation dominated period of the universe. The now operating LIGO facilities opens a new observational window onto the universe. If such facilities are upgraded they may permit an observation of the cosmic gravity wave background.

In this chapter various possible windows into quantum gravity and cosmology are examined. It is my opinion that nature cannot be structured in such a way that intelligent life within it are unable to probe its most basic foundations. This is not a particularly scientific statement, but one that ultimately scientific optimism is based upon. On this basis it is presumed that experimental detection of quantum gravity, however oblique, will occur in the not too distant future. If this does occur the results will likely change many of our current ideas about the foundations of physics.

3.2 Atomic Detection of Quantum Spacetime Fluctuations

Such detection is a theory of the coupling between a hydrogen atom and quantum gravity fluctuations. This is a more realistic theory than the two state atom model of atomic interactions with quantum spacetime fluctuations presented in the introduction. The basic theory holds that the Rabi oscillation of an atom with a photon in a state entanglement results in the evolution of a parameter which gives a Berry phase induced by quantum gravity fluctuations. The loop in parameter space marks an area with a quantum flux, due to spacetime fluctuations coupled to a quadrupole moment, through this loop that shifts the phase of the system. It is proposed that this mechanism may be used to detect quantum gravity fluctuations.

The underlying assumption is an atom in the first family such as Cesium with the outer electron in a high Rydberg level will behave approximately as a massive hydrogen atom. This permits a treatment of the energy level or wave packet orbit of this outer electron according to a quantum Kepler problem. The evolution of this Berry phase is then described according to the basis states of a hydrogen atom. From there the theory is modified to consider the experimental setup according to a one-atom laser. The basic theory is advanced further to develop the physics of an atomic laser. Matters of quantum and thermal noise are introduced so noise problems may be eliminated. Further, the proposed device is discussed that would use Bose condensates of hydrogen-like atoms.

The post-Newtonian form of gravity is considered for brevity. The post-Newtonian gravito-electromagnetic fields are similar to Maxwell fields and are written as a linear operator expansion without the difficulties of gravitational nonlinearity. The metric in Post-Newtonian gravity is expanded around a flat spacetime:

$$g_{\mu\nu} = \eta_{\mu\nu} + g^{(2)}_{\mu\nu} + g^{(4)}_{\mu\nu} + \ldots \quad \{\mu = \nu = 0, \text{ or } \mu, \nu = i, j\},$$

$$g_{i0} = g^{(3)}_{i0} + g^{(5)}_{i0} + \ldots. \tag{3.1}$$

$\eta_{\mu\nu}$ is the flat spacetime metric, where since the spacetime is source free $g^{(2)}_{\mu\nu} = -(2GM/r)\delta_{\mu\nu} = 0$. The next highest terms within the harmonic condition are $g^{(3)}_{0i} = \mathcal{A}_i$, where $\mathcal{A}$ is a vector potential and $g^{(4)}_{00} = -2\phi$ a scalar field. The connections to lowest order are [4]

$$\Gamma^{(4)i}{}_{00} = \frac{\partial \phi}{\partial x^i} + \frac{\partial \mathcal{A}_i}{\partial t}, \quad \Gamma^{(3)i}{}_{0j} = \frac{1}{2}\left(\frac{\partial \mathcal{A}_i}{\partial x^j} - \frac{\partial \mathcal{A}_j}{\partial x^i}\right). \tag{3.2}$$

For $\mathcal{E}_i = -\Gamma^{(4)i}{}_{00}$ and $\mathcal{B}_i = \epsilon_{ijk}\Gamma^{(3)j}{}_{0k}$ the fields are:

$$\mathcal{E}_i = -\frac{\partial \mathcal{A}_i}{\partial t} - \nabla_i \phi, \quad \mathcal{B}_i = \frac{1}{2}(\nabla \times \mathcal{A})_i. \tag{3.3}$$

In the harmonic condition $\nabla \cdot \mathcal{A} = 0$ it is easy to find the Maxwell type of equations

$$\nabla \times \mathcal{E} - \frac{2}{c}\frac{\partial \mathcal{B}}{\partial t} = 0 \tag{3.4a}$$

$$\nabla \times \mathcal{B} + \frac{1}{2c}\frac{\partial \mathcal{E}}{\partial t} = 0. \tag{3.4b}$$

Now expand the gravito-electromagnetic field into normal modes with the operators b and $b^\dagger$ analogous to QED. The linearized gravity vector potential expanded as spatial eigenmodes quantized in a box is,

$$\mathcal{A}_i = i\epsilon_i \sum_k \sqrt{\frac{\hbar}{2\omega\epsilon_g V}}\left(b^\dagger e^{-ikx} - b e^{ikx}\right), \tag{3.5}$$

where ϵ_g is a vacuum permittivity for gravity, analogous to the permittivity in electromagnetism. Now the electric and magnetic analogues of the gravity field are operators that describe linearized gravitons.

The Hamiltonian for the hydrogen atom is

$$H = \frac{p^2}{2m} - \frac{Ze^2}{r}, \tag{3.6}$$

which in the classical limit describes the orbit of a particle around a central potential that obeys Kepler's laws of planetary motion. The interaction of the electromagnetic field is included with $\mathbf{p} \to \mathbf{p} + ie\mathbf{A}$. Quantum mechanically this is treated with the wave function

$$\psi_{nlmsm_s} = R_{nl}(r)Y_{lm}(\theta,\phi)\Phi_{sm_s}(\phi), \tag{3.7}$$

where the radial, angular and azimuthal parts of the wave function obey their own independent Schrödinger equation. For the sake of simplicity the azimuthal wave function is ignored.

The Hamiltonian is time independent which ensures the constancy of the energy with time. Further, the rotational symmetry of H ensures that the orbit lies on a constant plane. To ensure the motion is defined by a closed orbit the Runge-Lenz vector is imposed as an additional constant of the motion. A Hermitian quantum operator corresponding to the Runge-Lenz vector is

$$\mathbf{R} = \frac{1}{2}(\mathbf{p}\times\mathbf{L} - \mathbf{L}\times\mathbf{p}) - \frac{Ze^2}{r}\mathbf{r}, \tag{3.8a}$$

with $\mathbf{p} = \frac{\hbar}{i}\nabla + ie\mathbf{A}$. The square of $\mathbf{R}$ is then

$$\mathbf{R}^2 = 2H(\mathbf{L}^2 + 1) + Z^2. \tag{3.8b}$$

That $\mathbf{R}$ is a constant of the motion is found by the commutator $[H, \mathbf{R}] = 0$, where with $\mathbf{A} = 0$ $\mathbf{R}$ is in the plane of motion $\mathbf{L}\cdot\mathbf{R} = 0$. If the Runge-Lenz is replaced with $\mathbf{R} \to -\frac{iZ}{\sqrt{2H}}\mathbf{R}$ the following operators are defined

$$\mathbf{J} = \frac{1}{2}(\mathbf{L} + \mathbf{R}), \quad \mathbf{J}' = \frac{1}{2}(\mathbf{L} - \mathbf{R}), \tag{3.9}$$

which satisfy the algebra,

$$[J_i, J_j] = i\epsilon_{ijk}J_k, \quad [J'_i, J'_j] = i\epsilon_{ijk}J'_k, \quad [J_i, J'_j] = 0. \tag{3.10}$$

This algebra is two copies of $SU(2)$ or $SO(4) \sim SU(2)\times SU(2)$. If $\mathbf{A} \ne 0$ the operators $\mathbf{J}$ contain elements of the $SU(2)$ algebra so the above commutators include the zero point energy of the QED vacuum. These operators as contained in the algebra $SO(4)\times_b U(1)$. This $\times_b$ notation is used to indicate that the operators $L_\pm$ will project onto $\mathbf{A}$ as a dipole interaction between the atom and the field, which breaks the degeneracy of the atom.

The perturbed Casimir operators in $SO(4)\times_b U(1)$ are

$$\mathbf{J}^2 = \frac{1}{4}(\mathbf{L}^2 + \mathbf{R}^2) + \frac{1}{2}(\mathbf{L}\cdot\mathbf{A} + \mathbf{A}\cdot\mathbf{L}), \quad \mathbf{J}'^2 = \frac{1}{4}(\mathbf{L}^2 + \mathbf{R}^2) - \frac{1}{2}(\mathbf{L}\cdot\mathbf{A} - \mathbf{A}\cdot\mathbf{L}). \tag{3.11}$$

The $SO(4)$ Casimir operators for this system are easily seen to be

$$\mathcal{J} = \mathbf{J}^2 + \mathbf{J}'^2 = \frac{1}{2}(\mathbf{L}^2 + \mathbf{R}^2|_{A=0}), \quad H_{int} = \mathbf{L}\cdot\mathbf{A} = \mathbf{J}^2 - \mathbf{J}'^2. \tag{3.12}$$

If $\mathbf{A} = 0$ everywhere the eigenvalues of $\mathbf{J}^2$ and $\mathbf{J}'^2$ are $j(j+1)$ and $j'(j'+1)$, where the condition $\mathbf{A}\cdot\mathbf{L} = 0$ insures that $j = j' = 0, \frac{1}{2}, 1, \ldots$. This atom is then n^2 fold degenerate. However, the electromagnetic field breaks the degeneracy of the energy eigenvalues. In a real atom this degeneracy is primarily broken by the magnetic moment of the nucleus of the atom. The electromagnetic field that interacts with the atom in a cavity is considered to be too small to break this degeneracy. Physically this is a splitting of the energy levels by the orientation of the electron spins with the magnetic field. It is the small ΔE that is a tool to perform measurements of quantum gravity fluctuations. For $j = j'$ the eigenvalues for $\mathcal{J}$ are $2j(j+1) = n^2$ and $\langle|H|\rangle = E$ in the redefinition of $\mathbf{R}$. This gives the result

$$\langle|\mathcal{J}|\rangle = 2j(j+1) = \frac{1}{2}\left\langle\left|\mathbf{L}^2 + \frac{Z^2}{2H}\mathbf{R}^2\right|\right\rangle = -\frac{1}{2} - \frac{Z^2}{4E}, \quad (3.13)$$

which leads to the $E = -Z^2/2n^2$.

The above calculation of the atomic energy levels illustrate that the n^2 degeneracy removes any energy dependency with the angular momentum states. This degeneracy is removed by considering the magnetic moment of the nucleus. Let $\mathbf{A} = \mathbf{r}\times\mathbf{B}$ so that $\mathbf{L}\cdot\mathbf{A}$ is a spin orbit interaction between the nucleus and the orbiting electron. The diagonal terms define the perturbing Hamiltonian according to the coupling with the magnetic moment M_z, $H_B = L_z M_z$ as

$$L_z M_z|l, m, n\rangle - m\hbar M_z|l, m, n\rangle. \quad (3.14)$$

There is now an energy difference with angular momentum states, where L_z may plays the role of σ_z in the two state atom. Now introduce a photon field that interacts with the atom and is tuned to the energy gap between two values of m. We consider the eigenstates $|l, m, n\rangle$ for the atom and the photon field. The last quantum number is $n = 0, 1$ for the photon field. Also the interaction Hamiltonian is written according to the ladder operators for the atom and photon states in a rotating wave approximation,

$$H_{int} = \kappa(L_+ a + L_- a^\dagger) \quad (3.15)$$

The action of these two operators on these states is

$$\mathcal{J}|l, m, n\rangle = l(l+1)\hbar^2|l, m, n\rangle \quad (3.16a)$$

$$H_{int}|l, m, n\rangle = \hbar\kappa(\sqrt{l(l+1) - m(m+1)}|l, m+1\ n-1\rangle + \sqrt{l(l+1) - m(m-1)}|l, m-1, n+1\rangle). \quad (3.16b)$$

The constant κ is determined by the splitting of the angular momentum states with an energy gap ΔE.

The time evolution of the atom and photon is governed by H_{int}, which requires work in the interaction picture. The degenerate Hamiltonian terms are irrelevant to this time evolution. The time evolution by H_{int} is then

$$-i\hbar\frac{\partial}{\partial t}H_{int} = [H_{int}, H_B] = i\hbar(L_- a - L_+ a^\dagger)M_z. \quad (3.17)$$

Similarly the evolution of the operators $L_\pm(t)$ are given by

$$L_\pm(t) = L_\pm(0)\cos(M_z t/\hbar) + iL_\mp(0)\sin(M_z t/\hbar). \tag{3.18}$$

This evolution involves the Rabi oscillation of the atom between the split atomic levels and the photon in a cavity.

Now introduce quantum fluctuations of spacetime into the atom. By conservation of momentum the gravity field must interact with the atom by a quadrupole interaction. This is an analog of the electric quadrupole interaction, now with the gravito-electric field $\mathcal{E}$. This Hamiltonian must then contain the quadrupole tensor and this interaction term is of the form

$$H_{Q\mathcal{E}} = g\sum_{ij}(3r_i r_j - \delta_{ij}r^2)\mathcal{E}. \tag{3.19}$$

The commutator $[L_i, r_j] = i\hbar\epsilon_{ijk}r_k$ demonstrates that

$$Q_{ij}|_{i\neq j} = 3r_i r_j|_{i\neq j} = -\frac{3}{4\hbar^2}\epsilon_{ikl}\epsilon_{jmn}L_k L_m r_n r_l. \tag{3.20}$$

The destruction and creation of a graviton requires that $\Delta j = \pm 2$. As with a, $a^\dagger$ the lowering and raising operators for photons, in analogue to the rotating wave approximation, the raising and lowering operators for the gravity field b, $b^\dagger$ must be associated with a product of raising and lowering operator of the atom and photon field $L_+ a^\dagger$, $L_- a$

$$H_{Q\mathcal{E}} = \frac{g}{2\hbar^2}\Big(ba^\dagger(L_z L_+ + L_+ L_z)(r_+ r_z - 3r_z r_+ + r_-^2) +$$

$$b^\dagger a(L_z L_- + L_- L_z)(r_- r_z - 3r_z r_+ + r_+^2)\Big) \tag{3.21}$$

where all constants are absorbed into g. It is noted that in this interaction there is a change in the electron spin by $\Delta s = \pm 1$ and a change in the orbital angular momentum by $\Delta l = \pm 1$ Hence the total change in the atomic angular momentum is $\Delta j = \pm 2$ in the absorption and emission of a linearized graviton. As the atom is defined to be in the x, y plane this Hamiltonian is then

$$H_{Q\mathcal{E}} = \frac{g}{2\hbar^2}\Big(ba^\dagger(L_z L_+ + L_+ L_z)r_-^2 + b^\dagger a(L_z L_- + L_- L_z)r_+^2\Big). \tag{3.22}$$

This gives the measure of a quantum flux through the area enclosed by the atom in the $x - y$ plane. From this Hamiltonian compute the Berry phase [5] on the wave function induced by quantum gravity fluctuations. The operators $L_\pm(t)$ are the parameters involved with this phase

$$\phi = \sum_\psi \frac{\langle|\nabla_\mathbf{L} H_{int}|\psi\rangle \times \langle\psi|\nabla_{\mathbf{L}'} H_{int}|\rangle}{(\Delta E)^2} =$$

$$(g'\hbar)^2\frac{\langle|[b, b^\dagger]a^\dagger a L_z^2(r_+ r_-)^2|\rangle \hat{n}_z}{(\Delta E)^2}, \tag{3.23}$$

92 *Quantum Fluctuations of Spacetime*

for $\mathbf{L} \neq \mathbf{L}'$. It is then evident that the Berry phase induced on the atomic and photon wave function is due to the zero point energy of the quantum gravity field.

The above theory is a starting point that needs to be further expanded. However currently it contains the core physics involved with this process. The Schrödinger equation for this Hamiltonian has been numerically integrated to illustrate the phase shift after a sufficient period of time.

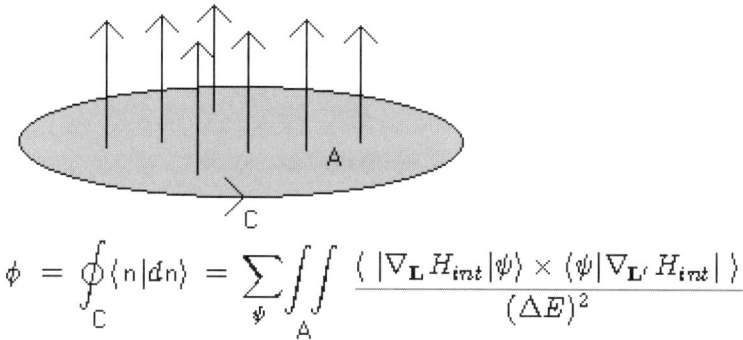

$$\phi = \oint_C \langle n|dn\rangle = \sum_\psi \iint_A \frac{\langle\, |\nabla_\mathbf{L} H_{int}|\psi\rangle \times \langle\psi|\nabla_{\mathbf{L}'} H_{int}|\,\rangle}{(\Delta E)^2}$$

Quantum flux through a parameter loop and
the geometric phase induced by gravity.
Figure 3.1

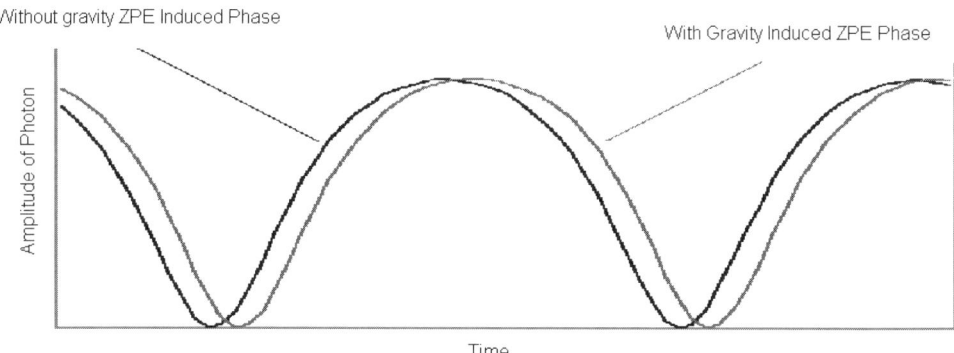

Numerical similation of the Berry phase induced by quantum gravity fluctuations.
Figure 3.2

This illustrates the influence of the quantum zero point energy of gravity on the phase of entangled states between an atom and a photon.

It is obvious this phase is extremely small, $\phi \sim 10^{-29}$, per Rabi oscillation of the atom plus photon field. It is not possible to maintain an atom in dressed states for the long period of time required to induce a measurable phase shift on the atom. It would then appear that an experiment designed to detect this quantum gravity induced phase will require identically prepared atoms be injected into a laser cavity. Each atom is prepared in the excited state and emits a photon into the cavity and exits the cavity before reabsorbing the photon. Each atom emits its photon by stimulated emission. The coherent state of the photon field is

$$|\alpha\rangle = \sum_n e^{|\alpha^2|^2/2} \frac{\alpha^n}{\sqrt{n!}} |n\rangle, \qquad (3.24)$$

where with $|n\rangle = ((a^\dagger)^n/\sqrt{n!})|0\rangle$

$$|\alpha\rangle = \sum_n e^{|\alpha|^2/2} \frac{(\alpha a^\dagger)^n}{n!} |0\rangle = e^{|\alpha|^2/2} \alpha^{a^\dagger} |0\rangle, \qquad (3.25)$$

where α is an arbitrary complex number. Now consider the product of $a^\dagger$ operators as due to the repeated emission of a photon by atoms injected into the cavity. The injection of a fresh excited atom into the cavity is treated by the application of an L_+ on the atomic state. So the interaction term for the electromagnetic field is

$$H_{int} = \kappa L_+ L_- \frac{(a^\dagger)^n}{n!}. \qquad (3.26)$$

Here the n is the number of times an atom is injected into the cavity. It is obvious that H_{int} is a constant of the motion. Since each atom completes half of a Rabi oscillation the induced phase is half that in equation 3.23. The photon state in the cavity is then

$$|\alpha\rangle = \sum_n e^{|\alpha^2|^2/2} \alpha^{a^\dagger} e^{in\phi/2} |0\rangle, \ \alpha = \alpha_r + i\alpha_i. \qquad (3.27)$$

Consider the case where there are two cavities with atoms injected through them. If the phase shift of the atom and photon field in the two cavities are opposite the overlap of the two resulting states is

$$\langle \beta | \alpha \rangle = \sum_n \langle 0 | e^{|\beta^2|^2/2} \frac{a^n}{n!} e^{in\phi/2} e^{|\alpha^2|^2/2} \frac{(a^\dagger)^n}{n!} e^{in\phi/2} |0\rangle$$

$$= e^{-(|\alpha|^2 + |\beta|^2)/2} e^{i\Phi} \beta^* \alpha, \qquad (3.28)$$

where Φ is the total phase accrued. For a small phase the overlap is approximately

$$\langle \beta | \alpha \rangle \simeq e^{-(|\alpha|^2 + |\beta|^2)/2} (\beta^* \alpha (1 + i\Phi), \qquad (3.29)$$

where the last term is the important data to be measured. The real part is the measured phase:
$$\mathcal{R}e\langle\beta|\alpha\rangle \simeq i\mathcal{I}m(\beta^*\alpha)\Phi e^{-(|\alpha|^2 + |\beta|^2)/2}. \tag{3.30}$$
This overlap is a distance in the complex plane and determines how orthogonal the two laser beams are. The Berry phase induced by quantum gravity fluctuations is measured by this overlap.

This experiment is a reasonable follow up to the measurement of gravitational Bohr orbits [6]. The Berry phase induced per Rabi oscillation is around 10^{-29}, which is too small to be directly measured. The above theory was derived under the assumption that a confined coherent photon field accrues this phase shift on an atom by atom basis. The Rabi oscillation occurs on a time frame of around 10^{11}Hz, so if this experiment is conducted for one second the phase shift in the coherent photon field will be $\sim 10^{-18}$ and if run for a year period this phase shift will be $\sim 10^{-11} - 10^{-10}$. The device will require the application of tools that are capable to measuring this phase shift. It will then be important to introduce a parametric amplifier to obtained squeezed states of coherent photons [7] in the cavity.

The phase in equation 3.30 is a measure of the distance between two states on the complex plane. The expectation of the operators a and $a^\dagger$ are
$$\langle\alpha|a|\alpha\rangle = \alpha, \quad \langle\alpha|a^\dagger|\alpha\rangle = \alpha^*. \tag{3.31}$$
These operators define the fields $E = E_0(a + a^\dagger)$ and $B = B_0(a^\dagger - a)$ with analogies to x, p. It is then obvious there are the expectations
$$\langle\alpha|E|\alpha\rangle = E_0\mathcal{R}e(\alpha), \quad \langle\alpha|B|\alpha\rangle = B_0\mathcal{I}m(\alpha) \tag{3.32}$$
$$\langle\alpha|E^2|\alpha\rangle = E_0^2((\alpha + \alpha^*)^2 + 1), \quad \langle\alpha|B^2|\alpha\rangle = -B_0^2((\alpha - \alpha^*)^2 - 1),$$
where $E_0 = \sqrt{\hbar\omega/2V\epsilon_0}$ and $B_0 = \sqrt{\hbar V\epsilon_0/2\omega}$ and are the electric and magnetic field associated with one photon. The quantum fluctuations in the electric and magnetic fields are determined to be
$$(\Delta E)^2 = \langle\alpha|E^2|\alpha\rangle - (\langle\alpha|E|\alpha\rangle)^2 = E_0^2$$
$$(\Delta B)^2 = \langle\alpha|B^2|\alpha\rangle - (\langle\alpha|B|\alpha\rangle)^2 = B_0^2 \tag{3.33}$$
so $\Delta E \Delta B = E_0 B_0 = \hbar/2$ and the coherent states are waves with minimum uncertainty.

These vacuum fluctuations are the sole contributor to noise in the amplitude and phase of the coherent states. For a coherent state this noise is contained in both the uncertainty in the electric and magnetic fields. While the uncertainty principle is absolute in quantum physics it is possible to arrange that $\Delta E = E_0/s$ and $\Delta B = B_0 s$ so this noise is minimized in the electric field component as the term $s \rightarrow \infty$. The squeeze state operator which permits this is [7]
$$S(z) = e^{(z(a^\dagger)^2 - z^*a^2)/2}, \tag{3.34}$$

where this acts on the states $|\alpha\rangle$ according to

$$S(z)|\alpha\rangle = e^{(z(\alpha^*)^2 - z^*\alpha^2)/2}|\alpha\rangle. \quad (3.35)$$

Now evaluate $\langle\beta|\alpha\rangle$ with the Berry phase

$$\langle\beta|\alpha\rangle \simeq e^{-(|\alpha|^2 + |\beta|^2)/2} e^{(z(\alpha^*)^2 - z^*\alpha^2)/2} e^{-(z(\beta^*)^2 - z^*\beta^2)/2} \beta^*\alpha(1 + i\Phi). \quad (3.36)$$

Let $z = \zeta + i\xi$ to find that

$$z(\alpha^*)^2 - z^*\alpha^2 = \zeta(\alpha^* + \alpha)(\alpha^* - \alpha) + (i/2)\xi((\alpha^* + \alpha)^2 + (\alpha^* - \alpha)^2)$$

$$= i\zeta\frac{\langle\alpha|E|\alpha\rangle\langle\alpha|B|\alpha\rangle}{E_0 B_0} + \frac{i}{2}\xi\left(\frac{\langle\alpha|E|\alpha\rangle^2}{E_0^2} - \frac{\langle\alpha|B|\alpha\rangle^2}{B_0^2}\right). \quad (3.37)$$

If $\zeta \to \infty$ is absorbed into $\langle\alpha|B|\alpha\rangle$ then ζ is a parameter for the expansion of this expectation. By the constancy of the phase $\langle\alpha|E|\alpha\rangle \to 0$, so that $\Delta B \to \infty$ and $\Delta E \to 0$. A parametric amplifier down steps the frequency $\omega \to \frac{1}{2}\omega$ by splitting the photon into two entangled photon states, which suppress the uncertainty in $\Delta E \propto 1/\sqrt{2}$. The entangled photons are absorbed by the atom and re-emitted as a photon with frequency ω. The parametric downshift is then repeated. This gives $\Delta E \to 0$.

The utilization of a parametric amplifier will permit quantum noise to be reduced to near zero. Assume the experiment measures the interference of the electric fields from photons in two cavities, where the phase is opposite in each cavity. The uncertainty in the electric field is

$$\langle\alpha(t)|E^2|\alpha(t')\rangle - \langle\alpha(t)|E|\alpha(t)\rangle\langle\alpha(t')|E|\alpha(t')\rangle = \Delta E^2 \delta(t - t'), \quad (3.38)$$

where $\Delta E = sE_0$ is squeezed to a very small value as $s \to 0$. The gravity ZPE induced phase to be measured increases by at a linear rate, but quantum noise in ΔE is suppressed which permits the detection of this phase.

Realistically there there will be a small quantum noise present in the system from the environment. Further noise from thermal fluctuations in mirrors will similarly have to be reduced by supercooling, but will never be reduced to zero. This thermal noise will be approximately $cKT\delta(t - t')$, where c is a constant. From an experimental point of view both of these noise contributions are minimized and are of comparable strength. This increase in the phase change will compete with the diffusion induced by quantum and thermal noise.

Consider the angular frequency for the evolution of the measured phase is Ω. The actual evolution rate of the phase $r(t)$ per atom as simply given by

$$\frac{dr(t)}{dt} + \Omega r(t) + F(t) = 0, \quad (3.39)$$

where the driving "force" $F(t) = 0$ if the phase evolves at a constant rate. For the case single particle motion this equation is just Newton's second law of motion. Equation 3.39 describes the evolution of the phase that deviates from a constant change due to external forces. The external force is given by the thermal noise part

$$\langle F_T(t)F_T(t')\rangle = \frac{2\Omega kT}{\rho}\delta(t-t'), \qquad (3.40)$$

where ρ is a constant of power. The contribution due to quantum noise is

$$\Delta F_Q(t)\Delta F_Q(t') = \frac{2\Omega\epsilon}{\rho'}\delta(t-t'), \qquad (3.41)$$

where $\epsilon = h\nu$ is the energy of the photon that interacts with the atom. Here $2\Omega kT/\rho$ and $2\omega\epsilon/\rho'$ define the diffusion constants for the Langevin equation [9]. It is know that these processes are guided by differential equations of the form

$$\dot{x}(t) = h(x(t),t) + g(x(t),t)F(t) = h(x(t),t) + D\Big\langle g(x,t)\frac{\partial}{\partial x}g(x,t)\Big\rangle, \qquad (3.42)$$

where $D = 2(kT/\rho + \Delta\epsilon/\rho')$. The Fokker-Planck equation associated with the Langevin equation is

$$\frac{\partial P}{\partial t} = \frac{\partial}{\partial \phi}\Big(P\frac{\partial V}{\partial \phi}\Big) + D\frac{\partial^2 P}{\partial \phi^2}. \qquad (3.43)$$

where $V = \Omega r(t)$ and ϕ is the phase. The first two terms appear similar to those in equations 3.39 and 3.41. Now expand the probability function as

$$P = \sum_{n=-\infty}^{\infty} a_n(t)r^n, \ a_n(t) = \frac{1}{2\pi}\langle r^n\rangle. \qquad (3.44)$$

and with the orthogonality of these cyclic functions then

$$\frac{d\langle a^n(t)\rangle}{dt} + in\Omega a_{n-1}(t) - Dn^2 a_n(t), \qquad (3.45)$$

for $\Omega \sim 10^{-18}sec^{-1}$. The physically relevant magnitudes for these quantities are given by

$$\Omega > \frac{2kT}{\rho} \simeq \frac{2\Delta\epsilon}{\rho'} \simeq e^{-s}\omega. \qquad (3.46)$$

The e^s is an estimate for the squeezing parameter due to parametric amplification with each Rabi oscillation. This gives the estimate $s > 60$. This also infers that the temperature must be approximately $10^{-1}K$.

This suggests required technology may be available in the near future. However, given the slow rate that the phase evolves this would require that the apparatus needs to operate for a prolonged period of time with strict tolerances. It is then apparent that this atomic laser may require more that just a single atom per Rabi flop. This will amplify the phase with each Rabi flop. To do this might require that the "atom" be a Bose-Einstein condensate [11] of hydrogen-like atoms. The Bose condensate of atom act as essentially one large atom. The "quantum flux" of the gravitational zero point energy coupled to the quadrupole moment contributes a phase from the sum of all parameterized loops in the atoms. In the condensate all the atom act coherently as one atom, where physically the phase of the whole system is given by the net parameterized loops in all atomic-photon interactions. The interaction of this Bose condensate with the photon will induce a phase $n\phi$, where n is the number of atoms that form the condensate. If a Bose condensate can be arranged with some 10^5 atoms then the rate of phase change Ω will increase by that factor. Liquid helium temperatures of $1°K$ are likely sufficient for enough atoms in the Bose condensate. Obviously this requires a type of B-E condensate "gun," where trapped Bose condensates are accelerated by phase modulation of the laser trap in order to inject them through the laser cavity.

With an ensemble of atoms their total angular momentum contributes to a sum $J_i = \sum_{a=1}^{N} J_i^a$. The measure of their state entanglement is

$$\xi^2 = \frac{N(\Delta \mathbf{n_1} \cdot \mathbf{J})^2}{\langle |\mathbf{n_2} \cdot \mathbf{J}| \rangle + \langle |\mathbf{n_3} \cdot \mathbf{J}| \rangle} > 1, \tag{3.47}$$

where $\xi^2 < 1$ is the condition for state entanglement or a squeezed spin state. Consider alkaline atom with states prepared in the form

$$|\psi\rangle \otimes \prod_{c=1}^{N} \otimes (|a\rangle + |b\rangle), \tag{3.48}$$

where $|\psi\rangle$ are the internal states of the atoms. Now let them interact weakly by a potential U_{ab} so the internal states of each atom are not significantly perturbed. The change in the squeezing parameter is then

$$\frac{d\xi}{dt} = \frac{(N-1)(U_{aa} + U_{bb} - 2U_{ab})}{2\hbar} \int d^3 r \langle \psi | \psi \rangle^2, \tag{3.49}$$

where condition that $U_{aa} + U_{bb} \neq 2U_{ab}$ gives the squeezed spin state. This can obviously be generalized for angles that deviate from n_1. When this condition is satisfied there is a strong measure of wave function overlap where $U_{aa} \simeq U_{bb}$ and that $\psi_a(N_a, t) \simeq \psi_b(N_a, t)$. The Berry phase induced on the wave function for the entire system due to quantum gravitational fluctuations is

$$\phi = \sum_a \sum_\psi \frac{\langle |\nabla_{\mathbf{L}} H_{int} |\psi\rangle|a\rangle \times \langle a|\langle \psi|\nabla_{\mathbf{L'}} H_{int}| \rangle}{(\Delta E)^2} = N\phi_{1\ atom}. \tag{3.50}$$

An experimental device will consist of two cavities where the quadrupole interaction of the atom with the zero point energy of gravity induces an opposite phase. This device will be an interferometer that detects the phase difference between coherent photons that interact with Bose condensates in the two cavities. The Bose condensates are created in a laser trap and accelerated towards a Helmholtz coil. A periodic sequence of Bose condensates are accelerated towards these coils with a frequency that is numerically commensurate with the Rabi frequency of the atom-photon interaction in the high-Q cavities. The spins of the atoms in the Bose condensate "packets" are randomly oriented. Each of these is effectively one giant atom with an undetermined spin. These pulses of Bose condensates enter a region with a magnetic field. This magnetic field then separates the atoms according to their spins. However, as these atoms exist in a Bose condensate this is not a separation of individual atoms so much as it is a separation of eigenstates in an entangled state. The splitting of the Bose condensates is analogous to the splitting of spin eigenvalues of an entangled state. The Bose-Einstein condensate is then split as a quantum state vector towards the two cavities. The different spin states of the Bose-Einstein condensates that enter the two cavities are then superpositions of a single wave function. This atomic analog of beam splitting then further puts the coherent photons in each cavity in a superposed state. The experimental setup would then be a quantum interferometer, where the detector measures the overlap of the two components of the total wave function associated with the two cavities.

The total phase shift measured is very small. This requires the quantum coherence of the atomic beam and the photons they interact with must have a high degree of fidelity. The above analysis shows this requires the application of parametric amplification and supercooling. This leads to the question of how to cool the mirrors to the temperatures required. This may then require that a reference beam be applied to measure the thermal noise of the mirrors. The detected thermal noise is used in real time to introduce photons into the cavity that are entangled with the photon-atom wave function. The introduced photons will have a wave form that is designed to interfere destructively with thermal noise in the system. This would amount to a specially designed quantum eraser system.

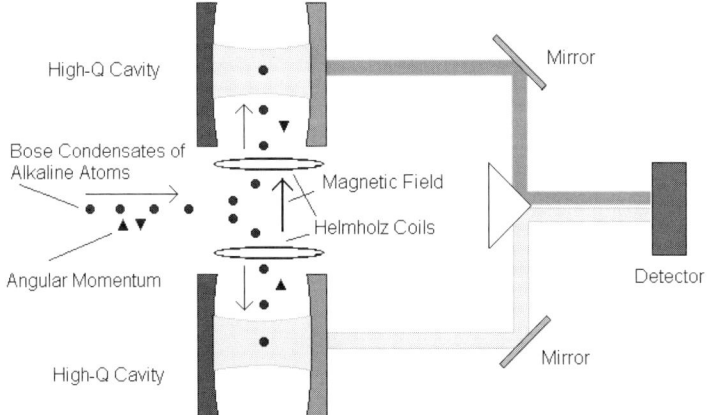

Proposed apparatus for measurement of quantum gravity induced phase.
Figure 3.3

Obviously other noise suppression techniques may have to be employed. Mechanical vibration in the system will have to be severely suppressed by techniques used in the LIGO machines. Either that or the entire machine might be lofted aboard a spacecraft, where any vibration due to thermal cycling of the craft as it orbits the Earth must be suppressed. The option here would be guided by the mass of the machine, costs and benefits. At this point serious experimental physics techniques and engineering would have to be employed. The subject is now beyond my expertise in these matters.

What has been presented is the outline of how quantum gravity fluctuations might be detected. It appears that quantum gravity should not be a subject that is completely sealed off from experimental study. It is my opinion that if quantum gravity is not experimentally probed in the next few decades the subject is destined to become a highly arcane mathematical topic very few can study, where those who study it are effectively removed from the scientific community. Such people would be in the position of the Hellenic philosophers who used "pure thought" to try to deduce the atomic nature of the world. This is a situation that may not need to occur.

3.3 Detecting Virtual Black Holes

The quantum gravity vacuum on a fine scale is probably filled with virtual black holes. These are thought to exist on such a small scale as to be impossible to detect. Yet this may not entirely be the case. Spacetime on a small scale contains a fluctuating arrangement of bubbles or loops that define quantum spacetime foam. The holographic principle was used by Ng [12] to deduce its structure. This result illustrates that its structure may be detected by gedanken experiments involving spacetime measurements. This thought experiment suggests a laser based atom interferometry techniques to look for spacetime fluctuations. This illustrates that the physics of quantum foam is defined by the existence of black holes.

The holographic principle indicates that the upper bound on the number of modes or degrees of freedom that can be put in a region of spacetime is determine by the area of the region, which is defined by the numbers of Planck units of area. The physics may be considered in a box of dimensions $L \times L \times L$, which may then be divided into the smallest units of measurement, which could be Planck scale units. Hence if the fluctuation in distance is δL the smallest unit of volume is $(\delta L)^3$. Within each unit of volume an independent degree of freedom is placed. The total number of modes is then $(L/\delta L)^3$ where the holographic principle requires $(L/\delta L)^3 \leq (L/L_p)^2$ for $L_p = (G\hbar/c^3)^{1/2}$. This gives the result

$$\delta L \geq L_p \left(\frac{L}{L_p}\right)^{1/3}. \quad (3.51)$$

This leads to the surprising result that the spacetime fluctuations sum to give a $\sim L^{1/3}$ dependency, which is considerably larger that usually considered with $\delta L \geq L_p$ Further, metric fluctuations in the volume L^3 scale as $\delta g_{\mu\nu} \geq (L_p/L)^{2/3}$. This then leads to fluctuations that can be considerably larger than the Planck scale.

Variants of standard spacetime measurements may be done in the light of these fluctuations. Explicit measurements to determine distances is central to general relativity, where coordinates do not have any meaning independent of observations. A coordinate system is established only by explicitly carrying out spacetime distance measurements. Wigner [13] proposed an experiment with a clock at one point and a mirror at another point. A light signal from the clock to the mirror in a timing experiment will determine the distance. The quantum uncertainty in the positions of the clock and the mirror will introduce an inaccuracy δL in the distance measurement. Hence a photon with a spread δL will exhibit an increase in spreading by $\delta L(2L/c) = \delta L + \hbar L/(mc\delta L)$, where the clock has a mass m. The uncertainty in position is then

$$\delta L \geq \sqrt{\frac{\hbar L}{mc}}. \tag{3.52}$$

The quantum mechanical relation may be given an added limit from general relativity. Assume the clock is a light-clock consisting of two parallel mirrors, where each has mass $m/2$. The two mirrors are placed a distance d apart, between which bounces a beam of light. For the uncertainty in distance measurement not to be greater than δL the clock must tick off time fast enough so $l/c \geq \delta L/c$. The size of the clock d must be larger than the Schwarzschild radius Gm/c^2 of the mirrors to read the time registered on the clock. From these it follows that

$$\delta L \geq \frac{Gm}{c^2}. \tag{3.53}$$

If the clock runs for a time T, where the smallest time measured by the clock is t, then from the time analog of the above relationship this gives

$$t^2 \geq \frac{\hbar T}{mc^2}. \tag{3.54}$$

Also the minimum is related to the time it takes light to cross the black hole's event horizon. $t \geq Gm/c^3$. If the black hole mass is the mass of our simple clock the total running time of the clock is

$$T = \frac{G^2 m^3}{\hbar c^4}, \tag{3.55}$$

which is the lifetime of a Black Hole predicted by Hawking's theory of black hole evaporation.

It is proposed in [12] that an interferometer experiment may detect these fluctuations. The proposed experiment requires an atomic beam be split along two arms of an interferometer. If the two arms of the interferometer are different then this is equivalent to two clocks that run for different times. When the two atomic beams are recombined by lasers the resulting phase may then be determined.

There is also another method that could be exploited. Assume an atom with the ground state $|g\rangle$ and a photon of frequency ν excites the atom to the state $|up\rangle$. Further, let the state $|up\rangle$ be split into the states $|a\rangle$ and $|b\rangle$ by the Stark effect or some other perturbation. If the atom is contained in a high-Q cavity there will be a Rabi oscillation. This is due to the level splitting, two "clocks" in a quantum superposition that sample the cavity length in according to different frequency intervals. The density matrices for the system obey

$$\dot{\rho}_{aa} = i(PE/\hbar)(\rho_{ab} - \rho_{ba})$$
$$\dot{\rho}_{bb} = -i(PE/\hbar)(\rho_{ba} - \rho_{ab}) \quad (3.56)$$
$$\dot{\rho}_{ab} = -i\omega_{ab}\rho_{ab} - i(PE/\hbar)(\rho_{bb} - \rho_{aa}),$$

where P is the magnitude of the polarization vector, E is the magnitude of the electric field vector and ω is the frequency $|E_a - E_b|/\hbar$. The Rabi oscillation frequency is $PE/\hbar$, where the time for a complete "flop" between the states $|a\rangle$ and $|b\rangle$ is $\hbar/(PE)$. The energy splitting is $PE = eEx$. The atoms pass through a region with an electric field induced by a capacitor. For an applied field of $10^6 volts/m$ the dipole Hamiltonian is $\sim 10^{-16} erg$ or $10^{-4} ev$. The wave length for the photons emitted by the split level is $\sim .3cm$. The atomic state acts as a clock that measures time over a time intervals $\hbar/(PE) \sim 10^{-11}s$ and the electron frequency is $mc^2/\hbar$, where m is the mass of the electron involved with the Rabi oscillation. The phase is $mc^2/(PE) \simeq 5.0 \times 10^9$. The uncertainty in time is

$$\delta T \geq \sqrt{\frac{\hbar^2}{PEmc}} = 1.4 \times 10^{-33} s \quad (3.57)$$

where m is the mass of the atom and $m_p = \sqrt{\hbar c/G}$. The atom will then exhibit a phase shift per Rabi oscillation of

$$\delta\phi = \frac{\delta T}{T}\frac{mc^2}{PE} \simeq 6.8 \times 10^{-13}. \quad (3.58)$$

It is possible to introduce atoms into a cavity in much the same manner as illustrated above, where the phase shift per Rabi oscillation that needs to be detected is $\delta\phi/\phi \sim 1.4 \times 10^{-22}$. Assume for the sake of argument that this atom is maintained in this dressed state for one second. The relative phase shift is then $\sim 10^{-11}$, which is within the capabilities of current technology. Current LIGO detectors are sensitive to a metric perturbation of $h \sim 10^{-21}$, which gives the phase on that order, where second generation upgrades should be capable of $h \sim 10^{-23}$. Assume there are two cavities with different energy level splitting PE. An optical interferometry measurement, which is simpler to arrange than an atomic interferometer, would then detect the phase difference between the two cavity photons.

A discrete beam of atoms must be injected into a cavity in much the same manner as illustrated above. This may be experimentally adjusted by the magnitude of a static electric field applied. An optical interferometry measurement, which is simpler to arrange than an atomic interferometer, would then detect the phase difference between the two cavity photons. Apparently the smaller the electric field applied in the Stark effect that the sensitivity of the time fluctuations can be greatly enhanced.

The holographic principle may be examined to look at the influence on curvature by quantum gravity fluctuations. Consider the volume tri-vector $V(\mathbf{e}_\alpha \wedge \mathbf{e}_\beta \wedge \mathbf{e}_\gamma)$, where V is the measure of this volume. The holographic principle indicates that the largest contribution will be from the term $O(\delta \mathbf{e}^3)$ so that

$$V\delta\mathbf{e}_\alpha \wedge \delta\mathbf{e}_\beta \wedge \delta\mathbf{e}_\gamma \geq L_p^2 A^{1/2} \epsilon^\mu{}_{\alpha\beta\gamma} * \mathbf{e}_\mu, \tag{3.59}$$

where A is the area that encloses the volume and $*$ is the Hodge star dual operator with $*^2 = 1$. The spacial three dimensional volume has its Hodge dual $*\delta V_{\alpha\beta\gamma} = V\epsilon^\mu{}_{\alpha\beta\gamma}\delta\mathbf{e}_\mu$. Apply the Hodge star dual operation on the last equation to define the fluctuation of a volume vector as

$$*\delta\mathbf{V}_\mu \geq L_p^2 A^{1/2} \mathbf{e}_\mu. \tag{3.60}$$

The magnitude of volume vector is reduced to a length by taking the cube root of its measure. This gives the fluctuation of a vectors as

$$\delta\mathbf{u}_\mu \geq (L_p^2 A^{1/2})^{1/3} \mathbf{e}_\mu. \tag{3.61}$$

By taking the coboundary operator on this with $d\mathbf{e}_\mu + \omega_\mu{}^\nu \mathbf{e}_\nu$ the curvature associated with this volume fluctuation is,

$$d^2\delta\mathbf{u}_\mu \geq (L_p^2 A^{1/2})^{1/3} \mathcal{R}^\nu{}_\mu \mathbf{e}_\nu, \tag{3.62}$$

which modifies the Riemann curvature tensor by

$$R^\alpha{}_{\beta\mu\nu} \rightarrow \left(1 + (L_p/L)^{2/3}\right) R^\alpha{}_{\beta\mu\nu}. \tag{3.63}$$

The modification is given by the magnitude of a metric fluctuation. For the Earth's gravity the curvature is approximately $\sim 10^{-28} cm^{-2}$ Consider three satellites in orbit around the Earth ranging one another with lasers. If the area defined by the triangular face of the positions of the satellites is on the order of $1 \times 10^{12} m^2$ the modification of the curvature by quantum gravity fluctuations is approximately 10^{-27}, or $\delta R \simeq 10^{-55} cm^{-2}$. Even close to a stellar mass black hole this factor is close to 10^{-24} and modifies the curvature near the event horizon by this factor. This is a very good result, for it indicates that the fluctuations in spacial lengths can be sufficient for empirical verification, but its impact of classical gravitation is almost infinitesimal.

3.4: Quantum Optical Detection of Quantum Gravity Induced Decoherence

A quantum system with on a very short time scale or a narrow pulse will couple with metric fluctuations of spacetime. If the holographic principle is correct fluctuations will occur on time intervals that are considerably longer than the Planck time. This coupling will induce phase information of the system to be absorbed by the quantum gravity vacuum. It should be possible to measure this with fempto-second pulsed lasers.

Quantum fluctuations of the metric will induce a wave function to exhibit a mixing of spacial and temporal correlations between paths in a sum over histories. A null geodesic on a small enough scale becomes uncertain due to metric fluctuations. A signal sent of probe sufficiently small region of space will have its information "processed" by such fluctuations. This results in a measure of decoherence due to the loss of quantum information to the quantum gravity vacuum. This means that a short pulse wave function will exhibit a measure of decoherence from metric fluctuations.

While this process is very subtle for quantum systems studied today, the holographic principle [14] indicates metric fluctuations are $\delta g_{\mu\nu} \simeq (t_p^2/t)^{1/3} g_{m\mu\nu}$ for the system with a period or time envelope t [15]. For $t \gg t_p$ fluctuations are small and these decoherences are small. For t large enough the wave function will largely cancel out the influence of these metric fluctuations. However, for t very small metric fluctuations have a significant impact to the wave function. The argument can be translated, or Fourier transformed, into one involving energy. Very high energy measurements will detect aspects of metric fluctuations in the fine grained details of a wave function. This physics deviates from the type of fluctuations and correlations expected from a two-point functional [16].

Quantum gravity has no experimentally measured datum to date, which are regarded as impossible by many due to the extreme scales involved. Future experimental tests of quantum gravity will invoke subtle physics induced by metric fluctuations. The electromagnetic field is probably the best probe to employ in such experiments. The electric and magnetic field satisfy an uncertainty principle $\Delta E \Delta B = \hbar/2$ for a minimal uncertainty photon wave packet. Metric fluctuations will then play a role in the physics of a photon if the wave packet has a small extent in space. The fluctuation of the electric field due to quantum gravity should then become apparent for a fempto-second laser pulse. This will be more pronounced if the quantum state of the photon field is parametrically amplified so the standard quantum uncertainty in the electric field is near zero.

The electric field and magnetic field associated with a photon in a volume V are $E_0 = \sqrt{\hbar\omega/2V\epsilon_0}$ and $B_0 = \sqrt{\hbar V \epsilon_0 / 2\omega}$, with the uncertainty $\Delta E \Delta B = E_0 B_0 = \hbar/2$ and ϵ_0 the electric permittivity. Parametric amplification of a photon converts a photon into two entangled photons with angular frequencies $\omega_1 + \omega_2 = \omega$. The uncertainty in the electric field for the two photons is then suppressed, where the uncertainty in the magnetic field component is correspondingly increased. If the two photons are absorbed by an atom and re-emitted as a single photon with frequency $\omega/2\pi$ then the process may be repeated enough times, such as with photons in a cavity, to cause $\Delta E \to 0$.

With the minimal uncertainty $\Delta E \Delta B = \hbar/2$ wave evolution of the operators $\hat{E}$ and $\hat{B}$ obeys [17]

$$(\hat{B} - \langle\hat{B}\rangle)|\psi\rangle = \frac{i\hbar}{2(\Delta E)^2}(\hat{E} - \langle\hat{E}\rangle) \times \hat{\mathbf{k}}|\psi\rangle, \quad (3.64)$$

where $\hat{\mathbf{k}}$ is the direction of photon propagation. The field fluctuations are $\delta E = \hat{E} - \langle\hat{E}\rangle$ and similarly $\delta B = \hat{B} - \langle\hat{B}\rangle$. The fluctuations in the electric field then contains the standard quantum fluctuations plus metric fluctuations,

$$\delta\hat{E} = \delta_0 \hat{E} + \frac{d\hat{E}}{dt}\delta t = \delta_0 \hat{E} + \left(\frac{\partial\hat{E}}{\partial t} + c\nabla \times \hat{E}\right)\delta t = \delta_0 \hat{E} + \omega(\hat{E} + \hat{k} \times \hat{E})\delta t. \quad (3.65)$$

The holographic fluctuation of the metric is $\delta g_{\mu\nu} = (t_p/t)^{2/3} g_{\mu\nu}$, for $t_p = \sqrt{G\hbar/c^3}$ and t_s the sample time or pulse width. The fluctuation in proper time from the interval is $\delta t = (t_p/t_s)^{1/3} t$, for t the sample time given by the periodicity of the photon. The fluctuation in the field then contains a term $\delta_0 E$ for the ordinary quantum fluctuation plus a term $\frac{d\hat{E}}{dt}\delta t$ with $\delta t = (t_p/t_s)^{1/3} t \ll 1/\nu$. Here t_s is the sample time for metric fluctuations, which is the time spread of the laser pulse and ν is the frequency of the photon field. The fluctuation in the electric field is then

$$\delta \hat{E} \simeq \delta_0 \hat{E} + \omega(\hat{E} + \hat{\mathbf{k}} \times \hat{E})(t_p/t_s)^{1/3} t, \tag{3.66a}$$

with

$$(\hat{B} - \langle \hat{B} \rangle)|\psi\rangle = \frac{i\hbar}{2(\Delta E)^2} \Big(\hat{\mathbf{k}} \times \delta_0 \hat{E} + 2\pi\omega \hat{\mathbf{k}} \times (\hat{E} \times \hat{\mathbf{k}} + \hat{E})(t_p/t_s)^{1/3} t \Big) |\psi\rangle, \tag{3.66b}$$

where $\hat{\mathbf{k}}$ indicates the direction of the photon. The analogous equation with the role of $\hat{E}$ and $\hat{B}$ interchanged is

$$(\hat{E} - \langle \hat{E} \rangle)|\psi\rangle \simeq \frac{i\hbar}{2(\Delta B)^2} \Big(\hat{\mathbf{k}} \times \delta_0 \hat{B} + \omega \hat{\mathbf{k}} \times (\hat{\mathbf{k}} \times \hat{B} - \hat{B})(t_p/t_s)^{1/3} t \Big) |\psi\rangle. \tag{3.67}$$

Now construct the Hamiltonian $\frac{1}{2}(E^2 + B^2)$ in order to find the contribution of metric fluctuations to the Schrödinger evolution of the wave function. For $t \sim 1/\omega$ the pulse width, equivalently the probe time t_s, is "refreshed" as a probe with each oscillation. For $\langle \hat{E} \rangle = \hat{E} - \delta \hat{E}$ and similarly for $\hat{B}$ these terms to first order in the fluctuations can be given according to the fields instead of their expectations. The square of the magnetic and electric fields are

$$\hat{B}^2 \simeq \hat{B}_0^2 + \frac{i\hbar}{(\Delta E)^2} \hat{\mathbf{k}} \cdot (\hat{E} \times \hat{B})(t_p/t_s)^{1/3}$$

$$\hat{E}^2 = \hat{E}_0^2 + \frac{i\hbar}{(\Delta B)^2} \hat{\mathbf{k}} \cdot (\hat{B} \times \hat{E})(t_p/t_s)^{1/3}. \tag{3.68}$$

The Hamiltonian density has a perturbation due to metric fluctuations.

$$\mathcal{H} = \frac{\epsilon_0}{2}(\hat{E}_0^2 + \hat{B}_0^2) + i\hbar\epsilon_0 \Big(\frac{1}{(\Delta B)^2} \hat{\mathbf{k}} \cdot (\hat{B} \times \hat{E}) + \frac{c^2}{(\Delta E)^2} \hat{\mathbf{k}} \cdot (\hat{E} \times \hat{B}) \Big)(t_p/t_s)^{1/3}. \tag{3.69}$$

The electric field operator is $\hat{E} = \mathbf{E}_0(\hat{a} e^{i\theta} + \hat{a}^\dagger e^{-i\theta})$, for $\theta = \hat{\mathbf{k}} \cdot \mathbf{r} - \omega t$ and similarly the magnetic field operator is $\hat{B} = i\mathbf{B}_0(\hat{a}^\dagger e^{-i\theta} - \hat{a} e^{i\theta})$. The perturbing Hamiltonian density is then written as

$$\mathcal{V} = i\hbar\epsilon_0 \Big(\frac{1}{(\Delta B)^2} - \frac{c^2}{(\Delta E)^2} \Big) E_0 B_0 (\hat{a}^{2\dagger} e^{-2i\theta} - \hat{a}^2 e^{-2i\theta} + [\hat{a}, \hat{a}^\dagger])(t_p/t_s)^{1/3}, \tag{3.70}$$

where the zero point energy term may be dropped. For an uncertainty in quadrature, $\Delta E = c\Delta B$ the squeezing Hamiltonian operator vanishes. However, if there is a "seed" parametric downshifting of the photon states they will contribute. Let $|z_s| > 1$ be the magnitude of the squeezing parameter. The uncertainty in the electric field is $\Delta E = E_0/|z_s|$, where correspondingly $\Delta B = |z_s|B_0$. The term $\hat{a}^2 e^{2i\theta} - (\hat{a}^\dagger)^2 e^{-2i\theta}$ is the generator of a squeeze state operator. The effective squeezing parameter in the second part of $\mathcal{V}$ is $|z| = (|z_s|^2 - |z_s|^{-2})(t_p/t_s)^{1/3}$. The metric fluctuations on the electromagnetic field induces a parametric amplification. The approximate Hamiltonian density that squeezes the vacuum state is then

$$\mathcal{V}_{sq} \simeq iE_0^2|z|\left(\hat{a}^2 e^{2i\theta} - (\hat{a}^\dagger)^2 e^{-2i\theta}\right) \quad (3.71)$$

In an interaction picture this Hamiltonian term gives the unitary evolution

$$U(t) = e^{-i\mathcal{V}t/\hbar}, \quad (3.72)$$

which for $ie^{2i\theta} = e^{2i\theta'}$ is

$$U(t) = exp\left(-iE_0^2|z|(t_p/t_s)^{1/3}(\hat{a}^2 e^{2i\theta} - (\hat{a}^\dagger)^2 e^{-2i\theta})t/\hbar\right) \quad (3.73)$$

The evolution of any observable is then $\mathcal{O}(t) = U(t)\mathcal{O}U^\dagger(t)$. Consider the evolution of Δx and Δp by $U(t)$. The evolution of the annihilation and creation operators are

$$a(t) = U(t)aU^\dagger(t) = a\, cosh(r) + a^\dagger e^{2i\theta} sinh(r),$$

$$a^\dagger(t) = U(t)a^\dagger U^\dagger(t) = a^\dagger cosh(r) + a e^{-2i\theta} sinh(r), \quad (3.74)$$

where $r = \epsilon_0 \mathcal{V} E_0^2 |z| t/\hbar = \omega t$. The uncertainties in the conjugate variables obey

$$\langle(\Delta x)^2\rangle = \frac{1}{2}(cosh(2r) - cos(2\theta)sinh(2r)), \quad \langle(\Delta p)^2\rangle = \frac{1}{2}(cosh(2r) + cos(2\theta)sinh(2r)). \quad (3.75)$$

Squeezed states have an entropy interpretation [19] with $S - \frac{1}{2}ln(\langle(\Delta x)^2\rangle\langle(\Delta p)^2\rangle)$ so that [19]

$$S = \frac{k_B}{2}ln\left(\frac{1}{4}\left(cosh^2(2r) - sinh(2r)(cos^2(2\theta) + sin^2(2\theta))\right)\right) = k_B ln(1/2). \quad (3.76)$$

This is to be expected for a pure state evolution with $\Delta S = 0$. However, this is an idealization. The laser pulse is assumed to be a femtosecond or shorter in wave length. This means that there is large spread of Fourier terms. For such a pulse the phase θ is difficult to ascertain, so an averaging over this phase in a coarse grained approach is required. The averaged uncertainties are such that

$$S = \frac{k_B}{2}ln\left(cosh^2(2r)/4\right), \quad (3.77)$$

where for r small $S = k_B(2r^2 + ln(1/2))$. Since $r = r(t)$ the system will exhibit an increasing decoherence with time.

The induced squeezing r for $t_s = 10^{-18}s$ and $|z_s| = \sqrt{2}$ gives $|z| \simeq 7.0 \times 10^{-9}$. This is a very small squeezing parameter. For a system with $\lambda = 500nm$ and a photon path of $\sim 1.0m$ the entropy associated with the decoherence will then be

$$S \simeq k_B|z|\omega t \simeq 8.4 \times k_B. \tag{3.78}$$

The Shannon-Khinchin formula for the entropy content for n bits of information lost is

$$S = -k_B n ln(n), \tag{3.79}$$

which for ΔS here corresponds to the loss of $n \sim 6$ bits, or q-bits of information. A quantum system consisting of a photon with a wave packet envelope width $\sim 10^{-8}cm$ will couple with approximately 24 virtual black holes in the vacuum, or spacetime foam. If femptosecond lasers are used the entropy drops by a factor of 10, so around two to three virtual black holes coupled to the pulse over a length of 1m.

A sufficiently short pulse of light will couple to the holographically induced fluctuations of the metric and produce detectable results. The shorter the pulse of light is the more it will exhibit perturbations due to the spacetime quantum foam and holographic effects. It is beyond the scope of this theoretical discussion of an "experiment in principle" to present explicit designs for this experiment. However, once atto-second pulse lasers are a practical tool for experimental quantum optics, it appears reasonable that this experiment could be configured. This and other related experiments [15][20][21] should be able to probe for some large scale physics potentially induced by quantum gravity fluctuations.

The decoherence of a wave function by quantum gravity fluctuations means that the gravity vacuum is an environment that prevents a complete Poincaré recurrence of a wave function. Quantum correlations or phase information is leaking into the quantum gravity vacuum. It is being irreversibly lost to this vacuum, but is not destroyed. This is a signature which indicates quantum gravity is a source of thermodynamics in the big bang induced in how the universe quantum tunnelled out of the vacuum state.

3.5: Ring Gyro Test of the Holographic Principle

The holographic principle states that matter-fields that compose a black hole exist on a membrane a Planck length above the event horizon according to a distant observer. This conjecture stems from the time dilation effect of clocks that fall onto black holes as observed at a distance [14]. Objects falling onto a Schwarzschild black hole are never observed to reach the event horizon at $r = 2M$, but appear to hover frozen by time dilation just above this membrane. To a distant observer the region just above the event horizon contains quantum amplitudes of everything that composes the black hole. The black hole also emits radiation and quanta by the Hawking effect. The Hawking radiation is the scatter by a black hole of quanta frozen above the event horizon. The Hawking radiation within a three dimensional volume results from fields pinned on a two dimensional membrane a Planck length above the black hole horizon. This is the origin of the term holography, for quantum amplitudes within a three dimensional region are determined by quantum amplitudes on a two dimensional surface.

The quantum gravity vacuum likely consists of virtual black holes near the Planck length scale $L_p = \sqrt{G\hbar/c^3}$. These virtual black holes determine the quantum modes for gravitation by the holographic principle. Consequently fields holographically projected from virtual black holes on a closed two dimensional surface determine the fields within the bounded three dimensional volume. Virtual black holes in the volume secondarily project fields from virtual black holes that are on this surface. This leads to fluctuations of lengths larger than the Planck scale $L_p = \sqrt{G\hbar/c^3}$ [15]

$$\delta L \geq \left(LL_p^2\right)^{1/3}, \qquad (3.80)$$

where L is the length scale of the volume. These fluctuations have a considerably larger length scale than expected from estimates of metric fluctuations by $\delta g \sim L_p/L$, or with straight forward calculations by two point functions [16]. The holographic principle has not yet been substantiated by experiments and is speculative at this time.

Atomic interferometry has been proposed as a method for detecting these fluctuations in [15]. There is a proposed quantum optical method of detection that employs atom-photon interactions, where atomic transitions defines a clock [22]. Atomic interferometry is in a comparatively early stage of development and more difficult than optical experiments. The photon-atom interaction approach proposed by the author avoids some of the difficulties of atomic interferometry, but is still comparatively complex. It is the purpose here to present another experimental architecture that may be simpler to construct and execute.

It is best if experimental physics employs the simplest method possible to detect an aspect of how nature behaves. Here it is demonstrated that δL may be detected with only photons. Sagnac interferometry has been proposed as a method for detecting gravity waves [23][24]. This may as well be used to detect the fluctuations due to the holographic principle. This eliminates the complexity of establishing atomic dressed states in an atom-photon interaction approach, where atomic states serve as a clock. Further, a rotating gyro has the added feature that noise due to thermal and quantum fluctuations of mirrors are cancelled out in this system. This type of experiment could well be performed soon as the detection of quantum gravity effects.

The Sagnac effect is due to the interference of light in a rotating interferometer [25]. A photon that reflects off a ring of mirrors on a rotating platform will return to its source, but where that source has been rotated slightly from its initial position. Consequently two counter directed photons will exhibit interference when they are recombined, as the round trip of the two photons is different. The ring gyro is a technology currently employed to detect rotations and used in the guidance of aircraft and rockets. The ring gyro is not sensitive to thermal and quantum uncertainties in the position of mirrors. This is because two counter directed photons will experience the same fluctuations. The identical phase differences of the two beams due to these fluctuations completely cancel. The ring gyro is then a very sensitive detector of rotations with a resulting Sagnac frequency $\Delta\nu$.

Suppression of noise in the ring gyro might at first thought mean this could not detect fluctuations due to spacetime fluctuations. However, there is a difference between fluctuations in the mirrors and spacetime fluctuations. With ordinary thermal and quantum fluctuations of the mirrors the photon responds only by reflecting off the mirrors. Hence counter directed photons will experience a phase shift that destructively interfere upon detection. Quantum fluctuations of spacetime due to the holographic principle gives a fluctuation of the spacetime metric $\delta g_{\mu\nu} = g_{\mu\nu}(L_p/L)^{2/3}$, which interact with the photons throughout the ring gyro. These fluctuations in the metric induce a net field effect through the area enclosed by the device and induce a measurable phase on the photons.

The frame of a rotating interferometer is noninertial. The metric has the following nonzero components up to first order in $\Omega L/c$

$$g_{00} = -g_{11} = -g_{22} = -g_{33} = 1, \; g_{0k} = -\left(\frac{\mathbf{\Omega} \times \mathbf{R}}{c}\right)\bigg|_k. \tag{3.81}$$

The rotation vector for this frame is then $g_k = -g_{0k}/g_{00}$, which in the $\Omega L \ll c$ approximation is given by

$$\mathbf{g} = \frac{\mathbf{\Omega} \times \mathbf{R}}{c}. \tag{3.82}$$

The Sagnac phase for two counter directed laser beams with wavelength λ in a ring gyro is determine by [26]

$$\Delta\Phi = \frac{4\pi}{\lambda c}\int_{\mathcal{A}} \nabla \times \mathbf{g} \cdot d\sigma, \tag{3.83}$$

where $\mathcal{A}$ is the area enclosed by the gyro. The vector $\mathbf{g}$ for the apparatus rotating with angular velocity $\mathbf{\Omega} = \Omega\mathbf{k}$, sitting in the $x - y$ plane, is $\mathbf{g} = \Omega(y\mathbf{x} - x\mathbf{y})/c$. This vector may be expanded according to the length fluctuation δL with $L = \sqrt{x^2 + y^2}$ as

$$\mathbf{g}(L + \delta L) = \mathbf{g}(L) + \frac{\partial \mathbf{g}}{\partial L}(L)\delta L = \mathbf{g}(L)\bigl(1 + \delta L/L\bigr). \tag{3.84}$$

where $\mathbf{g}(L)$ is the classical rotation vector. The integral in equation 3.83 for a square interferometer is then

$$\Delta\Phi = \frac{4\pi\Omega}{\lambda c}L^2\bigl(1 + (L_p/L)^{2/3}\bigr). \tag{3.85}$$

If the ring gyro has a length of 10cm, for a green laser with $\lambda = 500$nm, the Sagnac phase shift is then

$$\Delta\Phi = 8.4 \times 10^{-4}\Omega(1 + 2.9 \times 10^{-23})s. \tag{3.86}$$

The result in equation 3.85 illustrates how metric fluctuation enter directly into the Sagnac phase and equation 3.86 illustrates that this contribution is comparatively small. Clearly for large enough Ω, or base line L, the contribution from spacetime fluctuations are amplified. If the rotation of an $L = 10$cm apparatus is greater than 10^3Hz the fluctuation induced phase shift is within current technology similar to the detection of phase changes in LIGO facilities [22][28] with strain induced by metric variations $\delta g = \delta L/L \sim 10^{-21}$. The term $h = \frac{4\pi\Omega}{\lambda c}L^2\left(\frac{L_p}{L}\right)^{2/3}$ is analogous to the strain h in LIGO detection.

The above result is comparatively small and this problem has relativistic corrections beyond $\Omega \times \mathbf{R}/c$. A more general line element is needed of the sort found in "Roto-Stacey" coordinates [29]. The modified rotation vector $g_\theta = g_{t\theta}/g_{tt}$ is

$$\mathbf{g} = \frac{\Omega \times \mathbf{R}/c}{1 - \Omega^2 R^2/c^2}. \tag{3.87}$$

For a 10cm interferometer arm length the g_{tt} term introduces a first order correction to the Sagnac phase of $\sim \Omega \times 10^{-21}$s. The correction by g_{tt} to the δL fluctuation term is negligible.

With $\Delta\phi \simeq (4\pi\Omega L^2/(\lambda c))(1 + \Omega^2 R^2/c^2)^{-1}$, the Sagnac frequency shift is $\Delta\nu = \Delta\phi c/4L$ and the holographically induced frequency shift is $\delta\nu = \Delta\nu(L_p/L)^{2/3}$. Define E_{cw} and E_{ccw} so the counter directed fields are $E_{cw} = Ee^{ikx}$ and $E_{ccw} = Ee^{-ikx}$. A heterodyne measurement of the counter directed beams gives a correlation function for $\delta\nu$

$$\langle\,|\rho E_{cw}(t)E_{ccw}(t)|\,\rangle\Big|_{\delta\nu} = \langle\,|E_0|^2\,\rangle\frac{\delta\nu c}{2L}sin2\omega t. \tag{3.88}$$

A time integration will then define a power spectrum, which would be a signature for the holographic effect. A time integration of this will result in a relative term $\delta\nu c/(2L\omega) \sim 10^{-6}s^{-1}$.

The noise in the experimental system is going to be entirely due to photon shot noise and radiation pressure on the mirrors. In contrast any quantum noise motion in the mirrors will shift the mirror positions by $x_t = x_0 + \frac{p}{m}t$. A measurement of the mirror position in a heterodyne measurement would have a limit to its accuracy given by a commutator between the operators for position as different times. Ignoring the rotational motion of the apparatus this commutator gives

$$[\hat{x}_0, \hat{x}_t] = \frac{1}{m}[\hat{x}_0, \hat{p}]t = \frac{i\hbar}{m}t, \tag{3.89}$$

with the standard quantum limit $\Delta x_0 \Delta x_t \geq \hbar t/2m$. If the uncertainties have quadrature form then $\Delta x_0 \simeq \sqrt{\hbar t/2m}$. Let the angular frequency of the photons be ω, so the total Sagnac phase shift, with metric fluctuation data, be Ω. The quantum noise due to the mirrors then contributes to the electric field of the photons by

$$\mathbf{E}_{cw} = \mathbf{E}_0 cos(\omega t + \Omega + \mathbf{k}\cdot\mathbf{x}) = \mathbf{E}_0\big(cos(\omega t + \Omega) - \mathbf{E}_0(\mathbf{k}\cdot\mathbf{x})sin(\omega t + \Omega)$$

$$\mathbf{E}_{ccw} = \mathbf{E}_0 cos(\omega t + \Omega - \mathbf{k}\cdot\mathbf{x}) = \mathbf{E}_0\big(cos(\omega t + \Omega) + \mathbf{E}_0(\mathbf{k}\cdot\mathbf{x})sin(\omega t + \Omega), \tag{3.90}$$

where $|\mathbf{k}| = \omega/c$ and the sign change in the second equation emerges from the counter motion of this photon. The combined electric fields of the photons is then

$$\mathbf{E}_{cw} + \mathbf{E}_{ccw} = 2\mathbf{E}_0 cos(\omega t + \Omega). \quad (3.91)$$

Consequently quantum noise due to fluctuations of the mirror centers of mass cancels out. Thermal vibrations of the mirrors will similarly cancel out as well.

The contributing noise is then due to shot noise in photon production and the radiation pressure of photons on the mirrors. The contributing noise is entirely due to photons. Shot noise results from fluctuations in Poisson statistics of photon production [30]. Let N is the average number of photons produced in a time interval. There is also a fluctuation in photon number in that time interval. So in any give time interval there are $N + \delta N$ photons produced. In this time interval there is also a fluctuation $\delta\phi$ in the average phase ϕ. The Heisenberg uncertainty principle then dictates that $\delta N \delta\phi = 1$. For photons with energy $E = \hbar\omega$ produced by a laser with a power I the number of photons per unit time is I/E so in a time interval $N = It/E$, where E/t is a quantum limit for power I_{QL}. Hence the shot noise induced fluctuation in the position is

$$\Delta x_S = \Delta x_0 \sqrt{I/I_{QL}}. \quad (3.92a)$$

The uncertainty radiation pressure is a Heisenberg microscope result with

$$\Delta x_R = \Delta x_0 \sqrt{I_{QL}/I}, \quad (3.92b)$$

where the total noise induced fluctuation is $\Delta x = \Delta_0 \sqrt{(\Delta x_S)^2 + (\Delta x_R)^2}$.

For a 1 watt laser and a wavelength $.5\mu$m the photon production rate is $\dot{N} = 1.2 \times 10^{18} s^{-1}$. If the photon production obeys Poisson statistics then $\delta N = \sqrt{\dot{N} \times t} = 3.5 \times 10^{10} s^{-1}$, for $t = 1$s. Thus the shot noise induced phase is $\delta\phi = 3.5 \times 10^{-10}$. The effective strain in the interferometer is then

$$\delta h = \frac{.5 \times 10^{-6} cm \times 3.5 \times 10^{-10}}{2\pi 10 cm} = 2.8 \times 10^{-18}. \quad (3.93)$$

This is five orders of magnitude larger than the expected result from spacetime fluctuations. To reduce this noise a parametric down shifting of frequencies is required. This will induce $\Delta E \rightarrow 0$ and by correlation then $\Delta x_0 \rightarrow 0$. Experimentally this requires the use of a parametric amplifier in the laser, where the emission and absorption of photons process must occur > 20 times. This gives a net squeezing parameter of $z > 14$.

3.6: Measurement of Unruh-Hawking Physics With Bose-Einstein Condensates

The Einstein equivalence principle states that locally there is no difference between an inertial frame in flat Minkowski space and in a freely falling frame in a gravity field.

Similarly there is no local difference in physics between a frame on the surface of a gravitating body and an accelerating frame with the same local value of g. Lower order tidal physics, or Weyl curvature effects, are ignored if the frames are regarded as sufficiently small.

There is a similar effect with rotating reference frames. M. G. Sagnac sent light rays in a clockwise and counter clockwise directions on a rotating platform and found interference fringes [25]. By the invariance of the speed of light in an inertial frame, with the fact that light propagating in one direction will have a shorter distance to travel than the other, an phase interference was observed. This has lead to a "controversy," where a transformation to the rotating frame should also result in an invariance in the speed of light. However, the spacial extent of the rotating frame required introduces deviations analogous to low order tidal effects. In the same light this is why light is found to be frozen on an event horizon of a black hole as seen by a distant observer.

Hawking illustrated black holes quantum mechanically radiate in a blackbody distribution [31]. Analogously Unruh found that a detector on an accelerated reference frame will measure a thermal distribution of quanta from the particle horizon [32]. This is an application of the equivalence principle. Particle production by a black hole is equivalent to particle production on the split horizon bounding Rindler wedges. Here the observer on an inertial frame will find that some of the power required to drive this frame on an accelerated path is producing a thermal bath of quanta. Similarly a rotating frame with its artificial gravity should also produce such an effect. It is then reasonable to propose an experiment that detects this with low temperature systems on a rotating reference frame. The two possible physical systems to examine are low temperature superconductors and gaseous Bose-Einstein condensates.

To describe rotating coordinates consider the transformation between frame F and F'

$$x^{\mu'} = \Lambda^{\mu}{}_{\nu} x^{\nu}. \tag{3.94}$$

In cylindrical coordinates the radial and azimuthal coordinates are $r' = r$ and $z' = z$ and

$$t' = \gamma(t - \Omega \chi^t), \quad \theta' = \gamma(\theta - \Omega \chi^\theta), \tag{3.95}$$

where γ is the Lorentz factor. The angular momentum Ω defines the velocity tangential to the rotation $v = \Omega r$ as measured by a nonrotating observer. Now define the rotation in these coordinates by

$$\eta = \frac{d\theta}{dt}, \quad \eta' = \frac{d\theta'}{dt'}, \tag{3.96}$$

where the second rotation is given by

$$\eta' = \frac{d\theta - \Omega\left(\frac{\partial \chi^\theta}{\partial t} dt + \frac{\partial \chi^\theta}{\partial \theta} d\theta\right)}{dt + \Omega\left(\frac{\partial \chi^t}{\partial t} dt - \frac{\partial \chi^t}{\partial \theta} d\theta\right)}. \tag{3.97}$$

With these the velocities $v = r\eta$ and $v' = r\eta'$ are defined, where $\chi^t = \theta$ and $\chi^\theta = t$. This gives a formula for the addition of rotation

$$\eta' = \frac{\eta - \Omega}{1 - r^2\Omega\eta/c^2}, \tag{3.98}$$

which is analogous to the linear equation for velocity addition in special relativity.

The metric for a rotating frame is then of the form

$$ds^2 = dt^2 - r^2\left(d\theta + \frac{\Omega}{c}dt\right)^2 - dr^2 \tag{3.99a}$$

or

$$ds^2 = \left(1 - \frac{r^2\Omega^2}{c^2}\right)dt^2 - r^2d\theta^2 - 2r^2\frac{\Omega}{c}d\theta dt - dr^2. \tag{3.99b}$$

The horizon on this frame exists at $r = c/\Omega$. A solution to these coordinates according to the proper time s, with $dr = 0$ is then

$$t = g^{-1}\sinh(gs), \quad \theta = \frac{g^{-1}}{r}\cosh(gs) - g^{-1}\frac{\Omega}{c}\sinh(gs), \tag{3.100}$$

where $g = \frac{\Omega^2 r}{\sqrt{1 - \Omega^2 r^2/c^2}}$ is the acceleration parameter. Until later set $c = 1$. A Greene's function for the propagation of fields assumes the form

$$G(x, x') = -\frac{1}{4\pi^2((t - t' - i\epsilon)^2 - |\mathbf{r} - \mathbf{r}'|^2)}. \tag{3.101}$$

Now write the Greene's function as $G(x(s), x'(s')) = G(\Delta s)$, where massless particles the Greene's function is

$$G(\Delta s) = -\left(16\pi^2 g^2 \sinh(\Delta sg - i\epsilon g)\right)^{-1}. \tag{3.102}$$

This Greene's function then enters a Fourier integral for the propagation of the vacuum as observed by the shift in the atomic level $E \to E'$ of an atom on this rotation frame

$$\sum_{E'}|\langle 0', E'|0, E\rangle|^2 = \sum_{E'}|\langle E'|\mu|E\rangle|^2 \int d(\Delta s) e^{i(E' - E)\Delta s} G(\Delta s). \tag{3.103}$$

This integral then has the form [32]

$$\sum_{E'}|\langle 0', E'|0, E\rangle|^2 = \frac{1}{2\pi}\sum_{E'}\frac{(E' - E)|\langle E'|\mu|E\rangle|^2}{e^{2\pi(E' - E)/g} - 1}, \tag{3.104}$$

where $\mu = \mu(s)$ is a coupling Lagrangian between the vacuum and the atom or a monopole operator for a detector. This illustrates that the vacuum on a rotating reference frame will be a thermal vacuum. The temperature of this vacuum with units restored is $T = \hbar g/(2\pi c k_B)$.

It may be illustrated this is equivalent to a black hole. To explicitly see this in $U = u + v$, $V = u - v$ coordinates

$$U = \frac{1}{\Omega^2 r} g_{00}^{1/2} \sinh(gs), \quad V = \frac{1}{\Omega^2 r} g_{00}^{1/2} \cosh(gs) \qquad (3.105)$$

for $g_{00} = 1 - \Omega^2 r^2/c^2$. The motion of a point at a radial distance is equivalent to the motion of a particle in a gravity field. In general relativity the result is exactly Kepler's laws that $\Omega^2 r^3 = M$. Now perform the substitution that $r = 2M$ to arrive at a gravity force at the particle horizon as $g = \Omega^2 r = 1/4M$. The U, V coordinates are then of the form

$$U = 4M g_{00}^{1/2} \sinh(gs), \quad V = 4M g_{00}^{1/2} \cosh(gs), \qquad (3.106)$$

from which it is evident that $V^2 - U^2 = 16M^2 g_{00}$. This is equivalent to $1/g$ and the relationship $2\pi T = g$ gives

$$2\pi T = \frac{1}{\sqrt{16M^2 g_{00}}} = 2\pi g_{00}^{-1/2} T_0. \qquad (3.107)$$

The temperature is $T_0 = (8\pi M)^{-1}$, which is the temperature of a black hole with the mass parameter M, which with units restored is GM/c^2.

With units restored the temperature is dependent on the rotation of the frame by,

$$T = \frac{\hbar \Omega}{\pi k_B} g_{00}^{1/2} = 1.5277 \times 10^{-11} K - \sec \times \nu. \qquad (3.108)$$

For a gaseous Bose-Einstein condensate at $T \sim 10^{-9} K$ rotating at 100Hz the temperature change induced by the effects of quantum fields in spacetime would become measurable. The thermal deBroglie wavelength $\lambda_{dB} = h/\sqrt{2\pi m k_B T}$ defines the number of particles in the excited state within the Bose gas

$$N_e \leq \frac{3}{2} kT \frac{V}{\lambda_{dB}} \zeta\left(\frac{3}{2}\right). \qquad (3.109)$$

Those in the ground state are $N_0 = N - N_e$ [33], where for a large deBroglie thermal wavelength N_e can be made very small. For the quantity $z \simeq 1 - 1/N_0$, a shifted z due to an Unruh effect is then

$$\delta z = \frac{2V}{3h} \sqrt{\frac{2\pi m}{k_B}} \left(\frac{1}{\sqrt{T}} - \frac{1}{\sqrt{T'}}\right), \qquad (3.110)$$

where $T' > T$ is given by the Unruh effect in equation 3.08. The change in z is then a measure of the number of bosons removed from the ground state by interaction with thermal quanta in the Rindler wedge.

The particle horizon is defined where $g_{00} = 0$ at a radius of $r = c/\Omega$. For $\nu = 10s^{-1}$ the particle horizon on the rotating frame is located from the center of rotation as $r = 4.8 \times 10^8$m, which is nearly the radius of the Earth. The Bose-Einstein condensate then interacts with a heat bath defined by its acceleration $g = \omega^2 r$ for a horizon at this distance. Bose-Einstein condensates of Rb atoms have been induced to rotate by an evaporative spin-up technique. It is observed that any rotation rate faster than 16.6πHz will exceed the centrifugal limit for the system [34]. To examine the temperature change at the transition temperature for a Bose-Einstein condensate $T \sim 10^{-9} K$ requires an rotating frequency of $\sim 10^3 sec^{-1}$. It appears that the Bose Einstein condensate will have to be contained in some manner. Further, this has to be done so that the outer s shell electronic state is not effected. This material will have to permit laser light to penetrate it, give a rigidity required to contain the Bose Einstein condensate under rotation, at the same time not perturb the outer electron and its interaction with the laser light. It appears that such a material confinement is difficult or problematic. An better approach would be to use singly ionized barium that emulates cesium. The ionized gas is caught in the rotating egg carton potential and trapped, but in addition a magnetic field is applied to provide a Lorentz force that counters the centrifugal breakup of the condensate in the rotating reference frame.

A Ba^+ ion can be thought of as a Ce atom with an additional nuclear charge in computing some estimates. The energy state of the outer s electron is

$$E^Z_{1,0} \simeq \frac{2}{5} \frac{\sigma e^2}{a_0} \left(\frac{\rho(Z)}{a_0}\right)^2, \qquad (3.111)$$

where $Z = 55$ and 56 and $\sigma = 1$ and 2 for Ce and Ba^+[35]. The average radius factor is $\rho \simeq (Z-1)/2Z$. The shift in the energy of the outer s electron for an atom with Z and a singly ionized ion with $Z+1$ is then approximately for large Z

$$\Delta E^Z_{1,0} \simeq \frac{1}{5} \frac{e^2}{a_0}. \qquad (3.112)$$

As a percentage of the $E^Z_{1,0}$ this amounts to $\sim 1/Z^2$ and for the atomic and ionic species of interest this is about a $-.009 E^Z_{1,0}$ change. This indicates that the essential dipole interaction with light in the laser trap should not be significantly effected by this change.

3.7: Testing Gravitational Squeezing of the Vacuum State

In chapter 2 the phenomenology of spacetime fluctuations indicates that quantum gravitation infers a generalized uncertainty principle. This Uncertainty principle results ultimately from a unification of the well known Heisenberg uncertainty principle with the uncertainty in gravity on the amount of mass-energy a region of space contains. This uncertainty principle infers that the particle number does not have a one to one relationship with the eigenmodes defined by a diagonalizable Hamiltonian. This is due to the presence of a^2 and $(a^\dagger)^2$ terms in the dynamics. This further indicates that the vacuum state is squeezed by the self-parametric down conversion of the gravitational vacuum. In this section a brief outline is given for how this may be experimentally tested.

In the previous chapter the generalized uncertainty relationship

$$(\Delta p_j^b)^2(\Delta e_k^c)^2 \geq \frac{1}{4}\Big((1 + \langle e^{-2\pi\mathcal{H}/g}\rangle)\delta_{jk}\eta^{bc} - \frac{\pi}{g}\langle e^{\pi\mathcal{H}/g}(\alpha_j^{0b*}\alpha_k^{0b} + \alpha_j^{0b}\alpha_k^{0b*})e^{-\pi\mathcal{H}/g}\rangle\Big)^2, \quad (3.113)$$

was derived between the position and momentum measurements conducted on a particle with an acceleration g. The first experimental test that could be performed on this is to attempt to accelerate an electron to $g \geq 10^{22} cm/s^2$. The obvious method that exists today is to use high powered lasers. Such lasers, as those at the National Ignition Facility are capable to accelerate an electron to 150KeV in 10^{-10}s. This induces an acceleration of $\sim 10^{20} cm/s^2$. This corresponds to a temperature of .1K, which is currently too small for any feasible measurement. Yet the technology is beginning to reach a stage where this type of experiment could in the future be conducted.

Consider the situation where longitudinal electric fields exist in laser beams in nonlinear media. This requires the use of nonabelian electromagnetism, where the effects of nonlinear media may be modelled according to an extended electrodynamics with a $SU(2)$-like group structure or what is referred to as $O(3)$ structure [36]. This will lead to the photon as a soliton with a longitudinal field component. The nonlinear Schrödinger equation describes bunched photons. To start define the vector potentials as consisting of two parts $A^{(1)} = (A^{(2)})^*$. In addition use the definition of the auxiliary magnetic field $B^{(3)} = ie/\hbar A^{(1)} \times A^{(2)}$. Now compute Maxwell's equation, where $\mathcal{D} = \nabla + (ie/\hbar)(\mathbf{A}^{(1)} + \mathbf{A}^{(2)})$ is a covariant form of ∇

$$\mathcal{D} \times \mathbf{E}^1 = \nabla \times \mathbf{E}^{(1)} + \frac{ie}{\hbar}(\mathbf{A}^{(1)} + \mathbf{A}^{(2)}) \times \mathbf{E}^{(1)}. \quad (3.114)$$

Now compute $\mathcal{D} \times \mathcal{D} \times \mathbf{E}^{(1)}$ to find the covariant wave equation,

$$\mathcal{D} \times \mathcal{D} \times \mathbf{E}^{(1)} = \nabla^2 \mathbf{E}^{(1)} + \left(\frac{e}{\hbar}\right)^2 (|\mathbf{A}^{(1)}|^2 + |\mathbf{A}^{(2)}|^2)\mathbf{E}^{(1)}, \quad (3.115)$$

where **e** is a vector in the direction of Now use $|\mathbf{A}^{(1,2)}| = (1/\omega)|\mathbf{E}^{(1,2)}|$ to find

$$\mathcal{D} \times \mathcal{D} \times \mathbf{E}^{(1)} = \nabla^2 \mathbf{E}^{(1)} + \left(\frac{e}{\hbar\omega}\right)^2 |\mathbf{E}^{(2)}|^2 \mathbf{E}^{(1)}, \quad (3.116)$$

where $\hat{\mathbf{k}}$ is a unit parallel to the momentum. Now $\mathcal{D} \times \mathcal{D} \times \mathbf{E}^{(1)} = -(1/c^2)\mathcal{D}^2 \mathbf{E}^{(1)}/\partial t^2$ which leads to the nonlinear equation

$$\nabla^2 \mathbf{E} + \left(\frac{e}{\hbar\omega}\right)^2 |\mathbf{E}|^2 \mathbf{E} = -\epsilon\mu \frac{\mathcal{D}^2 \mathbf{E}}{\partial t^2}. \quad (3.117)$$

At this stage drop the superscript that indicates the extension into nonabelian electromagnetism. Now write the Fourier expansion for the electric field to express this according to the electric field displacement $\mathbf{D} = \epsilon \mathbf{E}$. This gives the amplitude fixed to the wavelength,

$(e/\hbar)A_0 \simeq \omega$, which is the case for solitons. This results in a cubic Schrödinger equation with a longitudinal field term:

$$\partial_x^2 \mathbf{E} + \left(\frac{e\mu}{\hbar c}\right)^2 |\mathbf{E}|^2 \mathbf{E} = -\frac{i\omega}{c^2} \frac{\partial \mathbf{E}}{\partial t}. \quad (3.118)$$

The solution to this equation is $A sech(kx)e^{i\omega t}$ which is a soliton solution.

The magnitude of the longitudinal term is $Ne\hbar\omega/2V\epsilon_0$, where N is the number of photons in the soliton. This quantity is related to the Poynting vector with units of $energy/area \times time$. If the soliton travels a distance d in the region with volume V this gives $Nde\hbar\omega/2V\epsilon_0$. For high powered lasers approximately $\sim 10^{33}$ photons are generated in a pulse. For a $10^9 cm^3$ cavity that traps the soliton, the longitudinal field magnitude will be approximately $\sim 10^{27} erg/cm^2 - s$. An electron accelerated to $v \sim !c$ has a power of $10^{27} watts$. The mass of a mole of electrons is $\sim 10^{-3}g$ and so the acceleration of 10^{-10} moles of electons would be on the order of $10^{30} cm/s^2$, with an Unruh temperature of $T \sim 10^9 K$. If the above conditions could be experimentally arranged the measurable impact of the Unruh effect may be highly significant.

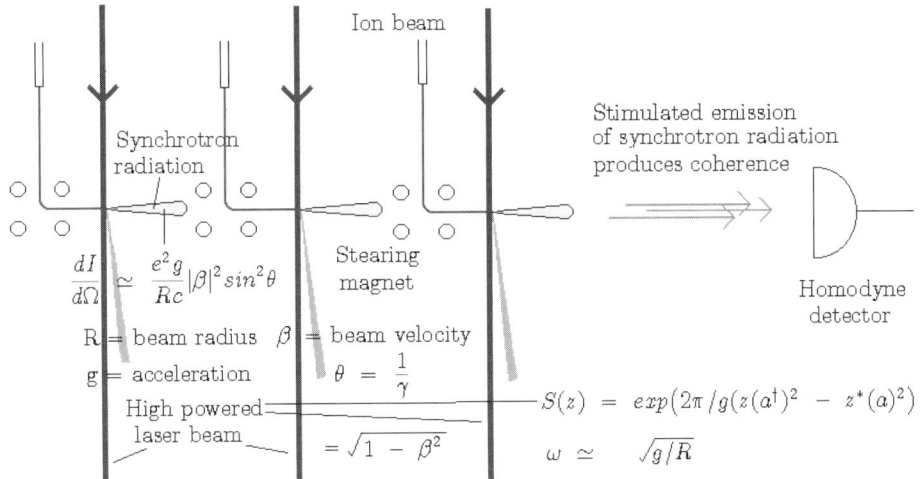

Proposed apparatus for detecting the acceleration induced squeezed state.
Figure 3.4

The ability to create nonlinear effects in intense laser field-matter interactions to create longitudinal electric fields would dramatically increase the ability to accelerate electrons. It is uncertain what are the requirements to induce this sort of behavior in a laser-matter interaction. Doubtless as this field of study is pursued that further developments will occur, where possibly a nonlinear interaction between lasers and materials will emulate the sort of nonabelian electrodynamic interaction above. If a longitudinal field component to a laser beam can be arranged then the potential to accelerate electrons with high g will be accomplished.

Figure 3.4 illustrates a possible experiment along these lines. A series of high powered laser beams accelerate the electrons from a cool stream of helium atoms. The highly accelerated electrons produce synchronized radiation with a lobe pattern

$$\frac{dI}{d\Omega} = \frac{e^2 g}{Rc}\left(\frac{v}{c}\right)^2 sin^2\theta, \ \theta = \frac{1}{\gamma} = \sqrt{1 - \frac{v^2}{c^2}}. \tag{3.119}$$

The X-rays produced in this array will induce a degree of stimulated emission. This array is analogous to the wiggler laser. The down stream X-rays produced should then have a degree of coherence as a X-ray laser. A homodyne detector could then be used to determine if there is any measure of parametric amplification of the X-ray beam. A homodyne detector for X-rays is technology that currently does not exist. Further, to "kick start" the gravitational squeezing of the vacuum state a parametric amplifier may have to be employed midstream. As yet a parametric amplifier or crystal for X-rays does not exist, but such technology might be developed in the future. In principle such an experiment could be performed if these devices exist in the future.

Such technologies envisioned in this experiment might be developed, for better or for worse, in connection with directed weapons developments. Currently with the push for exotic weaponry for space scenarios such efforts might be undertaken. The push for space weaponry will likely result in some type of arms race, where such arms races (again for better or for worse, as I will leave commentary on this aside here) do have a tendency to accelerate certain technological developments.

It is possible that this technology might be applicable in the distant future as a power source. The generalized uncertainty principle infers that there is a renormalization of L_p. If it were possible to generate highly coherent laser beams that could induce the parametric amplification of the quantum gravity vacuum so that $L_p \sim 10^{-13} cm$, then the virtual black holes associated with the virtual fluctuations may absorb the nucleus of an atom. The result will be the direct conversion of matter to energy. The complete conversion of matter to energy would amount to the ultimate energy source that could be obtained

This type of energy source might be the "final solution" to the energy problems we will be facing in the near future. Fossil fuels will deplete away as we currently are near the point where half of these resources have been consumed. Once consumption passes the half way point energy generation will peak out and begin to decline according to the Hubbert curve [37]. Currently the only economically viable alternatives the world has are wind turbines and nuclear energy. Solar photovoltaic (PV) energy currently is highly expensive, where the energy required in manufacture is a significant percentage of the energy produced. Thin film solar PV energy does hold some future promise. Without nuclear energy it is unlikely that alternative energy can provide anywhere near the energy equivalent of 80 million barrels of oil per day used today. Yet nuclear energy is fraught with considerable uncertainty over matters of radionucleide wastes and environmental and health concerns. Even with nuclear energy it is unlikely that we will enjoy the energy abundance currently available. Without nuclear energy it is doubtful that we could continue to live in a world

118 *Quantum Fluctuations of Spacetime*

at all resembling the one we live in today. The energy future of the 21st century is likely to be matrixed with various sources such as solar, wind, tidal and also nuclear.

So it is likely that the 21st century will be an age of nuclear energy, regardless of our positions on this energy source. If we exploit this energy source without significant to further ruin our world, at some point might come fusion energy, or possibly quantum gravity energy. If quantum gravity is used to annihilate matter completely then it is likely, or certainly prudent, that the material annihilated would be the radioactive wastes of the 21st century. If in the next century quantum gravity turns out to be a viable method for generating energy then humanity will have an energy source to not only bring about an economic renaissance of unprecedented proportions, but also to construct relativistic spacecraft and begin the exploration of extrasolar systems. Whether or not Homo sapiens is capable of the wisdom to harness this sort of energy is something the ethicists may ponder.

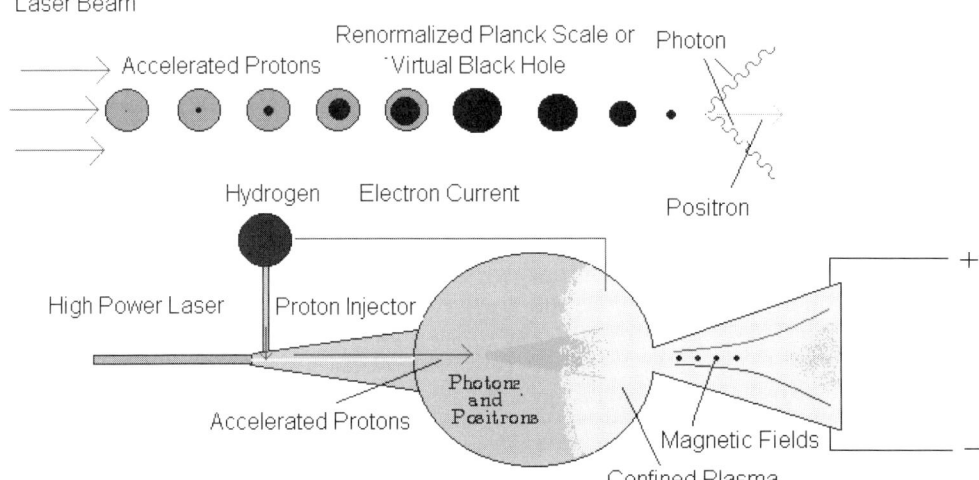

A futuristic method using the renormalization of the
Planck scale to convert mass to energy.
Figure 3.5

3.8: Cosmological Gravity Waves

The grandest physical system to observe is the universe itself. Observations of the universe at large record information about the early history of the universe. The cosmological horizon is the absolute limit of observation. In front of the horizon lies information that is piled up asymptotically towards the horizon. The observed Microwave radiation are photons released at the end of the radiation dominated period. This forms an electromagnetically opaque "wall" beyond which nothing can be observed. At later times we observe the early formation of quasars and galaxies all with red shifted atomic spectral lines that indicate their rapid motion. Observations beyond the opaque electromagnetic wall will have to employ neutrino astronomy and gravity wave detection. Neutrinos produced in the early inflationary phase of the universe may contain data on the inflationary phase and GUT symmetry breaking event in the early universe. Gravity wave interferometry [38] may give us a glimpse of the even earlier universe when gravity was as strong a force as the other gauge fields.

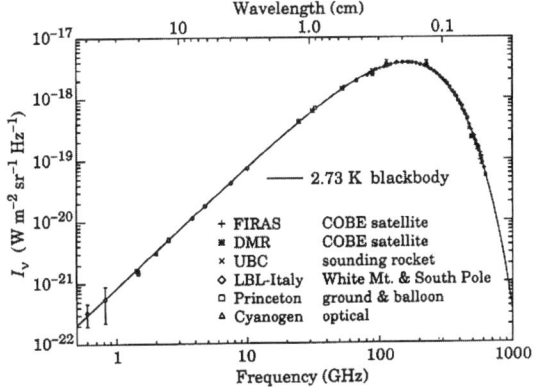

The cosmic background radiation.
Figure 3.6

There the prospect that "fossil" quantum spacetime fluctuations may be detected with gravity wave detectors. There are already indications that early quantum fluctuations existed in the early universe. The power spectrum is comparatively smooth and indicates little anisotropy in the universe. To a high degree of sensitivity it appears Gaussian. The observed microwave photons in the universe have a redshift of $z = 1000$, but at the time of the radiation dominated period were energetic enough to ionize hydrogen. These $z = 1000$ photons decoupled from matter at the end of the radiation dominated period.

Yet the microwave background exhibits fluctuations in its intensity by 3.5×10^{-5}K $\pm 1.0 \times 10^{-5}$K and may be seen in a map of the microwave sky. The appearance of small anisotropy are best understood by their appearance according to spherical harmonics. At smaller angular scales, or high numbers in spherical harmonics the fine details and features in the power spectrum also exist. These small anisotropy terms appear likely due to fluctuations in the very early universe.

The exact cause for these anisotropy terms on scales measured has several dependencies. These include acoustical effects and structure formation. The cosmic microwave background then has a number of competing affects and secondary affects due to reionization of matter at the end of the radiation dominated period. Various models are derived include parameters for the curvature, cosmological constant, matter-radiation ratio and the baryon-photon ratio. These models are run on computers and benchmarked against the measured data, such as those initially obtained by the WMAP spacecraft.

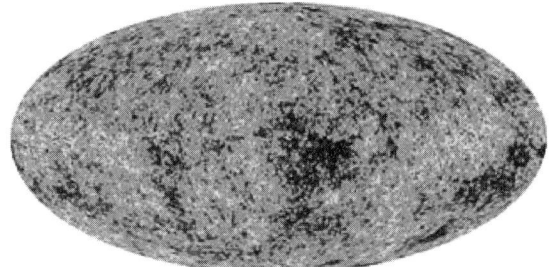

Anisotropy of the cosmic microwave
background detected by the WMAP probe.
Figure 3.7

Gravitational effects will modify the spectrum of photons emitted at the $z = 1000$ recombination period. Local variations in the spacetime metric between the recombination event at a certain region and the observer may induce redshifts or slight blueshifts in a photon. The entire spacial surface $\Sigma^{(3)}$ of the early universe may have local regions where it is contracted or expanded to lead to Doppler effects that deviate from an overall isotropic distribution of microwave photons. If there are deviations in the potential well photons scatter through as a function of θ, ϕ in spherical coordinates then there exist differential redshifts of photons that are observed as an anisotropy. Further, if there is a local gravitational potential well that is expanding, as the photon wave length is stretched along with the expanding spacial surface of evolution, the blue shift in the photon that fell into it will not be completely cancelled out by the redshift induced by the photon's exit from this region.

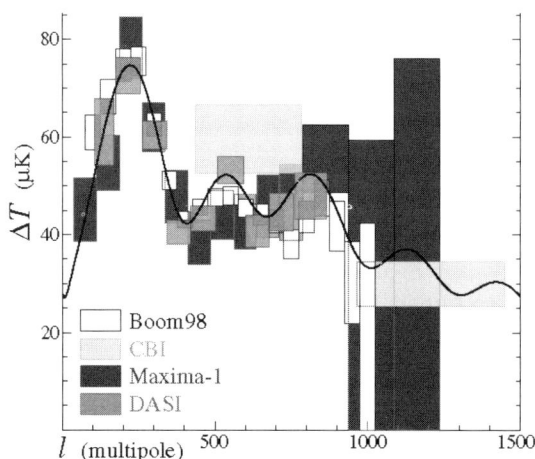

A multipole expansion of the anisotropy in the Microwave background of the universe.
Figure 3.8

For local regions with sufficiently large gravitational potentials a competing affect ensues, where photons that exert a pressure against the gravitational potential heat up baroyonic matter. This induces oscillations in this matter with regions of high and low pressures and temperatures. These oscillations are acoustical in nature. These regions of high and low temperatures associated with standing waves of acoustical energy will further induce a $\Delta T/T$ variation in the cosmic microwave background. This implies that the universe was not completely matter dominated during deionization. Once the potential region expands beyond the wave length of these acoustical oscillations the density fluctuations are frozen in.

There is considerable physical theory beyond this discussion, but the essential feature here is the existence of gravitational fluctuations. These gravitational fluctuations are the driving process behind the anisotropy of the cosmic background radiation. These fluctuations have to be seen as separate from gravitational lensing from large elliptical galaxies. These introduce anisotropy terms at various scales, which have the effect of smearing out the primary anisotropy. These contributions must be subtracted out by their association with optical effects of gravitational lensing. The perturbations are likely due to fluctuations associated with the inflationary epoch in the early universe, or are due to even earlier remnants of the quantum fluctuations in spacetime that are frozen in as the universe expands. Further, as indicated above, since the Planck energy is renormalized to $\sim 10^{15}$GeV these fluctuations could be the result of both. Possibly fluctuations associated with inflation are equivalent to quantum fluctuations. These quantum gravity fluctuations would be frozen into the classical background as stochastic gravity waves. For an overview of the detector physics of stochastic gravity waves L. S. Finn and A. Lazzarini provide an overview in reference [33].

The Roberston-Walker cosmology considers a spacial 3-sphere that evolves through space. It is the simplest of possible models that assumes the universe is isotropic and homogeneous. The metric for this universe is written as

$$ds^2 = dt^2 - R^2(t)\left(\frac{dr^2}{1 - kr^2} + r^2 d\Omega^2\right). \tag{3.120}$$

The term k in this metric is a constant that determines the spacial curvature of the cosmology. For $k = 1$ the cosmology is a closed spherical universe, for $k = 0$ the cosmology is flat, and for $k = -1$ the cosmology is open. The Einstein field equations give a constraint equation and a dynamical equation for the rate the radius changes with time. If the velocity is defined as $\mathbf{v} = (\dot{R}/R)H(t)\mathbf{r}$, where $H(t)$ is the Hubble parameter, a constant locally, the constraint equations is,

$$\frac{\dot{R}}{R} = \frac{8\pi}{3}G\rho - \frac{k}{R^2} \tag{3.121a}$$

and the dynamical equation is

$$\frac{\ddot{R}}{R} = -4\pi G\left(p + \frac{\rho}{3}\right). \tag{3.121b}$$

Apparently the evolution of this cosmology is then reduced to a problem common in classical mechanics. An integration of the dynamical equation with the constraint equation, a statement of energy conservation within a sphere of radius R, results in solutions for cosmological evolution.

For simplicity consider the universe to be radiation dominated. This is because the particles, even earlier with highly massive X, Y and Z GUT particles, have such large kinetic energies that they behave similar to photons or massless bosons. This situation in the early universe existed until the universe dropped to a temperature below $10^3 K$ 400,000 years into its evolution. For the radiation dominated period in the evolution of the universe the pressure and density were related by

$$p = \frac{1}{3}\rho. \qquad (3.122)$$

The density is related to the total number of bosonic and fermionic helicities,

$$\rho = 1/2 g(T)\rho_{bb} = 3TN(T), \quad g(T) = H_{boson} + \frac{7}{8} H_{fermion}, \qquad (3.123)$$

where H are the numbers of helicity states, T is the temperature and $N(T)$ is the number density of particles. For a partition function for the system of particles in the early universe there is a pressure determined by

$$p = -\left(\frac{\partial A}{\partial V}\right)_{N,T} = Z^{-1}kT\left(\frac{\partial}{\partial V}\ln Z\right)_{N,T}. \qquad (3.124)$$

For the partition function in a standard Boltzmann form $Z = \sum_n exp(-\beta E_n)$, $kT = \beta^{-1}$, the pressure is then

$$p = -Z^{-1}\sum_n \frac{\partial E_n}{\partial V} exp(-\beta E_n). \qquad (3.125)$$

The pressure density enters into the constraint equation 3.121a, which gives the conservation of energy density in a sphere

$$\frac{d}{dt}\left(\frac{4\pi}{3}\rho R^3\right) = -p\frac{d}{dt}\left(\frac{4\pi}{3}R^3\right). \qquad (3.126)$$

During the inflationary period of expansion the radius of the cosmology expands exponentially:
$$R(t) = R_0 exp(\gamma t), \qquad (3.127)$$

where

$$\gamma = \sqrt{\frac{8\pi\rho_0}{3M}} \simeq \frac{T_c^2}{M}. \qquad (3.128)$$

Here M is the mass of the particles in the universe. The universe then exhibits a scale change $\sigma = R(t + \Delta t)/R(t)$, where Δt is the duration of the inflationary period. The parameter for the geometry of the spacial universe is $\Omega = \rho/\rho_c$, for $\rho_c = (\dot{R}/R)3/8\pi G$. The condition for flatness $\Omega = 1$ in the early universe requires that $\sigma > 10^{27}$. The expansion parameter will reduce Ω by a factor of σ to thus guarantee cosmic flatness, or close to flatness.

At this stage in astronomy the value of ω based on luminous matter is $\Omega = .05$. There is dark matter and dark energy in the universe which changes this parameter to $\Omega \simeq .05$. The nature of dark matter and dark energy is an unresolved topic of research. Also the masses of elementary particles in the early universe may have played a role. The early universe near the GUT scale probably was more dominated by massive particles such as the flavor and quark-lepton transforming bosons, where this mass is now in the form of photons and neutrinos. The contribution by neutrinos to dark matter is another unresolved matter, but data from the Super-Kamiokande indicates nonzero neutrino masses. This means that with more complete observational understanding of the early the universe better bounds on the value of σ may be imposed with better empirical values for Ω. Further measurements of neutrino masses will also provide a benchmark for the density of matter in the universe. This will provide a constraint on the types of theories that can be proposed for the grand unification of gauge theories.

Consider the situation where the radius of the universe exhibits a variation

$$R(t) = R_0(t) + \delta R(\theta, \phi, t), \tag{3.129}$$

where variations in the radius are expanded into spherical harmonics

$$\delta R(\theta, \phi) = \sum_{l=0}^{\infty} \sum_{m=-l}^{l} A_{lm}(\kappa R)^l Y_{lm}(\theta \phi). \tag{3.130}$$

Here κ is a scale factor for the fluctuations with $1/\kappa \simeq L_p = 1.6 \times 10^{-33}$cm. These fluctuations will evolve according to

$$\delta \ddot{R} = 4\pi G \left(p + \frac{\rho}{3}\right) \kappa \sum_{l=0}^{\infty} \sum_{m=-l}^{l} (l+1) A_{lm}(\kappa R)^{-l-2} Y_{lm}(\theta \phi) + O(G^2)$$

$$= 4\pi G \left(p + \frac{\rho}{3}\right) \frac{\partial \delta R}{\partial R}. \tag{3.131}$$

Under inflationary expansion equation 3.127 illustrates that the influence of this term will become highly suppressed. Equation 3.131 gives the curvature terms that are relevant for the gravitational potentials that will induce the anisotropy in the cosmic background radiation.

The metric g_{ij} of $\sigma^{(3)}$ and its conjugate momentum metric π^{ij} exhibit an uncertainty of the form

$$\pi^{ij} - \langle \pi^{ij} \rangle = \frac{i\hbar \mathcal{F}(|z|)}{2(\Delta g)^2}(g^{ij} - \langle g^{ij} \rangle), \tag{3.132}$$

where $\mathcal{F}(|z|)$ is a function of the squeezing parameter. This gives the normalized wave functional

$$\Psi[g] = (2\pi(\Delta g)^2)^{-1/4} exp\left(i\frac{\langle\pi^{ij}\rangle g_{ij}}{\hbar} - \frac{(g_{ij} - \langle g_{ij}\rangle)^2 \mathcal{F}(|z|)}{(\Delta g)^2}\right). \qquad (3.133)$$

In general $\mathcal{F}(|z|) \simeq 1/2cosh(|z|)$. With equations 2.171 identify the raising and lowering operators a, $a^\dagger$ and b, $b^\dagger$ by the unitary operators

$$U = exp(-if(t)(c^\dagger c + d^\dagger d)),$$

$$S(|s|) = exp\left(\sum_{l=0}^{\infty}\sum_{l=-l}^{l} A_{lm}\omega_l(zY_{lm}(\theta, \phi)cd - z^*Y_{lm}(\theta, \phi)c^\dagger d^\dagger)\right), \qquad (3.134)$$

where the generator of the squeeze state operator is expanded according to spherical harmonics. Further the spacial indices are implied. This leads to

$$e_i = \int d\omega S(|z|)^\dagger (c_i \mathcal{F} + c_i^\dagger \mathcal{F}^*) S(|z|), \qquad (3.135)$$

so that for $g_{ij} = e_i e_j$

$$\overline{\langle |g_{ij}|\rangle} \simeq R\delta_{ij} e^{|z|} \sum_{l=0}^{\infty}\sum_{m=-l}^{l} A_{lm} Y_{lm}(\theta, \phi), \qquad (3.136)$$

where R is the radius of the spacial surface. This illustrates how the generator of the unitary squeezed state operator is decomposed into elements for the angular eigenstates of the wave function for the cosmology.

This argument illustrates that the wave function of the universe may be decomposed into eigenstates expanded into a complete set of functions, Legendre polynomials or Bessel functions or etc. This then opens the door to the question of how curvature in the universe is frozen into some set of spherical harmonics. It appears that as the radius of the universe becomes much larger than the Planck scale the cosmology has an increasingly classical spacetime structure. So the data on the inflationary bubble that tunnels out of the vacuum state is rapidly stretched in the classical spacetime as the whole spacetime expands. This indicates that quantum superpositions of the manifold are coupled to the vacuum and disentangled from the spacetime manifold. As such by the statistics in this process some residual spacetime curvature due to the quantum state of the nascent cosmology is frozen into the classical manifold. This physical data is hugely stretched further by the inflationary situation, which contributes to a small anisotropy to the observable universe.

It is possible a LIGO may detect quantum gravity fluctuations from the big bang. These curvature fluctuations would have to have a radius of curvature approximately $\sim$ 100km or less. With squeezed state operators and the entropy production in the early universe a quantum system may behave classically for long times. This suggests that quantum fluctuations in the spacetime manifold may have persisted and been frozen out for some time considerably longer than 10^{-43}s to 10^{-38}s, or those renormalized to 10^{-40}s to 10^{-35}s, where Planck scale fluctuations were predominant. Quantum gravity fluctuations should have been coherently coupled to the nascent universe during the inflationary period, as marked by the entropy production in the early universe. Planck scale fluctuations may have been frozen out as gravity waves currently a scale of $\sim 1 - 100$m by quantum fluctuations when the observable universe was 10^{-14}cm to 10^{-16}cm in radius. The renormalization of the Planck scale will push these gravity waves to a wave length of $\sim 10^3 - 10^5$ meters. The gravity waves should form a spectrum analogous to the microwave electromagnetic spectra from the end of the radiation dominated period of the universe. The curvature associated with these gravity waves should be on the order of $10^{-64}cm^2$, or approximately 10^{-36} that of Earth's gravity. Assume that the interferometer is on a plane with the **z** normal vector, the vibration detected would then be

$$\frac{d^2\mathbf{x}}{dt^2} = R_{x0x0}\mathbf{x} - R_{x0y0}\mathbf{y} \sim 10^{-64}cm^2(\mathbf{x}+\mathbf{y})$$

$$\frac{d^2\mathbf{y}}{dt^2} = R_{y0y0}\mathbf{y} - R_{y0x0}\mathbf{x} \sim 10^{-64}cm^2(\mathbf{x}-\mathbf{y}) \quad (3.137)$$

The motion of the detector will be approximately 10^{-22} the length of an interferometer arm. This is close to the performance of the LIGO machines.

There are difficulties due to how the "initial conditions" of the wave function of the universe are not determined by the no boundary condition. By random fluctuations there may indeed be spherical harmonics to the wave function of the universe, but determining them is difficult to ascertain, where they are probably completely random. So it is difficult to predict the spectra of a gravitational cosmological background radiation, yet a physical argument would suggest they are of a blackbody radiation form. Further, these measurable fluctuations left over from the early universe might be incommensurate than the size of the interferometer. As such detecting them will be difficult with current LIGO designs. Longer wave length gravity waves should exist, but the motion of the interferometer arms will be less than the minimum of the detectability limits of the LIGO designs.

3.9 The Universe as a $\sum(all) = 0$

The universe proposed to be a $\sum all = 0$ is discussed within the context of general relativity. Such a cosmology fits within data on large redshifted supernovae (SN1), the CMB data from the WMAP space probe and recent Hubble SN1 data [39][40]. This frugal property means this type of cosmology can spontaneously tunnel from the vacuum state and is consistent with the holographic principle of quantum gravity.

How did the universe emerge from nothing? If some deity capable of "anything," such as violating physical laws, is not invoked the universe then emerged by processes that did not violate conservation of basic physical observables. This would suggest that the entire universe is a $\sum all = 0$, where nothing was created in the big bang.

It is not possible to remove one's self from the universe and weigh it. Thus any notion of a net mass-energy to universe is meaningless, which suggests that the universe has no mass-energy. Weight is a force opposing geodesic motion in spacetime, something intrinsic to the universe. Given a particle of mass m and energy equivalent mc^2 its gravitational interaction with the rest of the matter in the universe is a negative quantity. If the total universe has zero net mass-energy content the total mass-energy of all fields is countered by its gravitational interaction with itself. An understanding on how the universe in its current state is ultimately a "zero" would aid greatly in our understanding of how it spontaneously emerged from the vacuum state.

The idea of a universe with zero net mass-energy is not new. A. Haas introduced the concept in 1936 [41] and expanded upon by E. P. Tyron in 1973 [42]. With the second reference the idea is advanced that the universe is the result of a quantum fluctuation of the vacuum. This means anything identified as any object with mass-energy is simply local deviation from a grand total that is zero. This removes the problem of any net creation process with the occurrence of the universe. This economy is quite compelling, where current data with $\Omega_{tot} \simeq 1$ indicates that this will emerge as one crucial aspect of cosmology.

Consider a particle with mass-energy mc^2 that interacts with the mass-energy of the universe with mass M. The Newtonian potential energy of this interaction is $V = -GMm/d$, where d is the distance between this mass and these other gravitational sources. The mass M of the universal gravitating sources is estimated from the critical density of the universe $\rho_{cr} = 3H/(8\pi G)$, where H is the Hubble constant. Since $\Omega = \rho_M/\rho_{cr}$ then $M = \frac{4}{3}\pi d^3 \rho_M = \Omega H^2 d^3/2G$. The gravitational potential on the mass m is

$$V = -\frac{\Omega}{2}\left(\frac{Hd}{c}\right)^2 mc^2. \qquad (3.138)$$

The distance to these sources is estimated as $c/H < d < 2c/H$ then $V \simeq -\Omega mc^2$. The current estimated value of Ω, with dark matter and energy is included, is $\Omega \simeq 1$. This suggests that the total energy is $mc^2 + V \simeq 0$. Thus from an elementary Newtonian argument it appears that the zero universe conjecture is reasonable. An extended theory along these lines is presented by J. Overduin and H. Fahr [43].

The $\sum all = 0$ cosmology is extended to the Robertson-Walker metric for a spherical cosmology written as

$$ds^2 = c^2 dt^2 - R^2(t)\left(\frac{d\rho}{1 - k\rho^2} + r^2 d\Omega^2\right), \qquad (3.139)$$

where $d\Omega^2 = d\theta^2 + sin^2\theta d\phi^2$ and ρ is a radial comoving coordinate $\rho = r/(1 + \frac{1}{4}kr^2)$. The term $R(t)$ represents the radial expansion of the spherical model cosmology. This metric is used in the Einstein field equation $G_{\mu\nu} = 8\pi T_{\mu\nu}$ with $G_{\mu\nu} = R_{\mu\nu} - \frac{1}{2}Rg_{\mu\nu} + \Lambda g_{\mu\nu}$, where Λ is the cosmological constant. The spacetime curvature information, or Einstein tensor, in the field equation is for this metric

$$G_{ii}/g_{ii} = -\frac{2\ddot{R}}{R} - \frac{\dot{R}^2}{R^2} - \frac{k}{R^2} + \Lambda, \; i = 1, 2, 3$$

$$G_{00}/g_{00} = -\frac{3\dot{R}^2}{R^2} - \frac{3k}{R^2} + \Lambda. \tag{3.140}$$

For the moment the constants G and c are set to unity. Due to the diagonal structure of the metric $G_{\mu\nu} = 0$ for $\mu \neq \nu$. The mass-energy source for the spacetime curvature defines a momentum-energy for "dust" $T_{\mu\nu} = \rho U_\mu U_\nu$,

$$T_{00} = \rho, \; T_{ii} = p_i, \; T_{\mu\nu} = 0 \; otherwise. \tag{3.141}$$

The cosmological constant has emerged as a subject of interest since the observations of type I-A supernovae [39] and microwave anisotropy data was received by the WMAP spaceprobe [40][44]. These datum indicate that the universe is not only expanding, but the expansion is accelerating. This means that Λ is dominated by a "cosmic force" that is repulsively pushing galaxies outward at greater velocities with time. This indicates that there is a cosmological Casimir type of effect such that the zero point of the quantum vacuum results in a negative pressure that induces a runaway expansion of the universe. There is an Ω_Λ associated with a cosmological term

$$\Lambda_{eff} = \frac{3H^2 \Omega_\Lambda}{c^2} \tag{3.142}$$

involved with the zero point energy of the vacuum with $\rho_{vac} = \Lambda_{bare}c^2/(8\pi G) + \rho_{vev}$. Here units are restored and H is the Hubble parameter. This gives $\Omega_\Lambda = \rho_{vac}/\rho_{cr}$. In general the momentum-energy tensor with this pressure and energy density is going to be $diag(T_{\mu\nu}) = (\rho c^2, p_i)$. The universe does not spontaneously create mass-energy within it. Thus it obeys the first law of thermodynamics $\Delta E = -\Delta W + \Delta Q$, where E, W and Q are the internal energy, the work on the system and the heat absorbed by the system. A closed cosmology demands that $\Delta E = 0$. The negative pressure of the universe is one that absorbs work, as opposed to a positive pressure that does work. This demands that for $\Delta W = -p$ and $\Delta Q = \rho_{vac}c^2$ that $p + \rho_{vac}c^2 = 0$ or $p = p_M + p_{vac} = -\rho_{vac}c^2$, where p_M may be ignored in a universe filled with "dust."

The equation of state defined by $w = p/\rho$ is $w = -1$ for the $\sum all = 0$ cosmology. The evolution of a cosmology according to the equation of state is

$$\left(\frac{\dot{R}}{R}\right)^2 = H_0^2 \left(\frac{\Omega_m}{R^3} + (1 - \Omega_m)R^{-3(1+w)}\right), \tag{3.143}$$

For $w > -1/3$ the cosmology recollapses and for $w \in [-1/3, -1]$ it expands according to a DeSitter metric. For $w < -1$ the dominant energy condition is violated, where the potential acts to pump negative energy into the system so the universe exponentially expands with a complete vengeance towards the "big rip" [45]. It is similar to a run away harmonic oscillator with an undamped driving force. For $w = -1$ the averaged weak energy condition is not violated and the cosmology satisfies $\sum all = 0$. Further, this cosmology favors "dark energy" as an aspect of spacetime with a cosmological constant as opposed to quintessense models [44]. Recent data from the Hubble space telescope analyzed by Reiss et al. indicates a cosmology with the equation of state $w = -1.02$, $\pm^{12}_{19}$ [45]. for $z = .46 \pm .13$ the transition between decelerated and accelerated expansion these datum indicate that $dw/dz = 0$. [45]

These results give the cosmological constant that may be inserted into equations 3.140,

$$\Lambda = \frac{8\pi G}{c^2}(\rho_{vac} + 3p_{vac}/c^2). \qquad (3.144)$$

The mass-energy of the universe is then a constant with a constraint written as

$$\frac{\dot{R}^2(t)}{R^2} = \frac{8\pi G}{3}(\rho + 3p) - \frac{k}{R^2}, \qquad (3.145a)$$

which gives the dynamical equation

$$\frac{\ddot{R}}{R} = \frac{4}{3}\pi G(\rho + 3p). \qquad (3.145b)$$

Consider the source of this acceleration as a potential function $\Phi(R)$ so that $\ddot{R} = -\partial\Phi/\partial R$. This potential is of the form

$$\Phi(R) = -\frac{2}{3}\pi G(\rho + 3p)R^2$$

$$= -\frac{H^2}{2}(\Omega_M + 2\Omega_\Lambda)R^2. \qquad (3.146)$$

It is interesting to note that this is similar to the potential derived for Newtonian models of the zero mass cosmology which is quadratic in the radial distance. In the Newtonian case the R is replaced by a radial distance to gravitational sources with $\nabla^2\phi = -4\pi G(\rho + 3p)$ [43]. This potential gives the gravitational binding of mass-energy in the universe. In a relativistic approach the gravitational binding energy of the universe is given by $\int_0^{d_c} dM\Phi$, where d_c is the distance to the cosmological event horizon.

$$V = \int dM\Phi = -\frac{3\Omega_M}{10}\left(\frac{Hd_c}{c}\right)^2\left(1 - \frac{2\Omega_\Lambda}{\Omega_M}\right)\int \rho_M\sqrt{-g}d^3r. \qquad (3.147a)$$

The kinetic analogue is the Tolman integral for the total mass energy,

$$E = \int (\rho c^2 + 3p)\sqrt{-g}d^3r = \left(1 - \frac{2\rho_{vac}}{\rho_M}\right)\int \rho_M\sqrt{-g}d^3r. \qquad (3.147b)$$

The total energy is then $E + V = 0$ so that

$$\left(1 - \frac{2\Omega_\Lambda}{\Omega_M}\right)\left(1 - \frac{3\Omega_M}{10}\left(\frac{Hd_c}{c}\right)^2\right) = 0. \qquad (3.148)$$

The unphysical solution is where $E = V = 0$ with $\Omega_M = 2\Omega_\Lambda$ and the other more realistic solution is with $\Omega_M = 10c^2/(3(Hd_c)^2)$.

This solution with $E + V = 0$ may be justified on more physical grounds. The surface gravity associated with the cosmological horizon κ_H is computed from the Killing vector $\xi = \frac{\partial}{\partial t}$ as $\xi_{\mu;\nu}\xi^\nu = \kappa_H \xi_\mu$. The Killing vector has the magnitude $\simeq \sqrt{\Lambda/3}\,r$ which gives a surface gravity $\kappa \simeq \sqrt{\Lambda/3}$ [47]. Since this result is computed from the Killing vector the isometries of spacetime define a frame dragging of galaxies in the universe. Physically this is similar to the Lense-Thirring effect or frame dragging of a mass orbiting a rotating black hole. This frame dragging also suggests the Mach principle, for here the inertia of galaxies is determined by the total gravitational field it interacts with.

The Hubble parameter is $H = 50 km/s/Mpc$ and the area of the cosmological event horizon is $12\pi\Lambda^{-1}$. The zero cosmology universe involves a ratio of the Hubble distance c/H and the radial parameter defined by $d_c = \sqrt{3/\Lambda}$ with Ω_M and Ω_Λ related by

$$\Omega_M = \frac{10}{9}\Omega_\Lambda. \tag{3.149}$$

Further the causal distance may be used to replace the distance to the cosmological horizon as $d_c \simeq 6/\sqrt{\Lambda}$ and so

$$\Omega_M = \frac{5}{9}\Omega_\Lambda. \tag{3.150}$$

This combined with $\Omega_M + \Omega_\Lambda = 1$ gives the result that $\Omega_\Lambda = 9/14 = .643$ and $\Omega_M = .357$. The observational range for these parameters is $(\Omega_M, \Omega_\Lambda) = (.27, .73)$ [44]. The Hubble data [48] indicates that $\Omega_M = .29^{+.05}_{-.03}$. The discrepancy with these theoretical values probably comes from the large characteristic distance for vacuum dominated cosmologies. For $d_c \simeq 12/\sqrt{\Lambda}$ then $\Omega_\Lambda = .78$ and $\Omega_M = .22$. These estimated results are within the bounds set by the SNI, WMAP and Hubble datum for supernovae and the cosmic background data.

The universe may have emerged from some type of quantum fluctuation in the vacuum state. The quantum superspace operators presumably exhibited fluctuations that resulted in a virtual occurrence of the spacial metric q_{ij} that exhibited inflationary pressures. Such a fluctuation is "frozen" into the runaway spacetime that exponentially expands. The existence of the Ω_Λ most likely reflects the vacuum physics and fluctuations that gave rise to the universe.

ADM relativity treats spacetime as foliated by spacial surfaces that are "pushed" forward by lapse functions. The spacetime manifold $\mathcal{M}$ is decomposed into $\Sigma^{(3)} \times \mathbf{R}$, where $\Sigma^{(3)}$ is a spacelike region and $\mathbf{R}$ is a real interval. On $\mathcal{M}$ we have the metric $g_{\mu\nu}$ that is covariantly constant with respect to the operator ∇_μ. The spacetime $\mathcal{M}$ is foliated into spacelike slices $\Sigma_t^{(3)}$, where the parameter is defined along $\mathbf{R}$ and defines the vector field $t_\mu = \partial_\mu t$. Thus for some $\Sigma_t^{(3)}$ there exists the future directed normal vector n_μ, with the Gauss fundamental forms

$$h_{\mu\nu} = g_{\mu\nu} + n_\mu n_\nu, \quad K_{\mu\nu} = -\frac{1}{2}\mathcal{L}_n h_{\mu\nu}. \tag{3.151}$$

130 Quantum Fluctuations of Spacetime

Given a projector operator $P_i{}^\mu$ its action on $h_{\mu\nu}$ is to define $h_{i\nu}$, which projects $h_{\mu\nu}$ from $\mathcal{M}$ to $\Sigma_t^{(3)}$ according to $P_i{}^\mu P_j{}^\nu h_{\mu\nu} = h_{ij}$. This defines the Gauss fundamental forms on $\Sigma_t^{(3)}$. The lapse and shift functions for the foliation are $N = -t^\mu n_\mu$ and $N^i = h^i{}_\mu t^\mu$ with $t^\mu = Nn^\mu + N^\mu$.

The momentum conjugate to the metric variable is determined by the extrinsic curvature tensor K_{ij} as $\pi_{ij} = g^{1/2}(K_{ij} - \frac{1}{2}Tr(K)g_{ij})$. This thus determines the Hamiltonian constraint

$$H = -\left(G_{ijkl}\pi_{ij}\pi_{kl} + g^{1/2}(R^{(3)} - 2\Lambda)\right), \quad G_{ijkl} = \frac{1}{2}g^{-1/2}(g_{ik}g_{jl} + g_{il}g_{jk} - g_{ij}g_{kl}), \tag{3.152}$$

which gives the dynamics by $H = 0$. Within the canonical quantization the momentum becomes the functional operator $\hat{\pi}^{ij} = -i\delta/\delta g_{ij}$ and the Hamiltonian constraint acts on a cosmological wave function so that $\mathcal{H}\Psi(g) = 0$ [49].

The minimal uncertainty fluctuations in the metric and momentum-metric observables obey $\Delta\pi^{ij}\Delta g_{kl} = (\hbar/2)\delta^i{}_k\delta^j{}_l$, and the wave function obeys the following equation [17][50]

$$(\hat{\pi}_{ij} - \langle\hat{\pi}_{ij}\rangle)\Psi(g) = \frac{i\hbar}{2(\Delta g)^2}(\hat{g}_{ij} - \langle\hat{g}_{ij}\rangle)\Psi(g). \tag{3.153}$$

The fluctuation in the momentum is $\delta\hat{\pi}_{ij} = \hat{\pi}_{ij} - \langle\hat{\pi}_{ij}\rangle$. The metric fluctuations are similarly defined. The wave function is then

$$\Psi(g) = \frac{1}{2\pi(\Delta g)^2}exp\left(-\frac{(g_{ij} - \langle g_{ij}\rangle)^2}{4(\Delta g)^2} - \frac{\langle\pi^{ij}g_{ij}\rangle}{\hbar}\right). \tag{3.154}$$

This wave function is defined for pure fluctuations of the metric and momentum metric observables.

A tensor field associated with this wave function might be written according to operators of the form $\hat{\pi}_{ij} = \sigma^a{}_i A^a{}_j$, where the index a is associated with an internal $SO(4)$ symmetry of spacetime tetrads [51]. The connection term $A^a{}_j$ is expanded according to raising and lowering operators so that a tensor field is

$$\phi(g)_{ij} = \frac{1}{2\pi(\Delta g)^2}exp\left(-\frac{(g_{ij} - \langle g_{ij}\rangle)^2}{4(\Delta g)^2}\right)\sigma^a{}_i\sum_\omega\left(a^a{}_j e^{-i\langle\pi^{ij}\rangle g_{ij}} + a^a{}_j{}^\dagger e^{i\langle\pi^{ij}\rangle g_{ij}}\right). \tag{3.155}$$

This field then acts upon a Fock space of states and determines the spinor valued tensor fields with Sen type of operators. The application of the operator $\hat{\pi}_{ij} - \langle\hat{\pi}_{ij}\rangle$ on equation 3.153 results in

$$(\hat{\pi}^{ij} - \langle\hat{\pi}^{ij}\rangle)(\hat{\pi}_{ij} - \langle\hat{\pi}_{ij}\rangle)\Psi(g) = \frac{i\hbar}{2(\Delta g)^2}(\hat{\pi}^{ij} - \langle\hat{\pi}^{ij}\rangle)(\hat{g}_{ij} - \langle\hat{g}_{ij}\rangle)\Psi(g). \tag{3.156}$$

The Gaussian envelope for the fluctuation is determined by $\delta\pi^{ij}\delta g_{ij}$ and has a mixed representation in the conjugate observables. This means that

$$\hat{g}_{ij} - \langle\hat{g}_{ij}\rangle = \frac{\sigma^a{}_i}{\sqrt{2\omega}}(\sqrt{1 - \beta}a^a{}_j + \sqrt{1 - \beta^*}a^a{}^\dagger{}_j) \tag{3.157}$$

and the Gaussian fluctuations will have the form [17][50],

$$\frac{(\hat{g}_{ij} - \langle \hat{g}_{ij}\rangle)^2}{4(\Delta g)^2} = \sum_\omega \Big(\frac{g^2 - \pi^2/\omega}{8(\Delta g)^2} +$$

$$\frac{\sqrt{(1+\beta)(1+\beta^*)}}{8(\omega\Delta g)^2}(\pi^2 + \omega^2 g^2) + \frac{1}{2}\big(|z(\omega)|e^{i\theta}(a^\dagger)^2 - |z(\omega)|^2 e^{-i\theta}a^2\big)\Big). \quad (3.158)$$

These fluctuations exhibit a squeezing of the vacuum state. Consider the spacial metric and its conjugate as spinor variables $g_{ij} = \sigma_i^a x_j^a$, $\pi^{ij} = \sigma^{ai}p^{aj}$. The electric field and magnetic field associated with a photon in a volume V are $E_0 = \sqrt{\hbar\omega/2V\epsilon_0}$ and $B_0 = \sqrt{\hbar V\epsilon_0/2\omega}$, with the uncertainty $\Delta E \Delta B = E_0 B_0 = \hbar/2$. By correspondence $p_0^i = \sqrt{\hbar\omega/2\epsilon_0}\hat{p}^i$ and $x_{i0} = \sqrt{\hbar\epsilon_0/2\omega}\hat{e}_i$, where ϵ_0 is a gravitational permittivity. The boosts of test particle momenta towards an event horizon induce $\omega \to \infty$ as the length constricts $L = V^{1/3} \to 0$. Metric fluctuation then decreases $\Delta x_{i0} \to 0$, in general agreement with $\delta L = (L_p^2 \times L)^{1/3}$. This is a squeezing of conjugate variables that conversely sends $\Delta p_0^i \to \infty$.

Consider the deSitter universe with the metric

$$ds^2 = -(1 - \Lambda r^2/3)dt^2 + (1 - \Lambda r^2/3)^{-1}dr^2 + d\Omega^2, \quad (3.159)$$

where there is a coordinate singularity at the cosmological horizon. This difficulty may be removed by switching to Kruskal coordinates. The delayed coordinate is then $r^* = \sqrt{3/\Lambda}\tanh^{-1}(\sqrt{\Lambda/3}r)$ and the coordinates $U = t - r^*$ and $V = t + r^*$ are used to define

$$u = -e^{-U\sqrt{\Lambda/3}} = -e^{-t\sqrt{\Lambda/3}}\sqrt{\frac{1+\sqrt{\Lambda/3}r}{1-\sqrt{\Lambda/3}r}},$$

$$v = e^{V\sqrt{\Lambda/3}} = e^{t\sqrt{\Lambda/3}}\sqrt{\frac{1+\sqrt{\Lambda/3}r}{1-\sqrt{\Lambda/3}r}}. \quad (3.160a)$$

There are further the coordinate conditions that

$$uv = -\frac{1+\sqrt{\Lambda/3}r}{1-\sqrt{\Lambda/3}r}, \quad u/v = -e^{-2t\sqrt{\Lambda/3}}. \quad (3.160b)$$

With each increment in time the momentum of a particle is boosted by $p \to p + \frac{1}{2}\delta t\sqrt{\Lambda/3}$ and similarly the momentum conjugate variables π_{ij} similarly increment. The motion of a particle is then carried along by the rapid separation of points in a spacial surface. For large boosting at $t \to (\Lambda/3)^{-1/2}$, $uv \to -\infty$ and $u/v \to -e^{-2}$. Both the u and v coordinates diverge as $r \to \sqrt{3/\Lambda}$. The ingoing and outgoing modes then intersect with a divergent γ, which is what is expected of fields on the cosmological event horizons, which is the boundary of observation up to the big bang.

As a particle according to an observer approaches within a Planck length of the cosmological event horizon at $r = \sqrt{3/\Lambda}$ then

$$uv \simeq \frac{2 - \sqrt{\Lambda/3}L_p}{\sqrt{\Lambda/3}L_p}, \quad u/v \simeq e^{-2}(1 + 2\sqrt{\Lambda/3}L_p). \tag{3.161}$$

The squeezing parameter is constrained to be $|z| < (\Lambda L_p^2/3)^{-1}$, which leads to

$$|z| \sim \frac{3c^5}{8\pi G^2 \rho_v} \tag{3.162}$$

The form of the wave function in equation 3.154 has an envelope with a representation in spacial and temporal coordinates. This may be extended to the u, v coordinates. Thus a probability density has the form

$$P(u, v) = (2\pi(\Delta x)^2)^{-1} exp(1/2(\alpha u^2 + \beta p_u^2 + 2\gamma u p_u)) \tag{3.163}$$

where $\gamma u p_u$ involves the squeezing parameters. The entropy of the wave function is [52]

$$S = \frac{1}{2} k_B ln(\langle(\Delta u)^2\rangle\langle(\Delta p_u)^2\rangle - \langle\Delta u \Delta p_u\rangle), \tag{3.164}$$

where the uncertainty spreads are evaluated with $S(z)aS^\dagger(z)$ and $S(z)a^\dagger S^\dagger(z)$. This gives the entropy

$$S = \frac{k_B}{2} ln\left(\frac{1}{4}(cosh^2(2|z|) - sinh^2(2|z|)[cos^2(2\theta) + sin^2(2\theta)])\right) = k_B ln\frac{1}{2}. \tag{3.165}$$

This indicates the evolution of pure states to pure states. Yet there is something physically wrong here. This assumes that phase information is preserved. However, this phase evolves with a group velocity $\theta(t) \simeq \theta_0 + t(k - k_0)d\omega/dk$, so the wave packet contains phase information in the region $r > \sqrt{3/\Lambda}$. This information can not be ascertained, and so the entropy involves a phase averaging with

$$\Delta S = \frac{k_B}{2} ln(cosh(2|z|)) \simeq k_B(|z| + (1/2)ln2) \simeq$$

$$k_B((\Lambda L_p^2/3)^{-1} + (1/2)ln(2)) = \frac{3c^5 k_b}{8\pi G^2 \rho_v} + (k_B/2)ln(2), \tag{3.166}$$

which is the entropy of the universe.

These indicate that the universe has an entropy determined by the cosmological event horizon. Further, this entropy is an indication of the total field theoretic information, or number of Q-bits, associated with the universe. This suggests that equation 3.118b is associated with a wave equation

$$i\frac{d\psi}{dt} = \mathcal{H}\psi, \tag{3.167}$$

where the Hamiltonian $\mathcal{H}$ is related to the Hamiltonian constraint in equation 3.152, where the wave function is expanded according to the operators a and $a^\dagger$, as in equation 3.155. Further there is a a fluctuation in the wave function

$$\delta\psi(t) = -i\mathcal{H}\delta t, \tag{3.168}$$

where the $\delta\psi(t)$ is expressed according to another basis. This is because the metric and momentum metric elements are themselves changed and are mixed as indicated in equations 3.156 and 3.163. Thus the wave function absorbs one vacuum and creates another one of the form

$$\psi(t) = a\phi(t) + b\phi^*(t). \tag{3.169}$$

However, for this to occur the wave function must both reflect off and transmit through the vacuum induced barrier. The above derivation of the cosmological entropy from the squeezed state of the cosmological wave functions suggests that conjugate variables do not exist in quadature phase. Further the tunnelling barrier is considered to be a parametric amplification. Consider the squeeze state operator

$$S(z) = exp\left(\frac{z}{2}a^{\dagger 2} - \frac{z^*}{2}a^2\right), \tag{3.170}$$

where the squeezing parameter z is $z = re^{i\theta}$. This operator when applied to these generalized operators $\alpha = S(z)aS^\dagger(z)$ gives

$$\alpha = S(z)aS^\dagger(z) = \left(a\,cosh(|z|) - a^\dagger \frac{z}{|z|}sinh(|z|)\right). \tag{3.171}$$

This is a Bogoliubov transformed annihilation operator [22][52]. This permits the wave function to be written according to the future scattered wave function

$$\psi(t) = c\phi(t) + d\phi^*(t). \tag{3.172}$$

Set $S(z) = exp(zcd - z^*c^\dagger d^\dagger)$ and $T(y) = exp(-iy(c^\dagger c + d^\dagger d)$ with $a = T^\dagger S^\dagger cST$ and $b = T^\dagger S^\dagger dST$. The Bogoliubov transformation on the operators

$$a = e^{iy}\left(cosh(|z|) - sinh(y)\right)$$

$$b^\dagger = e^{-iy}\left(cosh(|z|) - sinh(y)\right). \tag{3.173}$$

The tunnelling amplitude is $T^2 = c^\dagger c/(a^\dagger a) = 1/cosh^2(|z|)$. This also permits the calculation of the β term in equation 3.157 and equivalent the γ in equation 3.156 as

$$\beta = \frac{cosh(|z|) + e^{i\phi}cosh(|z|)}{cosh(|z|) - e^{i\phi}sinh(|z|)}. \tag{3.174}$$

The squeezing parameter determines the number of quantum states on the cosmological event horizon and is an expression of $(L/L_p)^2$, where $L = \sqrt{3/\Lambda}$. In addition the squeezing parameter is also dependent upon the mass-energy density of the vacuum. $|z| \sim 3c^5/8\pi G^2 \rho_v \sim (L/\delta L)^3$, where $\delta L \simeq 8\pi G^2 M/3c^5$, where M is the mass fluctuation of the vacuum associated with a volume V. This gives the holographic equation for the scale of fluctuations $(L/L_p)^2 \simeq (L/\delta L)^3$.

This suggests that the physics of the vacuum state and the cosmological constant responsible for the accelerated expansion of the universe are directly tied to the holographic principle. The dark energy is then a consequence of the holographic principle on the cosmological scale. There is a proposed uncertainty relationship between the volume of the universe and the cosmological constant $[\delta V, \delta \Lambda] \sim G$. Loop space gravity replaces classical spacetime with causal sets [53]. Since Λ is a measure of the number of Q-bits, a cosmology with a causal net of N Q-bits will exhibit Poisson statistics $\delta N = \sqrt{N}$ then $\delta V \sim G\sqrt{V}$. This is probably an indication of the unification between the local physics of quantum fields with the physics of the universe.

A net zero cosmology is the most economical one that can emerge spontaneously from the vacuum state. Such spontaneous big bangs require no input of some mass-energy, $G_{\mu\nu}$ or equivalently initial or boundary conditions from outside of physics. This satisfies the "no boundary" condition [49] for a cosmology. This is one of the more salient theoretical features that is thought to be required for a universe to tunnel from a pre-big bang or vacuum state. It appears that observational cosmology is bearing out some theory, and is leading to the most economical requirements for quantum cosmology.

The observational science of cosmology is rapidly revealing new aspects of the universe. The recent discovery that the universe is accelerating outward is the latest of important results, which further indicates that the universe could well be a net zero. This then indicates that the observed universe is the result of a fluctuation in the quantum gravity vacuum. This then should stimulate theoretical research into the nature of gravitational instantons that can result in the tunnelling of a cosmology out of the vacuum state. As this physics is tied to the holographic principle it is important to design experiments to measure the resulting large scale fluctuations $\delta L \sim (L_p^2 L)^{1/3}$. Various experimental schemes have been proposed to test for these fluctuations [37][38]. Such experiments would provide empirical weight to the holographic principle, plus lend support to theories of some spontaneous quantum process that generates a $\sum all = 0$ cosmology.

The de Sitter cosmology is an extension of Einstein's relativity to five dimensions. This results in a spacetime cosmology that exponentially expands. With a Wick rotation on the additional dimension the cosmology is an anti-de Sitter (AdS) cosmology. The structure of a de Sitter cosmology is obtains from a 15 dimensional $spin(6)$ as the conformal group, which contains the $spin(5) \sim SO(3,2)$ de Sitter group, which defines spacetime and gravitation. The AdS is the corresponding $SO(4,1)$ group.

The universe is a D3-brane. There are even and odd dimensional D-branes which type IIA and type IIB open string have their ends anchored to, eg. a cello string anchored to the instrument. Type IIA strings have right and left moving modes that transform under separate supersymmetries with opposite chiral symmetries and the modes of a type

IIB strings transform under two supersymmetries with the same chirality. Thus a type IIB string on a 3-Dbrane will have its supersymmetries with opposite chirality mapped into two supersymmetries with the same chirality if it becomes anchored to a D2-brane within the D3-brane. An example of this is a spacetime with an event horizon, such as on a black hole.

The quantum emission by black holes has been found to be consistent with the scattering of a string from a $D5$-brane. The theory for the black hole 3-brane is then $AdS_5 \times S^5$, where AdS_5 is the anti-de Sitter group. This points to a correspondence between conformal field theory and the anti-de Sitter group, or equivalently the cosmology. In 10-dimensional supergravity the D10-brane has two self dual components that decompose as $AdS^5 \times S^5$, which is the manifold of dynamics in the full 11-dim supergravity. Here AdS^5 is the antiDeSitter space of extended general relativity. Thus a 3-dim space, a $D3$-brane exists in AdS^5, which in turn is dual to a $D5$-brane. An event horizon is a mesh of fields that is a $D2$-brane in the $D3$-brane. So a string — a $D1$-brane in the D3-brane scatters off the black hole by transforming itself from a type IIB to a type IIA on the $D2$-brane and then back. However, this is dual to a picture on the $AdS \times S^5$ dual branes. By the mirror symmetry transformation (M-theory) the black hole scatters off all strings in a way that ultimately preserves unitarity.

The above discussion on the $\sum(all) = 0$ and the accelerated expansion of the observable universe indicates that spacetime cosmology may be de Sitter, or de Sitter-like. The AdS/CFT correspondence issue is of considerable importance in our understanding of the universe, in particular quantum cosmology.

The AdS/CFT correspondence is not a primary issue of this monograph. While it is of considerable importance, the subject today is one of considerable mathematical complexity based on current M-theory and superstrings. In chapters 7 and 8 a potential underlying structure of these are presented. It is possible that matters of the AdS/CFT correspondence may only be addressed once this or some other substrate to current theories are understood.

References

[1] P. Chen, http://www.slac.stanford.edu/slac/media-info/20000605/chen.html

[2] G. A. Mourou, D. Umstadter *Scientific American*, **286**, 5, (2002).

[3] G. A. Mourou, C. P. J. Barty, M. D. Perry, *Physics today*, **51**, 5, 1774-1785 (1998).

[4] S. Weinberg (1972), *Gravitation and Cosmology*, p. 215, John Wiley and Sons, New York.

[5] M. V. Berry (1984), *Proc. R. Soc. London* **A 392**, 45.

[6] V. Nesvizhevsky, H. G. Börner, A. K. Petukhov, H. Abele, S. Bae (17 Jan 2002), *Nature* **415**, 297-299.

[7] R. Munoz-Tapia, *Am. J. Phys.* **61**, 11 (1993).

[8] X. Hu and F. Nori, *Physical Review Letters* **79**, 4605 (1997).

[9] W. T. Coffey, Y. P. Kalmykov, J. T. Waldron, *The Langevin Equation* World Scientific, Singapore (1996).

[11] A. Sorensen, L. Duan, I. Cirac, P. Zoller, *Nature* **409**, 63 (2001).

[12] Y. J. Ng, http://www.arxiv.org/abs/gr-qc/0201022

[13] E. P. Wigner, *Rev. Mod. Phys.* **29**, 255 (1957); H. Salecker and E.P. Wigner, *Phys. Rev.* **109**, 571 (1958).

[14] G. 't Hooft http://www.arxiv.org/PS_cache/hep-th/pdf/0003/0003004.pdf

[15] Y. J. Ng, "Spacetime Foam," *Int. J. Mod. Phys.* **D11** (2002) 1585-1590.

[16] W. Beirl, H. Markum, J. Riedler, "Two-Point Functions of Four-Dimensional Simplicial Quantum Gravity," *Nucl. Phys. Proc. Suppl.* 34 (1994) 736-738.

[17] A. Kempf, G. Mangano, R. B. Mann, Phys. Rev. **D52**, 1108 (1995).

[18] B. L. Hu, Y. H. Zhang, Mod. Phys. Lett. **A8**, 3575 (1993).

[19] M. Hillery, R. F. Connel, M. O. Scully and E. P. Wigner, *Phys. Rep.* **106**, 121 (1984).

[20] G.Amelino-Camelia, *Int. J. Mod. Phys.* **D11**, 35 (2002).

[21] G. Amelino-Camelia, Phys. Lett. **B510**, 255 (2001). L. Crowell, "Quantum Optical Detection of the Holographic Principle," *Found. Phys. Lett.* **16**, 3, 281-286 (2003).

[22] P. Beyersdorf, R.Byer, M. Fejer, "Polarization Sagnac Interferometer with Postmodulation for Gravitational-Wave Detection," *Optics Letters* **23**, 16: 1112-1114 (1999), "Results From the Stanford 10 m Sagnac Interferometer," *Class. Quantum Grav* **24** 1585-1589 (2002).

[23] Y. Chen, "Sagnac Interferometer as a Speed-Meter-Type, Quantum-Nondemolition Gravitational-Wave Detector," http://www.arxiv.org/PS_cache/gr-qc/pdf/0208/0208051.pdf

[24] M. O. Scully, private conversation.

[25] G. Sagnac, "La Demonstration de l'Existence de l'Ether Lumineux á Travers le Measures d'un Interferometre En Rotation,"*C. R. Acad. Sci. Paris* **157**, 708-709 (1913) (1913).

[26] E. J. Post, "Sagnac Effect," *Rev. Mod. Phys.* **39**, 475-493 (1967).

[27] Barry C. Barish, "The Detection of Gravity Waves with LIGO," DPF'99, http://www.arxiv.org/pdf/gr-qc/9905026

[28] K. Thorne,"Quantum Noise and Quantum Nondemolition in Gravitational-Wave Interferometers," http://www.ligo.caltech.edu/docs/G/G000327-00.pdf

[29] M. Pauri, M. Vallisneri, "Märzke-Wheeler Coordinates for Accelerated Observers in Special Relativity," *Found. Phys. Lett.* **14**, 5 (2000).

[30] C.W.J. Beenakker, M. Patra, "Photon Shot Noise," *Mod. Phys. Lett.* **B 13**, 337 (1999).

[31] R. M. Wald, *Quantum Fields in Curved Spacetime and Black Hole Thermodynamics*, University of Chicago, Chicago (1994).

[32] W. G., R. M. Wald, "What Happens When An Accelerating Observer Detects A Rindler Particle?"*Phys. Rev.* **D29**, 1047 (1984).

[33] R. K. Pathria, **Statistical Mechanics** Pergamon Press (1972).

[34] V. Schweikhard, I. Coddington, P. Engels, V. P. Mogendorff , E. A. Cornell, "Rapidly Rotating Bose-Einstein Condensates in and near the Lowest Landau Level," http://arxiv.org/abs/cond-mat/0308582

[35] C. Cohen-Tannoudji, B. Diu, F. Laloë, *Quantum Mechanics*, p. 1145, 1410, John Wiley & Sons, New York (1977).

[36] M. W. Evans, L. B. Crowell, *Classical And Quantum Electrodynamics and the B(3) Field*, pages 175-179, World Scientific, Singapore (2001).

[37] K. S. Deffeyes, *Hubbert's Curve*, Princeton Univ. Pr., (2001).

[38] L. S. Finn, A. Lazzarini, *Phys. Rev.* **D64** (2001) 082002.

[39] S. Perlmutter *et al*, "Measurements of Ω and Λ from 42 high-redshifted supernovae", *Astrophys. J.* **517** 565, (1999).

[40] M. R. Nolta, *et al*, "First Year Wilkinson Microwave Anisotropy Probe (WMAP) Observations: Dark Energy Induced Correlation with Radio Sources", submitted to *ApJ*, http://www.arxiv.org/abs/astro-ph/0305097

[41] A. Haas, "An attempt at a purely theoretical derivation of the mass of the universe,"*Phys. Rev* **49**, 411 (1936).

[42] E. P. Tyron, "Is the universe a vacuum fluctuation?"*Nature* **246**, 396 (1973).

[43] J. Overduin, H. Fahr, "Vacuum energy and the economical universe,"*Found. Phys. Lett.* **16**, 2, 119-125, (2003).

[44] M. R. Nolta, *et al*, "First Year Wilkinson Microwave Anisotropy Probe (WMAP) Observations: Dark Energy Induced Correlation with Radio Sources", submitted to *ApJ*, http://www.arxiv.org/abs/astro-ph/0305097

[45] John N. Bahcall, Charles L. Steinhardt, David Schlegel, "A Dynamical Solution to the Problem of a Small Cosmological Constant and Late-time Cosmic Acceleration"*Astrophys. J.*. 600 (2004) 520.

[46] R. R. Caldwell, M, Kamionkowski, N. N. Weinberg, "Phantom Energy and Cosmic Doomsday ,"*Phys. Rev. Lett.* **91** (2003) 071301.

[47] G. W. Gibbons, S. W. Hawking, "Cosmological event horizons, thermodynamics and particle creation", *Phys. Rev.* **D 15** 10 (1977).

[48] A. G. Riess et al., "Type Ia Supernova Discoveries at z ¿ 1 From the Hubble Space Telescope: Evidence for Past Deceleration and Constraints on Dark Energy Evolution", Accepted *Astrophysical Journal*, 607 (2004) 665-687, http://www.arxiv.org/PS_cache/astro-ph/pdf/0402/0402512.pdf

[49] J. B. Hartle, S. W. Hawking, "Wave function of the universe," *Phys. Rev.* **D 28** 12 (1984).

[50] L. Crowell, "Holographic Bounds on Hilbert Space," *Int. J. Mod. Physics.* **D12**, 9 (2003) 1663-1668.

[51] A. Ashtekar, *Phys. Rev. Lett.* **57**, 2244 (1986).

[52] A. Feinstein, M. A. P. Sebastián, "The origin of entropy in the universe," *Found. Phys. Lett.*, **13** 2 (2000).

[53] L. B. Crowell, "Generalized Uncertainty principle for quantum fields in curved spacetime," *Found. Phys Lett.*, **12** 6 (1999).

[54] R. Sorkin *et al*, "'Observables'in causal set cosmology," *Phys. Rev.* **D67** (2003) 084031.

4: Quantum Gravity Fluctuations, Strings and Two Dimensional D-Branes

4.1 Quantum Gravity Fluctuations in Modern Physics

The current literature in physics is filled with a plethora of field theories near the Planck scale. The most dominant set of these are string theories. This has recently been extended into theories of membranes with Dirichlet boundary conditions, or D-branes, by an examination between closed and open strings and boundary conditions on the ends of open strings as Chan-Paton factors. This has lead to a variety of theories which involves the universe as D3-branes of three dimensions connected to others by strings and other highly exotic field theories. These intellectual trends will likely continue into the future.

The purpose here is to examine some aspects of these theories to determine what fluctuation physics may be derived from them or included into them. This is in order to lay a foundation for these theories from a physical viewpoint. This may lead to a prospect for further experimental tests according to methods similar to those discussed in chapter 3 or potentially by entirely different methods. Without this sort of examination of these theories they can only be considered as being mathematical games.

Much of what follows involves 't Hoof's horizon algebra [1], based on the Susskind holographic hypothesis. This posits that ingoing and outcoming fields with respect to an event horizon are projected onto the horizon and determined there. The upper bound on the number of degrees of freedom in a spacetime volume is given by the number of Planck units of area bounding this volume. Hence an event horizon contains all field theoretic information of the black hole. An $SO(2,1)$ Lorentz algebra is derived for operators that describe fields that scatter off event horizons. Since the total field theoretic information is pinned on event horizons this total information is defined by the entropy of the black hole. Boltzmann's H-theorem and the use of statistics lead to an understanding of the underpinning of thermodynamics. As black holes are thermodynamic the same should obtain with quantum gravity. A Planck scale black hole of mass $M = \sqrt{\hbar c/G}$ and temperature $T \simeq \sqrt{\hbar c^5/G}/k_b \sim 10^{35} K$, indicates that the maximum entropy in a region of radius R is $S = R^2/4Gk_B$. Given a partition function $Z = e^{E/k_BT}$ with

$$S = k_BT\frac{\partial Z}{\partial T}|_R, \tag{4.1}$$

the number of required states in this region must grow exponentially with R. This suggests a computation of quantum gravity according to pure states is intractable or NP complete. Hence there does not exist, except with quantum computers, a way to efficiently compute quantum gravity with N pure states for N large.

Consider the threading of the event horizon by spin states in a manner similar to some Ising type of model. The thermodynamic aspect results from looking at operators in the algebra as being of the form

$$a^\mu = b^\mu + e^{-H/g}c^\mu, \tag{4.2}$$

where c^μ are operators with support on the other side of the horizon and b^μ are those with support on the observer's side of the horizon. g is the "gravity" on the horizon determined by Killing vectors on the surface. This leads to generalized commutators for algebras. This would then permit the horizon algebra to be examined according to Yang-Baxter equations for statistical mechanics.

black hole and cosmological horizons contain a "history lesson" of the fields and particles that pile up on them. The conjecture is this braiding structure on event horizons is related to the braiding operations for Calabi-Yau spaces. As fields and particles pile up on an event horizon, or conversely tunnel out from behind these horizons, these fields are ultimately described by a whole statistical ensemble of Calabi-Yau spaces "glued" to the event horizon. Further, thermodynamics then emerges from these Calabi-Yau spaces glued to the horizon, which may be reshuffled around so the braiding describes a partition function over an ensemble of Calabi Yau spaces.

Field theory may be formulated with weighted projective spaces. Physically this describes a Bohr wave according to a Virasoro type of structure. From this construction an abstract gauge theory may be derived. Further, on a weighted CP^4 one could describe Calabi-Yau spaces according to braid operations under various constraints. From this are Toric varieties for these weighted projective spaces.

This raises a question whether D-brane theory at exceedingly high energy is able to recover quantum gravity according to pure states. While this is the case in formal theory it may not be that way with physical reality. Physical reality is likely to continue to surprise us. There is a measure of confidence that physics at $\nu \ll \sqrt{c^5/G\hbar}$ quantum gravity is a finite temperature theory, however approximate it may be. If experiments on quantum gravity are performed likely this finite temperature regime will be observed.

Pure state quantum gravity or string theory of gravity most probably involves physics at the Hagedorn temperature. This is a phase transition temperature or energy where from there might emerge some deeper "inner picture" from string theory. Currently in string theories the partition function for strings diverges as all highly excited states of the string contribute to the density of states. This occurs in the strong coupling limit for strings and the connection between the domain $> T_H$ and $< T_H$ should be T-duality in string theory. The strong coupling limit is also a domain with a Tachyon background, which gives a fundamental cutoff in observable physics. This may suggest that the Hagedorn temperature involves a 4th law of thermodynamics where nature can only be observed up to T_H, but not beyond. This is then analogous to the 3rd law of thermodynamics. This then infers that T-duality should be examined in a finite temperature situation where potentially physics is equivalent under the limits

$$lim_{T \to T_H} \sim lim_{T \to 0}. \qquad (4.3)$$

Consider the case of a rotating string. The finite temperature vacuum which interacts with this string is physically equivalent to the vacuum for a string coupled to a black hole. This black hole is then demonstrated to be due to the excited states of the string. From this a Regge trajectory for particles is derived that is equivalent to the condition for a rotating black hole. We next examine the nonlinear Lorentz transformations of Lee Smolin and its implications for quantum mechanics and gravity. These nonlinear transformations could be relevant for quantum gravity. The bulk of this chapter concerns topological quantum numbers and 't Hooft's horizon algebra. The topological indices associated with event horizons, computed by the Riemann-Roch and Donaldson theorems, show the existence of an internal gauge theory associated with the singularity in the moduli space determined by an event horizon. This internal gauge theory emerges from the "blow-up" of points in the supersymmetric moduli space. This structure suggests a self-similar structure of gauge fields, their moduli and the emergence of gauge fields. Potentially nature exhibits an infinite nesting of structures as measurement scales approach the Planck length.

4.2 Finite Temperature Theory of Rotating Strings

String theories parameterize fields along a one dimensional chord or loop. The string contains modes of vibration corresponding to quantum states or elementary particles. This esoteric study holds the theoretical promise of unifying all the fundamental forces of nature. However, the experimental prospects are far more problematic as the energies required to probe nature on this scale are far beyond current capabilities. Yet quantum operators for fields under intense gravity or acceleration interact with a thermal vacuum [2] and exhibit departures from the standard quantum uncertainty principle [3]. Sensitive studies with stimulated emission of photons by highly accelerated electrons induced by a high powered laser might reveal slight departures from the standard quantum statistics [3]

The spectrum of a string is interpreted as a one to one correlation between their modes and particles[4]. However, this construction may not necessarily be universally valid. The nature of relativistic quantum fields in curved spacetime has some bearing on this problem. These studies lead to the startling conclusion that there is no general definition of a particle as directly associated with quantum states within curved spacetime. The notion of a particle as identified with a quantum state is only approximately valid for a state with a wavelength much smaller than some local region that is approximately flat. A quantum state, with a wavelength that extends over a large enough region of curved spacetime, defines inner products and maps between quantum operators and classical variables where there is a lose of unitarity equivalence. This makes it impossible to define a particle as identical to a quantum state.

For a rotating open string, let the string couple with an achronal region or horizon at the end points. This horizon is associated with a thermal vacuum. As a horizon interacts with the string, quantum states are unable to propagate along the string as a pure state. The string is defined by a set of states with thermal distributions according to a Killing time parameter determined by an operator. This may then by related to the problem of a string that interacts with a quantum black hole. This leads to the prospect that black holes act as maps between string types according to transformations between thermalized states.

A string is an extended object in one dimensions with a position in spacetime given by [4]

$$X^\mu(\sigma,\tau) = x^\mu + p^\mu\tau + i\sum_{n=1}^{\infty}\frac{1}{n}\left(a^\mu{}_n - a^\mu{}_{-n}\right)\cos(n\sigma)e^{in\tau}, \tag{4.4}$$

where σ and τ are the spatial and temporal parameterizations of the string onto a string world sheet and $a^\mu_n \to a^\mu_n/\sqrt{n}$. This string is written for open ends so that the right and left moving modes combine into standing waves on the string. The Lagrangian density for the string is then

$$\mathcal{L} = \frac{1}{2\pi}\left((\dot{X}^\mu)^2 - (X^\mu{}_{,\sigma})^2\right)$$

$$= \frac{1}{2\pi}\partial_a X^\mu \partial^a X_\mu, \tag{4.5}$$

where the first term involves derivatives with respect to τ and the second with respect to σ. The action $S = \int d\tau d\sigma \mathcal{L}$ is proportional to the area of the string world sheet swept out in τ, σ space.

The string is naturally quantized, where the momentum of the string is $P_\mu = \partial\mathcal{L}/\partial\dot{X}^\mu$ or

$$P^\mu(\sigma,\tau) = \frac{1}{\pi}\left(p^\mu + \sum_{n=1}^{\infty}(a^\mu{}_n + a^\mu{}_{-n})\cos(n\sigma)e^{in\tau}\right). \tag{4.6}$$

The canonical variables are expanded according to harmonic oscillators with an equal string time commutator

$$[P^\mu(\sigma,\tau), X^\nu(\sigma',\tau)] = -ig^{\mu\nu}\delta(\sigma - \sigma'), \tag{4.7}$$

with oscillator modes naturally quantized as

$$[a^\mu{}_m, a^\nu{}_n{}^\dagger] = [a^\mu{}_m, a^\nu{}_{-n}] = g^{\mu\nu}\delta_{m,-n}. \tag{4.8}$$

The string has the Hamiltonian

$$H = \int_0^\pi \left(P_\mu \dot{X}^\mu - \mathcal{L}\right)d\tau = \sum_{n=1}^{\infty} n a^\mu{}_{-n} a_{\mu n} + \frac{1}{2}p^{\mu 2}. \tag{4.9}$$

This Hamiltonian is an infinite sum over free harmonic oscillator energy states. The spectrum of this Hamiltonian exist in the Fock space of all possible states

$$\prod_n a^\mu{}_n |0\rangle. \tag{4.10}$$

This set of states is then defined as rungs on a ladder, where each rung is identified with a specific string mode. At this point that departures may occur due to the centripetal acceleration on the frame of the rotating string.

To describe rotating coordinates consider the transformation between frame F and F'

$$x^{\mu'} = \Lambda^{\mu}{}_{\nu} x^{\nu}, \quad (4.11)$$

In cylindrical coordinates the radial and azimuthal coordinates are $r' = r$ and $z' = z$ and

$$t' = \gamma(t - \Omega \chi^t), \quad \theta' = \gamma(\theta - \Omega \chi^\theta), \quad (4.12)$$

where γ is the Lorentz factor. The angular velocity Ω gives the tangential velocity $v = \Omega r$ as measured by a nonrotating observer. The rotation in these coordinates is

$$\eta = \frac{d\theta}{dt}, \quad \eta' = \frac{d\theta'}{dt'}, \quad (4.13)$$

where in the rotating frame is given by

$$\eta' = \frac{d\theta - \Omega\left(\frac{\partial \chi^\theta}{\partial t} dt + \frac{\partial \chi^\theta}{\partial \theta} d\theta\right)}{dt + \Omega\left(\frac{\partial \chi^t}{\partial t} dt - \frac{\partial \chi^t}{\partial \theta} d\theta\right)}. \quad (4.14)$$

Define the velocities $v = r\eta$ and $v' = r\eta'$, for $\chi^t = \theta$ and $\chi^\theta = t$. This gives a formula for the addition of rotation

$$\eta' = \frac{\eta - \Omega}{1 - r^2 \Omega \eta/c^2}, \quad (4.15)$$

which is analogous to the linear equation for velocity additions.

The metric for a rotating frame is then

$$ds^2 = dt^2 - r^2\left(d\theta + \frac{\Omega}{c} dt\right)^2 - dr^2 \quad (4.16a)$$

or

$$ds^2 = \left(1 - \frac{r^2 \Omega^2}{c^2}\right) dt^2 - r^2 d\theta^2 - 2r^2 \frac{\Omega}{c} d\theta dt - dr^2. \quad (4.16b)$$

The horizon for any observer on the rotating frame exists at $r = c/\Omega$. A solution to these coordinates with $dr = 0$ according to the proper time s is found to be

$$t = g^{-1} \sinh(gs), \quad \theta = \frac{g^{-1}}{r} \cosh(gs) - g^{-1} \frac{\Omega}{c} \sinh(gs), \quad (4.17)$$

where $g = \frac{\Omega^2 r}{\sqrt{1 - \Omega^2 r^2/c^2}}$ is the acceleration parameter. At this point set $c = 1$. A Greene's function for the propagation of fields assumes the form

$$G(x, x') = -\frac{1}{4\pi^2 (t - t' - i\epsilon)^2 - |\mathbf{r} - \mathbf{r}'|^2}. \quad (4.18)$$

144 Quantum Fluctuations of Spacetime

If the Greene's function is expressed as $G(x(s), x'(s')) = G(\Delta s)$ and if the particles are massless the Green's function is

$$G(\Delta s) = -\left(16\pi^2 g^2 sinh(\Delta sg - i\epsilon g)\right)^{-1}. \tag{4.19}$$

This Greene's function then enters a Fourier integral for the propagation of the vacuum as observed by the shift in the energy level $E \rightarrow E'$ of a test particle or detector on this rotation frame

$$\sum_{E'} |\langle 0'E'|0E\rangle|^2 = \sum_{E'} |\langle E'|\mu|E\rangle|^2 \int d(\Delta s) e^{i(E' - E)\Delta \dot{s}} G(\Delta s). \tag{4.20}$$

This integral then has the form [5]

$$\sum_{E'} |\mu\langle 0'E'|0E\rangle|^2 = \frac{1}{2\pi} \sum_{E'} \frac{(E' - E)|\langle E'|\mu|E\rangle|^2}{e^{2\pi(E' - E)/g} - 1}, \tag{4.21}$$

where $\mu = \mu(s)$ is a coupling Lagrangian between the vacuum and the atom or a monopole operator for a detector. This then gives a vacuum on a rotating reference frame as a thermal vacuum. The temperature of this vacuum is then with units restored $T = \hbar g/(2\pi c k_B)$

Now transform to the coordinates

$$u = (1 - r\Omega/c)t - r\theta, \; v = (1 + r\Omega/c)t + r\theta, \tag{4.22}$$

or

$$u = g^{-1}(sinh(gs) - cosh(gs)), \; v = g^{-1}(sinh(gs) + cosh(gs)), \tag{4.23}$$

where the line element is $ds^2 = dudv$. The momentum components are then du/ds and dv/ds. It is apparent the Killing vector for spacetime in these coordinates is

$$\xi = \frac{1}{2}\left(\frac{du}{ds}\partial_u + \frac{dv}{ds}\partial_v\right), \tag{4.24}$$

where the off diagonal metric are

$$\xi^\mu P_\mu = \frac{du}{ds}\frac{dv}{ds} = 1. \tag{4.25}$$

This is a covariant constant along the momentum of a particle fixed to the rotating reference frame. For sufficiently large proper time or for $gs >> 0$ use the approximation that $du/ds \simeq gu$ and $dv/ds \simeq gv$ so that

$$\xi = g(u\partial_u + v\partial_v), \tag{4.26}$$

Now construct the gauged covariant derivatives for a string

$$\frac{\partial}{\partial u} \rightarrow \frac{D}{\partial u} = \frac{\partial}{\partial u} + \frac{i}{l}\sum_{n=1}^{\infty}\left(a^u{}_n + a^u{}_{-n}\right)cos(n\sigma)e^{-in\tau} \tag{4.27}$$

and
$$\frac{\partial}{\partial v} \to \frac{D}{\partial v} = \frac{\partial}{\partial v} + \frac{i}{l}\sum_{n=1}^{\infty}\left(a^v_n + a^v_{-n}\right)\cos(n\sigma)e^{-in\tau}, \qquad (4.28)$$

where l is the length of the string. This then gives the Killing vector defined in a covariant manner
$$\xi = g\left(u\frac{D}{\partial u} + v\frac{D}{\partial v}\right). \qquad (4.29)$$

With the Killing vector it is possible to calculate the thermal distribution of states for a rotating string. The rotation of the string is associated with a particle horizon at the ends, where the rotational speed is c. The tension on the string is the centripetal acceleration associated with the rotation is
$$g\xi^\mu = -\frac{1}{2}\nabla^\mu(\xi^\nu \xi_\mu), \qquad (4.30)$$

which diverges at the horizon when $r = \Omega/c$. The tension on the string is a measure of the force required to keep a point on the string a distance from the horizon.

This event horizon, split into the past and future horizons $h_I \cup h_{II}$, as observed by an inertial observer, is determined by the background metric as well as the graviton modes on the string. This horizon splits the Hilbert space according to states described on either side of the horizon [6]. The string states defined in the spacetime containing the string are in $\mathcal{H}_I$, while states in the region on the other side of the horizon span $\mathcal{H}_{II}$. The states are completely defined by the Hilbert space $\mathcal{H}_I \oplus \mathcal{H}_{II}$. However, causality conditions mean an observer in one region can only specify states in only that region with complete certainty. This means that initial data can be freely specified on h_I completely independently, as this region is not causally connected by the future of h_{II}. It is this that leads to the thermalization of states and the inability to define states as isomorphic to particles. This translates to an inability to define distinct string modes isomorphic to eigenstates of particles.

The projection of the Killing vector on to a vector determined by the covariant derivative is a constant $\langle \xi^\mu P_\mu \rangle = const$. This leads to the definition of the Killing time parameter w which satisfies
$$\xi^\mu \nabla_\mu w = const, \qquad (4.31)$$

where w is to be determined. The coordinates u, v define the two null directions. The Killing time evaluated on the null surface h_{II} $u = 0$, which gives the Killing vector
$$\xi = gv\frac{D}{\partial v}. \qquad (4.32)$$

With a normal ordering this Killing vector gives the expected equation of motion for the Killing time w
$$gv\left\langle\frac{Dw}{\partial v}\right\rangle = 1 + \frac{2v}{l}(ln|v| - 1)\sum_{n=1}^{\infty}\langle a^v_n a^v_{-n}\rangle\cos^2(n\sigma), \qquad (4.33a)$$

where

$$w = \frac{1}{g}ln|v| - \frac{iv}{gl}(ln|v| - 1)\sum_{n=1}^{\infty}(a^v{}_n + a^v{}_{-n})cos(n\sigma)e^{in\tau}. \quad (4.33b)$$

Here the last term in 4.33a contains the expectation of the Hamiltonian. The action of $gvD/\partial v$ on w is constant, determined by the modes of the string and the position on the string world sheet by the coordinates σ, τ. This is is a string version of the Unruh effect, where string modes contribute to the Killing time. The Killing time is then partially determined by the time operator

$$\hat{w} = -\frac{v}{l}(ln|v| - 1)\sum_{n=1}^{\infty}(a^v{}_{-n} + a^v{}_{-n})cos(n\sigma)e^{in\tau}, \quad (4.34)$$

with $w = w_0 + i\hat{w}/g$. $\hat{w}$ is evaluated on the Fock space of string states whose eigenvalues determine the physical Killing time. This Killing time operator is not the same as the putative time operator of Finkelstein[7].

Consider a string restricted to h_{II}, which are causally measured by an observer at $t = 0$. The Killing time for $v > 0$ introduces a new phase $exp(-i\omega_n w)$, where ω_n is the frequency with respect to the Killing time. The mode expansion of the string with $u = 0$ on h_{II} is then,

$$\mathcal{X}^v(\sigma, \tau) = x^\mu + p^\mu\tau + i\sum_{n=1}^{\infty}(a^v{}_n - a^v{}_{-n})cos(n\sigma)e^{i\omega_n w}e^{in\tau}. \quad (4.35)$$

For the oscillating term

$$\psi^v = \sum_{n=1}^{\infty}\psi^v{}_n e^{i\omega_n w}, \quad (4.36)$$

the Fourier transform of $\psi_n exp(i\omega_n w)$ with respect to the inertial time gives the field according to the modes n. This Fourier transformation according to spacetime variables is

$$\phi^v(n) = \psi^v_n \int_0^\infty e^{inv} e^{-i\omega_n w} dw. \quad (4.37)$$

Now extend this into the complex plane by setting $w_0 = ln(iz)$ then $ln(iz) = ln(z) + i\pi/2$ and the Fourier transform is

$$\phi^v(n) = i\psi^v_n exp\left(-\frac{\pi\omega_n}{2g} - \frac{\omega_n}{g}\hat{w}\right)\int_0^\infty e^{inv} e^{-i\omega_n ln(z)} dz. \quad (4.38)$$

Equation 4.38 illustrate how fields restricted to h_{II} define the fields in $\mathcal{H}_I$. For $\theta \to -\theta$ and $t \to -t$ this establishes the fields on h_{II} which pertain to the region on the other side of the horizon contained in $\mathcal{H}_{II}$. The same analysis above may also be performed for $v < 0$. In general this means that string modes contained in the region with the rotating string are entangled with states or vacuum modes within region II. These modes interact with the string as virtual strings. This means field modes on the string in region I, ψ_n^v, must be defined by the fields on the string plus the fields defined beyond the horizon nonlocally correlated to the string:

$$\psi_n^v = \psi_{nI}^v + exp\left(-\frac{\pi\omega_n}{2g} - \frac{\omega_n}{g}\hat{w}\right)\psi_{nII}^v. \qquad (4.39)$$

Further analysis with Bogoliubov operators indicates the pure vacuum state for the string is transformed into

$$U|0\rangle = |0\rangle_{n'} = \prod_n \left(\sum_{m=0}^{\infty} exp\left(-\frac{\pi\omega_n}{g} - \frac{\omega_n}{g}\hat{w}\right)\right)|0\rangle_{nI}|0\rangle_{nII}. \qquad (4.40)$$

With this vacuum the spectrum of the string states is defined as a product of all possible oscillator states,

$$\prod_{n'} a^{\mu}{}_n U|0\rangle = \prod_{n'}\prod_n a^{\mu}{}_{n'} \sum_{m=0}^{\infty} exp\left(-\frac{\pi\omega_n}{g} - \frac{\omega_n}{g}\hat{w}\right)|0\rangle_{nI}|0\rangle_{nII}, \qquad (4.41)$$

which means these string states are thermalized states of harmonic oscillators. This thermalization is due to the entanglement of states that are hidden by an event horizon.

This is equivalent to a string interacting with a black hole. To explicitly see this in the coordinates $U = u + v$, $V = u - v$

$$U = \frac{1}{\Omega^2 r} g_{00}^{1/2} sinh(gs), \quad V = \frac{1}{\Omega^2 r} g_{00}^{1/2} cosh(gs) \qquad (4.42)$$

for $g_{00} = 1 - \Omega^2 r^2$. The motion of a point at a radial distance is equivalent to the motion of a particle in a gravity field. In general relativity the result is exactly Kepler's laws that $\Omega^2 r^3 = M$. Perform the substitution $r = 2M$ to arrive at $\Omega^2 r = 1/4M$. The U, V coordinates take the form

$$U = 4M g_{00}^{1/2} sinh(gs), \quad V = 4M g_{00}^{1/2} cosh(gs), \qquad (4.43)$$

from which it is evident that in U, V coordinates $V^2 - U^2 = 16M^2 g_{00}$, which is equivalent to $1/g$. The relationship $2\pi T = g$ also gives

$$2\pi T = \frac{1}{\sqrt{16M^2 g_{00}}} = 2\pi g_{00}^{-1/2} T_0. \qquad (4.44)$$

This is a temperature $T_0 = (8\pi M)^{-1}$, which is the temperature of a black hole.

The thermal vacuum may be expressed in an abbreviated form as

$$|0'\rangle = \prod_n e^{W_n}|0\rangle_{nI}|0\rangle_{nII}, \qquad (4.45)$$

where the four amplitude

$$A_4 = \langle 0', k_1|V_0(k_2)\Delta V_0(k_3)|0', k_4\rangle \tag{4.46}$$

may be computed. Δ is a field propagator $\Delta = \int_0^\infty dt e^{-t\mathcal{H}} = \mathcal{H}^{-1}$, with the vertex operator $V_0(k)$. $\mathcal{H}$ is the Hamiltonian for the string, which gives $e^{-t\mathcal{H}}$ as the unitary operator for a string of length π along the direction t.

$$A_4 = \int_0^1 \frac{dx}{x} \langle 0', k_1|V_0(k_2,1)V_0(k_3,x)|0', k_4\rangle, \tag{4.47}$$

for $x = e^t$. This amplitude may be written as

$$\langle 0', k_1|V_0(k_2)\Delta V_0(k_3)|0', k_4\rangle =$$

$$\prod_n \prod_m \langle 0|_{mI}\langle 0_{mII}|e^{\mathcal{W}_m^\dagger} exp(-k_2 \sum_i^\infty \frac{a_i}{i})e^{-t\sum_i a_{-i} a_i} exp(-k_3 \sum_i^\infty \frac{a_{-i}}{i})e^{\mathcal{W}_n}|0\rangle_{nI}|0\rangle_{nII} \tag{4.48}$$

$$= \prod_n \prod_i exp((-k_{2\mu}k_3^\mu + \mathcal{W}_n^\dagger + \mathcal{W}_n)\frac{x^i}{i}) = (1-x)^{k_2 k_3 + \sum_n(\mathcal{W}_n^\dagger + \mathcal{W}_n)}.$$

From now let $\mathcal{W}$ be the sum of all $\mathcal{W}_n^\dagger + \mathcal{W}_n$. The final step in the calculation exploits the method of coherent states. Define the s parameter as $s = -(k_2 + k_3)^2$ so that $k_2 k_3 = s/2 + \epsilon$, for $\epsilon = (k_3^2 + k_4^2)$. The amplitude is then

$$A_4 = \int dx\, x^{-s/2 - 2}(1-x)^{s/2 + \epsilon + 2\mathcal{W}} \tag{4.49}$$

For a rotating string the tangential velocity of the string ends has a magnitude equal to the speed of light. For $2\mathcal{W}$ written as $-\omega/g \to -J$ this amplitude is

$$\langle 0'|V_2 V_3|0'\rangle = (1-x)^{-J + s/2 + \epsilon}. \tag{4.50}$$

This amplitude is a parameterized by

$$J \leq s/2 + \epsilon, \tag{4.51}$$

which for equality is a Regge trajectory. If $s = 2m$ this is precisely the condition imposed on rotating black holes:

$$J \leq m + \epsilon. \tag{4.52}$$

The rotating string is then physically equivalent to a string interacting with a black hole. Physically the term ϵ is interpreted as the zero point energy and $\epsilon = n\hbar\omega_{Planck}$, where $n = 1$ for an open string and $n = 2$ for a closed string. The doubling for a closed string accounts for the independent modes that traverse the string in opposite directions. The four-amplitude is then

$$A_4 = \int dx \, x^{\frac{1}{2}(k_3 + k_4)^2 - 2}(1 - x)^{k_2 k_3 + 2W}, \tag{4.53}$$

where for $a = -a/2 - 1$ and $b = s/2 + J/2 + 1$, A_4 will define a Veneziano amplitude, mathematically an Euler beta function:

$$A_4 = \int dx \, x^{a-1}(1 - x)^{b-1} = \frac{\Gamma(a)\Gamma(b)}{\Gamma(b)} \tag{4.54}$$

This exercise is important in considering where the string interacts with a quantum black hole. Assume the Killing vector and the acceleration parameter is defined by another spacetime configuration with a gravity field. The Killing vector field then defines a Killing horizon for a black hole as $\nabla^\mu(\xi^\nu \xi_\nu) = -2g\xi^\mu$ for a Killing time of a black hole. The state $a_{-1}^\mu a_1^\nu |0\rangle$ contains a traceless part that is a spin-2 graviton field, which enters $e^{a-n a_n}$ as the four point amplitude. So effectively the black hole that interacts with the string is also due to the spectrum of the string. However, this is associated with a finite temperature open string connected to a black hole. This suggests a black hole exciton which emerges on a closed string transforms that string into an open string with a thermal distribution of states.

Susskind states there is a preservation of information n systems that enter a black hole [8]. An observer on the string and within the black hole will record that the string is still a closed string with two distinct modes. Conversely an external observer will measure that these modes have been cut off at the Planck scale of energy, where a mixing of these modes gives thermally distributed modes of an open string. The relationship between these two points of view indicates how information is scrambled, so the thermal string is simply due to the statistical mixing of information not accessible to the observer external to the black hole. Further, the string will to the external observer appears to cover more and more of the event horizon. This suggests that if the black hole persists that the string will approach the Hagedorn temperature, but where the frequency modes observed are highly redshifted. This means the area of the black hole is proportional to the length of the string and that entropy dominates the density of states. This then points to a subtle connection between the lifetime of the black hole, determined by its mass and the Hagedorn temperature.

4.3 Nonlinear Spacetime Transformations, Quantum Theory and Gravity

Lee Smolin and João Magueijo [9] have proposed a nonlinear extension of the Lorentz transformation that includes the limiting factor for the energy of a particle as $E < E_{Planck}$. Physically it is expected that the wave length of a particle should be restricted to a value greater than the Planck length $\sqrt{G\hbar/c^3}$. Similarly the limiting energy a particle may be boosted to is less than the reciprocal of this, which is the Planck energy $\sqrt{c^5/G\hbar}$. This paper [9] makes reference to the fact this nonlinear extension is sensitive to only one particle. Physically this makes sense, as obviously macroscopic entities composed of many particles may have energies much larger than a Planck energy, such as the kinetic energy of motion for a planet or a star. Here this statement is justified and where a quantum underpinning to this nonlinear transformation is suggested. This result has some implications for quantum gravity.

The paper [9] sets up a postulate on the relativity of inertial frames, the equivalence principle and the observer independence of the Planck scale of length and energy. A nonlinear Lorentz group is then proposed with the generator of the standard Lorentz rotation generator J^i defined as

$$L_{ij} = p_i \frac{\partial}{\partial p^j} - p_j \frac{\partial}{\partial p^i}, \quad J^i = \epsilon^{ijk} L_{jk}, \tag{4.55}$$

where the boost generator modified with the dilaton operator

$$D = p^i \frac{\partial}{\partial p^i} \tag{4.56}$$

is

$$K^i = K_0^i + L_p p^i D. \tag{4.57}$$

These satisfy the standard commutation relationships for the Lorentz algebra

$$[J^i, J^j] = \epsilon^{ijk} J_k, \quad [J^i, K^j] = \epsilon^{ijk} K_k, \quad [K^i, K^j] = \epsilon^{ijk} J_k. \tag{4.58}$$

The addition of the dilaton operator means there is the inclusion of a p^i in the boost, which means the action is nonlinear in momentum space. The entire nonlinear Lorentz boost in the x direction then gives

$$\begin{aligned} p_0' &= \frac{\gamma(p_0 - vp_x)}{1 + L_p(\gamma - 1)p_0 - L_p \gamma v p_x} \\ p_x' &= \frac{\gamma(p_x - vp_0)}{1 + L_p(\gamma - 1)p_0 - L_p \gamma v p_x} \\ p_{y,z}' &= \frac{p_{y,z}}{1 + L_p(\gamma - 1)p_0 - L_p \gamma v p_x}. \end{aligned} \tag{4.59}$$

These transformations are nonlinear in momenta and derived for any Lorentz boost. Apparently these transformations are likely only to concern a quantum system or particle. There are plenty of macroscopic systems that have momenta and energy vastly larger than the Planck scale of momenta and energy. So the question to be asked is in what way these nonlinear extensions concern quantum wave functions and whether they are applicable to classical or macroscopic systems.

In the Lorentz group action consider the action of

$$U(p_0) = e^{-L_p p_0 D} \tag{4.60}$$

on the momentum p_i:

$$U(p_0)p_i = \frac{p_i}{1 - L_p p_0}. \qquad (4.61)$$

This extension of the Lorentz group illustrates there is a divergence when $p_0 = L_p^{-1}$ which points to the new invariant associated with the Planck scale.

The action of the dilaton operator on a quantum wave function suggests that these nonlinear Lorentz transformations operate on the quantum level. The dilaton operator

$$D = p^i \frac{\partial}{\partial p^i}, \qquad (4.62)$$

acts on a wave function $\psi \sim e^{ipx}$ to produce

$$D\psi = ipx\psi. \qquad (4.63)$$

The spacetime indices on the momentum and energy are suppressed for brevity. Now consider p and x written in raising and lowering operator form

$$a = x + ip, \quad a^\dagger = x - ip, \qquad (4.64)$$

where frequencies and constants are ignored for brevity. This means that

$$a^2 - (a^\dagger)^2 = 2i(xp + px), \qquad (4.65)$$

and $xp = px + [x,p] = px + i$ for $\hbar = 1$. So the dilaton operator on the wave function is equivalent to

$$D\psi = \frac{1}{4}(a^2 - (a^\dagger)^2 + 2)\psi. \qquad (4.66)$$

The operator $U(p_0) = e^{L_p p_0 D}$ acts on the wave function so that

$$U(p_0)\psi = e^{L_p p_0/2} e^{(L_p p_0/4)(a^2 - (a^\dagger)^2)}\psi. \qquad (4.67)$$

The dilaton operator acts as a squeeze state operator on the wave function. So this function measures the squeezing of the wave function. The squeezed state operator gives a measure on the phase space volume of a wave function, which is related to the Wigner function that gives a similar measure.

Consider the action on the density operator $\rho = |\psi\rangle\langle\psi|$ on the state variable

$$U(p_0)\rho = e^{iL_p p_0 p^i x_i}\rho. \qquad (4.68)$$

Two completeness sums $1 = \int dx |x\rangle\langle x|$ inserted on either side of $U(p_0)\rho$ gives

$$U(p_0)\rho = \int dx |x\rangle\langle x| e^{iL_p p_0 D} |\rho \int dx' |x'\rangle\langle x'|, \qquad (4.69)$$

with the eigenstate relation $\langle x|x_{op}\rangle = \langle x|x$ so that

$$U(p_0)\rho = \int dx \int dx' |x\rangle\langle x|e^{iL_p p_0 px}\rho|x'\rangle\langle x'|. \tag{4.70}$$

The Wigner function is formed by the substitution $x \to x + \frac{1}{2}q$ followed by $x' = x - \frac{1}{2}q$

$$U(p_0)\rho = \int dx e^{iL_0 px} \int dq |x + q/2\rangle\langle x + q/2|e^{iL_p p_0 pq/2}\rho|x - q/2\rangle\langle x - q/2|. \tag{4.71}$$

The expectation of this operator $\langle\psi|U(p_0)|\psi\rangle$ results in

$$\langle U(p_0)\rho\rangle = \int dx e^{iL_p p_0 px} \int dq \langle x + q/2|e^{iL_p p_0 pq/2}\rho|x - q/2\rangle\langle\psi|x + q/2\rangle\langle x - q/2|\psi\rangle. \tag{4.72}$$

With the Wigner operator

$$\langle\psi|x + q/2\rangle\langle x - q/2|\psi\rangle = e^{-ipq/2}|\psi(x)|^2 e^{ipq/2} \tag{4.73}$$

the expectation assumes the form

$$\langle U(p_0)\rho\rangle = \int dx e^{iL_p p_0 px}|\psi|^2 \int dq \langle x + q/2|e^{iL_p p_0 pq/2}\rho|x - q/2\rangle$$

$$= \int dx e^{iL_p p_0 px}|\psi|^2 \int W_{p_0}(p, q), \tag{4.74}$$

with the particular form of the Wigner function as [10]

$$W_{p_0}(p, q) = \int dq \langle x + q/2|e^{iL_p p_0 pq/2}\rho|x - q/2\rangle. \tag{4.75}$$

Consider the wave as existing in a region $[-a/2, a/2]$ with periodic boundary conditions,

$$\psi(x) = a^{-1/2}e^{ipx/\hbar}, \tag{4.76}$$

where $p = 2\pi n\hbar/a$ and the role of $\hbar$ is explicitly restored. The Wigner function assumes the explicit form

$$W_{p_0}(p, q) = \frac{\sin[(p - L_p p_0 p/2)q/\hbar]}{(p - p_0 p/2)q/\hbar}. \tag{4.77}$$

From here take the limit $\hbar \to 0$, with $\hat{p} \to 0$ to find that

$$\lim_{\hbar \to 0} W_{p_0}(p, q) = a^{-1}\delta(L_p p_0 p). \tag{4.78}$$

When equation 4.74 is integrated over the momentum this evaluates $U(p_0)$ over the phase space

$$\lim_{\hbar \to 0} \int dp \langle U(p_0)\rho\rangle = a^{-1} \int dp \int dx e^{iL_p p_0 px}|\psi|^2 \delta(L_p p_0 p)$$

$$= a^{-1} \int dx e^0 |\psi|^2 = 1, \qquad (4.79)$$

and the operator $U(p_0)$ in the limit that $\hbar \to 0$ is reduced to a unit operator.

The Wigner function defines a quasi-probability density in phase space for a quantum system. If the quantum unit of action is small compared to the total action for the system then the oscillations of the systems are far too rapid to be evaluated on the large scale of the system. When this is the case only an average is obtained, which is a classical probability density. In this situation the expectation of the operator $U(p_0)$ is reduced to a unit matrix. Now consider a density matrix for a quantum system with a large number of entangled states. The total action of this system, while large, is the minimal volume in phase space for the system. With disentanglement the phase space volume for this system enlarges and information is lost. It may be argued that a system with quantization on the large will exhibit this nonlinear momentum invariance principle. It is likely that if this nonlinear invariance is a correct description of nature they involve interesting and deep relationships between quantum mechanics and it correspondence with classical and macroscopic physics.

For a system with a large number of atoms or states in an entangled state it is then possible that deviations from the standard Lorentz transformation might be measurable. Such might in principle be the case with a Bose-Einstein condensate with a sufficient number of atoms. This condensate might then be accelerated, by variations of the phase in lasers that compose a trap as in the case of quantum fountains. Potentially deviations from the standard Lorentz boosts might be measurable.

The effect of the operator $U(p_0)$ on the element of the Lorentz group $g = exp(\omega_{\mu\nu} L^{\mu\nu})$ defines the nonlinear representation of the Lorentz group by

$$\mathcal{G}[\omega_{\mu\nu}] = U^{-1}(p_0) g U(p_0) = U^{-1}(p_0)(1 + \omega^{\mu\nu} L_{\mu\nu}) U(p_0) + \ldots =$$

$$1 + \omega^{\mu\nu} \frac{L_{\mu\nu}}{1 - L_p^2 p_0^2} + \ldots = exp\left(\omega^{\mu\nu} \frac{L_{\mu\nu}}{1 - L_p^2 p_0^2}\right) = e^{\omega^{\mu\mu} M_{\mu\nu}[p_0]}. \qquad (4.80)$$

This modifies the structure of general relativity. Let the vector e^u, where the index a indicates an internal space direction, define a tetrad basis by $e_\mu^a = \partial_\mu e^a$. The tetrad exhibits the nonlinear realization of the transformation according to

$$e_\mu^a \to e_\mu^{a\prime} = \mathcal{G}[\omega_{\alpha\beta}, p_0^b] e_\mu^a, \qquad (4.81)$$

where $p_0^b = e_0^b$. For $e^{a\mu}$ the transformation involves $\mathcal{G}^{-1}[p_0]$. Similarly the differential operator

$$D_\mu \to \mathcal{G}[e_0](\partial_\mu + \omega_\mu^{a\nu} e_\nu^a), \qquad (4.82)$$

transforms locally under the nonlinear Lorentz group. This then gives

$$D_\mu e_\nu^a = \mathcal{G}[p_0](\partial_\mu e_\nu + \omega_{\sigma\nu}^a)\mathcal{G}[e_0] + (\mathcal{G}[p_0]\partial_\mu \mathcal{G}^{-1}[p_0] e^a)_\nu, \qquad (4.83)$$

which for the local nonlinear transformation written according to indices gives the connection coefficients

$$\omega^{a\nu}{}_\mu[p_0] = \mathcal{G}_\alpha{}^\nu[p_0]\omega^{a\alpha}{}_\beta \mathcal{G}^{-1\beta}{}_\mu[p_0] + \mathcal{G}_\alpha{}^\sigma[p_0]\left(\partial_\mu \mathcal{G}^{-1\alpha}{}_\nu[p_0]\right)e^a_\sigma. \quad (4.84)$$

There is then an additional connection term. For $p_0 L_p \ll 1$ these additional connection terms are correspondingly small. Define these additional connection terms $\gamma^{a\mu}{}_\nu = \gamma^{a\mu}{}_{\nu\sigma}e^\sigma = \mathcal{G}^\mu{}_\rho \partial_\nu \mathcal{G}^{-1\rho}{}_\sigma e^a$. The curvature tensor is then

$$R^{a\alpha}{}_{\mu\beta\nu}[p_0] = \mathcal{G}^\alpha{}_\sigma[p_0] R^{a\sigma}{}_{\mu\rho\nu}\mathcal{G}_\beta^{-1\rho} + \partial_{[\beta}\gamma^{a\alpha}{}_{\mu\nu]} + \epsilon^{abc}\gamma^{b\alpha}{}_{[\beta}\gamma^c{}_{\mu\nu]}. \quad (4.85)$$

The standard curvature is homogeneously transformed by the nonlinear term, where the additional curvature term is labelled as $\rho^{a\alpha}{}_{\mu\beta\nu}$. This is an alternative approach to arriving at equation 2.74. Clearly the nonlinear transformation only becomes appreciable when the $p_0 L_p \to 1$.

The additional connection terms may then contain torsional terms, which might connect this to quantum fluctuations. Torsional construction in emerges below with the horizon algebra and topological indices. As showed in chapter 2 torsion is related to topological numbers. Before exploring this direction an overview of topological field theory is given first.

4.4: Overview of Basic Topological Field Theory

This is an introduction to the idea of topological field theory. Most of physics involves simply connected spaces and the equations of motion are tied to the geometry of the space. Even general relativity is most often worked this way. The classic solutions for black holes, Schwarzschild, Kerr, etc, involve an asymptotically flat spacetime. The topology of the total space is not considered. This is a convenient way to do physics, but in the problems of quantum cosmology and unification this luxury is not allowed.

The simplest approach to topological field theory uses Morse theory [11]. Morse theory involves the minimum, maximum and saddle points in a space. If for instance for a two dimensional 2-genus torus, where one end is is assigned as the bottom a potential minimum $V = gx$, there are two saddle points in the middle, a potential minimum as an attractor point at the bottom and a repelling point at the top. The geodesic of a particle on this space is then governed ultimately by these critical points.

Define a function f on this manifold that vanishes on the critical points

$$\frac{\partial f(X)}{\partial x_j} = 0, \quad (4.86)$$

where the behavior of this function at these critical points is determined by the second derivatives that defines the Hessian matrix

$$H_{ij} = \frac{\partial^2 f(X)}{\partial x_j \partial x_k}. \quad (4.87)$$

The range of the indices is over the dimension of the manifold, which in this example is $dim = 2$. Geometrically the eigenvalues of the Hessian matrix determines how many geodesics go "upwards" and go "downwards" by their signs. Consider the eigenvalues with negative sign. This defines the Morse index $\mu(X)$. The number with Hessian index 0 is $M_0 = 1$, the bottom, the number with index 2 is $M_2 = 1$ as the top and the number with index 1 is $M_1 = 4$. The Euler index is then defined accordingly as

$$\chi = M_0 - M_1 + M_2 = -2. \tag{4.88}$$

This also defines the Euler index for a multihandled torus with g handles,

$$\chi(M) = 2(1 - g). \tag{4.89}$$

In general the Euler index is defined by the Betti numbers b_n, where n is the dimension of the subspace of M with

$$\chi(M) = \sum_{n=1}^{dim M} (-1)^n b_n(M). \tag{4.90}$$

There is one weakness in this theory. The Morse index in this case above is $\mu_n = b_n(M)$, yet in general this may not always the case and in general $\mu_n \geq b_n(M)$. However, this theory originally developed by A. Floer [12] illustrates that this Morse theory approach does lead to a cohomological description of quantum fields. The power of Morse theory is extended through the use of physical operators, in particular supersymmetry generators.

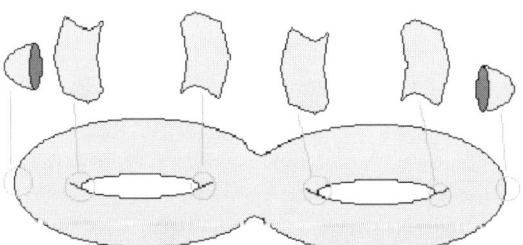

A g = 2 torus with critical points illustrated.
Figure 4.1

Let the coboundary operator $\mathbf{d}$ with $\mathbf{d}^2 = 0$ be parameterized according to t, which may be thought of as a time variable, where the Morse function f that describes the surface according to

$$\mathbf{d}_t = e^{-tf} \mathbf{d} e^{tf} = \mathbf{d} + t\mathbf{d}f. \tag{4.91a}$$

Also the Hodge dual star operator, $* *^2 = 1$ defines $\mathbf{d}^* = *\mathbf{d}*$ with

$$\mathbf{d}_t^* = e^{-tf} \mathbf{d}^* e^{tf} = \mathbf{d}^* + t\mathbf{d}^* f \tag{4.91b}$$

The momentum defined by $\mathbf{d}f$ and $\mathbf{d}^*f$ is then written according to the annihilation and creation operators a, $a^\dagger$ as

$$\mathbf{d}_t = \mathbf{d} + ta_i\frac{\partial f}{\partial x^i}\mathbf{d}x^i, \quad \mathbf{d}_t^* = \mathbf{d}^* + ta_i^\dagger\frac{\partial f}{\partial x^i}\mathbf{d}^*x^i. \tag{4.92}$$

The time developed Hamiltonian is then of the canonical form

$$H_t = \frac{1}{2}(\mathbf{d}_t\mathbf{d}_t^* + \mathbf{d}_t^*\mathbf{d}_t). \tag{4.93}$$

The Hamiltonian may then be written according to the coboundary operators as

$$H_t = \frac{1}{2}\Big(\mathbf{dd}^* + \mathbf{d}^*\mathbf{d} + t^2(a_ia_j^\dagger + a_j^\dagger a_i)\frac{\partial f}{\partial x_i}\frac{\partial f}{\partial x_j} + t[a_i, a_j^\dagger]\frac{\partial^2 f}{\partial x_i \partial x_j}\Big). \tag{4.94}$$

The operators obey $a_i a_j^\dagger + a_j^\dagger a_i = g_{ij}$ and $\mathbf{dd}^* + \mathbf{d}^*\mathbf{d} = -\partial^2/\partial x_i \partial x_j$. The Hamiltonian is then

$$H_t = \frac{1}{2}\Big(-\frac{\partial^2}{\partial x_i^2} + t^2\Big(\frac{\partial f}{\partial x_i}\Big)^2 + t[a_i, a_j^\dagger]\frac{\partial^2 f}{\partial x_i \partial x_j}\Big). \tag{4.95}$$

The last term is the Hessian for the critical points. Now expand the Morse function around the critical points with $f(x) = f(X) + \lambda_i(x - X_i)^2$. The Hamiltonian then takes the form of a harmonic oscillator perturbed by the Hessian term

$$H_t = \frac{1}{2}\Big(-\frac{\partial^2}{\partial x_i^2} + t^2\lambda_i^2 x_i^2 + t[a_i, a_j^\dagger]\lambda_i\Big). \tag{4.96}$$

This Hamiltonian act on a basis of harmonic oscillator state $\{|n\rangle\}$, $n \in \mathbb{Z}$ to give the energy

$$E_t = \frac{1}{2}\sum_i \big(|\lambda_i|(N_i + 1) - \text{sign}(\lambda)\lambda_i\big), \tag{4.97}$$

Evidently the vacuum is defined for $N_i = 0$, which is a vacuum with zero energy. The extension of Morse theory is connected to cohomology since the Betti number is

$$b_n(t) = 2(\dim \ker(H_t)). \tag{4.98}$$

The operators a_i, $a_i^\dagger$ pertain to fields that exist in spacetime. The modified Bogoliubov operators give a generalized uncertainty principle according to [3]

$$[\alpha_j^b, \alpha_k^{c*}] = \big(1 + e^{-2\pi\mathcal{H}/g}\big)\delta_{jk}\eta^{bc} -$$
$$\frac{\pi}{g}e^{\pi\mathcal{H}/g}\big(\alpha_j^{0b*}\alpha_k^{0c} + \alpha_j^{0b}\alpha_k^{0c*}\big)e^{-\pi\mathcal{H}/g}, \tag{4.99}$$

where $\alpha_j^b = \hat{\alpha}_j^b + \alpha_j^{0b}$, $\alpha_j^{0b} = a_i^b e^{-\pi\mathcal{H}/g}$ and g is an acceleration. Further, η^{bc} is the metric for the internal $SO(4)$ group. The Heisenberg uncertainty relationship between the momentum and position variables in quantum mechanics is

$$(\Delta p_j^b)^2 (\Delta e_k^c)^2 \geq \frac{1}{4}(i\langle[p_j^b, e_j^c]\rangle)^2. \tag{4.100}$$

From these modified operators the position and momentum operators are define $\alpha_j^b = 1/\sqrt{2}(e_j^b + ip_j^b)$ and $\alpha_j^{b*} = 1/\sqrt{2}(e_j^b - ip_j^b)$, which gives,

$$(\Delta p_j^b)^2 (\Delta e_k^c)^2 \geq \frac{1}{4}(\langle[\alpha_j^b, \alpha_k^{c*}]\rangle)^2. \tag{4.101}$$

The uncertainty relationship expressed according to thermalized creation and annihilation operators is

$$(\Delta p_j^b)^2(\Delta e_k^c)^2 \geq \frac{1}{4}\Big((1 + \langle e^{-2\pi\mathcal{H}/g}\rangle)\delta_{jk}\eta^{bc} - \frac{\pi}{g}\langle e^{\pi\mathcal{H}/g}(\alpha_j^{0b*}\alpha_k^{0b} + \alpha_j^{0b}\alpha_k^{0b*})e^{-\pi\mathcal{H}/g}\rangle\Big)^2. \tag{4.102}$$

For a vacuum spacetime the Einstein field equation reduces to $R_{\mu\nu} = \frac{1}{2}Rg_{\mu\nu}$. It is then apparent the above commutator for vacuum spacetime gives a correction to the Ricci curvature for Σ^3 as

$$\mathcal{R}_{jk} = \frac{1}{2}\Big(R(1 + e^{-2\pi\mathcal{H}/g})g_{jk} - \int_0^\infty d\omega \frac{\pi}{g}e^{\pi\mathcal{H}/g}(\alpha_j^{0b*}\alpha_k^{0b} + \alpha_j^{0b}\alpha_k^{0b*})e^{-\pi\mathcal{H}/g}. \tag{4.103}$$

From here evidently the Hamiltonian assumes the form

$$H_t = \frac{1}{2}\Big(-\frac{\partial^2}{\partial x_i^2} + t^2\lambda_i^2 x_i^2 + t[\alpha_i, \alpha_j^*]\lambda_i\Big)$$

$$= \frac{1}{2}\Big(-\frac{\partial^2}{\partial x_i^2} + t^2\lambda_i^2 x_i^2 + t((1 + e^{-2\pi\langle/g})\delta_{ij} + \mathcal{R})\lambda_i\Big). \tag{4.104}$$

It is evident that for $N_i = 0$ the vacuum state does not have zero energy.

This Morse structure has a correspondence with quantum mechanics. Parameterize the Hamiltonian according to a classical-like path of a particle that moves on the surface. A classical-like particle obeys a geodesic equation, where the Morse function is analogous to a potential,

$$\frac{dx^i(t)}{dt} = -g^{ij}[x(t)]\frac{\partial f[x(t)]}{\partial x^j}. \tag{4.105}$$

Yet the Hamiltonians in equations 4.95 and 2.104 are operators for quantum oscillators. The local critical points with a Morse index give harmonic oscillator states. Yet a quantum wave function will nonlocally exist throughout this manifold. Physically it is likely that a particle in one local critical point may tunnel to another critical point. As such given two critical points X_i and X_j compute the matrix element $\langle X_i | d_t | X_j \rangle$. With equation 4.105 a tunnelling amplitude due to the relative phase of a wave function, where that phase is given by the Morse potentials at the two critical points, is $\sim e^{-t(f(X_i) - f(X_j))}$. The tunnelling amplitude is

$$\langle X_i | d_t | X_j \rangle = n(X_i, X_j) e^{-t(f(X_i) - f(X_j))}, \tag{4.106}$$

where $n(X_i, X_j)$ is an index determined by the Morse function.

A quantum cohomology may now be constructed. Consider the set of eigenstates associated with the critical point A as $\mathcal{E}_a$ where $\mu(A) = a$. Let the operator Q that acts on the state $|A\rangle \in \mathcal{E}_a$ be defined by $n(A, B)$ for $B \in \mathcal{E}_a$ as

$$Q|A\rangle = \sum_{B \in \mathcal{E}_{a+1}} n(A, B) |B\rangle. \tag{4.107}$$

The operator Q takes the system from $\mathcal{E}_a$ to $\mathcal{E}_{a+1}$. Geometrically there exists a map from a set of eigenstates to the "boundary" of other eigenstates, e.g. $d|\ \rangle$, in other critical points. This operator Q satisfies

$$Q^2 |A\rangle = \sum_{C \in \mathcal{E}_{a+2}} \sum_{B \in \mathcal{E}_{a+1}} n(A, B) n(B, C) |C\rangle, \tag{4.108}$$

which geometrically is a map $\mathcal{E}_a$ to $\mathcal{E}_{a+2}$, according to the boundary of eigenstates as $d^2|\ \rangle$. Yet by the famous "boundary of a boundary is zero" truth this must be zero. Therefore $Q^2 = 0$ and $Q^{\dagger 2} = 0$ are cohomology operators. Now define the Betti number as the cohomology of Q with support on $\mathcal{E}_a$

$$b_p = dim((kerQ/imQ) \cap \mathcal{E}_a), \tag{4.109}$$

and this is a cohomological operator for quantum states.

Now consider the graded algebra with L_0 as the Poincaré algebra and L_1 as the space of supergenerators $span\{Q_\alpha\}$. Then for a in L_n and b in L_m there exists the bracket structure

$$\{[a, b]\} = ab - (-1)^{nm} ba. \{[\ ]\} = \text{either commute or anticommute} \tag{4.110}$$

L_0 is the Poincaré algebra consisting of the momentum and angular momentum operators p_μ and $M_{\mu\nu}$. The algebraic extension of the Poincaré algebra is then given by the bracket structures

$$[p_\mu, Q_\alpha] = 0, \quad [M_{\mu\nu}, Q_\alpha] = -(\sigma_{\mu\nu})_{\alpha\beta} Q_b, \tag{4.111}$$

with $\sigma_{\mu\nu} = \frac{i}{4}[\gamma_\mu, \gamma_\nu]$.

So the graded Lie algebra consists of operators in L_0 that act on elements $\{a\}$ of L_0 according to the map $a : L_0 \to L_0$. L_1 contains operators $\{a\}$ that elements of L_0 by the map $a : L_0 \to L_1$ and when they act on elements of L_1 they define the map $a : L_1 \to L_0$. Now the generators of the supersymmetric algebra can be written for $p_\mu = (m, 0, 0, 0)$ (in the rest frame of a particle of mass m) as

$$Q_\alpha = \sqrt{1/2m} a_\alpha, \bar{Q}_\alpha = \sqrt{1/2m} a_\alpha^\dagger \qquad (4.112)$$

where a and $a^\dagger$ are fermionic operators. So these are elements of L_1, where their application twice on either L_0 or L_1 is a map back into those respective spaces. This is in keeping with the algebra for fermionic operators that

$$\{a_\alpha^\dagger, a_\beta\} = \delta_{\alpha\beta} \qquad (4.113)$$

Therefore two supersymmetry (SUSY) transforms on an element on a multiplet returns that element under a Lorentz boost.

Topological field theory with supersymmetry makes the following substitutions:

$$\mathbf{d} \to Q, \; \mathbf{d}^* \to \bar{Q}, \; \mathbf{dd}^* + \mathbf{d}^*\mathbf{d} \to \{Q, \bar{Q}\} = i\gamma \cdot \partial, \qquad (4.114)$$

where the last one gives the supersymmetric Hamiltonian as $H = \frac{1}{2}\{Q, \bar{Q}\}$. There is also the replacement of the index $(-1)^p$ for the action of $\mathbf{d}$ on a p-form with the Witten index $(-1)^F$, where F is the fermion number. For the field

$$\Phi^i = \phi^i + \bar{\theta}\psi^i + \bar{\theta}\theta F^i \qquad (4.115)$$

it is possible to construct the action

$$S = \int d^2x d^2\theta \left(g_{ij} \bar{D}\Phi^i D\Phi^j + f(\Phi) \right), \qquad (4.116)$$

where an integration over the superfield θ gives

$$S = \int d^2x \Big(g_{ij} \partial_\mu \phi^i \partial^\mu \phi^j + ig_{ij} \bar{\psi}^i \gamma^\mu D_\mu \psi^j +$$

$$R_{ijkl}\bar{\psi}^i\psi^j\bar{\psi}^k\psi^l - g^{ij}\frac{\partial f}{\partial \phi^i}\frac{\partial f}{\partial \phi^j} - (D_i D_j f)\bar{\psi}^i\psi^j \Big). \qquad (4.117)$$

The gauged differential operator has the form $D_\mu \psi^i = \partial_\mu \psi^i + \Gamma^i{}_{jk}\partial_\mu \phi^j \psi^k$. This is an amazing result. An integration over the superfields is equivalent to a situation where the coboundary operators are replaced with Q and $\bar{Q}$

$$\mathbf{d}_t = \mathbf{d} + tQ_i \frac{\partial f}{\partial x_i} \mathbf{d}x^i, \; \mathbf{d}_t^* = \mathbf{d}^* + t\bar{Q}_i \frac{\partial f}{\partial x_i} \mathbf{d}^* x^i. \qquad (4.118)$$

4.5: Supersymmetry and Cohomology of the Horizon Algebra

The holographic principle [8] indicates that fields within regions bounded by an event horizon, such as the cosmological horizon, are "holograms." This infers that the quantity of data required to specify the state of a cosmology is significantly reduced. From a computer science point of view there are $N = 3^n$ data points that must be specified by a lattice representation of a spatial surface. Conversely only $N = 2 \times 2^n$ data points must be specified on two dimensional surfaces in spacetime. In fact, if duality is maintained then only half this quantity of data is required. If each data point is specified with a Planck distance separation the entropy per volume, from $n = V/l_p^3$ within the standard ADM approach will be

$$S = n\ln 3 \propto \frac{V}{l_p^3}\ln 3. \tag{4.119}$$

A cosmology based on 2-d surfaces, with $n = A/l_p^2$, has an entropy

$$S = n\ln 2 = \frac{A}{l_p^2}\ln 2, \tag{4.120}$$

which gives a formula for the entropy that is more commensurate with the Bekenstein-Hawking formula for entropy of a black hole $S = A/4G$, where A is the area of the event horizon. This indicates all the relevant data for fields in the universe may be described on surfaces of two dimensions, in particular on horizons. The horizon algebra is a way in which fields may be described according to ingoing and outgoing fields from event horizons.

The horizon algebra is a tool for constructing a supersymmetric membrane theory for the event horizon of a black hole. This theory is then extended to look at topological quantum numbers associated with event horizons and fluctuations. This then associates quantum eigen-numbers with time operators, such as in equation 4.34.

The horizon algebra describes ingoing and outgoing states according to states frozen on the event horizon [13]. The ingoing momentum $p^- = \frac{1}{\sqrt{2}}(p^1 - p^0)$ is a function of the transverse coordinates $x_t = (y, z)$ with $[p^-, x_t] = 0$. This momentum defines the ingoing states of a black hole $|i\rangle = |p^-(x_t)\rangle$. As p^- and x^- are not independent there is a change δp^- is associated with a δx^- according to

$$\delta x_o^-(x_t) = \int dx_t' G(x_t - x_t')\delta p_i^-(x_t). \tag{4.121}$$

$G(x_t - x_t')$ is a propagator that obeys the Poisson equation $\partial_t^2 G(x_t - x_t') = -\delta^2(x_t - x_t')$. There is a similar construction for outgoing momentum and position p^+, x^+ as functions of x_t. This results in the following commutators for these observables

$$[p^-(x_t), x^+(x_t')] = [p^+(x_t), x^-(x_t')] = -i\delta^2(x_t - x_t'). \tag{4.122}$$

Equation 4.121 then permits the momentum to be $\partial_{x_t}^2 x^+ = p^+$ and $\partial_{x_t}^2 x^- = -p^-$ so the commutators become

$$[\partial_{x_t} x^-(x_t), \partial_{x_t} x^+(x_t')] = i\delta^2(x_t - x_t'). \tag{4.123}$$

With $x^{\pm} = \frac{1}{\sqrt{2}}(x^1 \pm x^0)$ this gives

$$[\partial_{x_t} x^0(x_t), \partial_{x_t} x^1(x'_t)] = i\delta^2(x_t - x'_t). \tag{4.124}$$

Now let $x_t = x_t(\sigma)$, where σ is an arbitrary coordinate, which gives

$$[\partial_\sigma x^\mu(\sigma), \partial_\sigma x^\nu(\sigma')] = i\epsilon^{\mu\nu}{}_\rho \partial_\sigma x^\rho(\sigma') \delta^2(\sigma - \sigma'). \tag{4.125}$$

If the metric has the signature $(-,+,+)$ this algebra is $SO(2,1)$, which is a subalgebra of the Lorentz algebra. This is equivalent to the angular momentum algebra, with a signature change.

There exist connection one-forms determined by the parallel translation of the elements in the horizon algebra. Let $e^\mu = \partial_\sigma x^\mu$ be a basis element. The horizon algebra is then represented as

$$[e^\mu, e^\nu] = i\epsilon^{\mu\nu\gamma} g_{\gamma\rho} e^\rho. \tag{4.126}$$

With the further identification $e^1 = e^+$ and $e^2 = e^-$, the action of the coboundary operator on these elements is then

$$de^+ = \omega^- \wedge e^0 - \omega^0 \wedge e^-$$
$$de^- = \omega^0 \wedge e^+ - \omega^+ \wedge e^0 \tag{4.127}$$
$$de^0 = \omega^+ \wedge e^- - \omega^- \wedge e^+.$$

Under the previous 0, 1, 2 indices there is the identification

$$\omega^+ = \omega^0{}_2, \quad \omega^1 = \omega^0{}_1, \quad \omega^0 = \omega^1{}_2. \tag{4.128}$$

Now construct the action for gravitation. With these connections and the momentum conjugates π^0, π^1, π^2 the invariant action is

$$S_{grav} = \int (\pi_0 de^0 + \pi_+ de^+ + \pi_- de^-). \tag{4.129}$$

Now consider the supersymmetric pairs to the basis elements (e^+, e^-, e^0) as (ψ^+, ψ^-, ψ^0). Here the elements of the horizon algebra is a tetrad with an index for an internal space and the spin $\frac{3}{2}$ gravitino field also implicitly contains a spinor index. However, this is suppressed for brevity. The covariant coboundary operator $\mathcal{D}$ acts on these Rarita-Schwinger fields as

$$\mathcal{D}\psi^+ = d\psi^+ + \omega^- \wedge \psi^0 - \omega^0 \wedge \psi^-$$
$$\mathcal{D}\psi^- = d\psi^- + \omega^0 \wedge \psi^+ - \omega^+ \wedge \psi^0 \tag{4.130}$$
$$\mathcal{D}\psi^0 = d\psi^0 + \omega^+ \wedge \psi^- - \omega^- \wedge \psi^+.$$

From this the contribution to the action by this field is then

$$S_\psi = \int \left(\rho_0 d\psi^0 + \rho_+ d\psi^+ + \rho_- d\psi^-\right), \tag{4.131}$$

where ρ is the momentum conjugate $\partial\mathcal{L}/\partial\dot\psi$.

Spacetime curvatures may also be computed. The curvature two-forms for the elements (e^+, e^-, e^0) are

$$\begin{aligned} R^+ &= d\omega^+ + \omega^- \wedge \omega^0 \\ R^- &= d\omega^- + \omega^+ \wedge \omega^0 \\ R^0 &= d\omega^0 + \omega^- \wedge \omega^+. \end{aligned} \tag{4.132}$$

Here the cyclic nature of the $SO(2,1)$ group is apparent in these curvature forms. The invariant action may be found from the Ricci scalar derived from these components, which will provide the constraint required for the action in equation 4.129.

To further understand the physics the transformation properties of these fields are needed. To start consider the supersymmetric transformation of these fields given by the supergravity multiplet

$$\delta_\xi e^\mu = i(\psi^\mu \sigma \bar\zeta - \zeta \sigma \bar\psi^\mu) \tag{4.133a}$$

$$\delta_\xi \psi_\mu = \mathcal{D}_\mu \zeta - ie_\mu\left(2R(\sigma\bar\zeta) - b\zeta - \frac{1}{3}b(\zeta\sigma\bar\sigma)\right), \tag{4.133b}$$

where $\xi = 2i(\theta\sigma\bar\zeta - \zeta\sigma\bar\theta)$, θ, $\bar\theta$ are the Grassmann variables in the graded algebra and $\zeta = \zeta(x)$. The index $\mu \in (+, -, 0)$. The spacetime transformations of e^μ and ψ^μ are

$$(\delta_\xi \delta_\eta - \delta_\eta \delta_\xi) e^\mu = (\xi\eta R)^\mu{}_\nu e^\nu + (\xi\eta T)^\nu \mathcal{D}_\nu e^\mu. \tag{4.134a}$$

$$(\delta_\xi \delta_\eta - \delta_\eta \delta_\xi) \psi^\mu = (\xi\eta R)^\mu{}_\nu \psi^\nu + (\xi\eta T)^\nu \mathcal{D}_\nu \psi^\mu. \tag{4.134b}$$

Here T in equation 4.134b is the torsion term in supersymmetry.

BRST quantization employs Popov ghost fields in a field theory so these ghost fields cancel out states that are not physical. This theory determines a coboundary operator Q such that $Q^2 = 0$, so physical states are

$$|\psi\rangle = kerQ/imQ. \tag{4.135}$$

Hence physical states are those where $|\psi\rangle \neq Q|\phi\rangle$. It is the ghost field cancellations that permit the construction of this coboundary operator. The object is to then introduce a Lagrangian L^{bc} comprised of anticommuting ghost fields b, c as functions over the $U(1)$ line bundle variables z, $\bar z$,

$$L^{bc} = b_\mu \partial_{\bar z} c^\mu + \bar b_\mu \partial_z \bar c_\mu \tag{4.136}$$

that provides the proper number of fields so that $Q^2 = \frac{1}{2}\{Q, Q\} = 0$. c creates a ghost field out of the vacuum and b as annihilating the ghost field.

The ghost fields are anticommuting scalars $\{b^\mu, c_\nu\} = \delta^\mu{}_\nu$, which define the ghost number N as

$$N = \sum_\mu c^\mu b_\mu. \tag{4.137}$$

A Lie Algebra $\mathcal{G} = SO(2,1)$ with elements e^μ obey the commutation relation

$$[e^\mu, e^\nu] = C^{\mu\nu}{}_\rho e^\rho. \tag{4.138}$$

The operator which determines the cohomology of $\mathcal{G}$ is

$$Q = c_\mu e^\mu - \frac{1}{2} C^{\mu\nu}{}_\rho c_\mu c_\nu b^\rho \tag{4.139}$$

which satisfies $Q^2 = 0$ by the Jacobi identity on the structure constants $C^{\mu\nu}{}_\rho$. Then for ghost number $N = n$, Q defines the $H^n(\mathcal{G}, R)$ cohomology class.

The BRST variation of an element of $\mathcal{G}$ is

$$\delta_B e^\mu = [\xi Q, e^\mu] + [\bar\xi \bar Q, e^\mu]$$
$$= i\epsilon^{\mu\nu}{}_\sigma \xi c_\nu \partial_z e^\sigma - i\epsilon^{\mu\nu}{}_\sigma \bar\xi \bar c_\nu \partial_{\bar z} e^\sigma, \tag{4.140}$$

where $\xi, \bar\xi$ are variables of a $U(1)$ line bundle. On a component basis these transformations are

$$\delta_B e^+ = i(c^- \xi \partial_z e^0 - \bar c^- \bar\xi \partial_{\bar z} e^0) - i(c^0 \xi \partial_z e^- - \bar c^0 \bar\xi \partial_{\bar z} e^-)$$
$$\delta_B e^- = i(c^0 \xi \partial_z e^+ - \bar c^0 \bar\xi \partial_{\bar z} e^+) - i(c^+ \xi \partial_z e^0 - \bar c^+ \bar\xi \partial_{\bar z} e^0) \tag{4.141}$$
$$\delta_B e^0 = i(c^+ \xi \partial_z e^- - \bar c^+ \bar\xi \partial_{\bar z} e^-) - i(c^- \xi \partial_z e^+ - \bar c^- \bar\xi \partial_{\bar z} e^+)$$

This theory involves a graded algebra, which defines a supersymmetric transformation of the tetrad e^μ into the spin $\frac{3}{2}$ field ψ^μ,

$$\delta_B e^+ = i(c^- \xi \partial_z e^0 - \bar\xi \bar c^- \partial_{\bar z} e^0) - (c^0 \xi \partial_z e^- - i\bar\xi \bar c^0 \partial_{\bar z} e^-) + 2i(\psi^+ \sigma \bar\xi - \xi \sigma \bar\psi^+)$$
$$\delta_B e^- = i(c^0 \xi \partial_z e^+ - \bar\xi \bar c^0 \partial_{\bar z} e^+) - i(c^+ \xi \partial_z e^0 - \xi \bar c^+ \partial_{\bar z} e^0) + 2i(\psi^- \sigma \bar\xi - \xi \sigma \bar\psi^-) \tag{4.142}$$
$$\delta_B e^0 = i(c^+ \xi \partial_z e^- - \bar\xi \bar c^+ \partial_{\bar z} e^-) - i(c^- \xi \partial_z e^+ - \bar\xi \bar c^+ \partial_{\bar z} e^-) + 2i(\psi^0 \bar\xi - \xi \sigma \bar\psi^0).$$

Similarly there is a transformation of the ψ^μ field as

$$\delta_B \psi^\mu = [\xi Q, \psi^\mu] + [\bar\xi \bar Q, \psi^\mu] = \epsilon^{\mu\nu}{}_\sigma (\xi c_\nu \partial_z \psi^\sigma + \bar\xi \bar c_\nu \partial_{\bar z} \psi^\sigma) + \mathcal{D}^\mu(\xi + \bar\xi) -$$
$$i e^\mu \left(2R(\sigma\bar\xi) - \gamma\xi - \frac{1}{3}\gamma(\xi\sigma\bar\sigma + 2R(\bar\sigma\xi) - \gamma\bar\xi - \frac{1}{3}\gamma(\bar\xi\bar\sigma)\sigma) \right). \tag{4.143}$$

The fields γ exist to equalize the number of boson and fermion modes. Thus on a component basis there is

$$\delta_B \psi^+ = (\xi c^- \partial_z \psi^0 + \bar\xi \bar c^- \partial_{\bar z} \psi^0) - (\xi c^0 \partial_z \psi^- + \bar\xi \bar c^0 \partial_{\bar z} \psi^-) + \mathcal{D}^+(\xi + \bar\xi)$$

$$\delta_B \psi^- = \left(\xi c^0 \partial_z \psi^+ + \bar{\xi}\bar{c}^0 \partial_{\bar{z}} \psi^+\right) - \left(\xi c^+ \partial_z \psi^0 + \bar{\xi}\bar{c}^+ \partial_{\bar{z}} \psi^0\right) + \mathcal{D}^-(\xi + \bar{\xi})$$
$$ie^+\left(2R(\sigma\bar{\xi}) - \gamma\xi - \frac{1}{3}\gamma(\xi\sigma\bar{\sigma}) + H.C.\right)$$
(4.144)
$$ie^-\left(2R(\sigma\bar{\xi}) - \gamma\xi - \frac{1}{3}\gamma(\xi\sigma\bar{\sigma}) + H.C.\right)$$
$$\delta_B \psi^0 = \left(\xi c^+ \partial_z \psi^- + \bar{\xi}\bar{c}^0 \partial_{\bar{z}} \psi^-\right) - \left(\xi c^- \partial_z \psi^+ + \bar{\xi}\bar{c}^- \partial_{\bar{z}} \psi^+\right) + \mathcal{D}^+(\xi + \bar{\xi})$$
$$ie^0\left(2R(\sigma\bar{\xi}) - \gamma\xi - \frac{1}{3}\gamma(\xi\sigma\bar{\sigma}) + H.C.\right).$$

The variation of the ghost fields b, c gives

$$\delta_B b = \xi(c\partial_z b + (\partial_z c)b) = 2i\xi T^{bc}_{ghost}, \quad \delta_B c = \xi c \partial c, \quad (4.145)$$

where the variation in b defines the total ghost energy-momentum tensor $T_{ghost} = T^{(bc)}_{ghost} + T^{(\bar{b}\bar{c})}_{ghost}$.

At this point a constraint on the fields needs to be introduced as a Lagrange multiplier in the total Lagrangian. A coordinate condition on the spatial components e^+, e^- must be imposed to fix the spatial surface and synchronizes clocks at every point. The 1-forms are then

$$\mathbf{e}^+ = e^{\phi^+} dz, \quad \mathbf{e}^- = e^{\phi^-} d\bar{z}, \quad \mathbf{e}^0 = e^{\phi^0} d\tau, \quad (4.146)$$

where z, $\bar{z}$ are coordinates of $\Sigma^{(2)} \subset \Sigma^{(3)}$. Define the duality between these 1-forms with the vectors e_μ so that $\langle e_\mu | \mathbf{e}^\nu \rangle = \delta_\mu^{\ \nu}$. The duality transformation are defined as $\langle e_\mu | \mathbf{e}^\nu \rangle^c = \langle e_\nu | \mathbf{e}^\mu \rangle^c$. By duality under the Hodge star operator restricted to $\Sigma^{(3)} \subset \mathcal{M}^4$ the two dimensional subspace $\Sigma^{(2)} \subset \Sigma^{(3)}$ is dual to $\Sigma^{(1)} \subset \Sigma^{(3)}$. Hence, fixing the coordinate on e^+, e^- fixes the condition on e^0. The coordinate condition that fixes the Lorentz transformation is $\phi^+ = \phi^-$. Define $\phi = \phi^+ + \phi^-$ with $\phi = \phi(z, \bar{z})$ and perform the BRST transformation of these variables

$$\delta_B \phi^+ + \delta_B \phi^- = \langle e_+ | \delta \mathbf{e}^+ \rangle + \langle e_- | \delta \mathbf{e}^- \rangle =$$
$$-i(\xi c^- \langle e_+ | \partial_z \phi^0 \mathbf{e}^0 \rangle - \bar{\xi}\bar{c}^- \langle e_+ | \partial_{\bar{z}} \phi^0 \mathbf{e}^0 \rangle) + -i(\xi c^+ \langle e_- | \partial_z \phi^0 \mathbf{e}^0 \rangle - \bar{\xi}\bar{c}^+ \langle e_- | \partial_{\bar{z}} \phi^0 \mathbf{e}^0 \rangle). \quad (4.147)$$

Similarly the BRST transformation of $\mathbf{e}^0$ is

$$\delta_B \phi^0 = \langle e_0 | \delta \mathbf{e}^0 \rangle =$$
$$\left(-i(\xi c^- \langle e_+ | \partial_z \phi^0 \mathbf{e}^0 \rangle - \bar{\xi}\bar{c}^- \langle e_+ | \partial_{\bar{z}} \phi^0 \mathbf{e}^0 \rangle) + -i(\xi c^+ \langle e_- | \partial_z \phi^0 \mathbf{e}^0 \rangle - \bar{\xi}\bar{c}^+ \langle e_- | \partial_{\bar{z}} \phi^0 \mathbf{e}^0 \rangle)\right)^c.$$
(4.148)

which is $(\delta_B \phi^+ + \delta_B \phi^-)^c$. The terms $\partial_z \phi^\pm$, $\partial_{\bar{z}} \phi^\pm$. $\partial_z \phi^0$ and $\partial_{\bar{z}} \phi^0$ act as connection terms which induce basis rotations. The duality between the spatial surface with coordinates z, $\bar{z}$ and τ with $\mathbf{e}^+ \wedge \mathbf{e}^- = e^{\phi^+ + \phi^-} * d\tau$ is seen in a comparison of equations 4.127 and 4.148. Clearly then $\phi^0 = \phi^+ + \phi^-$.

The BRST variation includes the supersymmetric transformation of the fields into their pairs. The BRST variation in the $e^\pm$ field is then

$$\delta_B e^\mu = i(\psi^\mu \sigma \bar\zeta - \zeta \sigma \bar\psi^\mu) + i\big(\xi(c^+ - c^-)\partial_z \phi - i\bar\xi(\bar c^+ - \bar c^-)\partial_z \phi\big)e^\mu$$

$$\delta_B \psi^\mu = i\big(\xi(c^+ - c^-)\partial_z\phi - i\bar\xi(\bar c^+ - \bar c^-)\partial_z\phi\big)\psi^\mu + \quad (4.149)$$

$$i\big(\xi(u^+ - u^-)\delta_z\phi - i\bar\xi(\bar u^+ - \bar u^-)\partial_z\phi\big)e^\mu + ie^\mu\Big(2R(\sigma\bar\xi) - \gamma\xi - \frac{1}{3}\gamma(\xi\sigma\bar\sigma) + H.C.\Big),$$

where $\delta_B c^\mu = u^\mu + c^\nu \partial_\nu c^\mu$ and $\delta u^\mu = c^\nu \partial_\nu u^\mu - u^\nu \partial_\nu c^\mu$.

The Lagrangian or constraint for this field plus the Lagrangian for the field ϕ, with its Lagrange multiplier π, defines the Lagrangian for the bosonic sector as

$$\mathcal{L} = \int \pi \partial_z \partial_{\bar z} \phi + \int (b^\mu \partial_{\bar z} c_\mu + \bar b^\mu \partial_z \bar c_\mu). \quad (4.150)$$

The entire BRST transformation of the theory is

$$\delta_B e^\mu = \epsilon^\mu{}_{\nu\rho} \xi c^\nu \partial_z e^\rho + 2i\psi^\mu \sigma \bar\xi + H.C.$$

$$\delta_B \psi^\mu = \epsilon^\mu{}_{\nu\rho} \xi c^\nu \partial_z \psi^\rho + 2ie^\mu(R - \Gamma)$$

$$\delta_B c^\mu = \frac{i}{2}\epsilon C^\mu{}_{\nu\sigma} c^\nu c^\sigma = c^\nu \partial_\nu c^\mu \quad (4.151)$$

$$\delta_B b^\mu = \epsilon B^\mu,$$

with $\Gamma = \gamma\xi + \frac{1}{3}\gamma\xi\sigma\bar\sigma$. The Faddeev-Popov procedure on the path integral with the action S results in

$$\int \frac{\mathcal{D}[e]}{Vol_{gauge}} e^{-iS} \to \int \frac{\mathcal{D}[e, b, c, B]}{Vol_{gauge}} e^{-i(S + S_1 + S_2)}. \quad (4.152)$$

Here a gauge fixing action $S_1 = B_\mu \Gamma^\mu(c)$ and a Faddeev-Popov action $S_2 = b_\mu c^\nu \delta_\nu F^\mu(e)$ are added to the original gauge invariant action. The gauge fixing condition $F^\mu(e)$ and the ghost field b_μ determine these two actions by

$$\delta_B\big(b_\mu F^\mu(e)\big) = \epsilon(S_1 + S_2). \quad (4.153)$$

The overlap of two states ψ and ψ', determined by changes in the gauge fixing condition, gives

$$\delta\langle\psi|\psi'\rangle = \langle\psi|\delta_B(b_\mu \delta F^\mu)|\psi'\rangle = i\langle\psi|\{Q_B, b_\mu F^\mu\}|\psi'\rangle. \quad (4.154)$$

The variation in the gauge fixing condition is set to zero, which gives that the physical states are invariant under the BRST charge operator Q_B according to

$$Q_B|\psi\rangle = Q_B|\psi^\dagger\rangle = 0. \quad (4.155)$$

166 Quantum Fluctuations of Spacetime

Since the anticommutation of the BRST charge operator with $b_\mu F^\mu$ gives the gauge fixing and Faddeev-Popov actions, Q_B is a constant of the motion with the Hamiltonian and gives

$$[Q_B, \{Q_B, b_\mu F^\mu\}] = [Q_B^2, b_\mu F^\mu] = 0. \tag{4.156}$$

Thus Q_B defines a cohomology for all physical states, which must be states that are closed under the action Q_B modulo exact states $|\psi\rangle = Q_B|\phi\rangle$.

It is now possible to define a hierarchy of operators that are generators of the action, where the first one as $\chi_0 = \mathbf{e}^\mu(e_\mu)$. This defines the invariant interval for the spacetime. Consider the coboundary operator d that shifts the field from a point x to x' on this generator as

$$d\chi_0 = 2\mathbf{e}^\mu \wedge \mathcal{D}e_\mu = \frac{i}{2}\{Q, (\xi\sigma\psi^\mu - \bar\psi^\mu\sigma\bar\xi) \wedge e_\mu\} \tag{4.157}$$

so that

$$\chi_1 = \frac{1}{2}\left(\xi\sigma\psi^\mu - \bar\psi^\mu\sigma\bar\xi\right) \wedge e_\mu. \tag{4.158}$$

Now work through the hierarchy of $d\chi_i = i\{Q, \chi_{i+1}\}$, which gives

$$\chi_2 = \frac{1}{2}\gamma\left(\xi\sigma\psi^\mu - \bar\psi^\mu\sigma\bar\xi\right) \wedge e_\mu - \frac{1}{2}(\xi\sigma - \sigma\bar\xi)\mathcal{R}^\mu{}_\nu e^\nu \wedge e_\mu$$

$$\chi_3 = -\frac{1}{4}\gamma(\xi\sigma - \sigma\bar\xi)\mathcal{R}^\mu{}_\nu e^\nu \wedge \left(\xi\sigma\psi^\mu - \bar\psi^\mu\sigma\bar\xi\right) \tag{4.159}$$

$$\chi_4 = -\frac{1}{4}\mathcal{R}^\mu{}_\nu e^\nu \wedge \mathcal{R}^\sigma{}_\mu e_\sigma$$

If $\chi_2 = \chi_2^1 + \chi_2^2$ the second term defines the spacetime Lagrangian

$$\mathcal{L}(\chi_2^2) = *((\xi\sigma - \sigma\bar\xi)\chi_2^2 \wedge \Sigma^{(2)}), \tag{4.160}$$

where $\Sigma^{(2)}$ is dual to the 2-surface $e^\nu \wedge e_\mu$ and $*$ is the Hodge dual star operator on $\mathcal{M}^{(4)}$. The four form χ_4 defines a quantity when integrated over the whole spacetime gives the Euler-Poincaré characteristic for the whole manifold. This is a four dimensional extension of the Gauss-Bonnet theorem.

Since Q is a cohomological operator on a cocyle then for an integral over χ_n

$$\int_\alpha \{Q, \chi_n\} = -i\int_\alpha d\chi_{n-1} = i\int_{\partial\alpha} \chi_{n-1} = 0, \tag{4.161}$$

as $\partial\alpha = 0$. This implies the integration $\int_{\alpha + \partial\beta} \chi_n$ is invariant, where the addition of a boundary term keeps the integral unchanged due to the BRST anticommutator. This gives the topologically invariant path integral over all homological cycles α_j which evaluates all $W(\chi_{n_j})$

$$Z[\cup_j \alpha_j] = \int \mathcal{D}exp(i\prod_j \int_{\alpha_j} \mathcal{L}(\chi_{n_j})). \tag{4.162}$$

The gravitational Lagrangian evaluated over all homological cycles is a construction similar to loop space geometry. In four dimensions the invariants defined by equation 4.163 corresponds to the Donaldson polynomials.

4.6: Moduli Space Constructions of Horizons

An instanton is a local minimum of the action for a Euclidean path integral. Instantons are topologically nontrivial configurations, which physically correspond to tunnelling phenomenon. This ties into the Floer cohomology introduced above, where fields are locally expanded around minima in the manifold and states may tunnel from one minima to another. Spacetime instantons are nontrivial two-cycles on spacetime. String world sheets that wrap around these two-cycles define string world sheet instantons. Instantons then determine the moduli of a space or spacetime. The points of the moduli space are defined by parameters that label the geometry or topology of a manifold so each coordinate in the moduli space is a moduli. The moduli can then label degenerate vacua in quantum field theory, where the expectations for each vacuum are physically inequivalent. Here the moduli space would consist of inequivalent geometries. This has parallels with the nature of the Goldstone boson. In chapter 5 this is discussed in the light of renormalizing the Planck scale and large extra dimensions. This will lead to the surprising possibility that the degenerate vacuum of the Higgs field is correlated to topological obstructions on the large scale due to Planck scale physics. This raises the interesting prospect that high energy experiments may in the near future find that the Higgs field is related to quantum black holes on a large scale.

A moduli space is a space of solutions that defines gauge equivalent gauge vectors. The moduli space is the set of all self-dual connections (instantons) with degrees of freedom removed by gauge conditions. This space is $\omega_{SD}/\mathcal{G} = \mathcal{M}_{mod}$. The number of metric moduli and conformal Killing vectors of the moduli space is obtained with the Riemann-Roch theorem. Given an operator $\mathcal{O}$ that is covariantly constant on a geodesic,

$$\partial g_{ij} = (2\delta(\omega) - \nabla_k \delta\sigma))g_{ij} - 2(\mathcal{O}\delta\sigma)_{ij}, \tag{4.163}$$

where ω is the conformal factor, σ is a spatial parameter. Here the metric is defined on the complexified coordinates z, $\bar{z}$. $\mathcal{O}$ may be derived by a variation of the metric g_{ij}

$$\delta g_{ij} = 2\delta\omega g_{ij} - \nabla_i \delta\sigma_j - \nabla_j \delta\sigma_i, =$$

$$(2\delta\omega - \nabla_k \delta\sigma_k)g_{ij} - (\mathcal{O}\delta\sigma)_{ij}, \tag{4.164}$$

with $(\mathcal{O}\delta\sigma)_{ij} = \nabla_i \delta\sigma_j + \nabla_j \delta\sigma_i - \nabla_k \delta\sigma_k)g_{ij}$. $\mathcal{O}$ defines the conformal Killing equation $(\mathcal{O}\delta\sigma)_{ij}u^j = 0$. Thus $\mathcal{O}$ defines the number of conformal Killing vectors. The transpose of $\mathcal{O}$, $\frac{1}{2}(\mathcal{O}^t u)_i = -\nabla^j u_{ab}$ defines the metric moduli. The Riemann-Roch theorem states that the difference between the numbers of the metric moduli and the number of conformal Killing vectors is -3χ, where χ is the Euler characteristic for the manifold. This difference is then

$$\frac{1}{2}\big(dim\ ker(\mathcal{O}) - dim\ ker(\mathcal{O}^t)\big) = -3\chi = 6g(g-1) \tag{4.165}$$

168 Quantum Fluctuations of Spacetime

Here is where instantons come into play, for $\mathcal{O}$ determines the number of instanton in the theory. Killing vectors in general relativity define instantons.

To develop this sort of theory some requirements are needed. Primarily the space has to satisfy self-duality or antiself-duality with the Hodge dual-star operator (*),

$$*F_{ij}dx^i \wedge dx^j = \pm F_{ij}dx^i \wedge dx^j. \tag{4.166}$$

This gauge-like form of gravity is demonstrated with the use of spinors and a connection to the Dirac field. The massless Dirac field is described by the Lagrangian $\mathcal{L}_D = i\bar{\psi}\gamma^\mu \partial_\mu \psi$, where the Dirac field. The four component Dirac spinor field is

$$\begin{pmatrix} \psi_1 \\ \psi_2 \\ \psi_3 \\ \psi_4 \end{pmatrix} \rightarrow \psi_l = e^a{}_l \psi_a, \tag{4.167}$$

where the index l runs over the number of spinor components. The index a for the entries are related to an internal symmetry by $\sigma^{\hat{a}}{}_{ab}\gamma = \gamma^{\hat{a}}$, where $\hat{a}$ runs over the basis of an internal space. This Lagrangian is written according to basis elements $e^a{}_l$. The Dirac Lagrangian is then

$$\mathcal{L} = i\bar{\psi}_c e^c{}_k{}^\dagger \gamma^\mu{}_{kl}(e^a{}_l \partial_\mu \psi_a + (\partial_\mu e^a{}_l)\psi_a) =$$
$$i\bar{\psi}_c e^c{}_k{}^\dagger \gamma^\mu{}_{kl} e^a{}_l (\delta^b_a \partial_\mu \psi_a + \omega^{ab}_\mu)\psi_b). \tag{4.168}$$

Here ω^{ab}_μ is identified as a spinorial connection coefficient for gravity $\omega^c_\mu = \epsilon^{cab}\omega^{ab}_\mu$ and $\omega^\nu{}_\mu = e^\nu_c \omega^c_\mu$, where e^ν_c are spacetime tetrads.

The Dirac matrices may posses an infinite number of possible representations. The above formalism may be considered according to whether the representation of the Dirac matrices are local. If these representations vary continuously on a 4-manifold a gauge-like formalism of general relativity is seen to emerge. When this is the case the Dirac action assumes the more general form

$$L_D = i\bar{\psi}\gamma^\mu \partial_\mu \psi \rightarrow i\bar{\psi}\partial_\mu(\gamma^\mu \psi) = i\bar{\psi}_a \partial_\mu(\gamma^\mu{}_{ab}\psi_b) =$$
$$i\bar{\psi}_a \gamma^\mu{}_{ab}\partial_\mu \psi_b + i\bar{\psi}_a(\omega_{\mu ac}\gamma^\mu{}_{cb} + \omega_{\mu cb}\gamma^\mu{}_{ac})\psi_b, \tag{4.169}$$

where the curvature tensor is

$$R^\mu{}_{\beta\rho\alpha} = \{\omega_{[\alpha a}{}^b{}_{,\rho]} + \omega_{[\alpha a}{}^c \omega_{\rho]c}{}^b\}\gamma^\mu{}_{bd}\gamma_\beta{}^{ad}. \tag{4.170}$$

Spacetime curvature is due to an internal symmetry space with

$$F^{\hat{a}}_{\mu\nu} = \omega^{\hat{a}}{}_{[\nu,\mu]} + \epsilon^{abc}\omega^b{}_{[\nu}\omega^{\hat{c}}{}_{\mu]}. \tag{4.171}$$

If there are no sources for the gravity field there are the self-dual or antiself-dual equations

$$\partial^\mu F^{\hat{a}}{}_{\mu\nu} = 0, \; *\partial^\mu * F^{\hat{a}}{}_{\mu\nu} = 0, \quad (4.172)$$

With the horizon algebra example above the $SO(2,1)$ theory in three dimensions is embedded into $SO(3,1)$. So embed this theory where the extra spatial dimension is dual to $*e^0 \wedge e^+ \wedge e^-$ with a signature [-1, 1, 1, 1].

The Donaldson's theorem gives a polynomial that determines the dimension of the moduli space $\mathcal{M}_{mod}$ for a manifold $\mathcal{M}$ with an internal group structure $\mathcal{G}$ [14]. This formula is

$$dim(\mathcal{M}_{mod}) = p^1((adP)\mathcal{M}) - dim(\mathcal{G})(1 - b_1 + dim(H^2(\mathcal{M},\mathbf{R}))). \quad (4.173)$$

where p^1 is the 1st Pontrayagin class, which is a Chern class in a complex valued $\mathcal{M}$, adP is the adjoint action of the bundle $\mathbf{P}$ (or Lie algebra) and b_1 is the first Betti number of $\mathcal{M}$. This is evaluated on subspaces of CP^2 in $\mathcal{M}_{mod}$ corresponding to singular points. These subspaces of CP^2 are defined by equivalent lines in a cone. One can check that if $\mathcal{G}$ is a trivial group, such as a unit matrix, that this formula collapses to the one above. So the moduli space of $\mathcal{M}$ is determined by the disjoint union of these "cones," and that $\mathcal{M}_{mod}$ is a cobordant of them.

A remarkable fact of this is that the intersection for this disjoint union is determined by $[E_8] \oplus (\sigma_x)$, where $[E_8]$ is the Cartan matrix of the exceptional algebra E_8 and σ_x is the Pauli matrix. With self duality there is the construction

$$[E_8] \oplus [E_8] \oplus (\sigma_x). \quad (4.174)$$

Donaldson found that one cannot remove the $[E_8] \oplus [E_8]$, which demonstrates the existence of $\mathcal{M}^4$ spaces that are homeomorphic but not diffeomorphic. There are apparent connections with string theories here with the role of the exceptional group E_8.

Given 4-dimensional spaces, with the action of a p-bundle with a group $\mathcal{G}$, there are some curious results which emerge. One can define the above topological numbers, but one also finds that the total number of $\mathcal{M}^4$ spaces contains these weird "fake $\mathcal{M}^4$s," that are homeomorphic but not diffeomorphic. So consider spacetime, even though Donaldson works with Euclideanized $\mathcal{M}^4$, as sliced into $\mathcal{M}^2$ and $\mathcal{M}'^2$ so that the curvature on each of these is self dual. It appears that with the "fake" $\mathcal{M}^4$ spaces there is a problem. The number of these spaces is infinitely large, where these need to be removed from the path integral. The Polyakov integral measure is $\frac{\mathcal{D}[e]}{Vol_{gauge}}$ over the diffeomorphism symmetry. This means that the "fake" spaces are removed, as they are not diffeomorphic. This means the path integral is only a sum over the real $\mathcal{M}^4$ spaces.

The string causes the gravitational field due to the action of the string at a length larger than the string. Similarly consider the string as intersecting certain $\mathcal{M}^2$'s as D-branes with Dirichlet boundary conditions under a spacetime target map and that the

$\mathcal{G} = SU(N)$ is a Chan-Paton group action. Examples of this are the 2-dimensional surfaces of fields that hover one Planck unit above the event horizon of a black hole, as well as Lanzcos junction conditions. This 2-dimensional surface is then defined by the horizon algebra, where the above the supersymmetric extension of this leads to an invariant action which corresponds to the Donaldson polynomial. Strings interact with these 2-dimensional membranes and the nature of this interaction will involve the topological quantum numbers associated with the 2-dimensional membrane.

To illustrate that equation 4.162 corresponds to the Donaldson's polynomial consider the variation of the curvature and its dual

$$\delta(F_{\mu\nu} + *F_{\mu\nu}) = \partial_\nu \delta\omega_\mu - \partial_\mu \delta\omega_\nu + \delta(\omega_{[\mu}\omega_{\nu]}) = 0. \tag{4.175}$$

Now impose the gauge condition $\mathcal{D}_\mu \omega = 0$ to removed gauge redundancies. Clearly the action requires the input of the Rartia-Schwinger field as well

$$\mathcal{D}_\nu \delta\psi_\mu - \mathcal{D}_\mu \delta\psi_\nu + \mathcal{D}_{[\mu} \psi_{\nu]} = 0, \tag{4.176}$$

with a constraint equation $\mathcal{D}_\mu \psi^\mu = 0$. The χ equations and the constraints reduce the number of independent degrees of freedom in the fields ψ and ω, which is the dimension of the moduli space. This results in

$$dim(\mathcal{M}_{mod}) = \sum_j (4 - n_j). \tag{4.177}$$

Thus for the fields Ψ, which corresponds to gravity and its associated gravitino field, which are evaluated on cocycles α_j the path integral is then

$$Z(\alpha_1, \alpha_2, \ldots, \alpha_n) = \int \Psi_{\alpha_1} \wedge \Psi_{\alpha_2} \wedge \ldots \wedge \Psi_{\alpha_n}. \tag{4.178}$$

This defines the path integral for quantum cohomology in a form determined by the Donaldson polynomial.

A principal bundle which is split $p = p_1 \oplus p_2$ means the differential operator decomposes as $\mathcal{D} = d_1 \oplus d_2$. $\mathcal{D}$ has a nonvanishing kernal. For the operator $L = L^\dagger \in ker(D)$ that acts on an eigenvector e, then $Le = \pm i\lambda e$ and

$$L\mathcal{D}e = i(d\lambda)e + i\lambda \mathcal{D}e. \tag{4.179}$$

The differential of the eigenvalue is $d\lambda = Im(u\mathcal{D}e, e)$, which by manipulation is $\lambda(\mathcal{D}, e)$, vanishes by $(e, e) = 1$ The defines the splitting of the bundle and with $\mathcal{D}e = 0$ the differential operator and the connection are split. Under this splitting the bundle $SO(2,1) \sim U(1,1)$ is reduced to a $U(1)$ bundle. This is equivalent to the null condition on the elements of equation 4.146

$$0 = \mathbf{e}^+ \otimes \mathbf{e}^+ + \mathbf{e}^- \otimes \mathbf{e}^- - \mathbf{e}^0 \otimes \mathbf{e}^0 = e^{2\phi_+} dz^2 + e^{2\phi_-} d\bar{z}^2 - e^{2\phi_0} d\tau^2. \tag{4.180}$$

With the condition $\phi^+ = \phi^-$, and $\phi = \phi^+ + \phi^-$ the reduction of the bundle is seen with the condition that $\phi = \phi^0$. This null condition further occurs with the interchange in the signature in the metric. With the sourceless Einstein field equation the curvatures R^+, R^-, R^0 satisfy the following conditions

$$R^+ = R^- - e^{2(\phi^- - \phi^0)}R^0, \ R^- = R^+ - e^{2(\phi^- - \phi^0)}R^0$$

$$R^0 = \frac{1}{3}\left(e^{2(\phi^+ - \phi^0)}R^+ + e^{2(\phi^- - \phi^0)}R^-\right), \quad (4.181)$$

or

$$R^+ = R^-, \ R^0 = \frac{2}{3}e^{2(\phi - \phi^0)}R \to 0, \quad (4.182)$$

satsified on the horizon. As this approaches the event horizon $\phi^\pm \to 0$ and $\phi^0 \to \infty$, so $\partial_i e^{\pm,0} = e^{\pm,0}\partial_i \phi^{\pm,0} = 0$. It is evident that the connections satisfy

$$\omega^{\pm,0} = \mathcal{D}\chi \quad (4.183)$$

will satisfy $R^{(\pm),0} = d\omega^{(\pm),0}$. Thus the bundle is spilt into $\mathcal{D} = d^{(\pm)} \oplus d^0$ and is determined by the circle

$$\begin{pmatrix} e^{i\theta} & 0 \\ 0 & e^{-i\theta} \end{pmatrix}. \quad (4.184)$$

such that the bundle has been reduced to a $U(1)$ bundle. This splitting of the $SO(2,1)$ bundle for ingoing and outgoing states then defines the event horizon. This corresponds to a singularity in the moduli space. It is evident that the Lagrangians χ_i in equation 4.162 defines a topological invariant or charge on the horizon and is identified with a moduli "blow-up."

This split bundle then defines all fields "pinned" to an event horizon, such as with a black hole. Here fields are properly observed by an observer exterior to a black hole.

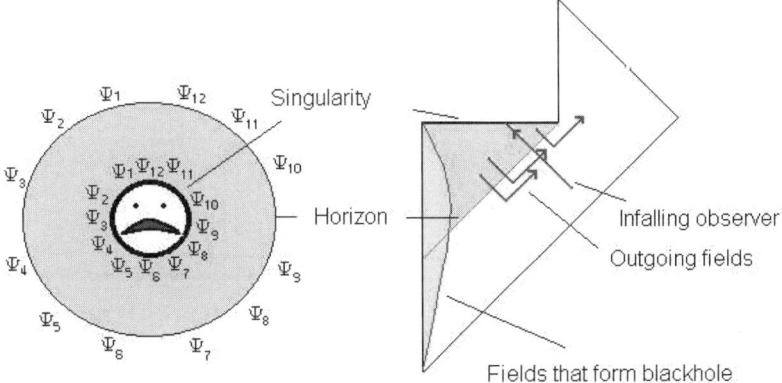

Duality of fields with a black hole.
Figure 4.2

This leads to a curious dichotomy. The exterior observer watches the interior observer get "pasted" onto the membrane of gravity field oscillations just above the horizon and appear to wink out, while the interior observer continues to persist into the spacelike region of the black hole only to then suffer a rapid disintegration upon approaching the interior singularity. The black hole quantum mechanically radiates the modes associated with the interior observer, as seen by the exterior observer. Conversely these modes may persist until the black hole finally radiates away in its final burst. So it appears that fields associated with event horizons have a dual description associated with singularities. For the black hole this is illustrated in figure 4.2. Here the past and future null direction $\mathcal{I}^-$, $\mathcal{I}^+$ bound the spacetime and the black hole is illustrated with its horizon and singularity. The singularity is a space-like surface, where the inner horizon on a Kerr black hole is likely a singularity. The inner horizon of a physical Kerr black hole presents an interior observer with a divergence of fields that pile up upon the observer's approach to the horizon. This leads to an infinite pulse of radiation and fields that impact on the inner horizon, which suggests this horizon predicted by a solution for an "eternal black hole," is physically the singularity. The interiors of black holes are unobservable, so this interpretation is considered without an examination of solutions that involve black hole interiors. The exterior observer then observes the infalling observer to flatten out near the horizon and measures quantum fields that scatter out of the black hole. The scattered field may contain quanta involved with the infalling observer, yet the infalling observer does not perceive anything particular upon crossing the event horizon.

This leads to the conjecture that the fields associated with the singularity are the same as those associated with the horizon membrane, given by a D2-brane under a spacetime target map. This implies some sort of nonlocality or entanglement associated with the fields on the event horizon and those with the singularity. Based on the Penrose diagram this implies fields pinned above the horizon are entangled with identical copies of fields associated with the singularity. In fact the membrane above the horizon is from a field theory point of view identical to the field modes of the singularity. However, this leads to a curious situation that appears similar to "quantum cloning," whereby an operator acts on a state, or field so that $\mathcal{U}\psi = 2\psi$. Quantum cloning suffers from the problem that if $\mathcal{U} = e^{-iO} \simeq 1 - iO$ then

$$\mathcal{U}\psi \simeq (1 - iO)\psi = \psi - i(O\psi), \qquad (4.185)$$

where $iO\psi = -\psi$. This means that $iO^\dagger\psi = \psi$ and so then $U^\dagger U$ acts on ψ so that

$$UU^\dagger\psi = 4\psi, \qquad (4.186)$$

and so U is not a proper unitary operator.

This black hole cloning is different however. From equation 2.112 any such two fields define a single field by a thermal distribution with a Bogoliubov transformation between these fields. This would infer that these fields are duplicated across a horizon, where an external observer is unable to gain access to information about both fields with complete accuracy. Further, the splitting of the bundle at the horizon indicates these two fields are not duplicated by an operator that is generated by elements of group operators in $SO(2,1)$. This splitting of the bundle generated by the horizon algebra is an indication of how there is not a quantum unitary evolutionary description between the fields associated with the singularity and those pinned on the event horizon.

Given the moduli space $\mathcal{M}_{mod}$ the singularities in this moduli space may be "blown up" into projective geometries, or subspace cones within a projective space as a "blow-up" of a point in these moduli spaces. If the metric is Euclideanized for the horizon algebra $(-,+,+) \rightarrow (+,+,+)$, the topology of the space is described by an elliptic complex

$$0 \rightarrow \Lambda^0(ad\ P) \rightarrow^D \Lambda^1(ad\ P) \rightarrow^{P\text{-}D} \Lambda^2(ad_P), \quad (4.187)$$

where $\pi_- D$ is the projection of the differential operator onto the anti-self dual forms and $ad\ P$ is the adjoint action of the group that describes the vector field for the principle bundle **P**. Near a singular point in the moduli space the group action defines the kernal of the operator $\pi_- D$ is in $\Lambda^2(ad\ P)$. By the application of the Fredholm alternative the operator may be split according to its image and kernal, where if both of these are finite the image of the operator defines cohomology of the operator D into the forms

$$H_D^1 = C^q,\ H_D^2 = C^p \oplus P_- H_{DR}^2(M), \quad (4.188)$$

where the cohomology in the second term is the deRham cohomology. Now if the second cohomology term vanishes then the first cohomology ring may be described are given by $H_D^1 \sim C^3$. Yet this is described within a neighborhood of the singularity, e.g., within S^1 and so the orbit space for the operator is

$$C^3/S^1 = CP^2 \times [0,1)/(x,0), \quad (4.189)$$

which is a cone of projective rays in CP^2.

This result is stunning, but it does not completely satsify what is needed. A hyperbolic metric, which is honestly more physical, needs similar short exact cohomological sequence without elliptic structure, but rather a hyperbolic structure. This sort of structure has not been well explored. However, the following physical argument is given for why this mathematical structure should still apply. It goes back to the old "high school" presentation on the relationship between the ellipse, parabola and hyperbola. A conical beam of light that passes through a plane will describe an ellipse. If one rotates the plane the light beam intersects the plane in a parabola. This is the critical angle where the plane is parallel to the edge of the beam. If the angle of this plane is rotated further the flashlight beam will define a hyperbola on its intersection with the plane. In the case of the ellipse the more distant endpoint on the ellipse will contain the same geometric information across that point. However, in the case of the parabola and hyperbola this condition is not assured. The two endpoints "at infinity" are free from each other and what topological information these two points have are independent of each other. This then demands some assignment of some sort of topological relationship between these endpoints. This requires that additional topological information must be imposed on the end points, to eliminate this additional freedom. This implies that elements $z \in CP^2$ will exhibit a discrete number involved with matching the two end points of the hyperbolas. So the elements of $z \in H_D^1$ when mapped to the complex space C^3 have a cyclicity or Bohr wave-like structure with end points that meet up in an integral number of "beats." The ansatz is then that the elements of the orbit space are given by coordinates $z \to z^n$ when mapped to the Euclidean geometry. This then means that the blow up of a singular point in the moduli space with hyperbolic metric structure has this "fanning" structure of a weighted projective space. This weighted structure is then carried into the cones at the blow up of a singularity in the moduli space. These cones will then have a weighted structure.

This argument is meant as a "physical" argument for a mathematical structure. The above argument is one that would need to be examined by serious mathematical research. However, here the interest is with physical insight over mathematical theorems.

The projective space CP^2 is the set of complex lines in the complex space C^3 with complex coordinates. Formally this is

$$CP^2 = (C^3 - \{0\})/(x_0 \simeq \lambda x_0), \qquad (4.190)$$

with the coordinates in the projective space determined by $z_j = x_j/x_0$. The weighted projective space defines the equivalence class on the complex coordinates in CP^2. The weighted projective space, $CP^2{}_w$, may be constructed from CP^2 with the map $CP^2 \to CP^2{}_w$ defined by the action on the coordinates in $CP^2{}_w$ by,

$$[z_1, z_2] \mapsto [z_1{}^{a_1}, z_2{}^{a_2}], \qquad (4.191)$$

or

$$[z_1, z_2,] \mapsto [r_1{}^{a_1} e^{i a_1 \theta_1}, r_2{}^{a_2} e^{i a_2 \theta_2}]. \qquad (4.192)$$

This establishes an identification between the points in the $[0, 2\pi r/a]$ "pie slices" of each complex line.

Now consider two maps:

$$f : CP^2 \to CP^2(a_1, a_2) = CP_w^2$$
$$g : CP^2 \to CP^2(b_1, b_2) = CP_{w'}^2, \qquad (4.193)$$

so that the weights for the two maps are unequal. Let $d\hat{z}_j$ and $d\hat{z}'_j$ be differential basis one-forms in CP_w^2 and $CP_{w'}^2$ respectively. Then under the two pull-back actions,

$$f^* : CP^2{}_w \to CP^2, \quad g^* : CP^2{}_{w'} \to CP^2. \qquad (4.194)$$

These basis one-forms are pulled back into CP^2 according to,

$$d\hat{z}_j \mapsto df_j = a_j z_j{}^{a_j - 1} dz_j$$

$$d\hat{z}'_j \mapsto dg_j = b_j z_j^{b_j - 1} dz_j. \tag{4.195}$$

The dual vectors, V_j, V'_j on CP^2 take the form,

$$V_j = a_j^{-1} z_j^{1-a_j} \frac{\partial}{\partial z_j}$$

$$V'_j = b_j^{-1} z_j^{1-b_j} \frac{\partial}{\partial z_j} \tag{4.196}$$

A commutator of the two vector fields gives,

$$[V_j, V'_j] = (a_j - b_j)(a_j b_j)^{-1} z_j^{1-a_j-b_j} \frac{\partial}{\partial z_j}. \tag{4.197}$$

The vector fields defined as

$$L^{a_j} = a_j V_j = z^{1-a_j} \frac{\partial}{\partial z_j}, \quad L^{b_j} = b_j V_j, = z^{1-b_j} \frac{\partial}{\partial z_j}, \tag{4.198}$$

satisfy the Witt algebra or Virasoro algebra without central extension [15] for each index value,

$$[L^{a_j}, L^{b_j}] = (a_j - b_j) L^{a_j + b_j}, j \in (1, 2). \tag{4.199}$$

This result begs for the existence of the map, $CP^2 \to CP^2(a_1 + b_1, a_2 + b_2)$ and so an infinite "ladder" of such maps exists. So the weighting of CP^2 results in two copies of the Virasoro algebra. Finally, the Virasoro algebra maybe extended by a one-dimensional center with the map

$$m : Vir \oplus C \to Vir, \tag{4.200}$$

where this central extension is the kernel of the map. The extended Virasoro group is then,

$$[L^{a_j}, L^{b_j}] = (a_j - b_j) L^{a_j + b_j} + c(a_j, b_j). \tag{4.201}$$

This prescription for weighted projective spaces permits the Virasoro generators to be expanded according to the Laurent expansion

$$L^{a_n} = \oint \frac{dz}{2\pi i z} z^{a_n + 2} T(z), \quad T(z) = \sum_{a_n = -infty}^{\infty} \frac{L^{a_n}}{z^{m+2}}. \tag{4.202}$$

The anomaly term may be evaluated if the Virasoro operators are extended by $T^{a_j} = L^{a_j} - \delta_{m,0} c/24$. Define central charges as

$$Q^{a_n} = \oint_C \frac{dz}{2\pi i} j^{a_n}, \tag{4.203}$$

for

$$j^{a_n} = z^{m+1}T(z), \quad T(z) = -\sum_{n=-\infty}^{\infty} e^{im(x+iy)}. \tag{4.204}$$

for $(x, y) \in C$. NOw compute the commutator between two charges to find

$$[Q^{a_n}, Q^{b_n}] = \oint_C \frac{dz_2}{2\pi i} res_{\bar{z}_1 \to z_2} j^{a_n} j^{b_n}, \tag{4.205}$$

where the residue is

$$res_{\bar{z}_1 z_2} j^{a_n} j^{b_n} = res_{\bar{z}_1 \to z_2} \left(\frac{c}{2|z_1 - z_2|^4} + \frac{2}{|z_1 - z_2|^2} T(z) + \frac{1}{|z_1 - z_2|} \partial T(z) \right)$$

$$= \frac{c}{12}(a_n^3 - a_n)z_2^{m+n-1} + (m-n)z_2^{m+n+1}T(z_2). \tag{4.206}$$

which gives the anomaly term in the commutation of Virasoro generators.

Commutators of the Virasoro generators L^{-1}, L^0, L^1 produce the $SL(2, R)$ algebra

$$[L^0, L^{-1}] = L_{-1}, \quad [L^0, L^1] = -L^1, \quad [L^1, L^{-1}] = 2L^0. \tag{4.207}$$

This is the same in form as the $SU(2)$ algebra for the angular momentum operators $L_{\pm}$, L_z, but is noncompact. This does indicate that the hyperbolic nature of the spacetime metric is codified into the moduli space singularities in an algebraic form.

This Virasoro algebra corresponds to field amplitudes of their own. The Virasoro generator L^n may then be written according to field operators α_m as

$$L^n = \frac{1}{2} \sum_{m=-\infty}^{\infty} \alpha_{n-m} \alpha_m. \tag{4.208}$$

This demands a Hamiltonian that contains field amplitudes associated with the two Virasoro algebras as

$$H = \frac{1}{2} \sum_{-\infty}^{\infty} (\alpha_{-n}\alpha_n + \alpha'_{-n}\alpha'_n). \tag{4.209}$$

Write the Virasoro algebra, without center, as,

$$[L^a, L^b] = C^{ab}{}_c L^c. \tag{4.210}$$

Consider the vector, $\xi^\alpha = \xi^\alpha{}_a L^a$, where ξ^α is an element of the Lie algebra $\mathcal{G}$. The commutator in $\mathcal{G}$ is then

$$[\xi^\alpha{}_a, \xi^\beta{}_a] = C_g{}^{\alpha\beta}{}_\gamma = [\xi^\alpha{}_a L^a, \xi^\beta{}_b L^b] = C_g^{\alpha\beta}{}_\gamma \xi^\gamma{}_{a+b} L^{a+b}. \tag{4.211}$$

A separation of variables, $\xi^\alpha{}_a(x, z) = P^n(z)\zeta^\alpha{}_a(x)$ for $x \in \mathbf{X}$ and $z \in CP^n$, gives the commutator,

$$[\xi^\alpha, \xi^\beta] = [\xi^\alpha{}_a, \xi^\beta{}_b]P_n^{-1}(z)\frac{\partial}{\partial z}P_n(z)z^{2-a-b}\frac{\partial}{\partial z}, \quad (4.212)$$

which requires $P_n(z) = z$, or z^{-1}. On the unit circle z and z^{-1} act as overall phases. The elements $\xi^\alpha{}_a$ act as n-trads, or vielbeins, $\xi^\alpha{}_a : Vir \to G$, which obey the Kac-Moody algebra [16],

$$[\xi^\alpha{}_a, \xi^\beta{}_b] = C_g{}^{\alpha\beta}{}_\gamma \xi^\gamma{}_{a+b}. \quad (4.213)$$

associated with the Lie algebra $\mathcal{G}$.

Now consider the $\mathcal{G}_{KM}$ valued principal bundle $\mathbf{P}$ over $\mathbf{X}$ with the fibration $\pi : \mathbf{P} \to \mathbf{X}$, where sections of the principal bundle are given by vielbeins $\Gamma^\alpha{}_a(\mathbf{P}) = \xi^\alpha{}_a$. On the principal bundle the Hermitian metric[16],

$$\eta^{\alpha\beta} = \eta(\xi^{\dagger\alpha}, \xi^\beta) = \xi^{\dagger\alpha}{}_a \xi^\beta{}_a, \quad (4.214)$$

is defined for $\mathbf{P}$ locally holomorphically trivial. This is the Killing form for the Lie algebra $\mathcal{G}$. The bracket structure for ξ and $\xi^\dagger$ are two commuting Virasoro algebras. Within this trivialization the connection coefficients are,

$$(\eta^{-1}\partial_\mu\eta)_\alpha{}^\beta = \eta_{\alpha\gamma}\partial_\mu\eta^{\gamma\beta}$$
$$= \xi^\dagger{}_{\alpha a}\xi^\gamma{}_a(\partial_\mu\xi^\dagger{}_\gamma \xi_b{}^\beta + \xi^\dagger{}_\gamma \partial_\mu\xi_b{}^\beta), \quad (4.215)$$

which equals,

$$(\partial_\mu\xi^\dagger{}_\alpha \xi_b{}^\beta + \xi^\dagger{}_\alpha \partial_\mu\xi_b{}^\beta) = A^\dagger{}_\alpha{}^\beta{}_\mu + A_\alpha{}^\beta{}_\mu = \mathcal{A}_\alpha{}^\beta{}_\mu. \quad (4.216)$$

The curvature tensors $\mathcal{F}_\alpha{}^\beta = F_\alpha{}^{\beta\dagger} + F_\alpha{}^\beta$, which consist of holomorphic and antiholomorphic curvatures, are then

$$\mathcal{F}_\alpha{}^\beta{}_{\mu\nu} = \partial_{[\nu}(\eta_{\alpha\gamma}\partial_{\mu]}\eta^{\beta\gamma})$$
$$= \partial_{[\nu}\mathcal{A}^{\beta\gamma}{}_{\mu]} + \mathcal{A}^\dagger{}_{\alpha\beta[\mu}\mathcal{A}^{\beta\gamma}{}_{\nu]}. \quad (4.217)$$

These elements may be seen to transform as connections. The gauge shift of these connection coefficients by a variation on the CP^2 coordinates $z, \bar{z}$ is,

$$A'_\alpha{}^\beta{}_\mu - A_\alpha{}^\beta{}_\mu = \delta A_\alpha{}^\beta{}_\mu = \delta(\eta^{-1}\partial_\mu\eta)_\alpha{}^\beta. \quad (4.218)$$

The identity $\delta\eta^{-1} = -\eta^{-1}\delta\eta\eta^{-1}$ and the chain rule gives,

$$\mathcal{A}_\alpha{}^\beta{}_\mu = \partial_\mu(\eta^{-1}\delta\eta)_\alpha{}^\beta + [\mathcal{A}_\mu, \eta^{-1}\delta\eta]_\alpha{}^\beta, \quad (4.219)$$

or $\mathcal{A}_\alpha{}^\beta{}_\mu = \mathcal{D}_\mu(\eta^{-1}\delta\eta)_\alpha{}^\beta$.

Since $\delta\eta = \eta' - \eta$ the gauge transformation is expressed as,

$$(\eta'^{-1}\partial_\mu\eta')_\alpha{}^\beta - (\eta^{-1}\partial_\mu\eta)_\alpha{}^\beta = \mathcal{D}_\mu(\eta^{-1}\eta')_\alpha{}^\beta, \qquad (4.220)$$

which is a solution generating transformation. Now write $\delta\eta$ according to vielbeins,

$$\delta\eta^{\alpha\beta} = \delta\xi^\dagger{}_a{}^\alpha \xi_a{}^\beta + \xi^\dagger{}_a{}^\alpha \delta\xi_a{}^\beta \qquad (4.221)$$

Let the variation of the vielbein components with respect to the coordinates $z\bar{z}$ be,

$$\xi'^\alpha{}_a = \xi^\alpha{}_a + \frac{\partial}{\partial z}\xi^\alpha{}_a \delta z$$

$$\xi^{\dagger\prime\alpha}{}_a = \xi^{\dagger\alpha}{}_a + \frac{\partial}{\partial\bar{z}}\xi^\alpha{}_a \delta\bar{z}. \qquad (4.222)$$

For $\xi^\alpha{}_a = z^{-1}\zeta^\alpha{}_a(x)$, $\xi^{\dagger\alpha}{}_a = \bar{z}\zeta^{\dagger\alpha}{}_a(x)$, $z = e^{i\theta}$ then,

$$\delta A^{\alpha\beta}{}_\mu = -2\delta\theta(A^{\alpha\beta}{}_\mu + [A^\alpha{}_{\gamma\mu}, \xi^\dagger{}_\gamma \xi^\beta]). \qquad (4.223)$$

This gauge transformation is equivalent to the standard expression for the gauge transform of a connection,

$$A^{\alpha\beta}{}_\mu = G^{-1\alpha}{}_\gamma A^{\gamma\delta}{}_\nu G_\gamma{}^\beta + G^{-1\alpha}{}_\gamma \partial_\mu G^\gamma{}_\beta \qquad (4.224)$$

Assume that $G^{\alpha\beta} = e^{i\epsilon\phi^{\alpha\beta}} \simeq \delta^{\alpha\beta} + i\epsilon\phi^{\alpha\beta}$, and define $\phi^{\alpha\beta} = \xi^{\dagger\alpha}\xi^\beta$ and $\epsilon = 2\delta\theta$, then the equivalence is apparent. Similarly the transformation of the gauge fields is,

$$\delta F^\alpha{}_{\beta\mu\nu} = [F^\alpha{}_{\gamma\mu\nu}, \xi^{\dagger\gamma}{}_a \xi^\gamma{}_a], \qquad (4.225)$$

which is equivalent to the standard expression.

In Chapter 5 a T-duality between fields on a D2-brane and a D5-brane is discussed. This means two gauge fields apparently exist. In order for this to be a proper description of a D-brane these must be introduced into the Dirac-Born-Infeld action

$$S = -T\int d^3\xi Tr\left(e^{-\Phi}(-det(G_{ab} + B_{ab} + 2\pi\alpha F_{ab} + 2\pi\alpha' F'_{ab})^{1/2}\right). \qquad (4.226)$$

Here G_{ab} and B_{ab} are the Neveu-Schwartz-Neveu-Schwartz (NS-NS) states and the two gauge fields are indicated. These are boson fields in type I and II superstring theories, which can include gravity and the diliton field. T is the membrane tension and ξ represent coordinates on the brane. The indices a, b run over the brane and the terms $\partial X^\mu/\partial\xi^a$ intertwine between spacetime coordinates and brane coordinates. It is reasonable to assume these gauge fields are identical copies of each other, but with possibly different coupling scales α, α'.

It is tempting to think these two gauge fields from the holomorphic coordinates z and $\bar{z}$ exhibit a duality, such as the Montenen-Olive duality [17]. In this sense consider that

$$\alpha' = \frac{2\pi}{\alpha}. \tag{4.227}$$

It is known that field theories exhibit duality between their electric and magnetic field components in the absence of sources. For the electromagnetic theory there is a duality between the electric and magnetic fields in the divergence equations and the Maxwell-Faraday equations for the electric and magnetic field in vacua. This can also be understood according to the duality of two-chains in four-dimensional manifolds. This lead Dirac, in his examination of the motion of an electric charge in the presence of a magnetic monopole field, to demonstrate that a wave function for such a particle will have proper phase construction if the product of the electric and magnetic charges are such that [18]

$$eq = 2\pi\hbar n. \tag{4.228}$$

Dirac's magnetic monopole consisted of an infinitely long solenoid "tail" that ended at the source of the longitudinal magnetic field. This result was generalized by 't Hooft and Polyakov who demonstrated that magnetic monopoles were entities that resulted from grand unified field theories. Further, these magnetic monopoles did not have the semi-infinitely long solenoid tail. This leads to the concept that sourceless gauge theories have a certain "internal" duality between the electric and magnetic fields, or their nonabelian analogs, where further results by Donaldson explores this duality according to the nature of four dimensional manifolds.

The fine structure constant of the electromagnetic theory, $\alpha = e^2/\hbar c$ is then associated with a magnetic fine structure constant $\alpha_m = q^2 c/4\pi\hbar$. This means that this magnetic fine structure constant is related to the reciprocal of the fine structure constant $\alpha_m = n^2 \pi \alpha^{-1}$. This observation leads to the conjecture that strong and weak coupling constants can be interchanged between dual gauge theories [19-22].

Montonen and Olive [17] suggested that the weak electric coupling constant in one gauge theory can be associated with a strong magnetic coupling constant in another. Similarly the strong electric coupling constant in one gauge theory is associated with a weak magnetic gauge coupling in another theory. As such one theory, where the gauge quanta couple to weak electric charges, is dual to a second field theory where the electric field coupling is strong, but where the quanta couple to weak magnetic charges. This means that in the case of QCD, where the coupling is very strong by the quark color field, the source of the chromo-electric field, there exists a dual theory with weak chromo-magnetic coupling. In the dual theory the quanta of the theory are determined by the magnetic charges.

These couplings then demonstrate T-duality. The mass of this two dimensional membrane above a black hole is

$$M = T_2 e^{-\Phi} 4\pi^2 R^2. \tag{4.229}$$

The T-duality on one of the directions x gives that this mass is also

$$M = 4\pi^2(\alpha + \alpha')T_1 e^{-\phi}R, \tag{4.230}$$

which is the form for the mass of a dilaton black hole tied to an open string. Thus the string tension T_1 and the membrane tensions are related by

$$T_2 = \frac{T_1}{2\pi\sqrt{\alpha + \alpha'}}. \tag{4.231}$$

From this duality consider that $F_{\mu\nu} = F'_{\mu\nu}$. Now expand the action in equation 4.226 to second order so the gauge field appears as

$$\frac{T}{4}\big(2\pi(\alpha + \alpha')\big)^2 Tr(F^{\mu\nu}F_{\mu\nu}), \tag{4.232}$$

in the action and recovers the standard action for gauge theories with Montonen-Olive duality

This gauge field then enters Chan-Paton factors for strings that interact with this two dimensional D-brane. Chan-Paton factors require oriented $U(n)$, unoriented $SO(n)$ and $Sp(n)$ theories. The above T-duality argument indicates that the gauge field on the membrane is determined by these endpoints of the string.

It is worth noting that this structure just illustrated leads to an infinite regression and a fractal type of structure. A gauge field emerges from the moduli space singularity at the horizon. This gauge field in turn will have a moduli space and singularities. Due to the pseudoEuclidean nature of the metric, the above argument should lead to an additional gauge field. This process is then apparently an infinite regression. In fact this appears to have a structure similar to that of Julia sets, discussed at the end of this chapter. For a manifold $\mathcal{M}$ with a principal bundle $\mathbf{P}$ then the moduli space $\mathcal{M}_{mod} = \omega_{SD}/\mathcal{G}$ is the input of $\mathbf{P}$ into a "machine", R that produces $\mathcal{M}_{mod}$ and a new principal bundle $\mathbf{P}'$

$$R : \mathbf{P} \rightarrow \mathbf{P}'. \tag{4.233}$$

Then for $\mathbf{P}$ evaluated at a point $x \in \mathcal{M}$, $R(\mathbf{P}(x)) = \mathbf{P}'(x)$. Obviously one can consider iterated applications of this R so that a sequence of gauge structures are generated. These gauge structures may have a self-similar nature to them, or they may correspond to a hierarchy or "tower" of structures. At this point the nature of this has not been determined. However, this leads to the conjecture that there exists an infinite number of structures associated with the two dimensional D-brane. Possibly as physics approaches the Planck scale nature exhibits an infinite amount of structure, where the underlying rules behind this involve fractal geometry and a form of holomorphic dynamics. Curiously due to the Bekenstein bound the amount of information associated with these structures must also approach unity as they converge to the Planck unit of length.

4.7: Brane Dynamics and Black Holes

The event horizon of a black hole contains all available quantum information about the black hole. This null surface is a subspace of a D3-brane. The quantum fields on this surface are correlated with internal fields according to the Bogoliubov transformation. The quantum operators for this D2-brane then have a general form as presented in chapter 3. The operators for the Horizon algebra may be generalized according to

$$\alpha_i^b = \hat{a}_i^b + a_i^b e^{-\pi\mathcal{H}/g}, \tag{4.234}$$

which obey the commutation

$$[\alpha_i^b, \alpha_j^c] = (1 + e^{-2\pi\mathcal{H}/g})\epsilon^{bcd} t_{ijk}\alpha_k^d -$$

$$\frac{2\pi}{g} e^{\pi\mathcal{H}/g}(\alpha_i^{b*}\alpha_j^c + \alpha_i^b \alpha_j^{c*})e^{-\pi\mathcal{H}/g}. \tag{4.235}$$

t_{ijk} is the structure constant for $SL(2,R)$. A departure from quadrature in the uncertainty in these quantum operators results from an amplified squeezing of the vacuum state. For $\beta_i = S(z)\alpha_j S^\dagger(z)$ a gravitational induced squeezed state S appears in the commutator

$$[\beta_i^a, \beta_j^b] = (1 + S(z)e^{2\pi\mathcal{H}/g}\epsilon^{bcd} t_{ijk}\alpha_k^d -$$

$$\frac{2\pi\omega}{g} e^{\pi\mathcal{H}/g}(\alpha_i^{b*}\alpha_j^c + \alpha_i^b \alpha_j^{c*})e^{-\pi\mathcal{H}/g}\left(\cosh^2|z| - \frac{z^2}{|z|^2}\sinh^2(|z|) - \right. \tag{4.236}$$

$$\left. \frac{4z}{|z|}\alpha_i^b \alpha_j^c \cosh(|z|)\sinh(|z|)\right).$$

$S(z)$ is the gravitational squeeze state operator

$$S(z) = \prod_j e^{2\pi(z(a_j^\dagger)^2 - z^* a_j^2)}. \tag{4.237}$$

Consider the antisymmetric torsional term $\rho^\alpha{}_{\mu\nu}$ defined in equation 2.67 as parameterized by an additional variable e^5, which is an additional spatial dimension, $\rho^\mu{}_{\nu\sigma}(e^5)$. This operator $\rho^\mu{}_{\nu\sigma} = e^\mu \omega_{\nu\sigma}$ is expanded according to operators of the form in equation 4.234. The torsion is expanded as,

$$\omega^a{}_{\mu\nu} = \epsilon^\beta{}_{\mu\nu} \sum_k \frac{1}{\omega_k}\left(\alpha_\beta^a f(k) + (\alpha^a)^*_\beta f^*(k)\right), \tag{4.238}$$

where the torsional component of the generator to the path integral in equation 2.69 now contains a gravitationally induced squeeze state operator. The modified Ricci curvature

$e^{\pi\mathcal{H}/g}(\alpha_\mu^{b*}\alpha_\nu^c + \alpha_\nu^b\alpha_\mu^{c*})e^{\pi-\mathcal{H}/g} = \mathcal{R}_{\mu\nu}^{bc}$ enters into the gravitational Lagrangian. The path integral then contains a term which defines a partition function

$$\int dE n(E) e^{-E/g} = Tr e^{-E/g}, \qquad (4.239)$$

where $n(E)$ is the number operator is written according to $\sum_b a^{b\mu} a_\mu^{b\dagger}$. For sufficiently large squeezing the following follows:

$$n(E) \simeq Tr e^{-E/g} \simeq \prod_k \exp\bigl(-\omega_k n(k)\bigr) \cosh(2\omega_k/g). \qquad (4.240)$$

This may be written as $\sqrt{L/L_p} = \cosh(2\omega_k/g)$, where the squeezed state operator effectively renormalizes the Planck scale. This term is the central charge for the string D5-brane interaction by $\sqrt{L/L_p} = 2\pi\sqrt{Q_1 Q_5}$ and $\omega/g = n_5$, where n_5 is the number of excitons of the modes for the 5th coordinate direction. Hence the D2-brane picture of the black hole recovers the D5-brane-string interaction picture, but where the reduction in dimension with the two dimensional D2-brane results in a loss of information which is recovered in the D5-brane picture.

4.8: Supersymmetric Moduli Space

The hierarchy of Lagrangians in equations 4.158 and 4.159 for the supersymmetrized horizon algebra include the fermionic partners and ghost fields. The construction of the moduli space needs to be extended to include the supersymmetric transformations. The Riemann-Roch theorem suggests the moduli space needs to incorporate the topology induced through the inclusion of these additional topological fields. The moduli space will then contain solutions for incoming and outgoing states, the ghost fields and the super-pairs. For a surface of genus g the Riemann-Roch theorem indicates the number of moduli minus the number of superconformal fields is $2g - 2$ and with the addition of boson and fermion numbers is

$$N = 2g - 2 + N_B + \frac{N_F}{2} \qquad (4.241)$$

To construct the super-moduli space consider the space as involving the variables x, θ, for coordinates in space and superspace. The supersymmetry algebra for a single particle with $p^2 = -m^2$ gives in the rest frame of this particle $p_\mu = (-m, 0, 0, 0)$. The supergenerators are then rescaled by the mass according to

$$a_a^\alpha = \frac{1}{\sqrt{m}} Q_a^\alpha, \quad (a_a^\alpha)^\dagger = \frac{1}{\sqrt{m}} \bar{Q}_a^{\dot{\alpha}}. \qquad (4.242)$$

These $2N$ fermionic operators satisfy the anticommuting algebra

$$\{a_a^\alpha, (a_b^\beta)^\dagger\} = \delta^{\alpha\beta}\delta_{ab}, \quad \{a_a^\alpha, a_b^\beta\} = \{(a_a^\alpha)^\dagger, (a_b^\beta)^\dagger\} = 0. \qquad (4.243)$$

The operators α_i^b and $\hat{\alpha}_i^b$ define a Clifford algebra of forms over a Clifford vacuum $|\Omega\rangle$. With the application of $(a_a^\alpha)^\dagger$ on the vacuum $|\Omega\rangle$ the various particle states of the theory are defined

$$|N_{i_1,i_2,\ldots i_n}^{a_1,a_2,\ldots a_n}\rangle = \frac{1}{\sqrt{n!}}(\alpha_{i_1}^{a_1})^\dagger(\alpha_{i_2}^2)^\dagger\ldots(\alpha_{i_n}^{a_n})^\dagger|\Omega\rangle.$$

$$|\hat{N}_{i_1,i_2,\ldots i_n}^{a_1,a_2,\ldots a_n}\rangle = \frac{1}{\sqrt{n!}}(\hat{\alpha}_{i_1}^{a_1})^\dagger(\hat{\alpha}_{i_2}^2)^\dagger\ldots(\hat{\alpha}_{i_n}^{a_n})^\dagger|\Omega\rangle. \tag{4.244}$$

The state vectors $|N\rangle$ and $|\hat{N}\rangle$ define a general number operator similar to equation 4.97.

These operators α_i^b and $\hat{\alpha}_i^b$ are included in a graded algebra. The cohomology of the supergenerators is by definition the BRST cohomology, where n is the number of hypermultiplets in the $N = 1, 2, 4, 8$ supersymmetries. It is possible to consider the Polyakov path integral according to quaternions. A hyper-Kähler metric is replaced in this manner so that an $SU(2)$ holonomy exists. This is then a $4n$ manifold with the group $Sp(n) \times SU(2)$ that embeds into $SO(4n)$. The moduli space is Kähler and its solutions given by a Kähler potential of the form

$$K(A) = im(A^b \partial_b F(A)) \tag{4.245}$$

So this moduli space defines $3 + n$ $SO(4n)$ instantons. The moduli space in the neighborhood of a singularity is then

$$C^{2n+3}/S^1 = CP^{2n+2} \times [0, 1)/(0, x). \tag{4.246}$$

The dimension of this moduli space near a singularity, that corresponds to the black hole horizon, for $n = 1$ is the same as for a system involving the interaction between a single D1-brane (string) with a single D5-brane system. This is the moduli corresponding to the interaction of quanta (a string) with the black hole.

The process of quantum tunnelling is where the generator of the path integral becomes real valued. black hole radiance as seen in equation 4.240 involves a thermal distribution of states. Equation 4.240 shows that for a sum over all states associated with the black hole this gives a partition function for the thermal distribution of energy. Further, the D5-brane approach shows that appearance of the partition function is due to a sum over all quantum states associated with the horizon.

The next chapter illustrates the nature of the two dimensional D-brane with an experimental interest in mind. With this it is possible to examine the nature of large extra dimensions, the violation of topological indices and the potential connection between virtual black holes in the quantum gravity vacuum with physics that could be probed by future high energy experiments with facilities such as the LHC. This will then take us to another experimental possibility for the detection of physics associated with quantum gravity and spacetime fluctuations. The breakdown of topological quantum numbers have consequences with large extra dimensions. These implications include CP violation and the prospect that the Higgs field is correlated with a quantum black hole.

4.9: Moduli Spaces With Non-Hausdorff Zariski Topology

The general theory of relativity has an $SO(3,1) \sim SL(2,\mathcal{C})$ group structure that locally defines Lorentz transformations. This is equivalent to a theory of $SU(1,1)$ gauge connection on a four-manifold. The moduli space $\mathcal{M}/\mathcal{G}$ of gauge equivalent connections is well examined for the compact case $SU(2)$, but is less well known for the noncompact case with $SU(1,1)$. In the compact case this moduli space is an infinite dimensional space except at a measure zero set of singular points where the bundle splits and the holonomy group is a $U(1)$ subroup. The moduli space for the noncompact case has a nonHausdorff structure, which exists on a restricted or measure zero set of open sets. The moduli space for $SL(2,\mathcal{C})$ and $SU(1,1)$ should then share a duality with the bundle splitting of $SU(2)$ on singular points. Physically this connection is seen with the nature of imaginary time where $t \to it$, or in general when there exists an argand angle θ for complex time $e^{i\theta}t$. There is then a $U(1)$ group structure which slits a $SU(2)$ bundle $d \to d_1 \oplus d_2$, where this $U(1)$ bundle is determined by the Cartan center

$$\begin{pmatrix} e^{i\theta} & 0 \\ 0 & e^{-i\theta} \end{pmatrix} \tag{2.247}$$

and is the circle in the group. These singular points are dense and result in the blowup of a point on the moduli space for $SU(2)$. In both cases the set of gauge connections with split bundles have measure-ϵ, where with the elliptic complex singular points are a dense set and in the noncompact gauge group case the space is nonHausdorff with a Zariski topology that is itself dense.

It is not well known how to quantize fields on nonHausdorff configuration spaces. Yet with the notion of imaginary time a connection between a gauge theory with Hausdorff configuration space, "modulo a measure zero set," with one that is completely nonHausdorff will define a map between the two. For a σ_3 gauge component in $SU(2)$ there is a $U(1)$ rotation into an imaginary part so that $A_3 \to \mathcal{R}eA_3 + i\mathcal{I}mA_3$. Here the "smoothness"of this rotation will permit the quantization of the noncompact group connections. The extension of the elliptic complex to include $U(1)$ transformations $SU(2) \to SU(1,1)$ is examined. This rotation has the effect of superposing a nonsingular point with a singular point on a manifold where $t \to it$. If the two manifolds are considered as the superposed states of a single wave function this rotation is then a dual between an accumulation of an instanton at a point in the Euclideanized manifold, by Taubes theorem, around a singular point and the topological charge for a noncompact manifold of $SU(1,1)$ with moduli having Zariski topology. By treating these two manifold descriptions as dual the moduli space for the noncompact $SU(1,1)$ is a Wick rotation of $\mathcal{C}^3 \to \mathcal{C}^2 \times \mathbf{R}$, where the singular point then defines a 5 dimensional cone on $\mathcal{C} \times \mathbf{R}^4 \to \mathcal{C} \times \mathbf{R}^3 \times [0,1)/x \sim x' \subset \mathcal{C} \times \mathbf{R}^4$. This space is then the 5-dimensional space for the deSitter cosmology defined according to projective coordinates.

Gauge theory posits the existence of nonphysical connection one-forms A and their associated fields dA or $dA + A \wedge A$ in the nonAbelian case. The connection one-forms are arbitrarily defined on a section of the principal bundle E over a manifold $\mathcal{M}$. So consider a section s and a transformation of this section $s' = gs$, where g is an element of a Lie group $\mathcal{G}$. The action of d on s', with $ds = As$, is then

$$ds' = d(gs) = (dg)s + gds, \tag{4.248a}$$

and so

$$ds' = (dg)g^{-1}s' + gAg^{-1}s'. \tag{4.248b}$$

The gauge transformation of the connection one-form A is

$$A' = gAg^{-1} + (dg)g^{-1}. \tag{4.249}$$

Since this transformation is inhomegeneous these gauge connections are physically unobservable. The field two-form $F = dA + A \wedge A$ transform homogenously as $F' = gFg^{-1}$. For the group element $g = e^{i\chi}$ then

$$ds' = (id\chi + A + i[\chi, A])s' \tag{4.250}$$

and gauge connections transform as $A \to A + id\chi + i[\chi, A]$. If this commutator is zero this is the gauge transformation of the electromagnetic vector potential $A \to A + \nabla \chi$, where the magnetic field $B = \nabla \times A$ is gauge invariant.

Consider the gauge theory as a group action on a manifold. For $g \in \mathcal{G}$ these act on $x \in \mathcal{M}$ and the pair x, gx defines a graph $\Gamma = \{(x, gx) : g \in \mathcal{G}, x \in \mathcal{M}\}$, where $\Gamma \subseteq \mathcal{M} \times \mathcal{M}$. The moduli space $\mathcal{M}/\mathcal{G}$ then has quotient topology which defines the orbit space. The moduli space is then Hausdorff if and only if Γ is closed. A slice of the action may also be defined as an open manifold $\mathcal{N}$ by the tangent spaces at each $x \in \mathcal{N}$ so that $T_x(\mathcal{M}) = T_x(\mathcal{G}x) \oplus T_x$, $\forall x \in \mathcal{N}$. If this obtains there exists an injective map from the moduli space to $\mathcal{N}$, where then $\mathcal{M}/\mathcal{G}$ is a manifold. If both these conditions obtain $\mathcal{M}/\mathcal{G}$ is a Hausdorff manifold. This theorem is proven by considering the orbit space $\mathcal{S}_{n-1}$ with a Sobolev norm up to n for the space of sections with derivatives $\leq n$. This orbit space is given by the action of group elements $\mathcal{G}$ on gauge connections $\mathcal{A}_{n-1}$. The tangent to the group action is $\mathcal{G} \cdot D$ where D is the gauge invariant operator one-form. For the group $\mathcal{G}$ the Lie algebra is $\Omega^0(ad\,\eta)$ with $\eta = P \times U$, $U \subset \mathcal{M}$. For the gauge connection $A \in \Omega^1(ad\,\eta)$ there is the map $\Omega^0(ad\,\eta)_n \xrightarrow{D} \Omega^1(ad\,\eta)_{n-1}$, with the Sobolev index noted. A gauge condition that defines the moduli space is then for $v \in \Omega^0(ad\,\eta)$, where the action Dv that satisfies $\langle A, Dv \rangle = 0$, equivalent to $\langle D^*A, v \rangle = 0$. This defines the slice or moduli

$$\mathcal{S}_D = \{A \in \Omega^1(ad\,\eta)_{n-1} : D^*A = 0\}, \tag{4.251}$$

which is the tangent space to $\mathcal{S}$ defined by a gauge condition. Here the differential D is defined in a space that is diffeomorphic the kernal of D^* tensored for group action with Sobolev index up to n, $\mathcal{A}_{n-1} \xrightarrow{diff} Ker(D^*) \times \mathcal{G}_n$.

On a neighborhhood $\mathcal{O}_D$ in $\mathcal{A}_{n-1}$ there is then the map $g : \mathcal{O}_D \to \mathcal{G}_n$ with $g(D^*)D' \in D + \mathcal{S}_D$, for $D' = D + A$. This map is seen with the differential equation

$$D^*(g^{-1}Dg + g^{-1}Ag) = 0, \tag{4.252}$$

which defines the map

$$\mu_D : \Omega^1(ad\ \eta)_{n-1} \oplus \Omega^0(ad\ \eta)_n \to \Omega^0(ad\ \eta)_{n-2} \tag{4.253}$$

that sends $A \oplus g \mapsto D^*(D(g) + A)$ for the two elements on the right hand side. This then defines the second order partial differential D^*D. This means that the inverse diffeomorphism between $\mathcal{O}_D \to \mathcal{G}_n$ exists on the pair of elements (A, g) on $\Omega^1(ad\ \eta)_{n-1} \oplus \Omega^0(ad\ \eta)_n$ as the gauge transformation

$$(A,\ g) \mapsto gA'g^{-1} - (Dg)g^{-1}. \tag{4.254}$$

From this construction it is possible to show that $\mathcal{S}_{n-1}$ is Hausdorf for $n > 2$. For the orbit space Γ let $x \to D$ and $M = \mathcal{A}_{n-1}$ Further, let there be a succession of unitary maps $g_i : i \in \{1,\ \ldots\ \infty\}$ so that $\lim_{i \to \infty} A_i \to A'$. For each Sobolov index let $||Dg_i||_{m \leq n}$ be bounded in the each $H^k(End\ \eta \times T^*M)$. This insures convergence for each term, where there exists an $\epsilon > 0$ so that the sequence of elements in $H^k(End\ \eta \times T^*M)$, c_k, $c_{k'}$ converge with $c_k - c_{k'} < \epsilon$. This insures that points of the moduli space are separable almost everywhere.

The elliptic complex is the exact sequence

$$0 \to \Omega^0(ad\ \eta)_n \xrightarrow{D} \Omega^1(ad\ \eta)_{n-1} \xrightarrow{P_-D} \Omega^2(ad\ \eta)_{n-2} \to 0, \tag{4.255}$$

where P_-D is a projection onto antiself-dual two-forms. Define (A, x) the tangent to the operator D as

$$T_D = \{(A, x) \in \Omega^1(ad\ \eta) \oplus c : P_-DA + \delta x = 0\}. \tag{4.256}$$

This is on the slice $\mathcal{S}$ if the tangent is taken on $D_{n-1}/\mathcal{G}_n$, with $D^*A = 0$. This defines the map $\pi : E \to \mathcal{C}$ as

$$ker(\delta\pi) = \{(A, x)|P_-A + \delta x = 0,\ D^*A = 0\}, \tag{4.257a}$$

and

$$im(\delta\pi) = (\delta^{-1}(im(P_-D))) \tag{4.257b}$$

so that π is a Fredholm map with the index $h_0 - h_1 + h_2 = -5$ by the Atiyah-Singer index theorem. This means that the dimension of the moduli space is 5.

If X is a topological space then X is a set of points and there is a set of subsets of X, called "open sets," satisfying the following conditions:

1. X itself and the empty set are both open.

2. Any union of open sets is open.

3. Any finite intersection of open sets is open. If X has a metric a topology is formed with open sets consisting of unions of interiors of balls of any radius $\epsilon > 0$ and any center. In physics such metric constructions are quite common.

The natural topology in algebraic geometry is Zariski topology [24]. Let X be embedded in $\mathcal{C}^n$ as the intersection of the zeroes of some algebraic equations in the homogeneous coordinates of $\mathcal{C}^n$. If f is any algebraic equation in the homogeneous coordinates of $\mathcal{C}^n$, the intersection of $f = 0$ with X will be an algebraic subspace of X. The Zariski topology on X is defined as the open sets consisting of the null set, X, unions of U_f's and finite intersection of U_f's, for any f's. Hence the open sets consist of complements of algebraic subspaces of X.

So within these U_f's there is a set of equivalent algebraic functions where the function is zero. This means that for any group action $\mathcal{G}$ on U_f and $U_{f'}$ that

$$GU_f = GU_{f'}, \tag{4.258}$$

where this obtains on two open sets. Now define the function F so that

$$f = F + e^{-x}(f' + F). \tag{4.259}$$

Then for $x \to \infty$ the function F exists in the closure of GU_f, called $\bar{G}U_f$. A similar argument may be made for $\bar{G}U_{f'}$. For this case then any measure μ gives

$$\mu(\bar{G}U_f \cap \bar{G}U_{f'}) \neq 0. \tag{4.260}$$

This obtains no matter how small these neighborhoods are made, where as a result the Zariski topology is nonHausdorff.

The $SL(2, \mathbf{C})$ and $SU(1, 1)$ groups are defined on four and three dimensional spacetime respectively. The generators of these groups exist on the principal bundles $P(\Sigma, \mathcal{G})$, where Σ is a Cauchy surface (spatial surface) of dimensions 3 and 2 for $SL(2, \mathbf{C})$ and $SU(1, 1)$. The set of connections A on the principle bundle define Wilson loops $\oint Adx$ are maps $\mu : [0, 1] \to \sigma$ for $\mu(0) = \mu(1)$. An element of $\mathcal{G}$ assigned to μ by the holonomy map $H(\mu, A)$ defines a function

$$F_\mu(A) = \frac{1}{2} Tr\, H(\mu, A) \tag{4.261}$$

that is invariant with respect to the group action of $\mathcal{G}$. Thus $F_\mu(A)$ is an element of $\mathcal{M} = A/\mathcal{G}$ or the moduli space. The rotation $\sigma_3 \to i\sigma_3$ will carry the separability condition to the noncompact case. Such separability is discussed in reference [25].

The group $SU(1,1)$ is related to $SU(2)$ by the signature change on the basis elements σ_1, σ_2, σ_3 of $SU(2)$. For $\sigma_\pm = \sigma_1 \pm \sigma_2$, the basis for $SU(1, 1)$ are then σ_+, σ_-, $\tau_3 = i\sigma_3$. Now consider a connection one-form

$$A = A^+ \sigma_+ + A^3 \sigma_3 \tag{4.262}$$

and a gauge transformation determined by the group action of $g \in \mathcal{G}$, $g = e^{i\lambda \tau_3}$. The gauge transformed connection is then

$$A' = g^{-1}Ag + g^{-1}dg = e^{-2\lambda}A^+\sigma_+ + A^3\sigma_3, \tag{4.263}$$

188 Quantum Fluctuations of Spacetime

where $d\lambda = A^3$. Thus λ is a parameterization of the gauge orbit for this connection. This leads to the observation

$$\lim_{\lambda \to \infty} A(\lambda) \to A^3 \sigma_3, \qquad (4.264)$$

where $A^+\sigma_+ + A^3\sigma_3$ and $A^3\sigma_3$ have distinct holonomy groups and thus represent distinct points in the moduli space $\mathcal{M}$. However by the last equation this gives

$$F_\mu(A^+\sigma_+ + A^3\sigma_3) = F_\mu(A^3\sigma_3), \qquad (4.265)$$

which obtains similarly for any gauge invariant function. Hence there exist two distinct points in the moduli space that define the same set of gauge invariant functions. Hence there does not exist a measure over these two points that separates them and $\mathcal{M}$ is then nonHausdorff with a Zariski topology. The above statement that $f_\mu(A^+\sigma_+ + A^3\sigma_3) = f_\mu(A^3\sigma_3)$ obtains for any gauge invariant function is proven in [26].

Physically this is a manifestation of the degeneracy of the vacuum state for quantum fields in curved spacetime, which results in a blackbody distribution of quanta associated with the Unruh effect and the curved spacetime result of black hole quantum radiance of Hawking. The group structure is given by a noncompact Bogoliubov transformation. Consider the gauge connections defined on a flat spacetime bases $|0\rangle$ ω_+, ω_- for states entering and exiting a black hole or an event horizon. These are related to the Minkowski connections A_+ and A_-, according to the Bogoliubov transformations by

$$A_+ = \alpha \omega_+ - \beta \omega_-, \quad A_- = \alpha \omega_- - \beta \omega_+, \qquad (4.266)$$

such that $\alpha = cosh^2(g)$ and $\beta = sinh^2(g)$, for g the generator or rapidity of the group. The noncompact nature of the Bogoliubov transformation leads to the nonclosure of the orbit space Γ, where the orbit space $\mathcal{M}/G$ is then nonHausdorff.

4.10: Modularity and Fractal Aspects of Moduli with Zariski Topology

The next two sections are largely due to R. Betts.

The above construction with the elliptic sequence is pertinent for the instanton where $g \to ig$. For the tunnelling of states time becomes imaginary time. For an $-i$ rotation the basis elements are mapped to σ_+, σ_-, $i\sigma_3 \to \sigma_+$, σ_-, σ_3, where the elliptic sequence pertains to the tunnelling states. For the conservation of quantum information the same topological information must obtain for both of these. In the Euclideanized version an event horizon is a fixed point. This fixed point then contains all the relevant quantum information, where the orbit space for the connection form is closed. The degeneracy of gauge connections for the hyperbolic case and the preservation of quantum information physically indicates the degeneracy results in no loss of quantum information. In other words there exist a discrete set of hyperbolic orbits that correspond to the orbit space of the elliptic complex. This connects with the ansatz of section 4.6 on a cyclic correspondence between elliptical cycles and hyperbolas. Other hyperbolic orbits outside this discrete set are then degenerate with respect to this set, where by the Zariski topology they have the same moduli as some hyperbola in this discrete set. The following construction is offered as a model set for this.

The projective lines in R^3 defines the Riemannian sphere $S^2 \simeq CP^1$. CP^1 has connections to the Lorentz transformations. The conformal transformations of the Riemannian sphere [27]

$$z \to z' = \frac{az + b}{cz + d}, \tag{4.267}$$

is isomorphic to the projective Lorentz group $PSL(2, C)$, or $SL(2,C)$ modulo the length of null vectors $x^\mu x_\mu = 0$. These null vectors define the heavenly sphere. Let $z' = c/d$ where $ad - bc = 1$ defines a discrete subgroup $SL(2, Z) \subset SL(2, C)$. The 2×2 matrices $A \in SL(2, C)$, $A = \begin{pmatrix} a & b \\ c & d \end{pmatrix}$, are such that

$$det(A) = ad - bc = 1. \tag{4.268}$$

The matrices A are elements of $SL(2, C)$, in a subset of the modular group, which in turn is a subgroup of the Möbius group [27]

The Farey Sequence of order n is the set of rational numbers l/m, for $0 \leq l \leq m \leq n$ and $gcd(l, m) = 1$ [28]. The Farey sequences are are nested as $F_1 \subset F_2 \subset \ldots \subset F_n \subset \ldots$. The fractions are arranged in ascending orders of magnitude. For the three successive terms a/b, c/d, e/f in a Farey sequence, $c/d = (a + e)/(b + f)$. The fraction c/d is the mediant of a/b and e/f with $ad - bc = -1$, which gives the modularity condition if $a, b, c, d \to ia, ib, ic, id$.

The modular transformation is isomorphic to $SL(2, Z) \subset SL(2, C)$. Now these are embedded into $SL(2, C)$ with the matrices

$$\sigma_z = \begin{pmatrix} 1 & 0 \\ 0 & -1 \end{pmatrix}, \sigma_y = \begin{pmatrix} 0 & i \\ -i & 0 \end{pmatrix}, \sigma_x = \begin{pmatrix} 0 & 1 \\ 1 & 0 \end{pmatrix}, \sigma_0 = \begin{pmatrix} 1 & 0 \\ 0 & 1 \end{pmatrix} \tag{4.269}$$

so the matrix A is a linear combination such as

$$M_1 = x\sigma_z, \quad M_2 = y\sigma_y, \quad M_3 = w\sigma_x, \quad M_4 = w\sigma_0, \tag{4.270}$$

where $A = M_1 + M_2 + M_3 + M_4$ or

$$A = \begin{pmatrix} x + w & iy + w \\ -iy + w & w - x \end{pmatrix}, \tag{4.271}$$

where $x + w = a$, $iy + w = b$, $w - iy = c$, $w - x = d$.

There exists a constraint condition on the values of x, y, w to produce the two successive terms of a/b, c/d. in some particular Farey sequence. Let $x + w = ia$, $w + iy = ib$, $w - iy = ic$, $w - x = id$. Then one may eliminate w by

$ia - x = x + id$ and $ib - iy = ic + iy$. This leads to $x = (1/2)i(a - d)$, $y = (1/2)(b - c)$. x is imaginary and y is a rational number. This defines the constraint equation

$$y^2 - x^2 = (1/4)[(b - c)^2 + (a - d)^2], \qquad (4.272)$$

which becomes the transformed "hyperbolic" equation $Y^2 - X^2 = 1$, for $F_n/\{(b = c) \vee (a = d)\}$ if

$$X = 2x/\sqrt{(b - c)^2 + (a - d)^2}, Y = 2y/\sqrt{(b - c)^2 + (a - d)^2}. \qquad (4.273)$$

X and Y are parameterized as $Y = cosh(at)$, $X = sinh(at)$ for $at = cosh^{-1}[(b - c)/\sqrt{(b - c)^2 + (a - d)^2}]$ so that $cosh^2(at) - sinh^2(at) = 1$. Thus the Farey sequence gives a discrete set of hyperbolic orbits that satisfy an $SL(2, Z)$ modularity and is a discrete Minkowski group. This discrete set then gives quantum numbers that define the degeneracy of gauge connections, or a superselection rule.

4.11: Julia Sets and NonHausdorff Measures over Moduli Spaces

The Farey sequence has relationships with fractal geometries. The number pairs of the sequences between $[0, 1]$ defines a Julia set. The connection to fractal geometry is with the nonHausdorff nature of the moduli space. The Hausdorff dimension over a space X measures the growth of the number of sets of diameter ϵ required to cover X as $\epsilon \to 0$. If $n(\epsilon)$ is the number of sets required to cover X and $n(\epsilon) \propto \epsilon^{-D}$ for $\epsilon \to 0$ then X has Hausdorff dimension D. The Hausdorff dimension is determined by a measure over the set X as

$$\mu_d(X, \epsilon) = inf\{\sum_{i \in Z} diam(S_i)^d : diam(S_i) < \epsilon\} \qquad (4.274)$$

$$\mu(X) = \lim_{\epsilon \to 0} \mu_d(X, \epsilon). \qquad (4.275)$$

The Hausdorff dimension is then the $d = D$ that separates the condition for $\mu_d(X)$ finite and infinite. This means that

$$H(X) = sup\{d \in R : \mu_d(X) = \infty\}. \qquad (4.276)$$

A Hausdorff topology is one that defines a unique limit point in a net. This means that if X is Hausdorff then $\mu(X)$ is a unique limit point for $\epsilon \to 0$. However, if the measure is infinite then the limit point is not defined and the space X is not Hausdorff.

The bundle curvature $F_\mu(A^+\sigma_+ + A^3\sigma_3) = F_\mu(A^3\sigma_3)$ is defined by a coboundary operator in a sequence analogous to the elliptic complex. The degeneracy of the gauge connection and their occurrence according to a discrete set or Farey sequence defines a set of points that converge to zero. This is then equivalent to a Julia set of points that converge to zero. Further, the Atiyah-Singer index for this system no longer has an integer dimension. For $c = 1/7 + i3/4$ and $c = -1/7 + 3/4$ the Julia sets appear in figure 4.3.

It is known that Farey sequences have correspondences with fractal geometries and the period doubling towards chaotic dynamics. The pairs of numbers given by a Farey sequence then define various Julia sets which are a measure of the nonHausdorfness of various moduli. It is interesting that a geometry with connections of dynamics systems should appear in the context of quantum gravity.

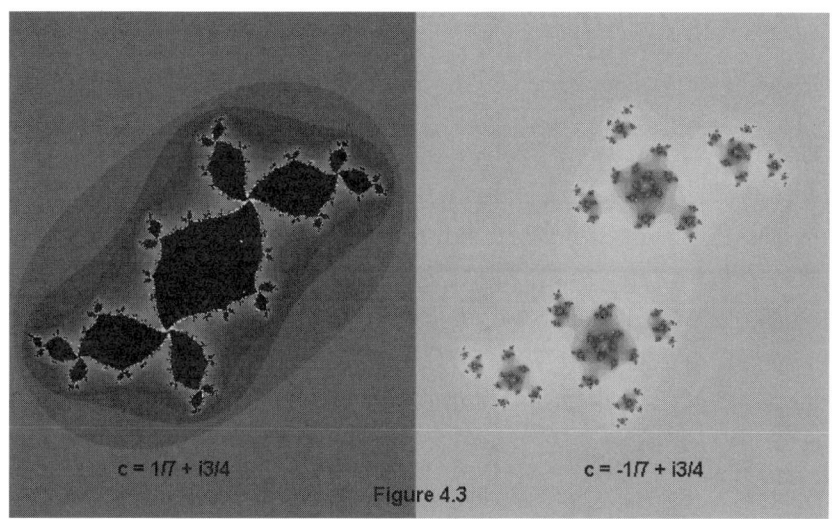

Julia sets associated with nonHausdorff moduli space.

References

[1] G. 't Hooft, *Class. Quant. Grav.* **16** (1999) 395-405.

[2] W. G. Unruh, R. M. Wald, *Phys. Rev.* **D29**, 1047 (1984).

[3] L. B. Crowell, *Found. Phys. Lett.* **13**, 2 (1999).

[4] M. B. Green, J. H. Schwarz, E. Witten, *Superstring Theory* Cambridge press (1987).

[5] N. D. Birrell, P. C. W. Davies, *Quantum Fields in Curved Space*, Cambridge University Press (1982).

[6] R. M. Wald, *Quantum Fields in Curved Spacetime and black hole Thermodynamics*, University of Chicago, Chicago (1994).

[7] D. Finkelstein and W. Hallidy, *Int. J. Theo. Phys.* **30**, 1991.

[8] L. Susskind, *J. Math. Phys* **36**, 11, (1995) 6377.

[9] L. Smolin, J. Magueijo, http://www.arxiv.org/abs/hep-th/0112090

[10] E. Wigner, *Phys. Rev.* **40**, 749 (1932).

[11] J. Milnor, "Morse Theory," Princeton University Press (1963).

[12] A. Floer, *Comm. Math. Phys.*, **118**, 215 (1988).

[13] G. 't Hooft, *Class. Quant. Grav.*, **16**, (1999) 395-405.

[14] S. K. Donalsdon, P. B. Kronheimer, "The Geometry of Four Manifolds," Clarendon Press, Oxford (1990).

[15] V. G. Kac, Infinite Dimensional Lie Algebras, Cambridge University Press, Cambridge, (1983).

[16] P. Goddard, A. Kent, *Physics Letters*, **152b**, 393 (1985).

[17] C. Montonen, D. Olive, *Nucl. Phys.*, **B110**, 237, (1976).

[18] J. Polchinski, *HEP-th/6907050 v2*, July 9, 1996.

[19] N. Seiberg, *Phys. Lett.*, **206B**, 75, (1988).

[21] N. Seiberg, E. Witten, *Nucl, Phys.*, *B426*, 19, (1994).

[22] N. Seiberg, E. Witten, *Nucl, Phys.*, *B431*, 484, (1994).

[23] E. Witten, *Phys. Lett.*, **86B**, 283, (1979).

[24] Kahn, D. W. "Topology: An Introduction to the Point-Set and Algebraic Areas." Dover, Dover, 1995.

[25] W. Fairbairn, C. Rovelli, "Separable Hilbert space in Loop Quantum Gravity" *J. Math. Phys.* **45** (2004) 2802-2814.

[26] A. Ashtekar, J. Lewandowski, "Completeness of Wilson loop functionals on the moduli space of $SL(2,C)$ and $SU(1,1)$-connections, " *Class. Quant. Grav.* **10** (1993) L69-L74.

[27] T. M. Apostol, "Modular Functions and Dirichlet Series in Number Theory," Springer-Verlag, NY (1996).

[28] A. A. Gioia, "The Theory of Numbers," Markham Publishing, Chicago.

5: Topology, Extra Large Dimensions and The Higgs Field

In chapter 2 quantum fluctuations of spacetime with have gauge-like properties were developed. This means potentially that gauge fields in our universe have a mirror version as the generators of spacetime fluctuations. This raises an interesting prospect that the Higgs field for the spontaneous breaking of the standard model has a mirror in the physics of spacetime which breaks the symmetry of the mirror gauge field. If so spacetime physics may play a role in the broken symmetry of the universe.

This might suggest a connection between the Higgs field and the black hole. After considerable development of quantum information theory of black holes in chapter 9 theory on the phase structure of black holes. For a black hole with a small number of Planck units of mass information flow through such a quantum black hole is unitary. In fact in general this is the case for even very massive black holes. However, for semi-classical black holes there is a phase change in the structure of spacetime that prevents an exterior observer from extracting information from a black hole in a way that reflects unitarity. This phase change is shown to have a Landau-Ginsburg structure, which emulates the Higgs field. If this mirror gauge field, say an $SU(n)$ field, reflects $SU(n)$ gauge fields for nongravitational forces in the universe, the net $U(n)$ theory will give a $U(1)$ symmetry between the Higgs field and the phase structure of black holes.

Physically this means that under the right conditions a Higgs field may have some small physical aspects of a quantum black hole. This has some parallels with theories that invoke large extra dimensions. Indeed in what follows some of the physics of these theories is invoked. The major departure is that the connection between the scale for the Higgs in the standard model and the Planck scale is given according to the renormalization of the Planck scale by Bogoliubov transformed fields and squeezed states. This departure from theories of large extra dimensions requires for Planck scale physics to appear at the TeV range in energy special care must be taken to do such experiments with particles in coherent or entangled states. If this can be done it is possible a scattering experiment may transfer an entanglement from two colliding particles to an entanglement between a Higgs field and a quantum black hole renormalized to a length scale far beyond $L_p = \sqrt{G\hbar/c^3}$.

5.1 Extra Dimensions and the Horizon D2-Brane and D5-Brane

In this chapter D2-branes are found to have consequences for gauge theories and the standard model. There is the breakdown of topological quantum numbers and connections to physics that could potentially be measured in high energy accelerators in the near future. These considerations come from the theory that there exist seven extra large dimensions which involve gravity. These extra dimensions in supergravity have the affect of making gravity appear weak on large spacetime scales, but on a sufficiently small scale where these extra dimensions exist gravity becomes stronger [1]. It is thought that these extra large dimensions can exist on scales much larger than the Planck scale, where the above result on the renormalization of the Planck scale does show (at least under some circumstances) this might be a possibility for squeezed states which renormalize the Planck scale. The particle physics possibilities explored here are that CP violations may have their origin in quantum gravity and the big bang, where further quantum gravity fluctuations or virtual black holes may on a large scale manifest themselves as aspects of the Higgs field. Experimentally this could mean that the Higgs particle would exhibit some decay processes that are anomalous. Effectively a Higgs field would be a superposition of the Higgs state with a small amplitude for a quantum black hole due to extra large dimensions.

This chapter is largely a review of theoretical results on extra large dimensions, but with the caveat that these extra dimensions only become large under a coherent squeezing of the vacuum state. If these compactified dimensions are large under ordinary conditions, then it would be likely that proton decay would be far more prevalent than what is observed. It is also likely that the cores of neutron stars, even more in the cores of newly discovered quark stars, would have sufficient energy to permit "soft" black holes to exist and decay away these objects. Yet neutron stars are known to be highly stable. Further, these ideas could turn out to be completely erroneous. However, the interest here is to make some contact with other current developments in theoretical physics.

A major thrust of this approach is with the intention of illustrating ways quantum gravity could be an experimental science. Without some experimental data on Planck scale physics it is hard really to call this subject physics or a science. It is likely that nature does have quantum gravity effects that are accessible to our methods of experimentation. It is our job to figure out where these might be and to examination them. In this chapter these connections between the renormalization of the Planck scale and "large extra dimensions" [2] are discussed.

The additional dimensions in nature are thought to be compactified into spaces on a scale near the Planck length. In a basic Kaluza-Klein setting the extra 5th dimension is curled up into regions $\sim 10^{-33}$cm as Calabi-Yau spaces, orbifolds or orientifolds. These compactified spaces form the basis for the existence of gauge forces in the world. The one difficulty is that the couplings for these gauge fields are dimension dependent $g_5 \propto 1/mass$ and the coupling $g_5^2 E$ is divergent for $E \gg 1/g_5^2$. This renders such theories unrenormalizable. This is one reason that stringy quantum gravity is appealed to for it provides a length scale to cut off physics with an effective Lagrangian. As a rule the ultraviolet divergences of compactified field theories has not been satisfactorily understood. However, the insight of Arkani-Hamed that extra dimensions can be understood dynamically has shed some light on this matter [3]. This reciprocates the usual picture of things, where now a four dimensional theory emerges that is asymptotically free at high energies. Consider the gauge groups as existing in a lattice, with two dimensional plaquettes, where the lattice is described by an $SU(m)^N \times SU(n)^N$ group. As each Lie algebra defines a vector space the total lattice is a "quiver" of vector spaces. This requires the gauge theory posses a Weyl fermion sector connected to each algebra. Each $SU(n)$ then has either a ϕ or $\bar{\phi}$ attached marked by an arrow either leaving or entering a node. These algebras then define a chain connected by the fermion sectors. A closed chain with $SU(m) \rightarrow SU(n) \rightarrow SU(m)$

on a side will define an N sided polygon. This chain is then a representation of the additional 5th dimension at low energy, which has been given the rather odd designation as a "moose." [4] This chain may exist within a compactification scheme on a torus, orbifold or orientifold.

Assign ϕ_i as the field that connects $SU(n)$ and $SU(m)$ at the ith side and $\psi_{i,i+1}$ as the field that attached $SU(m)$ at the ith node to the $SU(n)$ at the $i+1^{th}$ node. The S matrix is then defined as

$$S_{i,i+1} = g_s \langle |\phi_i \psi_{i,i+1}| \rangle. \tag{5.1}$$

A local gauge transition on this matrix is then determined by the $SU(m)$ groups at the vertices of the edge link by $g_i^{-1} S_{i,i+1} g_{i+1}$ and $S_{i,i+1}$ is an $m \times m$ matrix of bosons. These bosons are then "link variables" for the chain. When the gauge coupling g_s becomes large there is a confinement process that defines a mass, which by necessity breaks any chiral symmetry. The renormalization cut offs for confinement are set by the two groups defined as Λ_n and Λ_m, where free fermions and their gauge bosons (e.g. quarks and gluons) are free from confinement for $E \gg \Lambda_n, \Lambda_m$. Under this situation, where the strength of the $SU(n)$ is small, the differential of the scattering matrix in a nonlinear sigma model is,

$$\mathcal{D}_\mu S_{i,i+1} = \partial_\mu S_{i,i+1} - ig_s A_{\mu i} S_{i,i+1} + ig S_{i,i+1} A_{\mu i+1}, \tag{5.2}$$

where the effective Lagrangian for the field theory is [3]

$$\mathcal{L}_{eff} = -\frac{1}{2g^2} \sum_i F^a_{\mu\nu i} F^{a\mu\nu}{}_i + g_s^2 \sum_i Tr |\mathcal{D}_\mu S_{i,i+1}|^2. \tag{5.3}$$

This is the Lagrangian for a five dimensional $SU(m)$ theory, where the additional dimension has been placed on the N-polygon. The last term in the Lagrangian determines a mass Lagrangian of the form

$$\mathcal{L}_{mass} \sim g_s^2 \sum_i (A_i - A_{i+1})^2. \tag{5.4}$$

This Lagrangian can be written as $\bar\psi_i \mathbf{M}^2_{ij} \psi_j$ and the mass matrix is

$$\mathbf{M}^2 = \begin{pmatrix} 2 & -1 & 0 & -1 & \cdots \\ -1 & 2 & -1 & \cdots & 0 \\ & \cdot & & & \\ & \cdot & & & \\ -1 & \cdots & & \cdots -1 & 2 \end{pmatrix}. \tag{5.5}$$

At this point the problem is reduced to the simple problem of a spring loaded string with eigenfunctions for the Weyl field $\psi = exp(2\pi imk/N)$ with eigenvalues $E_k = 4g_s \sin^2(\pi k/N)$. These eigenvalues are related to the radius of the compactified 5th dimension by

$$E_k = \frac{2}{a^2} \sin^2(2\pi ka/R), \tag{5.6}$$

where a is the length of the edge links. For low energy this gives the situation where $2\pi k/R \ll a^{-1}$ and the momentum eigenvalues are $p_5 = 2\pi|k|/R$. For low energy, where the momentum in five dimensions satisfies $p_5 \ll a^{-1}$, this results in a coupling constant in four dimensions on the diagonal states which satisfies [5]

$$\frac{1}{g_4^2} = \frac{N}{g^2} = \frac{R}{g_5^2}, \tag{5.7}$$

where $g_5 = g/g_s$. Here the theory appears completely five dimensional for $R \gg 1/k \gg a$. The interaction distance is then smaller than the compactified radius and larger than the distance a, $R \gg r \gg a$, with a potential of the form $V(r) \simeq g_5^2/r^2$ [6]. The distance r is given by $r = \sqrt{(x-x_0)^2 + r_5^2}$ and the coupling constant is determined by the size of the compactified fifth dimension $g_5^2 \sim R$. Yet for $r \ll \Lambda_5^{-1} = \Lambda_m^{-1}$ the theory appears four dimensional.

There are two types of gauge groups $\mathcal{G} = SU(n)$ which act on four dimensions and $\mathcal{G}_5 = SU(m)$, respectively with parameters Λ and Λ_5. There is an obvious duality present here with the cases $\Lambda \ll \Lambda_5$ and $\Lambda \gg \Lambda_5$. In this matter there are two different ways to represent the fifth dimension in this manner. If $m = n$ the duality between the two gauge groups is similar to the Bohr-Sommerfeld duality. Given the connection form A from one $SU(n)$ the quantum phase is given by

$$exp\left(ie\oint_{C=\partial S} A\right) = exp\left(ie\int_S F\right) = exp(ieg), \tag{5.8}$$

so under the quantization condition $e\mu = 2\pi n$. Define the duality operator $*m$ that interchanges the electric field components of one $SU(n)$ with the magnetic field components of the other $SU(n)$: $*m(e_1) \to \mu_2 \sim \mu_1$. If $*m$ has Liebniz property of a differential operator $*m(e_1 e_2) = \mu_2 e_2 + e_1 \mu_1$ the duality between the gauge group is similar to the quantization condition between an electric charge and the topological magnetic monopole, or as a dyon

$$*m(e_1 e_2) = g_2 e_1 + e_1 g_2 = (e_1 g_1 + e_2 g_2) = 2\pi n. \tag{5.9}$$

There are then two situations where $\Lambda \gg \Lambda_5$ and $\Lambda \ll \Lambda_5$. A transition from one to the other reflects the compactification of one extra-dimension and the decompactification of another. For $|\Lambda/\Lambda_5 - 1| \sim O(1/n)$ there is the situation where $r <, \Lambda, \Lambda_5^{-1}$, where the theory is four dimensional. At this point the momentum of fermions and gauge bosons (quarks and gluons) is large enough to be free fields. For Λ/λ_5 large there is the explosion of fermions and their bosons associated with $SU(m)$ and the ring or lattice view of the 5th dimension vanishes. The vanishing of the 5th dimension provides a UV cutoff for a sigma model realization for both situations with an unbroken phase. In the 5-d phase there is an $SO(4,1)$ theory and in the 4-d phase an $SO(3,1)$ theory. The Lagrangian then consists of a gauge component plus a potential term that is a function of fields ϕ evaluated at each i, $i+1$ edge link

$$\mathcal{L} = \mathcal{L}_{gauge} + \sum_i V(\phi_{i,i+1}), \tag{5.10}$$

with $\langle \phi_{i,i+1} \rangle = S_{i,i+1}$. If the potential function is Higgsian there is a symmetry breaking and the expectation value,

$$\langle \phi_{i,i+1} \rangle = \mu e^{\lambda_i \sigma_i}, \tag{5.11}$$

where λ_i, σ_i are Goldstone and goldstonino fields.

At this stage gravity is included. It appears that gravity exists purely in the four dimensional world. Yet the radius of the 5th dimension is compactified in 4 dimensions. The radius of the 5th dimension is renormalizable by gravity. For the gravitational squeezing parameter z the Planck unit of action is adjusted and the Planck scale should shift according to $L'_p \sim L_p cosh(|z|)$. For gravitational self-squeezing the parameter is $z \simeq kz_0/g$, where z_0 is a random squeeze state parameter for the departure from quadrature in a fluctuation. So gravity is not effected by the gauge theoretic phase transition. Gravity holds for the domain $E = E_P \sim \lambda >> 1/Rcosh(|z|)$, where in the middle region the gauge theory is 5 dimensional. In this modified setting this middle region is near the Planck scale, but the occurrence of 5th dimensional physics at low energy occurs as $\lambda_{min} \sim exp(-z|)/R$, where in the high energy domain pure gravity obtains λ also decreases. As such this "window" of the 5th dimension is lowered towards the TeV range in energy.

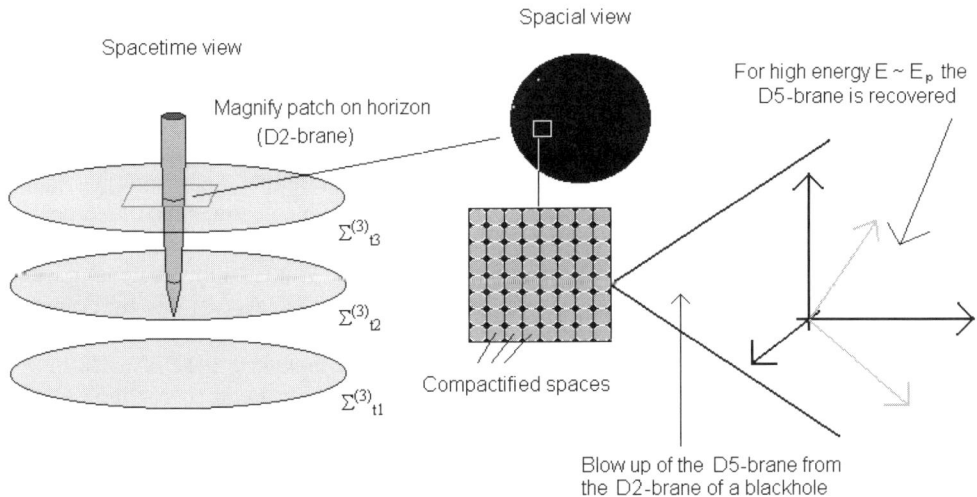

D-2 brane and D-5 brane duality with a black hole.
Figure 5.1

The two dimensional D2-brane construction above indicates there are ties to the D5-brane. Physically consider the following situation. In a low energy domain the D2-brane contains compactified three dimensional spaces curled up. This additional compactified dimension contains all the data on each $\Sigma^{(3)}$, by the holographic principle, with support determined by all the fields in spacetime. A special case of a D2-brane is the event horizon of a black hole. However, if one performs an experiment at high enough energy, or if the black hole is sufficiently small, the event horizon plus its compactified space uncurls into a five dimensional surface: a D5-brane. In 10 dimensions this is the Poincaré dual to the T-dual 5D-brane with the monopole charge corresponding to the gravity (electric-like) charge of the D2-brane. The event horizon contains all quantum field information on each $\Sigma^{(3)}$, dual to information on a three dimensional uncurled space within the D5-brane. The regions outside the black hole contains all quantum fields projected holographically by the event horizon. If one spatial dimension is ignored the horizon algebra $SO(2,1)$ is recovered for all the datum available to an observer.

Information on an event horizon is equivalent to information contained on D5-brane associated with the ensemble of states for a black hole. In 11 dimensions the dynamical surface of evolution is 10 dimensional. This means that any D5-brane for a black hole horizon has a Poincaré dual surface in five dimensions. As the D5-brane and its Poincaré dual contain all the quantum information in a 10 dimensional surface of evolution for supergravity, it is then evident that the information associated to a black hole is not really lost, but is only seriously buried from the perspective of an external observer with a coarse grained view of nature.

Embedding the horizon D2-brane into five dimensions, the uncompactified space, defines a D5-brane, which is demonstrated through M-theory [7][8]. The D-branes of two and five dimensions define "black branes [9]." In 11 dimensional supergravity the manifold of "evolution" has 10 dimensions. On this manifold a Poincaré duality exists between surfaces of two and eight dimensions. For type IIA supergravity a 1-form A_1 the field strengths are defined by $F_2 = dA_1$ The Hodge star duality operator then maps the field strength to $*F_2 = F_8 = *dA_1$. The Bianchi identities on these $dF_2 = 0$, $dF_8 = 0$ leads to the condition $d \wedge *dA_1 = d \wedge dA_7 = 0$. These are the dual Yang-Mills equations for dualities between electric and magnetic fields. The even dimensional nature of the type IIA fields indicates there is a similar duality between $F_4 = *F_6$ by Poincaré duality. Hence a D4-brane has a potential A_3, where this duality a D6-brane has a gauge connection A_5. Then by duality of fields the electric and magnetic field of a D6-brane is dual to the magnetic and electric field a D4-brane. A Dp-brane is then coupled to a $(p+1)$-form potential A_{p+1} that under a gauge determines the field by $\int A_{p+1}$. A type IIB field theory has $F_5 = *F_5$

Type IIA string theory requires odd valued-form potentials and Type IIB string theories require even valued-form potentials. For a type IIA electric potential on a p-brane the dual magnetic field is then defined on a $(6-p)$-brane. The magnetic flux through a

$p+2$ sphere is then

$$\int_{S_{p+2}} F_{p+2} = 2\kappa_{10}^2 g_{6-p} = e_p g_{6-p}, \qquad (5.12)$$

where κ_{10} is the coupling for the entire 10-dimensional surface or D10-brane. There exists a connection between the additional dimension and topological monopoles in equations 5.8-9 and the existence of these charges on D-branes.

The action for a Dp-brane interacting with an NS-NS string is given by the Dirac-Born-Infeld action

$$S_p = -g_p \int d^{p+1}\xi Tr\left(e^{-\Phi}\sqrt{-det(G_{ab} + B_{ab} + 2\pi\alpha' F_{ab})}\right), \qquad (5.13)$$

such that G_{ab} and B_{ab} are the NS-NS fields parallel to the D-brane and F_{ab} are the gauge fields on the D-brane [10]. The Latin indices are for coordinates on the D-brane and ξ are coordinates on the D-brane. The NS-NS fields are given by

$$G_{ab} = \frac{\partial X^\mu \partial X^\nu}{\partial \xi^a \partial \xi^b} G_{\mu\nu}, \; B_{ab} = \frac{\partial X^\mu \partial X^\nu}{\partial \xi^a \partial \xi^b} B_{\mu\nu}. \qquad (5.14)$$

A reasonable gauge condition to impose is $A_\nu = X^\mu F_{\mu\nu}$. For the left and right components of the string on the D-brane the Neuman condition is

$$\partial_n X^\mu = -i\partial_t X'^\mu, \; X'^\mu = X_L^\mu - X_R^\mu, \qquad (5.15)$$

which under T-duality becomes a Dirichlet condition. Further, as the D-brane is wrapped or compactified the endpoints satisfy $X'^\mu = e^\mu R = 2\pi\alpha' A^\mu$. The action is then given by a Chern-Simons Lagrangian

$$S_p = g_p \int_{p+1} Tr\left(exp(2\pi\alpha' F_2 + B_2) \wedge \sum_q A_q\right). \qquad (5.16)$$

The order of the term A_q is then determined by the order of the expansion of the exponential and the value of p. For $p = 4$ this action then contains the first order term

$$S_1 = 2\pi\alpha' \int Tr(F_2 \wedge A_3). \qquad (5.17a)$$

The second order term is

$$S_2 = 2(\pi\alpha')^2 g_4 \int Tr(F_2 \wedge F_2) \wedge A_1. \qquad (5.17b)$$

The second order action eqn. 5.17b contains the instanton number on a D4-brane as

$$8\pi^2 n = \int_{M^4} Tr(F_2 \wedge F_2). \qquad (5.18)$$

For $A_3 = e \wedge e \wedge A_1$ in a gauge-like form of gravity equation 5.17a is compared to the χ_2 equation of equation 4.159. Similarly equation 5.17b is to be compared to the χ_4 equation of eqn. 4.159.

This is curious 3-form potential with a charge defined on a D2-brane. The magnetic dual to this is then a monopole charge on a D5-brane. The D2-brane in M-theory this membrane has a tension

$$T_2 = \frac{1}{(2\pi)^2(\alpha')^{3/2}} \tag{5.19}$$

for the surface in the Dirac-Infeld-Sommerfeld action. This action in three dimensions in spacetime is then

$$S = T_2 \int d^3r \Big(\big(-det(g_{\mu\nu} + \partial_a X^\mu \partial^a X_\nu + 2\pi\alpha' F_{\mu\nu} \big)^{1/2} + \frac{\lambda}{2} \epsilon^{\mu\nu\rho} \partial_\mu F_{\nu\rho} \Big), \tag{5.20}$$

where the last term is a Lagrange multiplier that gives the Bianchi identity on the surface of charge. If spacetime has curvature the middle term will involve information about that curvature. S. Y. Chu derived weak gravity fields as an adjunct field that emerges from a string at distances larger than the string length [11]. A supersymmetric extension of this broken symmetry gives a breakdown in fermion-bosnic pairing. A symmetry breaking of the theory can result in a universe with an asymmetry in the particles over particles and an inflationary scenario in the early universe.

5.2: Time Symmetry and Violations Of Topological Quantum Numbers

Topological violations must describe a Lagrangian of broken symmetry for this field that acts on a degenerate vacuum state. The action for all bosonic strings and Majorana spinors as their supersymmetric partners is

$$S = -\frac{1}{2\pi} \sum_k \int d^2\sigma (h_k^{\alpha\beta} \partial_\alpha X_k^\mu \partial_\beta X_{k\mu} - i\bar{\psi}_k \gamma^\alpha \partial \psi_{k\mu}). \tag{5.21}$$

The sum is over all strings with the proper summation over equal numbers of advanced and retarded gravitational potentials associated with the strings. This action is not completely invariant under local supersymmetry. This requires the spin $\frac{3}{2}$ field, or gravitino ϕ_α, summed over every k in the theory with $\delta\phi_{k\alpha} = \nabla_\alpha \xi_k$ where ξ_k be Grassmann variables. Also introduce a zweibien (2-trad) to the theory on the string world sheet that transforms into the gravitino. The two actions

$$S' = -\frac{1}{\pi} \sum_k \int d^2\sigma \bar{\phi}_{k\alpha} \gamma^\alpha \gamma^\beta \psi^\mu \partial_\beta X_{k\mu}$$

$$S'' = -\frac{1}{4\pi} \sum_k \int d^2\sigma \bar{\psi}_{k\mu} \psi_k^\mu \bar{\phi}_{k\alpha} \gamma^\alpha \gamma^\beta \phi_\beta \tag{5.22}$$

are added together to obtain a total action invariant under local supersymmetric transformations

$$\delta X_k^\mu = \bar{\xi}\psi_k^\mu, \quad \delta\psi_k^\mu = -i\gamma^\alpha\xi(\partial_\alpha X^\mu - \bar{\psi}_k^\mu \phi_{k\alpha})$$
$$\delta e_\alpha^a = -2i\bar{\xi}\gamma^a \phi_{k\alpha}, \quad \delta\phi_{k\alpha} = \nabla_\alpha \xi_k. \tag{5.23}$$

Consider the string metric under the conformal gauge $h_k^{\alpha\beta} = e^{\Omega_k}\eta^{\alpha\beta}$, where Ω is a conformal factor and $\eta_{\alpha\beta}$ is the metric for a flat string world sheet. Every string is then similar up to a conformal factor. Consider the string under compactification. The bosonic variable for the string is a function of the position, where that position is by a target map parameterized by the string variables $\sigma_k^\alpha = (\tau_k, \sigma_k)$ as $X_k^\mu[x_k^\mu(\sigma_k^\alpha)]$. This gives

$$\partial_\alpha X_k^\mu = \frac{\partial X_k^\mu}{\partial x_k^\nu}\frac{\partial x_k^\nu}{\partial \sigma_k^\alpha} = \delta^\mu_\nu \frac{\partial x_k^\nu}{\partial \sigma_k^\alpha}, \tag{5.24}$$

so the bosonic string action may be written as

$$S = \frac{1}{2\pi}\sum_k \eta^{\alpha\beta}(g_k)_{\mu\nu}\partial_\alpha x_k^\mu \partial_\beta x_k^\nu. \tag{5.25}$$

At a large distance the position of the string separates into two components

$$x_k^\mu(\sigma_k^\alpha) = x_k^{0\mu}(\tau_k) + y_k^\mu(\sigma_k), \tag{5.26}$$

where $x_k^{0\mu}(\tau)$ gives the path of a particle-like entity parameterized by τ and $y_k^\mu(\sigma_k)$ is the fluctuation about that path due to small stringy behavior [11]. The relationship between these under a target map gives the classical flow of spacetime plus the torsional fluctuation. Consider the path according to an expansion around $x_k^{0\mu}$ by the stringy fluctuations

$$\partial_\alpha x_k^\mu = \partial_\alpha x_k^{0\mu} + y^\nu \nabla_{y^\nu}(\partial_\alpha x^{0\mu}) + \frac{1}{2}y^\nu y^\rho \nabla_{y^\nu}\nabla_{y^\rho}(\partial_\alpha x^{0\mu}) + \ldots. \tag{5.27}$$

These terms enter the bosonic action and each vanishes independently. To zeroth order in y this recovers the boson kinetic energy term according to $\partial_\alpha x_0^\mu$. To first order in y this defines a geodesic equation for y^μ and so that term vanishes. To second order in y this obtains the curvature

$$S_{O(y^2)} = -\frac{1}{2}\sum_k \int d^2\sigma \eta^{\alpha\beta}\left(g_{\mu\nu}(\nabla_\alpha y)^\mu(\nabla_\beta y)^\nu + \Omega R_{\mu\rho\nu\sigma}y^\rho y^\sigma(\partial_\alpha x_0^\mu)(\partial_\beta x_0^\nu)\right). \tag{5.28}$$

The first term is a kinetic energy for these fluctuations. The average of the string fluctuations in the second term and the conformal gauge indicates $\langle y^\sigma y^\rho \rangle = g^{\sigma\rho}\Omega$. The term $(\partial_\alpha x_0^\mu)(\partial^\alpha x_0^\nu) = const \times g^{\mu\nu}$ and so the second term is a constant times the Ricci scalar. The constant has units of cm^{-2}. This gives a target map so the integral over the string sheet is replaced by an integration over spacetime variables to arrive at the Hilbert-Palatini Lagrangian for the gravity field.

The action S' for scales much larger than the string length must be written according to the x^μ variables

$$S' = -\frac{1}{\pi}\sum_k \int d^2\sigma \bar{\phi}_{k\alpha}\gamma^\alpha\gamma^\beta \psi^\mu y^\nu \nabla_{y^\nu}\partial_\beta x_{0k\mu} +$$

$$-\frac{1}{\pi}\sum_k \int d^2\sigma \bar{\phi}_{k\alpha}\gamma^\alpha\gamma^\beta \psi^\mu y^\nu \nabla_{y^\nu}\partial_\beta x_{0k\mu}\left(g_{\mu\nu}(\nabla_\alpha y)^\mu(\nabla_\beta y)^\nu + \frac{1}{2}R_{\mu\rho\nu\sigma}y^\rho y^\sigma(\partial_\alpha x_0^\mu)(\partial_\beta x_0^\nu)\right). \tag{5.29}$$

It is evident that with T-duality a form of the action in equation 5.16 will yield two terms analogous to equations 5.17a and b that explicitly contain the χ_2 and χ_4 Lagrangian terms in equation 4.159. This derivation derives a form of gravitation that is analogous to a derivation of electromagnetism by considering fluctuations of scalar variables that define potentials under the action of a coboundary.

This heory leads to a breakdown of time symmetry, with zeta function representations of this breaking. The fermions in this theory will interact with each other through a Yukawa type of interaction of the form

$$\mathcal{L} = \kappa(\psi^\dagger M^2 \psi) + H.C., \tag{5.30}$$

where M is a matrix of the form in equation 5.5. This interaction from node to node on the chain is of the form

$$\sum_{i=1}^N gY\phi_i^*(\lambda_i - \lambda_{i-1})\psi_i + H.C.), \tag{5.31}$$

where Y is the Yukawa coupling constant that can be absorbed into g. With the coupling constant g define the Jarlskog parameter $J = g^N e^{2i\theta}$ that is a measure of CP violations and the breaking of supersymmetry[3]. This angle will provide a phase and a mixing angle that is the basis for the flavor mixing and Kobayashi-Maskawa mixing matrix. This then determines a fermion mass of the form

$$m \sim \frac{g_5^2}{\sqrt{2}}M_H. \tag{5.32}$$

CP symmetry braking may be tied to the breaking of supersymmetry [12], which gives partial supersymmetry. This construction illustrated below will be important in the discussion on partial breaking of supersymmetry and compactified dimensions. In the rest frame of a fermion with mass m the supergenerators Q_α and $\bar{Q}_{\dot\alpha}$ may define the fermion operators as

$$b_\alpha = \frac{1}{2m}Q_\alpha, \quad b^\dagger{}_\alpha = \frac{1}{2m}\bar{Q}_{\dot\alpha}. \tag{5.33}$$

The Fermi number operator is defined as
$$N_f = b^\dagger{}_\alpha b_\alpha. \tag{5.34}$$

The Fermi number operator acts as
$$N_F|\ \rangle = n_f|\ \rangle, \quad n_f = \{0,\ 1\}. \tag{5.35}$$

It is simple to demonstrate that N_f anticommutes with the fermion operator according to
$$b_\alpha(-1)^{N_f}|\ \rangle = (-1)^{N_f-1}b_\alpha|\ \rangle = -(-1)^{N_f}b_\alpha|\ \rangle, \tag{5.36}$$

which gives
$$Tr\big((-1)^{N_f}\{b_\alpha,\ b^\dagger{}_\alpha\}\big) = Tr\big((-1)^{N_f}b_\alpha b^\dagger{}_\alpha\big) - Tr\big(b_\alpha(-1)^{N_f}b^\dagger{}_\alpha\big)$$
$$= Tr\big((-1)^{N_F}\big)\big(Tr(b_\alpha b^\dagger{}_\alpha) - Tr(b_\alpha b^\dagger{}_\alpha)\big) = 0. \tag{5.37}$$

This is also equal to
$$Tr\big((-1)^{N_F}b_\alpha b^\dagger{}_\alpha\big) = Tr(-1)^{N_f}. \tag{5.38}$$

The trace of $(-1)^{N_f}$ is over both fermion and boson states and so $N_b = N_f$. The factor $(-1)^{N_f}$ is the Witten index, which vanishes when evaluated over both fermion and boson states [13].

This situation may be seen without appealing to asymptotic arguments applied to gravitation. The total Hamiltonian is then $\mathcal{H} = N\mathcal{H}_B + N\mathcal{H}_F$ for N strings with the bosonic operators a, $a^\dagger$ and the fermionic operators b, $b^\dagger$ for N strings is:
$$\mathcal{H} = \frac{N}{2}\sum_k \big(\omega(k)a^\dagger_k a_k\big) + \frac{N}{2}\sum_k \omega(k)b^\dagger_k b_k. \tag{5.39}$$

There exists an operator Λ so that $\omega(k)/2 = \omega log(\lambda_k)$. Thus by supersymmetry write the Hamiltonian as
$$\mathcal{H} = N\sum_k (log(\lambda_k)a^\dagger_k a_k) + N\sum_k log(\lambda_k)b^\dagger_k b_k. \tag{5.40}$$

Now Wick rotate the integrand in the path integral e^{-itH_B} under the substitution $it \to \beta$ to find that
$$Tre^{-NH_B\beta} = \sum_{i=1}^{\infty}\lambda_i^{-N\omega\beta} = \zeta(N\omega\beta). \tag{5.41}$$

This is a generalized zeta function. For both bosons and fermions
$$Tr\big((-1)^{N_F}e^{-N\mathcal{H}_B\beta}e^{-N\mathcal{H}_F\beta}\big) = e^{-N(N_B - N_F)\omega\beta} = 1, \tag{5.42}$$

since $N_B = N_F$ in unbroken supersymmetry. This then gives

$$Tr\big((-1)^{N_F} e^{-N\mathcal{H}_F \beta}\big) = \frac{1}{\zeta(N\omega\beta)}, \qquad (5.43)$$

where the fermion partition function may be written with the Möbius function μ as

$$Tr\big((-1)^{N_f} e^{-N H_F \beta}\big) = \sum_i \frac{\mu(\lambda_i)}{\lambda_i^{N\omega\beta}}. \qquad (5.44)$$

At this point supersymmetry is partially broken, where only half the stringy bosons have superpairs. The bosonic Hamiltonian takes the form

$$\mathcal{H}_B = 2N \sum_k log(\lambda_k) c_k^\dagger c_k + N \sum_k log(\lambda_k) f_k^\dagger f_k, \qquad (5.45)$$

where $c_k = a_k a_k$ and $c_k^\dagger = a_k^\dagger a_k^\dagger$. This is analogous to Cooper pairing of electrons in superconductivity. The operators f_k and $f_k^\dagger$ are fermionic operators "recovered" from the bosonic Hamiltonian. Now compute $Z_F(\beta)$ to find that

$$Z_F(\beta) = Tr\big((-1)^{N_F} e^{-N\sum_k log(\lambda_k) f_k^\dagger f_k \beta}\big) \times 1$$

$$= Tr\big((-1)^{N_F} e^{-N\sum_k log(\lambda_k) f_k^\dagger f_k \beta} e^{-2N\sum_k log(\lambda_k) c_k^\dagger c_k \beta} e^{-2N\sum_k log(\lambda_k) b_k^\dagger b_k \beta}\big)$$

$$= Tr\big((-1)^{N_f} e^{-\big(2N\mathcal{H}_F + N\mathcal{H}_B\big)\beta}\big), \qquad (5.46)$$

where the unity involves the old fermion and boson operators for unbroken supersymmetry. Now $Z_F(\beta)$ is

$$Z_F(\beta) = \sum_{n=1}^\infty \frac{\mu(n)}{n^{2N\omega\beta}} = \frac{\zeta(N\omega\beta)}{\zeta(2N\omega\beta)}. \qquad (5.47)$$

Here the c, $c^\dagger$ boson fields have superpartners and the f, $f^\dagger$ fields do not.

As this physically is associated the occurrence of gravity on a scale larger than the string, gravity leads to the violation of CP and T symmetry. The breakdown of T symmetry is due to the existence of black holes and event horizons that give a time asymmetry and entropy to the universe. This loss of T symmetry is on a larger scale, where it is likely that quantum information is ultimately preserved and this represents a large scale inability to access this information. The CP operator acts on a state $\psi_q(x, t)$ so that

$$CP\psi_q(x, t) = \psi_{-q}(-x, t). \qquad (5.48)$$

By CPT symmetry this is equivalent to $CPT\psi_q(x, t) = \psi_{-q}(-x, -t)$. However, the breaking of CP symmetry [12] and the occurrence of gravity on a scale larger than the

string length leads to a violation of T symmetry. Hence the Hamiltonian in equation 5.46 is such that
$$[T,\ \mathcal{H}] \neq 0. \tag{5.49}$$
As such the T operator is not a constant of the motion.

The D5-brane is now examined. The IIA string shares boundary conditions with a D2-brane and IIB string has boundary conditions on a D5-brane, which contains the NS-NS magnetic $B_{\mu\nu}$ charge. The type IIA strings are closed superstrings, where the clockwise and counterclockwise modes transform under separate spacetime supersymmetries with opposite chiralities. The type IIB strings are again closed but where the clockwise and counter clockwise modes transform under different supersymmetries, but with the same chiralities. These two string types are related by T-duality on the compactification of the X^9 coordinate compactification, where as $R \to 0$ for the Type IIA string is a type IIB with R large. The broken supersymmetry on the type IIA string separates the energy spectra of the two supersymmetries, which in carries to the Type IIB string sector. The type I string corresponding to a Type IIB string is on an odd dimensional p-brane, and under T-duality the Type I string corresponding to the Type IIA string is on an even dimensional p-brane. As the D5-brane contains the magnetic charge dual to the D2-brane electric charge, which means the endpoints of a string or D1-brane are contained on the D5-brane. This defines a point source for the $U(1)$ gauge field existing on the D5-brane. This is consistent with the existence of a $U(1)$ bundle found in equation 4.180. This $U(1)$ is then associated with the bundle splitting of the horizon algebra. Further, that it exists in the Type IIA theory is related to the S-duality on the $U(1)$ in the Type IIB theory. There is then a duality between Type IIA string on a D2-brane and a Type IIB string on a D5-brane. The D2-brane contains the quantum information emitted by Hawking radiation. [9]

We have the supersymmetrized horizon algebra in equations 4.157 through 4.160. In the NS-NS sector the transformation of the Rarita-Schwinger gravitinos contain the NS-NS 3-form field strength terms that must be added to the $\mathcal{D}(\xi + \bar{\xi})$ terms. Now extend the connection form ω as $\omega^{\pm} = \omega \pm \frac{1}{2}H$. These define two domains of unbroken supersymmetries in the D-brane world volume. In ten dimensions the noncompact group $SO(9,1)$ on the D5-brane is decomposed into
$$SO(9,1) \to SO(5,1) \times SO(4). \tag{5.50}$$
In $d = 10$ the Dirac representation for spinors is $\mathbf{32} = \mathbf{16} + \mathbf{16'}$. Under the decomposition in eqn. 5.50 the $\mathbf{16}$s decompose as $\mathbf{16} \to (\mathbf{4},\mathbf{2}) + (\mathbf{4'},\mathbf{2'})$:
$$\mathbf{16} \to (\mathbf{4},\mathbf{2}) + (\mathbf{4'},\mathbf{2'})$$
$$\mathbf{16'} \to (\mathbf{4},\mathbf{2'}) + (\mathbf{4'},\mathbf{2}). \tag{5.51}$$
The decomposition $SO(4) = SU(2) \times SU(2)$ the $\mathcal{D}^{\pm}$ correspond to unbroken supersymmetries with the transformations according to a $\mathbf{2}$ and $\mathbf{2'}$ of $SO(4)$. As a result there is a chirality here with left and right handed supersymmetries that are a $\mathbf{4}$ in $SO(4)$.

The ension of the D5-brane in NS-NS is

$$T_{D5} = \frac{1}{(2\pi)^5 g^2 \alpha'^3} = \frac{T_{D2}^2}{2\pi}. \tag{5.52}$$

These tensions are independent of the compactified dimension R_{10}, where this dimension is then compactified on the D5-brane to give the dimensional reduction to the D4-brane with the tension

$$T_{D4} = 2\pi R_{10} T_{D5}. \tag{5.53}$$

The interpretation of the D4-brane is a wrapped D5-brane. The Dirac quantization condition on the electric-like and magnetic-like charges $e_2 \mu_5 = 2\pi n$ and the dyon theory to identify a magnetic monopole field on the D2-brane show that $\mu_2 = \mu_5$. The two D2-branes with e_2 and μ_2 then determine the D4-brane with the tension in equation 5.53. The quantization condition on the D2-brane gives the coupling constants on the D4-brane, which on this D4-brane determines the instanton number for the gauge field, as well as the gravitational curvature coupling constant. The coupling constants g_5 and g_4 are related to each other by

$$\frac{1}{g_4^2} = \frac{R_{10}}{g_5^2}. \tag{5.54}$$

This is identical in form to the compactification relationship in equation 5.7.

5.3: TEV Physics From D-Branes and the Higgs Field

The compactification of an extra dimension is due to the wrapping of a D5-brane into a D4-brane with the resulting coupling constant for a gauge theory in four dimensions. From here a string (D1-brane) interacting with a black hole reproduces Hawking radiation, where if an extra dimension is compactified this is the scattering of a string from a D4-brane with quantum information concealed in the compactified dimension. In the dual view this means that quantum information is buried in the compactified space curled up just outside the event horizon of the black hole.

That this is connected to the Higgs field can be seen in equation 5.28. If this expansion is carried out this gives a topological term quadratic in the curvature plus a quartic term in ∇y^α

$$S_{O(y^2)} = -\int d^4x |-g| R + \frac{1}{2} \sum_k \int d^2\sigma \eta^{\alpha\beta} \big(g_{\mu\nu} (\nabla^\gamma \nabla_\alpha y)^\mu (\nabla_\gamma \nabla_\beta y)^\nu + (\nabla_\alpha y)^\mu (\nabla_\beta y)^\nu -$$

$$\frac{1}{2} \eta^{\alpha'\beta'} g_{\mu\nu} g_{\sigma\rho} (\nabla_\alpha y)^\mu (\nabla_\beta y)^\nu (\nabla_{\alpha'} y)^\sigma (\nabla_{\beta'} y)^\rho \big), \tag{5.55}$$

which has the form of the Higgs field if $\eta^{\alpha\beta} g_{\mu\nu} (\nabla_\alpha y)^\mu (\nabla_\beta y)^\nu = -\phi^2$. In this model gravity emerges from the small wiggling of the string, which results in the gravity field on a large scale. This emergent Higgs-like field results from small wiggly vibrations of the string. With the Born-Dirac-Infeld action the same is the case, gravity emerges from small vibrations in the D2-brane-string interaction. The cut-off for this compactification with respect to gravity occurs at higher energy than the compactification for gauge fields. This energy cut off is likely at or very close to the Hagedorn temperature. Yet this compactification gives a Higgs type of field associated with the quantum black hole. If this is the case quantum black hole at energies below the Planck scale, or longer than $\sim 10 - 100 L_p$, should act to break the symmetries of gauge theories. This then leads to several curious possibilities, such as the existence of a light Higgs in the TeV domain with a residual black hole nature.

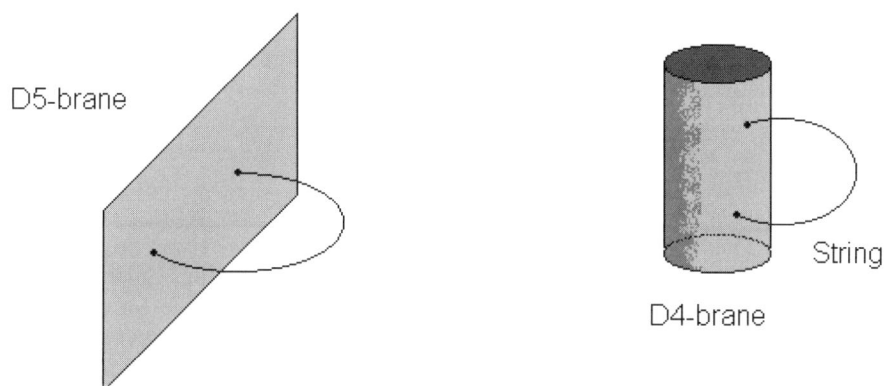

Blow up of the D5-brane from the D2-brane of a blackhole.
Figure 5.2

Consider the prospect that a light Higgs field emerges from this process. This gives new TeV physics of the standard model that is an effective or perturbative field theory. The emergent Higgs field is a form of the Landau-Ginzburg theory. We examine the above insight in the light of the supersymmetrization of the horizon algebra. The fundamental quantum conditions $[D_\alpha, X^\beta] = -\delta_\alpha^\beta$, determines the dynamic structure of the functional space. In group theory, this is realized in graded Lie algebra, where, $p_\mu = -2(cg_\mu)^{-1}\{\bar{Q}^\alpha, Q_\beta\}$, $[Q^\alpha, p_\mu] = 0$. The commutation relations $[T^\alpha, T^\beta] = if^{\alpha\beta\gamma}T^\gamma$, determines the group structure of the one-form D_l. In general with $N > 1$ supersymmetry a gauge connection from the supergenerator is

$$A_\mu = -2(cg_\mu)^{-1}\{Q^\alpha, Q_\beta\} \qquad (5.56)$$

with $p_\mu \to p_\mu + A_\mu$ in a gauge covariant form. This gauge theory is a component of a superfield.

Let Φ^k be the superfield expanded as $\Phi = \phi + \theta\psi + \bar{\theta}\bar{\psi} + \ldots$ that enter the superLagrangian

$$\mathcal{L} = \int d^2\theta d^2z D(\Phi, \bar{\Phi}) + V(\Phi). \qquad (5.57)$$

Consider the potential terms as a polynomial

$$V(\Phi) = \sum_i^N \Phi_i^{a_i} \tag{5.58}$$

The contribution to the path integral from this potential term is

$$Z[\Phi] = \int \prod_i^N D\Phi_i exp\left(\int d^2z d^2\theta \sum_i \Phi_i^{a_i}\right). \tag{5.59}$$

Now set $z^{a_i} = \Phi^{a_i}/\Phi_1^{a_1}$ in order to write this path integral terms as

$$Z[\Phi] = \int \prod_i^N D\Phi_i exp\left(\int d^2z d^2\theta \sum_i z_1^{a_1}(1 + z_i^{a_i})\right), \tag{5.60}$$

where the variation in the path integral $\delta Z[\Phi] = 0$ leads to the condition that $\sum_i \delta z_i^{a_i} = 0$. This is a constraint on the coordinates of the weighted projective space. This is the same sort of constraint that is imposed on CP^4 to obtain Calabi-Yau spaces. The simplest Calabi-Yau manifold is $Y_{4,5}$ on CP^4 determined by the constraint $\sum_i^5 z_i^5$. The identification is apparent for $z_i \rightarrow z_i/z_1$. It then becomes possible to examine Calabi-Yau spaces as determined by weighted projective spaces and the gauge fields they naturally contain.

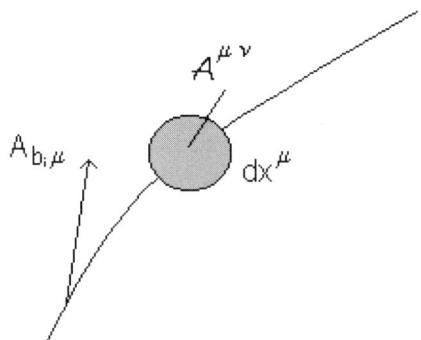

Variation of a path due to quantum fluctuations.
Figure 5.3

We then consider the variation in the variable $\delta z_i^{a_i} = (\zeta^\alpha)^{-1}\frac{\partial \xi^\alpha}{\partial z_i}\delta z_i$ determines a gauge connection, which exists in the projective space "cone" in the moduli space. This variation may be written as

$$\delta z_i^{a_i} = z^{a_i}\xi_{\alpha a_i}^\dagger \delta \xi_{a_i}^\alpha = z^{a_i}\xi_{\alpha b_i}^\dagger \delta \int^x dx^\mu A_{b_i\mu}^\alpha. \tag{5.61}$$

For the variation in the connection coefficient determined by a change in the path this gives the constraint condition

$$\sum_{i=1}^{N} \delta z^{a_i} = \frac{1}{2} \sum_{i=1}^{N} z^{a_i} \xi_{ab_i}^{\dagger} \int \int dx^\mu \wedge dx^\nu F_{b_i \mu\nu}^{\alpha}, \qquad (5.62)$$

where for small enough a variation $\int \int dx^\mu \wedge dx^\nu = \mathcal{A}^{\mu\nu}$. Physically the variation in the path corresponds to a quantum fluctuation.

Now write a path integral with the constraint condition as

$$Z[\Phi] = \int \mathcal{D}[\phi] e^{i\Gamma} exp\left(\frac{i}{2} \int d^2 z d^2 \theta \sum_{i=1}^{N} z^{a_i} \xi_{ab_i} \mathcal{A}^{\mu\nu} F_{b_i \mu\nu}^{\alpha}\right). \qquad (5.63)$$

The Chern-Simons Lagrangian $\Gamma = \frac{k}{4\pi} \int d^3 x \epsilon^{\mu\nu\rho} Tr(A_\mu \partial_\nu A_\rho + \frac{2i}{3} A_\mu A_\nu A_\rho)$ determines the field strength tensor to be

$$F_{b_i \mu\nu}^{\alpha} = \frac{k}{8\pi} \epsilon_{\mu\nu\rho} \frac{\delta \Gamma}{\delta A_\alpha^{b_i \rho}}. \qquad (5.64)$$

The Chern-Simmons action may be then used in the formation of the path integral. The expectation of a time ordered set of observables $TO_1 O_2 \ldots O_m \ldots$ up to $O(\mathcal{A}^2)$ is

$$T \langle O_1 O_2 \ldots O_m \ldots \rangle =$$

$$\int \mathcal{D}[\phi] \left(1 - \frac{4i\Gamma}{k} \epsilon_{\mu\nu\rho} \mathcal{A}^{\mu\nu} \int d^2 z d^2 \theta \sum_{i=1}^{N} z^{a_i} \xi_{ab_i} \frac{\delta}{\delta A_\alpha^{b_i \rho}}\right) e^{i\Gamma} (O_1 O_2 \ldots O_m \ldots). \qquad (5.65)$$

The action of the operator $\xi_{ab_i} \frac{\delta}{\delta A_\alpha^{b_i \rho}}$ on $O_1 O_2 \ldots O_m$ is to return the trace over the group constants

$$Z[\Phi_i] = \int \mathcal{D}[\phi] \left(1 - \frac{4i\Gamma}{k} \delta^3(x-y) \epsilon_{\mu\nu\rho} \mathcal{A}^{\mu\nu} dy^\rho \left\langle \sum_{i=1}^{N} z^{a_i} T_\alpha T_\alpha \right\rangle\right). \qquad (5.66)$$

The variation in this path integral leads to

$$\delta Z[\Phi_i] = \frac{2i\Gamma}{k} \delta^3(x-y) \epsilon_{\mu\nu\rho} \mathcal{A}^{\mu\nu} dy^\rho \left\langle \sum_{i=1}^{N} z^{a_i} T_\alpha T_\alpha \right\rangle, \qquad (5.67)$$

which is interpreted as the knot polynomial, or skein relation

$$Z[\Phi_+] - Z[\Phi_-] = -\frac{2i}{k} \left\langle c(t) \sum_i z^{a_i} \right\rangle. \qquad (5.68)$$

This braiding of compactified spaces on the event horizon is a representation of the segmentation of the event horizon.

This braiding of the event horizon is identified with a quantum group of q-deformed operators. The q-deformed algebra is

$$[L_3, L_\pm] = 2L_\pm, \quad [L_+, L_-] = \frac{q^{L_3/2} - q^{L_3/2}}{q^{1/2} - q^{1/2}}, \tag{5.69}$$

where for $q = e^{|z|\omega/g}$ this q-deformation can be seen to be due to the parametric amplification of the vacuum state when the angular momentum L_3 eigenstates are the Hamiltonian for the spin network. This indication of a braided structure for the D2-brane by equation 5.68, with the compactified radius R_{10} illustrates a quantum group structure to the renormalizion of the Planck scale to large values.

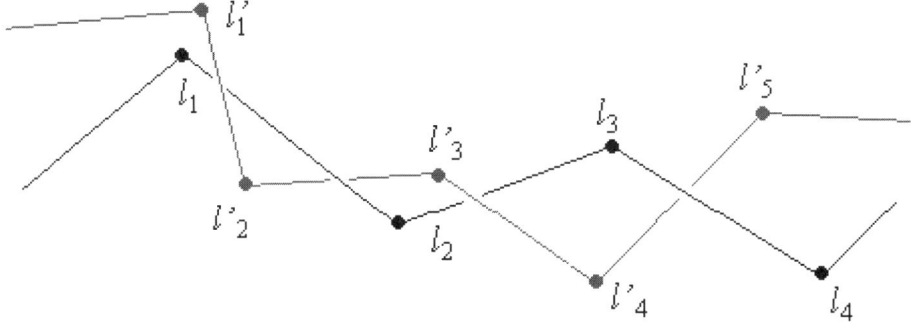

Braiding of compatified Calabi-Yau spaces on an event horizon.
Figure 5.4

The D5-brane and D1-brane (the string) have their respective charges Q_5 and Q_1. Define the coordinates m, n as tangent to the D5-brane, the coordinates i, j transverse to both D-branes and μ, ν and coordinates tangent to both D-branes. The momentum p_5 is defined for the wrapping of the D5-brane to give the D4-brane $p_5 = 2\pi n_5/R$. This gives a metric which defines a six dimensional space that embeds the D5-brane of the form [9]

$$ds^2 = \alpha_1^{-1/2} \alpha_5^{-1/2} \left(\eta_{ij} dx^i dx^j + (\alpha_n - 1)(dt + dx_5)^2 \right) +$$
$$\alpha_1^{1/2} \alpha_5^{1/2} g_{\mu\nu} dx^\mu dx^\nu + \alpha_1^{1/2} \alpha_5^{-1/2} \eta_{mn} dx^m dx^n, \tag{5.70}$$

where the metric coefficients are defined as

$$\alpha_1 = \frac{r^2 + r_1^2 - A/\pi}{r^2 - A/\pi}, \quad r_1^2 = \frac{(2\pi)^4 g Q_1 \alpha'^3}{V_4}$$

$$\alpha_5 = \frac{r^2 + r_5^2 - A/\pi}{r^2 - A/\pi}, \quad r_5^2 = g Q_5 \alpha' \tag{5.71}$$

$$\alpha_m = \frac{r^2 + r_m^2 - A/\pi}{r^2 - A/\pi}, \quad r_m^2 = \frac{(2\pi)^5 g^2 Q_1 \alpha'^3}{RV_4}$$

where A is the area of the black hole defined as [9]

$$A = 2\pi^2 L V_4 r_1 r_5 r_m = \kappa^2 (Q_1 Q_5 n_5)^{1/2}, \tag{5.72}$$

where the black hole entropy for this area larger than the Planck length is $S = 2\pi A/\kappa^2$. As indicated the expectation for the number operator for this quantum black hole is $n(E) \simeq exp(2\pi\sqrt{Q_1 Q_5 n_5})$. With this metric there is the prospect that a "zero mass" black hole may exist. This zero or low mass black hole may then be connected to a Higgs field.

The Chan-Paton factor for the string D5-brane involves an interaction with a monopole term. Such monopoles are topological defects that have the affect of breaking a global symmetry. If the theory describes fields far removed from the black hole the energy density of these fields asymptotically converge to a finite constant or zero value. Similar analysis has been done for Schwarzschild black hole, but here the "core mass" is found to be negative. This negative mass is measured from an asymptotic region that is flat, but with a deficit angle. This implicit residual core mass is different from the standard general relativistic, or ADM, mass. Here we apply a similar analysis with the metric in equations 5.70 and 5.71 to find that there is a residual core mass. This residual core mass is due to the topological defect of Chan-Paton factors on the D5-brane.

Equation 5.55 is written according to the field ϕ as

$$S = \int d^4 x \big(|-g| R - (1/2) \nabla^\mu \phi^m \nabla_\mu \phi_m + V(\phi) \big), \tag{5.73}$$

where the potential is a Higgs quartic potential and the Latin index is over the coordinates of a D5-brane. The Higgs field action is represented in spacetime coordinates rather than string coordinates. The Ricci curvature scalar and its Ricci tensor satisfy the Einstein field equation. The stress-energy tensor for the Higgs field is

$$T_{\mu\nu} = \nabla_\mu \phi^m \nabla^\mu \phi_m - g_{\mu\nu} \big(\nabla^\sigma \phi^m \nabla_\sigma \phi_m + V(\phi) \big), \tag{5.74}$$

where the field obeys the equation of motion $\nabla^\mu \nabla_\mu \phi_m = \partial V(\phi)/\partial \phi_m$. This indicates an ADM-like construction where the space is decomposed into a five dimensional surface $\Sigma^{(5)}$ and the time variable. On $\Sigma^{(5)}$ are further decomposed into a five dimensional ball $B^{(5)}$ and the remainder of the space that represents the asymptotic region.

The field is assumed to have an asymptotic form $\phi_m \simeq f(r) r_m^2/(r - A/\pi)$, for $f(r) \sim \sigma(1 - 1/r^2)$, $\sigma = const$. The 5-ball the metric coefficient is then

$$\alpha_m = \frac{r^2 + r_m^2 \sigma(1 - f(r))}{r^2 - A/\pi}. \tag{5.75}$$

Now transform coordinates so that $r_m \to \sqrt{1 - \sigma} r_m$ with the length and time contractions $t \to (1 - \sigma f(r))^{1/2} t$, $r \to (1 - \sigma)^{-1/2} r$. Then in that limit that $r, r_m, \to \infty$

$$\alpha_1 \to 1, \; \alpha_i \to 1, \alpha_m \to (1 + \sigma). \tag{5.76}$$

The spacetime part of the metric is then of the form

$$ds_{st}^2 = -(1 - \sigma)dt^2 + \sigma^{-1}dr^2 + \sigma\eta_{ab}d\Omega^2. \tag{5.77}$$

Define $\sigma = \sigma\sqrt{1 - \alpha}$ and reparameterize $(1 - \sigma)t \to t$ and $\sigma ds \to ds$ to obtain the metric

$$ds_{st}^2 = -dt^2 + dr^2 + (1 - \alpha)d\Omega^2. \tag{5.78}$$

Now a mass may be defined by the ADM formalism of gravitation [14]. This involves a technical distinction between a space that is asymptotically flat with a deficit angle (AFDA) and one that is standard asymptotically flat with a deficit angle (SAFDA). The SADFA may be considered as a reference space for differentials on the second space. The SAFDA may be considered as having no deficit angle at all. The ADM mass is computed as

$$16\pi(1 - \alpha)m_{ADM} = \int_{\partial\Sigma} dS_i \big(g^{0ij}g^{0kl} - g^{0ik}g^{0jl}\big)\mathcal{D}_k^0 g_{jl}. \tag{5.79}$$

The superscript 0 corresponds to the SAFDA and $\mathcal{D}_k^0$ is a covariant derivative with respect to this space. This corresponds to the Hamiltonian, which is then interpreted as the mass or energy associated with a spacetime referenced from an asymptotic region or the SAFDA. This mass may be computed if the coefficients of the theory are inserted into the Schwarzschild metric and the dynamical equations derived. The energy is found to be [14]

$$E = \frac{\alpha}{8\pi}\bigg(\frac{(\partial\tilde{r})^2}{2}\bigg(1 - 2\frac{\tilde{m}}{\tilde{r}}\bigg) + \frac{f^2}{(1 - \alpha)\tilde{r}^2} + \frac{(f^2 - 1)^2}{4}\bigg), \tag{5.80a}$$

where

$$\tilde{r} = r\sqrt{\alpha\lambda/8\pi}, \quad \tilde{m} = m\sqrt{\alpha\lambda/8\pi}, \quad \tilde{E} = \frac{E}{\alpha\lambda/8\pi}, \tag{5.80b}$$

where the last two terms correspond to the quartic part of the Higgs potential since $f < 1$ and approaches zero for $r \to \infty$. The ADM mass of the black hole is determined from the energy $\tilde{E}$ in the limit that the radius approaches zero with

$$\tilde{m}(0) = -\frac{\alpha}{2(1 - \alpha)}. \tag{5.80c}$$

This give positive masses only for $\alpha \in (0, 1)$, a zero mass for $\alpha = 0$, a singularity for $\alpha = 1$ and negative values elsewhere. The standard asymptotic mass of the black hole is however found to be positive

$$M = \sqrt{\frac{A}{16\pi(1 - \alpha)}}. \tag{5.80d}$$

This negative mass for a quantum black hole is due to the electric-like monopole field associated with the D2-brane. This negative mass emerges from the Higgs field due to small stringy vibrations, which also give rise to the gravity field. This also leads to an interesting observation on the nature of the vacuum. Equation 4.52 illustrates that for $J = 0$ that the mass of the black hole is $m = -2\epsilon$. The vacuum can have a large value $\sim 246 GeV$ in the standard model, but gravitationally the vacuum has a small mass-energy due to negative mass virtual black holes near L_p. A negative mass black hole is in fact a wormhole. This suggests a solution to the apparent contradiction between the large vacuum vev for the standard model and the small vacuum energy required to give the small cosmological constant. So the Higgs sector and the gravitational sector that emerges from the same stringy physics have near zero gravitational contribution, but can act on gauge theories with a large vev as seen in the light Higgs field in the standard model of electroweak interactions.

Now consider the nature of the Higgs potential. The problem that continually plagues the Higgs field is the quadratic divergences of radiative corrections of this field have suggested that the Higgs field is ultimately derivative in some manner. Equation 5.80a is suggestive of such a derivation. This new physics is expected to emerge at the TeV scale to stabilize the divergence at the electroweak scale. Even if this soft black hole as a light Higgs is a tiny portion of the field, it may still be able to bury these divergences into the Planck scale. This entails the construction of a chain of nodes with gauge actions connecting them. Assume that the above Higgs field that emerges from stringy vibrations compactifies two extra dimensions. Let the nodes in this toroidal compactified region be connected by matrices S_i and U_j in the two orthogonal directions. For a two dimensional plaquette with coordinates (i, j), $(i+1, j)$, $(i+1, j+1)$ and $(i, j+1)$ on this compactified surface to give the operator

$$-\sum_{i,j} \lambda_{i,j} S_{i,i+1} U_{j,j+1} S^\dagger_{i+1,i} U^\dagger_{j+1,j}, \quad (5.81)$$

where the compactified components of the gauge field A_5 and A_6 are absorbed into a Higgs field. Since the operators $S_i = exp(iT^a \phi_i^a / f)$, the mass contributed to the Higgs field by the gauge terms, $\phi = \sum_i \phi_i / f\sqrt{N}$ can be evaluated by a Wilson loop of these gauge fields wrapped into this compactified space [3].

The symmetries of this theory are chiral and the matrices S and U transform by left and right operators R_i and L_i

$$S'_i = L_i U_i R^\dagger_{i+1}. \quad (5.82)$$

Yet the introduction of masses breaks chiral symmetry. Now introduce a spurion field, q_i that contributes a term into the covariant derivative $\mathcal{D}_\mu \rightarrow \mathcal{D}_\mu + S_i q_i A_{\mu i+1} q_i^\dagger$ if the derivative is along 5th dimension, where U replaces S along the 6th dimension. Hence the connection terms transform exclusively by the left handed transformations. The spurion field transforms by a $U(1)$ group as well as by $q_i \rightarrow R_i q_i L^\dagger_{i+1}$. So any operator formed by product $\prod_i S_i q_i U_i q_i$ will account for the mass of the field ϕ. Once q_i has been exploited to give the "spurious symmetry" the value of this field may be set to unity. This then suggests that the quartic term in equation 5.81 corresponds to the mass term of the Higgs

field. This mass will be on the order of the energy computed in equation 5.6 and in terms of ϕ, which is
$$m(\phi) = 4gf\,\sin^2(2\pi ka/R + \phi/\sqrt{N}). \tag{5.83}$$
Additionally a Coleman-Weinberg effective potential may be written with the cutoff in energy at $\sim gf/N$. Now write the operators S and U according to generators s and u
$$S_{ij} = e^{is_{ij}/fN}, \; U_{ij} = e^{iu_{ij}/fN}. \tag{5.84}$$
A Taylor expansion of the operator in equation 5.81 leads to
$$const + \frac{1}{N^2}\sum_{ij}\lambda_{ij}tr|[s,\,u]|^2 + \ldots, \tag{5.85}$$
where the divergences are removed for $N > 2$. For $N > 3$ there now implicitly exist extra dimensions.

We then advance the argument that the chain of gauge fields stabilizes the ADM mass of the black hole and determines the parameter α. The results from equations 5.81-5.85 suggest that for $N = 1, 2, 3$ there is no compactified dimension, but there is for $N > 3$. For $N = 1$ there is a simple cutoff, for $N = 2$ there is a logarithmic divergence, but for $N > 2$ amplitudes are finite. This gives a Coleman-Weinberg potential of the form
$$V(\theta) = \frac{8g^4 f^4}{(N^2 - 1)(N^2 - 4)} \sum_n \frac{\cos(n\theta)}{n^5}. \tag{5.86}$$
The leading term before the sum gives m_ϕ and there is further cancellation of divergent diagrams. Then for N states the scales g/N are bounded by $m_\phi \sim g^2/(16\pi^2 N)\Lambda$, $\Lambda \sim 1 TeV$.

We then make the physical identification that the chain of fields that compactifies the extra dimension are those identified on the D2-brane of a black hole. This then indicates that for a small black hole its mass is renormalized by these fields braided on its horizon. The conjecture that the Higgs field is a form of a quantum black hole, with extra dimensional gauge connections threaded on it, means that this is the determinant of the parameter α in the ADM mass of the black hole. The Higgs nature of the quantum black hole stems from the wrapping of these extra dimensional gauge fields. This freezes the remaining 4-d gauge fields into one fixed gauge configuration, which breaks the gauge group $\mathcal{G}$ and introduces a mass term. This suggests the conjecture that a Higgs particle is in a superposition with a quantum black hole. Similar ideas are found in reference [15].

So apparently there exists a Higgs field that is the result of gauge fields in compactified dimensions associated with black hole horizons. These fields are likely the gauge-like fields illustrated in Chapter two, which may by a $U(1)$ symmetry be rotated into their mirror gauge fields familiar to physicists. Thus a quantum black hole may be related to the Higgs field by the mirror image gauge fields in the total $U(n)$ symmetry. The black hole is this curious beast with a potentially zero or negative mass, where the addition of gauge fields pinned on its horizon in extra dimensions give rise to a particle that is "Higgsian" in nature. There are then experimental prospects for this. This may be seen with the electroweak process in figure 5.5 below.

Consider a TeV fermion that scatters off the Higgs field, where the fields ψ_i and ψ_f have the initial and final momenta

$$p_i = (m_f/2)(1,\ 0,\ 0,\ 1),\ p_f = (m_f/2)(1,\ 0,\ 0,\ -1). \tag{5.87}$$

These enter into the invariant amplitude $\mathcal{M} = im(G_F\sqrt{2})^{1/2}\bar{u}(p_i)v(p_f)$, G_F being the Fermi coupling constant for weak interactions. The modulus square of this amplitude is then

$$|\mathcal{M}|^2 = \sqrt{2}m_f^2 G_F Tr(\gamma p_i \cdot \gamma p_f) = 2\sqrt{2}G_F m_f M_H^2. \tag{5.88}$$

This is the amplitude for the straight standard electroweak model. This may then be used to compute the differential decay rate

$$\frac{d\Gamma}{d\Omega} = \frac{|\mathcal{M}|^2}{64\pi^2 M_H} = \frac{G_F M_H m_f^2}{16\pi^2\sqrt{2}}. \tag{5.89}$$

Now assume that this process involves the scattering with a Z boson then the total differential decay rate is

$$\frac{d\Gamma}{d\Omega} = \frac{G_F^2 M_H M_w^2 m_f^2}{2\pi^2\sqrt{2}}. \tag{5.90}$$

We may then "rotate" this diagram to describe the annihilation of a proton and antiproton that then gives rise to the pair of Z bosons.

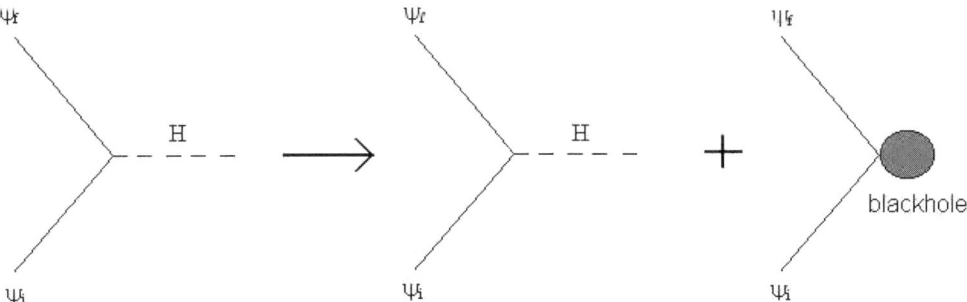

Amplitude for the Higgs field possessing quantum correlation with a black hole.
Figure 5.5

216 Quantum Fluctuations of Spacetime

There is the additional contribution from the black hole. This interaction is more problematic. An interaction between a proton and antiproton pair may produce a proton or antiproton plus a shower of additional particles. The black hole will violate conservation of baryon number. It is easiest to treat the whole process in a manner similar to a parton model "of yore" in describing inelastic scattering processes. The momenta of the particles in the frame of one of the proton are

$$p_i^\mu = (E,\ 0.\ 0, E),\ p_f^\mu = (E',\ E'\sin\theta,\ 0,\ E'\cos\theta). \tag{5.91}$$

and work with the invariants

$$s = (p + P)^2,\ Q^2 = -q^2 = -(p - p')^2,\ v = q \cdot P/M. \tag{5.92}$$

If M is the mass of the proton the invariant mass of the hadronic system is $U^2 = M^2 + 2Mv - Q^2$ for $v = E - E'$. This spray of hadrons or mesons can include the decay of the proton by its interaction with the black hole.

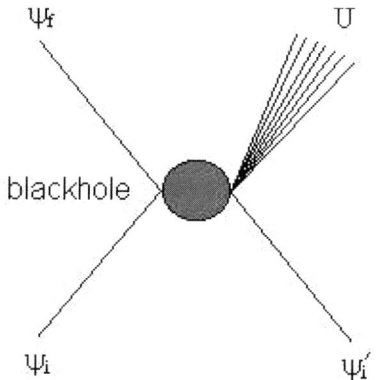

Blackhole contribution to a
scattering amplitude.
Figure 5.6

This invariant describes the momenta of the particle produced in this process is diagrammatically represented in figure 5.6. Here the entering states have momenta E/c and P, the scattering state on the left has the momenta p'. The momenta of the meson spray is identified with U, which gives

$$Q^2 = 4EE'\sin^2(\theta/2) = 2Mv + M^2 - M^{*2}, \tag{5.93a}$$

where M^* is the mass of an excited particle. The elastic scattering limit is $Q = 2Mv$, where the black hole does not absorb the proton and $U^2 = M^2$. The destruction of the proton by the black hole then modifies equation 5.93a by

$$Q^2 = 4EE'\sin^2(\theta/2) = 2Mv - M^{*2}, \tag{5.93b}$$

where the mass of the proton is subtracted from M^{*2}. This will result in a spray of mesons with $2Mv > Q^2$, along with a detectable violation of baryon number conservation. In a particle experiment this event should be readily detectable. This event would take place, based on a direct interpretation of the Planck energy, with a relative frequency from the standard model processes by a ratio of $\sim (G_F\sqrt{2})^{-1/2}/E_p \simeq 10^{-12}$. The probability of getting a micro-black hole with a Planck mass to fluctuate far from its ADM mass induced state as a field entangled with a Higgs particle, or as a Higgs particle, is likely small.

It is then apparent that to get the quantum black hole to "pop out" of the Higgs state that a renormalization of the Planck mass by induced squeezed states is required. For a squeezing parameter of $|z| > 9$ the production ratio $(G_F\sqrt{2})^{-1/2}/E_p$ could be increased by a factor of 4 orders of magnitude. Such a squeezing of the state of the protons would act as the catalyst for gravitational self-squeezing of the vacuum. Virtual black holes are associated with accelerations and teleparallel connections, where the squeezing of the wave functions of the protons will act as a catalyst to increase the gravitational squeezing of the vacuum state. This requires that a process analogous to parametric amplification with birefringent crystals be employed in high energy physics.

At high energy it is difficult to generate coherent and entangled states. For a proton collider the two counter rotating beams of protons and antiprotons have to be put into a coherent state. Coherent states for fermions is different than with bosons. For the operators $\hat{\psi}^\dagger|\psi\rangle = \psi|\psi\rangle$ the Grassmann algebra is such that $\psi^n = \psi^{\dagger n} = 0$ for $n > 1$. This means on the vacuum

$$e^{\hat{\psi}\hat{\psi}^\dagger}|0\rangle = (1 + \hat{\psi}\hat{\psi}^\dagger)|0\rangle. \tag{5.94}$$

Thus any path integral will be of the form

$$\int \mathcal{D}[\bar{\psi}]\mathcal{D}[\psi]e^{iS - \bar{\psi}\psi}. \tag{5.95}$$

If so then it is possible a parametric amplification in uncertainties in $\bar{\psi}$ and ψ are increased and decreased proportionately is possible.

To achieve this coherence a beam spitter on γ radiation is required. Assume γ rays can be split in a manner similar to a half mirror with optical photons. The entangled photons interact with the protons and antiprotons by "light catch up." In the rest frame of the proton or antiproton the γ ray is red shifted to a low value. Hence a proton and antiproton may interact with a split γ ray by a soft electromagnetic "jiggling." By this manner the entanglement of the split γ rays are transferred to the proton-antiproton beams. With a repeated cycling some proportion of the protons and antiprotons will enter into a Fermi coherent state. It is conceivable that some technology analogous to birefringence could be developed to induced squeezed states on γ rays. If so squeezed coherent states of proton antiproton beams at high energy are possible. This could further be enhanced if beam cooling is capable of cooling the beam into something closer to a Fermi condensate. By doing this it is possible that in these interactions the squeezed state entanglement of colliding beams will enter into the products of these interactions, which include the Higgs field. If so the Planck energy scale may then renormalized to a much lower energy. If any of these technologies are realistic it might be possible to observe a quantum black hole as a mirror of the Higgs particle.

It should also be mentioned that in this view of quantum black holes, compactified dimensions and Higgs particles most interactions with the Higgs particle do not violate baryon number. Proton decay by interacting with a virtual black hole would be a very rare event. So far proton decay has not been observed. This infers that both GUT predictions of proton decay and the occurrence of black holes on a large scale have bounds on their prevalence in the world. It should then be suspected that TeV black holes do not ordinarily occur, but if there is a measure of coherence in wave functions and an increase in gravitational self-squeezing of the vacuum state the physics of quantum black holes may become more pronounced at lower energies. This would mean that accelerators such as the LHC will not likely observe black hole physics without some type of technical upgrade that would involve as yet to be developed technology.

The idea that black holes may play a role in TeV physics has gained a measure of popularity. This type of physics is a possibility, but by no means certain. Much of the above construction involves the D-brane and string theory. The central departure is that the fluctuation of spacetime may be induced to exhibit squeezing of the vacuum state that renormalizes the Planck scale. So the decay of protons is not dominant under ordinary conditions, but where this decay at high energy might be induced, with state coherence and parametric amplification. Yet it has to be illustrated that this invokes D-branes and strings in very highly baroque theories and constructions that could in the end turn out to be little more than mathematics. Yet, this is offered as a way in which these types of theories could be tested soon.

5.4: Quantum Group Structure With Ricci Flows and Three-Manifold Surgery

Perelman's papers [16][17] lead to a possible proof of the Poincaré conjecture. Surgeries of a handles on a three-manifold can be performed so as to avoid the singularity at the "pinch off" of a handle. This means that the homotopy invariants of three-manifolds can be categorized in a nonsingular manner, which is likely a proof of the Poincaré Conjecture that any manifold homotopy equivalent to the three-sphere is topologically equivalent (homeomorphic) to the three-sphere. For the three-sphere plus a tube attached from one region to another, the cinching off of this handle creates a singular point where the Ricci curvature is divergent. The heat kernal generated by the general Laplacian operator acting on a functional F

$$\Delta F = \nabla_i \nabla^i F + g_{ij} R^{ij} = 0 \tag{5.96}$$

blows up as the Ricci scalar $g_{ij}R^{ij}$ goes to infinity. Perelman introduces a soliton on the small region around the cinching of the handle that varies with the scale of the small ϵ region in the limit the cinching closes off. This changes the definition of the Ricci scalar as

$$g_{ij}R^{ij} \rightarrow g_{ij}R^{ij} + \phi(x)/x, \tag{5.97}$$

so the Ricci scalar is finite as x approaches zero. This means that the Ricci flow can be described continuously across the cinching off of a handle, where the singular cusp is replaced by smooth caps.

General relativity describes spacetime with a single metric. A metric evolution is described as a classical "pushing" of a three-dimensional space through time. However, Einstein's field equations only work for a model with a single topology. Thus the quantum variant of the Hamiltonian constraint on a three-space only holds for a single topology. This Wheeler-DeWitt equation is then best applicable on the "tree level," where there are difficulties for single and multiple loop quantum corrections and fluctuations, which correspond to superpositions with topologically nontrivial spacetimes. The above renormalization of Ricci flow has an underlying quantum group structure. Perelman's results can lead to a model descriptions of topology changes in space, which physically correspond to virtual wormholes and so called quantum foam.

Consider on the cinched region the geodesic flow given by $dg_{ij}/dt = -2R_{ij}$, where R_{ij} is the Ricci curvature. The continuity of the Ricci curvature across the pinching off of a handle indicates how topological surgeries can be described analytically. Consider a geodesic flow on the squashed tube. The geodesic flow are given by the generator of group elements g. The group element then defines the Wilson line integral

$$W[C] = exp(i \int_C A \cdot dx), \tag{5.98}$$

for $A = gA'g^{-1} + gdg^{-1}$. Consider a curve that enters as a loop through a handle and re-emerges in the region near its entry. The cinching of the handle leaves a loop on one of the resulting horns and a remaining nonlooped curve on the other. When this handle is squashed to within a Planck unit of length, $L_p = \sqrt{G\hbar/c^3}$, there is an uncertainty and as to which the crossing of the loop within the handle. This results in an uncertainty in the crossing structure as seen with the Jones polynomial. Let g^+ and g^- be the group elements for these two loop crossings that fall into this undetermined situation. This leads to a quantum superposition of these two paths such that

$$g^+ - g^- = something, \tag{5.99}$$

which describes the existence of two crossings for the loop as the cinching action clamps the handle off. What is to be avoided are situations where $\infty - \infty = something$ or $g^+ - g^- = \infty$, where the infinities are due to a divergent Ricci flow at a singular point. However, these infinities can be removed. This equation is the Jones polynomial [18][19]. Given the over cross αL_+ and the under cross $\alpha^{-1} L_-$

$$\alpha L_+ - \alpha^{-1} L_- = zL_0, \tag{5.100}$$

where L_0 means no cross in the disconnected situation (manifold separation upon complete cinching), which is a quantum uncertainty in the breaking of topology. Perelman's result has profound implications for quantum gravity and topological change.

The Hamilton equation for Ricci flow for a Riemannian metric g_{ij} is $dg_{ij}/dt = -2R_{ij}$, where this is derived from a functional $F[g, \omega]$ such that [16]

$$\frac{dF[g, \omega]}{dt} = \int_M (R + |\nabla\omega|^2)e^{-\omega} dV. \tag{5.101}$$

This functional acts so that

$$\frac{dg_{ij}}{dt} = -2(R_{ij} + \nabla_i\nabla_j\omega), \quad \frac{d\omega}{dt} = -\Delta\omega - R, \tag{5.102}$$

which leads to

$$\frac{dg_{ij}}{dt} = -2(R_{ik} + \nabla_i\nabla_k\omega)g_{kj}, \tag{5.103}$$

an eigenvalued equation for the Ricci flow for $k = i$. The time derivative of the functional $F[g, \omega]$ is the low energy effective action in string theory and ω is a dilaton field, similar to that over a string world sheet [20]. The metric may then be rewritten according to a conformal gauge $g'_{ij} = exp(2\omega)g_{ij}$ and the Ricci curvature changed by $g'^{1/2}R' = g^{1/2}(R - 2\nabla^2\omega)$. By an appropriate diffeomorphism, or gauge choice, the governing equation for the metric can recover the Hamilton equation for Ricci flow. This eigenvalued equation then evolves a metric $g_{ij}(t)$ to $g_{ij}(t')$ as a Ricci soliton or breather. A trivial breather evolves the metric so it differs by scale factors and diffeomorphisms. The eigenvalued equation is bounded below according to infimum of the functional F, which prevents the occurrence of nontrivial breathers. Thus, under the shrinking, the infimum of the generating functional gives a bounded set of eigenvalues and the evolution of the metric is smooth.

This theory has an analog with statistical mechanics with a partition function $Z = \int exp(-\beta E)d\epsilon(E)$, for $\beta = 1/kT$ and $\epsilon(E)$ a density of states, that determines the average energy, entropy and fluctuation as [16]

$$\langle E\rangle = -\frac{\partial}{\partial\beta}Z, \; S = \beta\langle E\rangle + log\, Z, \; \langle(E - \langle E\rangle)^2\rangle = \frac{\partial^2}{\partial\beta^2} log\, Z. \tag{5.104}$$

These expectations can be defined according to Ricci flows if the functional is generalized to

$$F[g, \omega, t] = -\int_M \left(t(|\nabla\omega|^2 + R) + \omega - n\right)(-4\pi t)^{-n/2}e^{-\omega}dV), \tag{5/105}$$

so the evolution equations for the metric and ω are

$$\frac{dg_{ij}}{dt} = -2R_{ij}, \; \frac{d\omega}{dt} = -\Delta\omega + |\nabla\omega|^2 - R - \frac{n}{2t}. \tag{5.106}$$

The partition function for $log\, Z = \int(-\omega + n/2)(4\pi t)^{-n/2}e^{-\omega}dV$, where $t = 1/T$ is then

$$\langle E\rangle = -t^2\int_M \left(R + |\nabla\omega|^2 - \frac{n}{2t}\right)e^{-\omega}dV$$

$$S = -\int_M \left(t(R + |\nabla\omega|^2) + \omega - n\right)e^{-\omega}dV \qquad (5.107)$$

$$\langle(E - \langle E\rangle)^2\rangle = 2t^4 \int_M |R_{ij} + \nabla_i\nabla_j\omega - g_{ij}/2t|^2 e^{-\omega}dV$$

The boundedness of the the functional and the nontrivial breather result guarantees that these quantities are positive for $t > 0$. This construction is applicable to Euclideanized path integrals.

The Ricci curvature for a region being cinched will increase. The cinching off of a three-dimensional cylinder glued to a three-sphere results in a larger Ricci curvature. If the handle is cut the two cusps from the bifurcation point are commonly thought to have a divergent Ricci curvature. Given the Ricci flow $\dot{g}_{ij} = -2R_{ij}$ is κ collapsed at (x_0, t_0) on a scale $r > 0$ and $t \in [t_0 - r^2, t_0]$ if $|R(x, t)| \leq r^{-2}$, for all (x, t) with $d_{t_0}(x, x_0) < r$, so the volume of a ball $B(x, r^2)$ at a time t_0 has a volume less than κr^2. It is proven that for a smooth evolution of the metric by the Ricci flow a divergence may be avoided if the volume of the three-ball is kept at a bound at r_0. For $\kappa = \kappa(A) > 0$, $d_{T_0}(x, x_0) > r_0$ and the volume of the ball at least $A^{-1}r_0^n$, the metric g_{ij} is not κ-collapsed on scales less than r_0 at (x, r_0^2). The proof of this is an analysis of soliton solutions in this small region and a demonstration that this cinching of a multiply connected region is such that on one scale curvatures on a smaller scale may be cutoff. This is analogous to a renormalization procedure in quantum field theory. This theorem indicates that one can perform a topological surgery on a manifold by removing a small region with curvatures that are renormalized away. In this manner a neck connected to a three-manifold is described by the Ricci flow of g_{ij} as $\mathcal{M}^3 \times t$ for $t \in [0, T)$ that goes singular on the neck as $t \to T$, where the neck is surgically replaced by horns. The singular tips of the horns are replaced by smooth caps with finite Ricci curvature.

The physical interpretation is apparent. The quantum uncertainty principle indicates that the metric near the Planck scale should exhibit fluctuations. These fluctuations most likely contain virtual black holes and wormholes, where at a sufficiently small scale the manifold should fluctuate with topology changes. The Wheeler DeWitt equation for quantum gravity exhibits difficulties for topology changes, for the equation is established for a particular metric with a fixed topology. The construction of a partition function above is a Euclideanized path integral over an ensemble of the Ricci flows when $t \to it$. It is of interest to ascertain whether there exist quantum group structures that underlie this mathematics for surgeries on three-dimensional manifolds

The action principle may be written according to the minimal length of a path on a space. Let $\lambda(t) = \int_0^T \sqrt{g_{ij}U_iU_j}dt$ be an arc length on an small region. Now parameterize the velocities U_i according to all possible flows on the manifold as $U_i = exp(aX(t))\dot{\gamma}_i$. Here $X(t) = X\nabla_\mathbf{x}$ are for nearby paths and subject to variations. Under infinitesimal variations from a fiducial path $\dot{\gamma}_i$, or as $a \to 0$, $U_i = \dot{\gamma}_i(1 + aX\nabla_\mathbf{x} + O(\epsilon^2))$. Up to a^2 the integrand for $\lambda(t)$ is $\sqrt{a^2R + |\dot{\gamma}(t)|^2}$ and a summation over all ϵ^2 for each patch gives path length as

$$\lambda(t) = \int_0^T \sqrt{R + |\dot{\gamma}(t)|^2} dt. \tag{5.108}$$

The variation in this path length leads to the constraint on the geodesic equations

$$\nabla_X X = \frac{1}{2}\nabla R + 2Ric(X, \_) + X/2t. \tag{5.109}$$

An action $\mathcal{S} = \int(R + d\omega \wedge d\omega)$ is now defined, where R is a curvature two-form. The term $d\omega$ is a pure gauge term, where a gauge potential A is transformed as $A' = A + d\omega$. In general the action may then be written as

$$\mathcal{S} = \int (R + A \wedge A). \tag{5.110}$$

The Euler-Lagrange equation will give the above geodesic equation 14. Under an appropriate gauge choice this action may just then be $\mathcal{S} = \int R$.

A path integral

$$Z[g] = \int D[g, \omega] exp(-i\mathcal{S}) \tag{5.111}$$

is then constructed from the partition function. A gauge connection $A = \nabla \omega$ is found by the above conformal transformation, where this connection has a phase term

$$W(C) = exp\Big(i \int_C A \cdot dx\Big). \tag{5.112}$$

Now evaluate an expectation of this Wilson loop

$$\langle W(C) \rangle = \int D[g, \omega](W(C) exp(-i\mathcal{S})). \tag{5.113}$$

These expected values have a knot topological content when evaluated on a pinched off handle with $|R| \rightarrow |R_{cutoff}| < \infty$. This satisfies the Jones polynomial, which is the basis for quantum groups. This means that topological changes (surgeries) have an underlying quantum interpretation.

The curvature defines a two-form $\mathbf{R} = (R_{ij} + A_i A_j) dx^i \wedge dx^j$ where the Bianchi identify the Ricci two form gives

$$d\mathbf{R} + \mathbf{A} \wedge \mathbf{R} = 0. \tag{5.114}$$

Hence $\mathbf{R}$ is not closed under the co-boundary operator d, so there is no characteristic or first Chern class associated with $\mathbf{R}$. However, the Kähler four-form $\Omega = \mathbf{R} \wedge \mathbf{R}$ is closed $d\Omega = 0$. So Ω defines an element in $H^4(\mathcal{M}, R)$, where $\mathcal{M} = M^3 \times R$ is the four space of Ricci flows of three-manifolds, which is a second Chern class. The Kähler form defines a topological instanton number

$$I(\mathcal{M}) = \frac{1}{8\pi^2} \int_{\mathcal{M}} \Omega. \tag{5.115}$$

There exists the Kähler form on a topologically trivial space with the Chern-Simons three-form

$$\omega_3 = \mathbf{A} \wedge \mathbf{R} + \frac{2}{3} \mathbf{A} \wedge \mathbf{A} \wedge \mathbf{A}, \tag{5.116}$$

so that $\Omega = d\omega_3$, with $H^4(\mathcal{M}, R) = 0$. For a nonzero second Chern class the instanton charge is due to $ker(d)/im(d)$. From this the Chern-Simons Lagrangian may be defined on the three-manifold M^3 as

$$\mathcal{L} = \frac{1}{4\pi} \int_{M^3} \omega_3, \tag{5.117}$$

with the action on the Ricci flows determined by $\mathcal{S} = \int dt \mathcal{L}$, for t the parameter of Ricci flow evolution defined above.

Consider the cinching of a handle with a curve that winds around it. This handle consists of two balls $B_1(x, t, \epsilon, r)$ and $B_2(x', t, \epsilon, r)$, such that the points on ∂B_1 and ∂B_2 are identified. Physically this is a wormhole, which constitutes the quantum spacetime foam. As $\epsilon \to 0$ the two balls collapse to a point and the handle on the manifold is closed off. Let $\dot{\gamma}$ represent a curve crosses ∂B_1 from outside, then orbits B_2 in a region containing B_2 and then reemerges in the spatial region near B_1. There is then a loop that encloses the handle. The Ricci flow is divergent upon pinch off, with $R(x, t) \to R(x_0, t_0) + r^{-2}$ as $r = \epsilon r_0 \to 0$ [17]. Now impose the renormalization cutoff in the Ricci curvature, which physically would correspond to $R_{cutoff} \simeq L_p^{-2}$. The Wilson line integral with a small loop on this pinched off handle will then be

$$W(x_1, x_2) = W(x_1, x) i \mathcal{A}_{ij} R^a_{ij} T^a W(x, x_2). \tag{5.118}$$

An internal symmetry index has been imposed and $\mathcal{A}_{ij}$ is the small area enclosed by the loop. Now evaluate the Ricci curvature according to the Chern-Simons action $R^a_{ij} e^{iS} = i\epsilon_{ijk} \delta e^{iS}/\delta A^a_k$ at the cutoff for a time ordered product of fields

$$\langle T^a R^a_{ij} \phi_1 \phi_2, \ldots, \phi_n \rangle = Z^{-1} \int D[A] \kappa T^a \epsilon_{ijk} \left(\frac{-i\delta e^{iS}}{\delta A^a_k} \right) \phi_1 \phi_2 \ldots \phi_n, \tag{5.119}$$

which by integration by parts gives

$$\langle T^a R^a_{ij} \phi_1 \phi_2 \ldots \phi_n \rangle = -i Z^{-1} \epsilon_{ijk} \int D[A] \kappa T^a e^{iS} \frac{\delta}{\delta A^a_k} \phi_1 \phi_2 \ldots \phi_n, \tag{5.120}$$

where κ is a constant. By the computation of a matrix element of two gauge connections

$$\langle A^a_i, A^b_j \rangle = \frac{i}{k} \delta^{ab} \epsilon_{ijk} \frac{(x-y)^k}{|x-y|^3}, \tag{5.121}$$

224 Quantum Fluctuations of Spacetime

which is a topological invariant due to Gauss and defines the linking number. This gives the constant $\kappa = -4\pi i/k$. The functional derivative on the various fields $\phi = \phi_0 exp(i \int T^a A_j^a dx_j)$ brings down an additional structure constant, with

$$\langle T^a R_{ij}^a \phi_1 \phi_2 \ldots \phi_n \rangle = \frac{4\pi i}{k} \Big\langle W(x_1, x) T^a W(x, x_3) T^a W(x_3, x_4) \ldots W(x_n, x_2) +$$

$$W(x_1, x) T^a W(x, x_3) W(x_3, x_4) T^a W(x_4, x_5) \ldots W(x_n, x_2) + \ldots \Big\rangle. \quad (5.122)$$

By the renormalization of Ricci flows, consider this term as evaluated over the cutoff Ricci curvatures. Now evaluate the expected Wilson line $\langle W(L) \rangle$ removed from the cutoff by $R_{cutoff} \to R_{cutoff} - \epsilon$, for $\epsilon \to 0$ as

$$\langle W(L) \rangle = \langle W(L)^{-\epsilon} \rangle - \frac{4\pi i}{k} \Big\langle W(x_1, x) T^a W(x, x_3) T^a W(x_3, x_4) \ldots W(x_n, x_2) +$$

$$W(x_1, x) T^a W(x, x_3) W(x_3, x_4) T^a W(x_4, x_5) \ldots W(x_n, x_2) + \ldots \Big\rangle. \quad (5.123a)$$

The Fierz identity on the structure constants gives

$$T_{ij}^a T_{kl}^a = \frac{1}{2} \delta_{il} \delta_{jl} - \frac{1}{2N} \delta_{ij} \delta_{kl}, \quad (5.123b)$$

which leads to the relationship

$$\alpha \langle W(L) \rangle - \alpha^{-1} \langle W(L)^{-\epsilon} \rangle = z \langle W(L)^{cut} \rangle, \quad (5.124)$$

where $\alpha = 1 - 2\pi i/(kN) + O(k^{-2})$ and $z = -2\pi i/k + O(k^{-2})$. This last relationship is the skein polynomial for knot topology. Consider $W(L)$ as the over crossing loop, where $W(L)^{-\epsilon}$ as the under crossing loop, so this skein relation is then

$$\alpha \langle W(L^+) \rangle - \alpha^{-1} \langle W(L^-) \rangle = z \langle W(L_0) \rangle. \quad (5.125)$$

This then illustrates that the topology change in the three-manifold is accompanied by a change in the $SO(n)$ or $SU(n)$ topology of the path that loops through the neck that is pinched off. This then illustrates an underlying quantum group structure to topology changes in three dimensions.

The Chern-Simons Lagrangian in equations 5.16 and 5.117 define path integrals as in equation 5.63. Equation 5.125 is the same knot polynomial as the skein relation in equation 5.68. The changing in the topology of a spatial surface is then associated with an underlying quantum group structure, where in 5.3 this represents a threading of fields on an event horizon. This suggests that quantum foam, with its chaotic change in topology by quantum-morphisms, is tied to quantum indices on Calabi-Yau spaces. The collapse of a wormhole bridge in the quantum foam results in a virtual black hole. The quantum information for fields on the spatial manifold across the bridge is then preserved on the virtual black hole. Virtual black holes then preserve quantum information by holographically projecting field theoretic information into the space outside the event horizon.

The ADM approach to relativity is a form of Ricci flow for a three-manifold subject to the Hamilton and momentum constraints [21]. The Ricci flow is then a special case of Perelman's result, where the manifold of evolution has $H = 0$ and $N_i H^i = 0$. For the operator definition $\hat{\pi}^{ij} = -i\delta/\delta g_{ij}$ the connection terms in the Chern-Simons Lagrangian become field amplitudes for spacetime. A topology change due to a quantum fluctuation should change the overall instanton index given by the Kähler form. Physically it appears that topological changes in the spatial manifold under quantum-morphisms results in no net index change due to the underlying quantum group structure that absorbs this change. Thus quantum information associated with topology changes in the spatial manifold is simply transferred to the internal group structure for the braiding of the manifold.

A tie between quantum groups and topology changes identified by renormalized singularities implies that quantum information should be preserved. This is in contrast to the "no hair" theorems for black holes. The cut-off of divergent curvatures in a spatial cap, instead of a cusp or singularity, permits quantum information to thread through them. This conservation of quantum information is inferred by the holographic principle. There is also a growing suspicion that black hole thermodynamics does not involve the loss of information [22]. black hole thermodynamics probably involves hidden information later released in a scrambled form. Stephen Hawking recently conceded a bet by admitting that quantum information is preserved in black hole quantum radiance [23]. A Euclidean path integral over states for a deSitter cosmology with and without a black hole indicates that the topological entanglement between the initial and final states do not change the amount of quantum information in this cosmology. As yet a full report has not been written.

The quantum foam of the vacuum likely consists of virtual black holes and wormhole bridges. These two virtual quanta probably "blink" in between each other as superposed states. This likely means that the preservation of information across a wormhole bridge that pinches off into horns, or two black holes, implies that large black holes consisting of a large number of Planck mass units also conserve information.

The computation of Wilson lines on a three-manifold under Ricci flow the cutoff established on the cinching off of a handle produces a Jones polynomial if the metric variables are given an internal group description. The gauge connection is defined from a generator of a conformal factor $A' = A + \nabla \omega$. Such conformal field theories have Chern-Simons gauge field constructions [24] and these gauge potentials will exist in a space of conformal blocks. The conformal blocks are operators that determine a four point function between fields [25]. Further, these conformal block operators satisfy the Yang-Baxter relations for a braid group.

Knot theory, conformal field theory and the Yang-Baxter relations have an intimate relationship. Knot theory is a tool from which to analyze conformal field theory and the braid group structure of conformal blocks is used to tied the the fields together. The above underlying relationship between three-manifold topology and knot theory indicates that three-brane models of quantum cosmology can be analyzed topologically. From this perspective general relativity can be treated without reference to a background metric. There is then a "stringy" and "braney" structure that can be studied with these results on topology with three-manifolds when an internal group structure is assigned to the conformal gauge fields.

References

[1] L. Randall, R. Sundrum, "A Large Mass Hierarchy from a Small Extra Dimension," *Phys. Rev. Lett.* **83** (1999) 3370-3373.

[2] N. Arkani-Hamed, S. Dimopoulos, G. Dvali, N. Kaloper, *Phys. Lett.* **B480** (2000) 193-199 *Phys. Rev. Lett.* **84** (2000) 586-589.

[3] N. Arkani-Hamed, A. G. Cohen, H. Georgi, *Phys. Lett.* **B513** (2001) 232-240.

[4] N. Arkani-Hamed et al., http://www.arxiv.org/abs/hep-ph/0206020

[5] N. Arkani-Hamed, T. Gregoire, J. Wacker, *JHEP* **0203** 055 (2002).

[6] N. Arkani-Hamed, S. Dimopoulos, G. Dvali, N. Kaloper, *Phys. Rev. Lett.* **84** (2000) 586-589.

[7] A. Giveon, E. Witten, "Mirror Symmetry as a Gauge Symmetry," *J. Math. Phys.* **35** (1994) 5101-5135.

[8] E. Witten, "Five-branes and M-Theory On an Orbifold," *Nucl. Phys.* **B463** (1996) 383-397.

[9] A. Strominger, C. Vafa, *Physics Letters* **B379**, 99 (1999).

[10] A. A. Tseytlin, *Nucl. Phys.*, **B501**, 41 (1997).

[11] S. Y. Chu, http://www.arxiv.org/abs/gr-qc/9802070

[12] J. Ellis, G. G. Ross, *Phys. Lett.* **117B** (1982) 397.

[13] E. Witten Nucl. Phys. **B202**, 253 (1982).

[14] U. Nucamendi, D. Sudarsky, "Black Holes with Zero Mass," *Class. Quant. Grav.* **17** (2000) 4051-4058.

[15] S. Dimopoulos, G. Landsberg, *Phys. Rev. Lett.*, **87** (2001) 161602.

[16] G. Perelman, "The Entropy Flow For the Ricci Flow and Its Geometric Interpretation".
http://www.arxiv.org/PS_cache/math/pdf/0211/0211159.pdf

[17] G. Perelman, "Ricci Flows With Surgery on Three-Manifolds".
http://www.arxiv.org/PS_cache/math/pdf/0303/0303109.pdf

[18] L. Kauffman, *On Knots* Princeton Univ. Press, Princeton, NJ (1987).

[19] C. C. Adams, "The Knot Book," W. H. Freeman & Co. (1994).

[20] J. Polchinski, "String Theory," **1** pg 85, Cambridge University Press (2000).

[21] C. Misner, K. Thorne, J. Wheeler, "Gravitation,"Freeman press, San Francisco, (1973) 520-543.

[22] C. Adami, G. L. Ver Steeg, "black holes conserve information in curved-space quantum field theory," http://www.arxiv.org/PS_cache/gr-qc/pdf/0407/0407090.pdf

[23] http://math.ucr.edu/home/baez/this.week.html

[24] E. Witten, *Comm. Math Phys.* **121** 351 (1989).

[25] G. Moore, N. Seiberg, *Lectures on RCFT*, 1986 Summer Trieste.

6: Zeta Functions, Topological Quantum Numbers and M-Theory

It is common lore that the symmetry of the very early universe at very high energy was broken as the universe tunnelled out of the false vacuum state and entered its inflationary stage. The universe now is in its maximal state of broken gauge symmetry. As the universe expands this trend continues with thermodynamic phase changes as the universe become colder. This is seen in phase transitions, such as the breaking of the $U(1)$ gauge symmetry of electromagnetism in superconductivity. Compactified dimension have a supersymmetric interpretation, where the numbers of fermions and bosons have topological indices. The Witten index [1] is defined as,

$$Tr\big((-1)^{N_F} e^{-(H_F + H_B)\beta}\big) = 1 \qquad (6.1)$$

The trace is over both fermion and boson states so that $N_b = N_f$. The factor $(-1)^{N_f}$ is the Witten index and over the trace gives zero when evaluated over both fermion and boson states in unbroken supersymmetry. This index differs from unity for broken supersymmetry.

Twisted supersymmetry is a model of broken supersymmetry on a discrete ring of sites (i), $i = 1, \ldots n$ and links $(i, i-1)$ between sites [2][3]. At each site there is a gauge theory $\mathcal{G}_i$ and the total gauge theory is $\prod_{i=1}^{n} \mathcal{G}_i$. Assign a scalar field ϕ_i at each node and assume that $\phi_i = \phi$, $\forall i$, where the coupling constant at each site is g. This ring defines a compactified region of radius $R = N/2\pi g|\phi|$, with N sites. The Higgs scalar ϕ_i and fermion field ψ_i at each site is coupled to a gaugino field $\lambda_i - \lambda_{i-1}$ on a link within the Yukawa interaction

$$\sum_{i=1}^{N} gY\phi_i^*(\lambda_i - \lambda_{i-1})\psi_i + H.C.), \qquad (6.2)$$

where Y is the Yukawa coupling constant that can be absorbed into g. With the coupling constant g the Jarlskog parameter $J = g^N e^{2i\theta}$ is a measure of CP violations and supersymmetry breaking [2]. Supersymmetry is unbroken for $\theta = 0$. The twist is a measure of the breaking of supersymmetry around the ring lattice defined as $J - g^N$. The vacuum energy and the mass of the scalar field is then

$$E_{vev} = \kappa \sin^2\theta \left(\frac{g}{4\pi}\right)^N \frac{\Lambda^4}{16\pi^2}, \quad m_\phi^2 = \kappa' \sin^2\theta \left(\frac{g}{4\pi}\right)^N \Lambda^2, \qquad (6.3)$$

where Λ is the renormalization cutoff in the physics and κ and κ' are constants.

Fermions are coupled to the scalar field through the gaugino field. The Yukawa interaction according to the annihilation and creation operators a_k, $a_k^\dagger$, b_k, $b_k^\dagger$ and d_k, $d_k^\dagger$ for the scalar, fermion and gaugino field at each site is,

$$V = \sum_{i=1}^{N}\sum_{k,k'}(gY)(k,k')\big(a_k^{i\dagger}d_{k-k'}^{i,i-1}b_{k'}\mathcal{F} + a_k^i d_{k-k'}^{i,i-1\dagger}b_{k'}^{i\dagger}\mathcal{F}^* + H.C.\big), \qquad (6.4)$$

where $\mathcal{F} = e^{2i\theta}/N$. Consider a model where the scalar field is composed of two fermions, such as electrons in a Cooper pair or quarks in a meson. Within this model the absorption of the gaugino field by the fermion results in a bound state of two fermions. Similarly the decay of a boson into a fermion is equivalent to a transformation of one of the fermions in a bound state into a gaugino, where this gaugino escapes to "infinity" and leaves the fermion behind. The above interaction term becomes

$$V = \sum_{i=1}^{N}\sum_{k}\omega(k)\left((b_k^{i\dagger}b_k^{i\dagger})(b_k^i b_k^i)\mathcal{F} + (b_k^i b_k^i)(b_k^{i\dagger}b_k^{i\dagger})\mathcal{F}^*\right). \tag{6.5}$$

The creation and annihilation operators are the same at each site on the lattice so remove the sum and let $\mathcal{F} \to N\mathcal{F}$.

Consider the field ϕ as written in an eigenbasis $\phi = \sum_i a_i \phi_i$, with the operator Λ that acts on ϕ_i and returns the eigenvalue λ_i,

$$\Lambda \phi_i = \lambda_i \phi_i. \tag{6.6}$$

The action as the quadratic term of ϕ with the operator Λ is

$$S(\phi) = \frac{1}{2}\int \phi^\dagger \Lambda \phi. \tag{6.7}$$

This action gives a path integral

$$Z[\phi] = \int \mathcal{D}[\phi] e^{-iS(\phi)}. \tag{6.8}$$

For the integration measure $\mathcal{D}[\phi] = \frac{1}{\sqrt{\pi}}\prod_i da_i$ the path integral is

$$Z[\psi] - \frac{1}{2\sqrt{\pi}}\prod_i \int da_i e^{-\lambda_i a_i^2} - \frac{1}{2}\prod_i \sqrt{\lambda_i}, \tag{6.9}$$

which is then equal to $\frac{1}{2}\sqrt{det\Lambda}$. By taking the logarithm of the path integral the generator is

$$log(Z[\phi]) = \frac{1}{2}\sum_i log\left(\frac{\lambda_i}{2}\right). \tag{6.10}$$

Consider the case where supersymmetry is unbroken. Here the total Hamiltonian $\mathcal{H} = \mathcal{H}_B + \mathcal{H}_F$ for N lattice sites is:

$$\mathcal{H} = N\sum_k \omega(k) a_k^\dagger a_k + N\sum_k \omega(k) b_k^\dagger b_k. \tag{6.11}$$

By the above result there exists an operator Λ with eigenvalues λ_k that determine the frequencies by $\omega(k)/2 = \omega log(\lambda_k)$. Thus by supersymmetry the Hamiltonian is

$$\mathcal{H} = N\omega \sum_k log(\lambda_k) a_k^\dagger a_k + N\omega \sum_k log(\lambda_k) b_k^\dagger b_k. \tag{6.12}$$

Consider the integrand in the path integral $e^{-it H_B}$ under the substitution $it \to \beta$

$$Tr e^{-N\mathcal{H}_B \beta} = \sum_{i=1}^\infty \lambda_i^{-N\omega\beta} = \zeta(N\omega\beta), \tag{6.13}$$

which is the form of a generalized zeta function. This may be replaced with the Riemann zeta function if the eigenvalues are replaced with prime number in a Gödel numbering procedure. For unbroken supersymmetry the Witten index then gives,

$$Tr\big((-1)^{N_F} e^{-N\mathcal{H}_B\beta} e^{-N\mathcal{H}_F\beta}\big) = e^{-N(\mathcal{H}_B - \mathcal{H}_F)\beta} = 1 \tag{6.13}$$

since $N_B = N_F$. The Fermion sector must by the trace property $Tr(AB) = Tr(A)Tr(B)$ be

$$Tr\big((-1)^{N_F} e^{-N\mathcal{H}_F\beta}\big) = \frac{1}{\zeta(N\omega\beta)}. \tag{6.14}$$

The fermion partition function is then written with the Mø"bius function μ as

$$Tr\big((-1)^{N_F} e^{-N\mathcal{H}_F\beta}\big) = \sum_i \frac{\mu(\lambda_i)}{\lambda_i^{N\omega\beta}} = \frac{1}{\zeta(N\omega\beta)}. \tag{6.15}$$

Now let supersymmetry be broken. If the chain with N nodes have only n bosons with superpairs then the bosonic Hamiltonian for the whole chain then assumes the form

$$\mathcal{H}_B = n\omega \sum_k log(\lambda_k) c_k^\dagger c_k + 2N\omega \sum_k log(\lambda_k) f_k^\dagger f_k, \tag{6.16}$$

where $c_k = b_k b_k$ and $c_k^\dagger = b_k^\dagger b_k^\dagger$. The operators f_k and $f_k^\dagger$ are fermionic operators. This is tantamount to saying that some of the bosonic operators do not have fermionic superpartners with $\theta = 2n\pi a/R$, where a is the spacing between lattice sites with broken supersymmetry and R is the radius of the ring. The partition function $Z_F(\beta)$ is then

$$Z_F(\beta) = Tr\big((-1)^{N_F} e^{-N\omega \sum_k log(\lambda_k) f_k^\dagger f_k \beta}\big) \times 1$$

$$= Tr\big((-1)^{N_F} e^{-N\omega \sum_k log(\lambda_k) f_k^\dagger f_k \beta} e^{-2n\mathcal{H}_F\beta} e^{-2n\mathcal{H}_B\beta}\big)$$

$$= Tr\big((-1)^{N_f} e^{-\big((N + 2n)\mathcal{H}_F\beta + 2n\mathcal{H}_B\big)\beta}\big) \tag{6.17}$$

where the unity involves the old fermion and boson operators for unbroken supersymmetry. IN the fermion sector $Z_F(\beta)$ is

$$Z_F(\beta) = \sum_{m=1}^{\infty} \frac{\mu(m)}{m^{(N+2n)\omega\beta}} = \frac{\zeta(2n\omega\beta)}{\zeta((N+2n)\omega\beta)}, \quad (6.18)$$

We further see that $Z_{F,B}(\beta)$ is given by

$$Z_{F,B}(\beta) = Tr\left((-1)^{N_F} e^{-N\omega \sum_k log(\lambda_k) f_k^\dagger f_k \beta} e^{-n\omega \sum_k log(\lambda_k) a_k^\dagger a_k \beta}\right) \times 1$$

$$= Tr\left((-1)^{N_F} e^{-(N+2n)N \sum_k log(\lambda_k) f_k^\dagger f_k \beta} e^{-3n \sum_k log(\lambda_k) c_k^\dagger c_k \beta} e^{-N\mathcal{H}_F \beta} e^{-N\mathcal{H}_B \beta}\right)$$

$$= \sum_m \frac{\mu'(m)}{m^{(N+2n)\beta}} = \frac{\zeta(3n\omega\beta)}{\zeta((N+2n)\omega\beta)}, \quad (6.19)$$

which equal unity only for $n = N$. This illustrates that the Witten index is no longer an invariant under twisted supersymmetry, where an explicit demonstration of this with the Jarlskog parameter [5].

In effect the Witten trace operator is an inexact symmetry in broken or twisted supersymmetry. This means that the Dirac operator no longer has the same topological index as the boson operator. Bosonization is the equivalent to the topological breaking of Chern indices for the Dirac operator as given by the Atiyah-Singer theorem [5]. This index is associated with the spin 1/2 bundle S^+ and S^- with action by the Dirac operator is $D: C^\infty(S^+) \to C^\infty(S^-)$ in four dimensions

$$D: C^\infty(S^+ \times S^-) \to C^\infty(S^- \times S^+). \quad (6.20)$$

The Atiyah-Singer index is a Chern character of the Dirac operator over the manifold $\mathcal{M}$ defined according to [5]

$$A(\mathcal{M}) = \prod_i^{\frac{1}{2}dim\mathcal{M}} \left(\frac{y_i/2}{sinh(y_i/2)}\right), \quad (6.21)$$

where y_i are variables or roots of the Chern class over the manifold $\mathcal{M}$.

With supersymmetry there is a set of states with $j = 1/2$

$$|j_3 = 1/2, \ s = 1/2\rangle = 1|0\rangle$$

$$|j_3 = 1/2, \ s = 1, 0\rangle = \bar{Q}^{\dot{2}}|0\rangle$$

$$|j_3 = 1/2, \ s = 1\rangle = \bar{Q}^{\dot{1}}|0\rangle \quad (6.22)$$

$$|j_3 = 1/2, \ s = 1/2\rangle = \bar{Q}^{\dot{1}}\bar{Q}^{\dot{2}}|0\rangle,$$

so the number of fermion and boson states are equal with $Tr(-1)^{N_F}e^{N_B - N_F} = 1$, and the index number of fermions and bosons are equal in unbroken supersymmetry. With the set of observables $\mathcal{O}_i^b$ over bosons the correlation of these observables is

$$\langle \mathcal{O}_1^b \ldots \mathcal{O}_m^b \rangle = \sum_{\mathcal{S}_n^b} \int A(\mathcal{S}_n) \omega_1 \wedge \ldots \wedge \omega_m, \tag{6.23}$$

for $\mathcal{S}_n$ are solution spaces to $\Phi| \rangle = 0$. For the fermion sector similar observables $\mathcal{O}_i^f$ satisfy the correlation [4]

$$\langle \mathcal{O}_1^f \ldots \mathcal{O}_m^f \rangle = \sum_{\mathcal{S}_n^f} \int A(\mathcal{S}_n) \omega_1 \vee \ldots \vee \omega_m, \tag{6.24}$$

where $\vee$ denotes anticommuting forms and $\mathcal{S}_n^f$ are the space of solutions for $\Psi|0\rangle = 0$. Here $\sum_n A(\mathcal{S}_n^f)$ will be the Atiyah-Singer index. By the Witten index $\sum_n A(\mathcal{S}_n^f) = \sum_n A(\mathcal{S}_n^b)$, but with twisted supersymmetry $\sum_n A(\mathcal{S}_n^f) \neq \sum_n A(\mathcal{S}_n^b)$. It is then apparent that the zeta function realizations for fermion and boson fields for twisted supersymmetry violates

$$Tr\big((-1)^{N_F} e^{-N\mathcal{H}_B \beta} e^{-N\mathcal{H}_F \beta}\big) = 1. \tag{6.25}$$

The zeta functions for broken supersymmetry are then a measure of the topological violations or violations of topological quantum indices.

This physically has much to do with the nature of the vacuum energy. With unbroken supersymmetry the Witten index indicates that the total vacuum energy is zero. However, with broken or twisted supersymmetry the vacuum energy will become positive $H|0\rangle = \frac{1}{2}\sum_i \bar{Q}_i Q_i |0\rangle \geq 0$. The violation of topological indices associated with quantum field theory is then seen to be an aspect of setting the value of the vacuum energy. The vacuum energy is then set through the above zeta function regularization procedure.

6.1: Wrapped Spaces and Internal Symmetries

Equation 6.3 is similar to equation 5.6. Twisted supersymmetry is apparently a way of arriving at a similar type of wrapping of extra dimensions. The loss of supersymmetry is then made up for by the wrapping of supersymmetry in these compactified dimensions. By wrapping supersymmetry into compactified dimensions only partial supersymmetry can exist in spacetime. Assign the fields ϕ_i that connects a $SU(n)$ with a $SU(m)$ at the ith node and the fields $\psi_{i,i+1}$ that connects the $SU(m)$ with the next $SU(n)$ at the $i+1^{th}$ node. Again define operators that link group elements at the ith and $i+1^{th}$ nodes g_i, g_{i+1} respectively by $g_i^{-1} S_{i,i+1} g_{i+1}$ [6]. The differential operator on this matrix is the same as equation 5.2, but with an additional supersymmetry wrapped in the chain. The differentials of the scattering matrix are then extended to

$$\mathcal{D}_m S_{i,i+1} = \partial_\mu S_{i,i+1} + ig_s A_{mi} S_{i,i+1} + ig S_{i,i+1} A_{mi+1} \tag{6.26a}$$

$$\mathcal{D}_\alpha S_{i,i+1} = \frac{\partial S_{i,i+1}}{\partial \theta^\alpha} + i\sigma^\mu{}_{\alpha\dot\alpha}\bar\theta^{\dot\alpha}\mathcal{D}_m S_{i,i+1} \tag{6.26b}$$

$$\bar{\mathcal{D}}^{\dot\alpha} S_{i,i+1} = \frac{\partial S_{i,i+1}}{\partial \bar\theta_{\dot\alpha}} + i\theta^\alpha \sigma^\mu{}_\alpha{}^{\dot\alpha}\mathcal{D}_m S_{i,i+1}. \tag{6.26c}$$

The index m runs through the spacetime index μ plus the index for the compactified region. Now operate in the n-trad or vielbein basis of supersymmetry. The coordinates in superspace defined as $z^M = (x^m, \theta^\alpha, \bar\theta^{\dot\alpha})$, where Latin letters m runs over spacetime indices and Greek letters over spinor variables. This infers an effective Lagrangian that includes supersymmetric terms involved with the differential of the scattering matrix,

$$\mathcal{L}_{eff} = -\frac{1}{2g^2}\sum_i F^a_{\mu\nu i}F^{a\mu\nu}{}_i + g_s^2 \sum_i Tr\bigl(|\mathcal{D}_\mu S_{i,i+1}|^2 + |\mathcal{D}_\alpha S_{i,i+1}|^2 + |\mathcal{D}_{\dot\alpha} S_{i,i+1}|^2\bigr), \tag{6.27}$$

where μ, ν have been restored to their role as a spacetime indices. This illustrates a measure of supersymmetry trapped within the wrapped dimension of space.

The operator on a chain $\mathcal{O} = |tr S_1 S_2 q_2 \ldots S_n|^2$, where the trace operator of scattering operator is identified with

$$Tr\bigl((-1)^{N_F} e^{-N\mathcal{H}_B \beta} e^{-N\mathcal{H}_F \beta}\bigr) = e^{-N(\mathcal{N}_B - \mathcal{N}_F)\beta}. \tag{6.28}$$

The states on this chain are identified with the function $Z_{F,B}(\beta) = \frac{\zeta(3n\omega\beta)}{\zeta((N+2n)\omega\beta)}$ and the states are a ladder of eigen-numbers of the zeta functions. The states for the extra dimension exist in the range $g_f/N \to g_f$ as given by these eigen-numbers. Here for N large enough this physics may extend to the TeV range of physics.

This is an outline of this type of theory. It has the aesthetic structure of zeta function realizations which in recent years have come to play an increasingly important role in quantum physics. In addition this approach could be extended to the supersymmetric version of the horizon algebra with a gravitino sector.

6.2: Instantons Between String Types, Topology and Vacuum Ambiguity

This section is not about any derivation of a theory, but more of a conjectured occurrence as string theories approach the strong coupling limit. Consider the Born-Dirac-Infeld action [7]

$$S = T_2 \int d^3r \biggl(\bigl(-det(g_{\mu\nu} + \partial_a X^\mu \partial^a X_\nu + 2\pi\alpha' F_{\mu\nu}\bigr)^{1/2} + \frac{\lambda}{2}\epsilon^{\mu\nu\rho}\partial_\mu F_{\nu\rho}\biggr). \tag{6.29a}$$

Consider the limit the momentum p_{10} of a particle in the compactified 10 dimension is larger than the momentum in the additional dimensions q. So the energy is then

$$E = \sqrt{p_{10}^2 + q^2 + m^2} \simeq p_{10} + \frac{q^2 + m^2}{2p_{10}}. \tag{6.29b}$$

The momentum in the compactified region is then $P_{10} \sim n/R_{10}$, for n the winding number and R_{10} the radius of compactification. Therefore as $n \to \infty$ the momentum in the compactified region becomes far larger than the momentum in the uncompactified space. Now consider the NS-NS field as lying in the compactified region with flat connections. The coupling strengths according to the M-theory parameters M_{11} and R_{10} gives the effective Hamiltonian

$$H = R_{10} Tr\left(\frac{1}{2}p_{10}^2 + \frac{M_{11}^6}{16\pi^2}[A^\mu, A_\nu]^2 + \frac{M_{11}^3}{4\pi}\lambda\epsilon_\mu{}^{\nu\rho}\gamma^0\gamma^\mu[A_\nu, A_\rho]\right). \tag{6.30}$$

This equation is similar to equation 4.94. Equation 6.30 then contains topological information on the winding of strings and compactified supersymmetry on orbifolds.

One goal of M-theory is the unification of the various string types. The various string types are:

Type I: Superstring theory with open and closed unoriented strings with gauge group $SO(32)$. The opposite directed modes transform under the same supersymmetry. For the closed type I string there are no D0-branes, or a "braneless" theory.

Type IIA: A closed oriented string theory with D0, D2, D4, D6, D8-Branes that its corresponding open type I strings under T-duality satisfy Dirichlet or Neuman boundary conditions on. The right and left handed fields transform under separate supersymmetries and have opposite chirality

Type IIB: A closed oriented string theory with D1, D3, D5, D7, D9-Branes that its corresponding open type I string satisfies boundary conditions on. This right and left handed fields transform under separate supersymmetries and have the same chiralilty.

Heterotic: $E_8 \times E_8$ string which has no brane structure but contains both a 10 and a 26 dimensional structure.

Heterotic: $SO(32)$ string with a structure similar to $E_8 \times E_8$ string.

The Hamiltonian in equation 6.30 then describe the tunnelling of a NS-NS field between type I and II string types. The momentum for the NS-NS field is $\mathbf{p}_t = e^{-tf}\mathbf{p}e^{tf} = \mathbf{p} + t d\mathbf{f}$, where e^{tf} is the evolution operator due to the occurrence of the potentials $[A^\mu, A_\nu]^2$ and $\gamma^\mu[A_\nu, A_\rho]$. The momentum defined by $\mathbf{p}$ and $\mathbf{p}^*$ is then written according to the annihilation and creation operators a, $a^\dagger$ as

$$\mathbf{p}_t = \mathbf{p} + ta_i\frac{\partial f}{\partial x_i}\mathbf{d}x^i, \quad \mathbf{d}_t^* = \mathbf{p}^t + ta_i^\dagger\frac{\partial f}{\partial x_i}\mathbf{d}^*x^i. \tag{6.31}$$

The time developed Hamiltonian is defined in the canonical sense as

$$H_t = \frac{1}{2}(\mathbf{p}_t\mathbf{p}_t^* + \mathbf{p}_t^*\mathbf{p}_t). \tag{6.32}$$

The role of the potential due to the connection terms is absorbed into the definition of this Laplacian. The Hamiltonian is then written according to our coboundary operators as

$$H_t = \frac{1}{2}\left(\mathbf{pp}^* + \mathbf{p}^*\mathbf{p} + t^2(a_i a_j^\dagger + a_j^\dagger a_i)\frac{\partial f}{\partial x_i}\frac{\partial f}{\partial x_j} + t[a_i, a_j^\dagger]\frac{\partial^2 f}{\partial x_i \partial x_j}\right), \quad (6.33)$$

where $a_i a_j^\dagger + a_j^\dagger a_i = g_{ij}$ and there is the Laplacian $\mathbf{pp}^* + \mathbf{p}^*\mathbf{p} = -\partial^2/\partial x_i \partial x_j$ identified with p_{10} in equation 6.29. The Hamiltonian then has the form

$$H_t = \frac{1}{2}\left(-\frac{\partial^2}{\partial x_i^2} + t^2\left(\frac{\partial f}{\partial x_i}\right)^2 + t[a_i, a_j^\dagger]\frac{\partial^2 f}{\partial x_i \partial x_j}\right). \quad (6.34)$$

$\partial f/\partial x_i$ is identified with the momentum q_i in additional dimensions, with $\lambda_i = M_{11}^6/16\pi^2$. This internal momentum is associated with a connection term by $f = \mathbf{A}_\mu(e_\mu)$, where with flat connections $d\mathbf{A}_\mu(e_\mu) = [\mathbf{A}_\mu, \mathbf{A}_\nu]dx^\mu$. Similarly $\partial f^2/\partial x_i \partial x_j$ is $\partial q_i/\partial x_j = \lambda_j q_i$. Hence the Hessian is then identified with the last term in equation 6.30. The Lagrange multiplier is now the Hessian for the critical points. The Morse function expanded around the critical points with $f(x) = f(X) + \lambda_i(x - X_i)^2$ gives the Hamiltonian of a harmonic oscillator perturbed by the Hessian term

$$H_t = \frac{t}{2}\left(-\frac{\partial^2}{\partial x_i^2} + t^2\lambda_i^2 x_i^2 + t[a_i, a_j^\dagger]\lambda_i\right). \quad (6.35)$$

This Hamiltonian acts on a basis of harmonic oscillator state $\{|n\rangle\}$, $n \in \mathbb{Z}$, with the energy eigenvalue

$$E_t = \frac{1}{2}\sum_i (|\lambda_i|(N_i + 1) - \lambda_i n_i), \quad (6.36)$$

where N_i is the state number and n_i is the sign of the λ, the vacuum is defined for $N_i = 0$ at each critical point. There is then the Betti number for this "master space," or D-brane space that links string types together,

$$b_n(t) - 2(\dim \ker(H_t)) \quad (6.37)$$

The topology of this space is given by quantum numbers at each critical point. Further the vacuum states are defined under compactification of dimensions. Tunnelling between string types occurs as instantons since the energy linking these types is $E \sim E_p$ and the energy of this harmonic oscillator is below Planck energy cut-off. This tunnelling is modelled by the motion of a particle determined on this manifold according to

$$\frac{dx^i(t)}{dt} = -g^{ij}[x(t)]\frac{\partial f[x(t)]}{\partial x^j}. \quad (6.38)$$

Classically this describes the motion of a particle in a critical point with a harmonic oscillator potential. This potential is given by the terms in equation 6.30. Yet if this equation is quantized then there is a probability of tunnelling. As such given two critical points X_i and X_j the matrix element $\langle X_i|p_t|X_j\rangle$ will define the tunnelling probability. A tunnelling amplitude is due to the relative phase of a wave function, where that phase is given by the Morse potentials at the two critical points $\sim e^{-t(f(X_i) - f(X_j))}$. The tunnelling amplitude is

$$\langle X_i|d_t|X_j\rangle = n(X_i, X_j)e^{-t(f(X_i) - f(X_j))}, \qquad (6.39)$$

where $n(X_i, X_j)$ is an index determined by the Morse function.

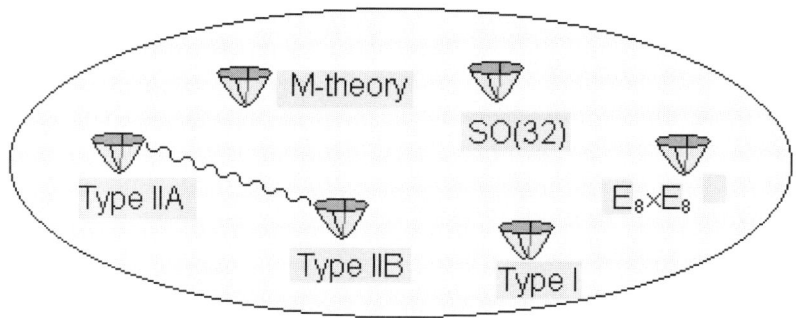

Tunnelling between Types IIA and Types IIB strings.
Figure 6.1

The above approximation breaks down when the momentum in the uncompactified dimensions approaches the momentum in the compactified dimension. At equality the compactified dimension unwraps. This suggests that at this point the wave functions of the NS-NS is no longer tunnelling across the potential barrier, but has an energy equal or larger than the barrier. It then appears that the number of symmetries associated with the theory diverges. One might be able to write this as a matrix theory with a myriad number of elements, where at low energy most of the terms are "buried away." However, this indicates that the vacuum structure above is not at the Planck energy, but rather a vev defined at lower energy. This further tends to suggest that the high energy vacuum of quantum gravity is elusive within this theory.

The tunnelling between a Type IIA and a Type IIB string is illustrated in figure 6.1. This tunnelling then illustrates the T-duality between Type IIA string and a Type I string on a D2-brane and a Type IIB string and its Type I on a D5-brane. The tunnelling is related to the Hawking radiation, for here the black hole has an event horizon with a large enough number of Planck units to make the theory operate. This tunnelling event between string types is associated with a transition between D-brane types in the M-theory sector. However, in the transition between string types the state is a Euclidean instanton, which in M-theory is a wave function that has an imaginary phase. In effect this is not the "final theory."

It appears that the primary problem is we do not know exactly what the vacuum state is for quantum gravity and strings. In the strong coupling limit in string theories all possible modes of the string become equally excited. This leads to a divergence of a partition function at the Hagedorn temperature. In effect we do not understand the nature of the quantum gravity vacuum as the scale approaches the Planck unit of distance and energy.

6.3 String Types, M-Theory and Octonions

The different string types are related to each other through M-theory. M-theory extends strings into entities of higher dimensions. There are both higher and lower dimensional objects called p-branes where p denotes their spatial dimensionality, such as a 1-brane is a string, a 2-brane a membrane and so forth. Higher dimensional objects existed previously in superstring theory but could vot be be studied properly before the "Second Superstring Revolution"[8] because of their non-perturbative nature.

The non-perturbative properties of p-branes of a class of p-branes assigned Dirichlet boundary conditions to type I superstring attached to them. These are designated Dirichlet p-branes (Dp-branes). Similarly there are Np-branes that hold the endpoints of open strings with Neumann boundarly conditions. Type I string theories for open strings have endpoints which satisfy the Neumann boundary condition the endpoints of strings These endpoints are free to move on the p-brane, where momentum does not flow into or out of the end of a string. The T duality infers the existence of open strings with positions fixed on the p-brane that are T-transformed strings with the Neumann boundary conditions. Type II theories are closed strings off a p-brane which correspond to Type I strings with end points on on a p-brane of a dimension. A Type I string is found by orientifolding a Type IIB string. The T duality of a Type I string with an equal number of left and right chiral waves transforms this into a Type IIA string in odd dimensions and a Type IIB string in even dimensions. This appears to break Lorentz invariance by assigning a special position. Yet the quantum physics of the Dp-brane restores this invariance.

The quantum states of the are found from the renormalizable 2D quantum field theory of the open string instead of the non-renormalizable world-volume theory of the D-brane itself. Hence perturbative methods are used to compute non-perturbative physics of the p-brane. Most of p-branes identified are Dp-branes, and where Np-branes are transformed into Dp-branes by duality symmetries. In this manner these are brought under mathematical control.

Mirror symmetry is a relationship between two Calabi-Yau manifolds. A Type IIB string on a orientifold is equivalent to a string without orientation, which is a Type I string. Similarly a a Type IIA string compactified in the X^9 coordinate, where this radius is small, is equivalent to a Type IIB string with this radius large. Thus an orbifold wrapping gives a T-duality between these string types. An orbifold with the Hodge structure $h^{1,1}$ is exchanged for an orientifold with Hodge structure $h^{1,2}$ by the wrapping of strings. This Hodge structure is

$$h^{p,q} = dim H^{p,q}(M), \qquad (6.40)$$

for M a complex valued manifold. The Dolbeault cohomology ring $H^{p,q} = H^p(M, \Omega^q)$, for Ω^q a sheaf of holomorphic q-forms. Duality obviously means that $h^{p,q} = h^{q,p}$.

A manifold that is Ricci-flat and Kähler has an $SU(3)$ holonomy. The Ricci form $\mathbf{R} = R_{i\bar{j}} dz^i \wedge dz^{\bar{j}}$ is holomorphic and closed $d\mathbf{R} = 0$, which defines a first Chern class c_1. For a Calabi-Yau manifold if $c_1 = 0$ $\mathbf{R}$ is exact. This manifold also has a covariantly constant holomorphic 3-form $\omega^{3,0}$, where $h^{3,0} = 1$. Further $h^{p,0} = h^{3-p,0}$. An example of this is the Eguchi-Hanson space EH^3 and the projective blow-ups of the T^6/Z_3 orbifold singularities. If the manifold is exactly $SU(3)$ holonomic then $h^{1,0} = 0$.

A way at arriving at this Calabi Yau mainfold may be seen in the following. Consider the spacetime metric $g_{\mu\nu}$ in four dimension with additional coordinate directions assigned for fluctuations. In this manner the fluctuations in each μ spacetime direction is associated with a coordinate in a unique direction in another four dimensions. This defines an eight dimensional vector of the form

$$X = (ct, x_i e^i, \hbar \xi_i \tilde{e}^i, c\hbar \omega \tilde{e}^8), \qquad (6.41)$$

for $i = 1, 2, 3$ and ω the angular frequency of the fluctuation. In eight dimensions there is the metric

$$\begin{pmatrix} g_{\mu\nu} & g_{\mu\tilde{\nu}} \\ g_{\tilde{\mu}\nu} & g_{\tilde{\mu}\tilde{\nu}} \end{pmatrix}. \qquad (6.42)$$

If the cross metric terms between the two four dimensional parts are zero a line element in flat spacetime is

$$ds^2 = -c^2 dt^2 + g_{ij} dx^i dx^j + \hbar^2 (g^{\tilde{i}\tilde{j}} d\xi_{\tilde{i}} d\xi_{\tilde{j}} + c\omega g^{\tilde{i}8} d\xi_{\tilde{i}} d\lambda + c^2 \omega^2 g^{88} d\lambda^2), \qquad (6.43)$$

for $d\lambda = de^8$. The $O(\hbar^2)$ terms are fluctuations in the line element.

Now consider these fluctuations on a black hole metric. The Schwarzschild metric in eight dimensions is then

$$ds^2 = -(1 - 2M/r) dt^2 + (1 - 2M/r)^{-1} dr^2 + r^2 d\Omega^2 +$$

$$\hbar^2 (g^{\tilde{r}\tilde{r}} d\xi_{\tilde{r}} d\xi_{\tilde{r}} + g^{\tilde{\theta}\tilde{\theta}} d\tilde{\theta}^2 + g^{\tilde{\phi}\tilde{\phi}} d\tilde{\phi}^2 + c\omega g^{\tilde{i}8} d\xi_{\tilde{i}} d\lambda + c^2 \omega^2 g^{88} d\lambda^2). \qquad (6.44a)$$

For $d\lambda = \frac{\partial \lambda}{\partial t} dt$ and $g_{88} = 1$ these terms may be grouped as

$$ds^2 = -[(1 - 2M/r) + \hbar^2 c^2 \omega^2 \partial\lambda/\partial t)^2] dt^2 + [(1 - 2M/r)^{-1} + \hbar^2 (\partial \xi_r/\partial r)^2] dr^2$$

$$+ [r^2 d\Omega^2] + \hbar^2 c\omega g_{\tilde{i}8} d\xi^{\tilde{i}} d\lambda. \qquad (6.44b)$$

The term $[r^2 d\Omega^2]$ represents the classical plus quantum metric terms grouped together. The eight dimensional metric has been mapped to the usual spacetime metric. The metric term $\hbar^2 c^2 \omega^2 (\partial \lambda/\partial t)^2$ is the fluctuation in the black hole mass. Hence this is assigned as

$$\hbar^2 c^2 \omega^2 \left(\frac{\partial \lambda}{\partial t}\right)^2 = 2m_p/r, \qquad (6.45a)$$

and similarly

$$\hbar^2 \left(\frac{\partial \xi_r}{\partial t}\right)^2 = \left(1 - \frac{2m_p}{r - 2M}\right). \qquad (6.45b)$$

The term $g_{\tilde{\phi}8} \partial \xi_{\tilde{\phi}}/\partial \lambda$ is a fluctuation in the angular momentum of the black hole

$$\hbar^2 g_{\tilde{\phi}8} \frac{\partial \xi_{\tilde{\phi}}}{\partial \lambda} = \sqrt{\frac{2m_p}{r}} \delta J/\hbar, \qquad (6.45c)$$

which contributed a $d\phi dt$ term in the spacetime metric. This is how an eight dimensional space can model the metric fluctuations in spacetime.

The Hopf fibration $S^7 \hookrightarrow S^{15} \to S^8$ gives the octonions, $\mathcal{O}$, with $\mathcal{O}P^1 \sim \mathcal{R}P^8$. The $\mathcal{O}P^1$ group is $spin(8)$. With the S^7 there is the exceptional group G_2 with $G_2 \to spin(7) \to S^7$. G_2 is also given by $SU(3)$ with the addition of two roots and $SU(3) \to G_2 \to S^6$

The group G_2 group contains two long and short roots a_1 and a_2 that satisfy

$$\frac{2a_1 a_2}{a_2^2} = -1 \text{ and } \frac{2a_1 a_2}{a_1^2} = -3, \qquad (6.46)$$

and so one representation might be

$$a_1 = (0, 1/\sqrt{3}), \quad a_2 = (1/2, -\sqrt{3}/2). \qquad (6.47)$$

The other roots are combinations of a_1 and a_2, call that B, which satisfies

$$\frac{2Ba_2}{a_2^2} = -1 \text{ and } \frac{2Ba_2}{a_1^2} = -3. \qquad (6.48)$$

These combinations are found from B as

$$a_3 = a_1 + a_2$$

$$a_4 = 2a_1 + a_2$$

$$a_5 = 3a_1 + a_2$$

$$a_6 = 3a_1 + 2a_2, \qquad (6.49)$$

where a_6 is the highest weight with $a_6 = (1,0)$. This weight satisfies

$$\frac{2a_1\mu^1}{a_2^2} = 1 \text{ and } \frac{2a_2\mu^1}{a_2^2} = 0. \tag{6.50a}$$

The second weight is found by demanding that

$$\frac{2a_1\mu^2}{a_2^2} = 0 \text{ and } \frac{2a_2\mu^2}{a_2^2} = 1. \tag{6.50b}$$

From these fundamental weights the others are then

$$\mu^2 - a_1$$

$$\mu^2 - a_1 - a_2$$

$$\mu^2 - 2a_1 - a_2$$

$$\mu^2 - 3a_1 - a_2, \tag{6.51a}$$

where with Weyl reflections the additional weights are found to be

$$\mu^2 - 3a_1 - 2a_2$$

$$\mu^2 - 4a_1 - 2a_2. \tag{6.51b}$$

The states corresponding to these weights can be built up from $|\mu^2\rangle$ by

$$E_{-a_1}|\mu^2>$$

$$E_{-a_2}E_{-a_1}|\mu^2>$$

$$E_{-a_1}E_{-a_2}E_{-a_1}|\mu^2>$$

$$(E_{-a_1})^2 E_{-a_2}E_{-a_1}|\mu^2>$$

$$E_{-a_2}(E_{-a_1})^2 E_{-a_2}E_{-a_1}|\mu^2>$$

$$E_{-a_1}E_{-a_2}(E_{-a_1})^2 E_{-a_2}E_{-a_1}|\mu^2>, \tag{6.52}$$

where these 7 states determine the 7 dimensional representation of the G_2 group. Along the H_1 and H_2 Cartan lines these states form a hexagon. The roots on the other hand form a 14 dimensional representation that appears as a large hexagon with a smaller hexagon with the short roots oriented 30 degrees relative to the large one.

Above the long and short roots are labelled a_1 and a_2. The label for the long root is maintained, but dropped for the short root. The fundamental or simple roots for $SU(3)$ are

$$a_1 = (1/2, \sqrt{3}/2), \ a_2 = (1/2, \sqrt{3}/2) \tag{6.53}$$

The weights are labelled as

$$\mu_1 = (1/2, 1/2\sqrt{3}), \quad \mu_2 = (1/2, 1/2\sqrt{3}), \tag{6.54}$$

where these weights define the additional weight

$$\mu_3 = (0, -1/\sqrt{3}) \tag{6.55}$$

that is not simple as it is determined by the first two. μ_1 is the highest weight. The generators are then defined as $T_i = \lambda/2$, for λ_i the Gell-Mann matrices, and the Cartan matrices are $H_1 = T_3$ and $H_2 = T_8$. According to the weights the off diagonal matrices define the E_i as

$$E_{\pm 1,0} = \frac{1}{\sqrt{2}}(T_1 \pm iT_2)$$

$$E_{1/2,\pm\sqrt{3}/2} = \frac{1}{\sqrt{2}}(T_4 \pm iT_5) \tag{6.56}$$

$$E_{1/2,\pm\sqrt{3}/2} = \frac{1}{\sqrt{2}}(T_6 \pm iT_7)$$

From the above definition of the roots the short roots for G_2 are computed as,

$$E_{a_1} = E_{0,1/\sqrt{3}}$$

$$E_{a_1+a_2} = E_{1/2,\mp 1/2\sqrt{3}} \tag{6.57}$$

$$E_{a_1+2a_2} = E_{-1/2,\mp 1/\sqrt{3}}.$$

Here the Weyl reflection is included as an abuse of notation. The matrix representations may be computed as

$$E_{a_1} = E_{0,1/\sqrt{3}} = \frac{1}{3}(E_{-1/2,\pm\sqrt{3}/2} + E_{1/2,\pm\sqrt{3}/2})$$

$$= \frac{1}{3\sqrt{2}}(T_6 \pm iT_7) + \frac{1}{3\sqrt{2}}(T_4 \pm iT_5)$$

$$E_{a_1+a_2} = E_{1/2,\pm\pm 1/2\sqrt{3}} = \frac{1}{3}(E_{-1,-} + E_{1/2,\pm\sqrt{3}/2}) \tag{5.58}$$

$$= \frac{1}{3\sqrt{2}}(T_1 \pm iT_2) + \frac{1}{3\sqrt{2}}(T_4 \pm iT_5)$$

$$E_{a_1+2a_2} = E_{-1/2,\pm 1/2\sqrt{3}} = \frac{1}{3}(E_{-1,0} + E_{-1/2,\pm\sqrt{3}/2})$$

$$= \frac{1}{3\sqrt{2}}(T_1 \pm iT_2) + \frac{1}{3\sqrt{2}}(T_6 \pm iT_7).$$

These are all the short roots, and are not simple for they are derived from the the simple roots of $SU(3)$. These roots plus the $SU(3)$ algebra define the G_2 algebra. It is evident that for the $SU(3)$ holonomy that in G_2 the Hodge structures $h^{1,0} = h^{0,1} \neq 0$.

The S^7 in the Hopf fibration then embeds the exceptional group G_2, which in turn embeds $SU(3)$. This naturally gives the octonions an orbifold structure. This also suggests that the octonions are the natural system from which M-theory should be derived. The above model also has an interesting feature to it. A Dp-brane with an attached type I string is equivalent to a closed Type II string off the brane. The metric fluctuations in 8 dimensions, with the correspondence to G_2 correlate the metric fluctuations with gauge-like fields. The graviton is a Type II string that is seen to be in a T-duality or entanglement with a closed type I string on a Dp-brane. This corresponding Type I string is as a gauge field content that is the mirror gauge field for gravitational fluctuations.

References

[1] E. Witten Nucl. Phys. **B202**, 253 (1982).

[2] N. Arkani-Hamed, A. G. Cohen, H. Georgi, http://www.arxiv.org/abs/hep-th/0109082

[3] N. Arkani-Hamed, A. Cohen, H. Georgi, Report-no: HUTP-01/A040, BUHEP-01-21, LBNL-48727.

[4] L. Baulieu, L. Losev, N. Nekrasov, *Nucl. Phys.* **B522** (1998) 82-104.

[5] S. K. Donaldson, P. B. Kronheimer, "The Geometry of Four-Manifolds," Oxford Science Publication, Oxford (1990).

[6] N. Arkani-Hamed, A. G. Cohen, H. Georgi, *Phys. Lett.* **B513** (2001) 232-240.

[7] A. A. Tseytlin, *Nucl. Phys.*, **B501**, 41 (1997).

[8] E. Witten, "Bound States of Strings and p-branes," *Nuclear Physcs* **B460**, 335.

[9] J. Baez, "The Octonions", http://www.arxiv.org/abs/math.RA/0105155

7: The Generalized Uncertainty Principle

7.1: The Vacuum Problem

The vacuum structure is a slippery concept. Virtual wormholes infer the parametric amplification of the vacuum state. There are operators similar to the squeeze state operator that act on the vacuum to determine structure outside what is defined by the standard Hamiltonian. The negative ADM mass for small black holes also illustrates something curious about the structure of the vacuum state. Finally the apparent difficulty in defining the high energy vacuum for strings and D-branes illustrates how elusive the vacuum structure is. If the universe is the result of a self-squeezing of the vacuum state it is best to have a reasonable description of that vacuum. The idea of *Creati Ex Nihilio* by quantum processes dates back to Lemaitre who proposed some sort of radioactive decay as the source of the big bang. Yet nobody really completely understands the "atom" here, which turns out to be the vacuum state.

Quantum mechanics infers complementarity between conjugate observables. This complementarity determine the uncertainty principle. General relativity has a sort of uncertainty principle of its own. Given a configuration of matter-fields in a region of space it is in general impossible to put a finger on that region and say "such and such" mass-energy resides here. The reason is this region of space exhibits diffeomorphisms according to the Einstein field equations. There is an uncertainty in the mass-energy contained in any region with compact support. Some researchers bang their heads against a wall to "solve" this apparent "problem" with general relativity. In fact it is a stunning and wonderful result that general relativity has this uncertainty. The first attempt at getting gravity and quantum mechanics to work together in the Hawking and Unruh effect illustrate that there is not a one to one connection between a particle number and a "rung" on the ladder of states. This includes the vacuum. These two uncertainties appear to have a commonality.

An important aspects of quantum gravity is this inability to identify exactly what the vacuum state is. The reason is that traditional attempts to make this identification are simply inadequate. The marriage of gravity and quantum mechanics involves a generalized uncertainty principle, where it is now uncertain by observers just what the vacuum state is! This generalized uncertainty involves the self-parametric amplification of the vacuum state. In laser physics this is where a photon enters a birefringent crystal with frequency ν and split into two photons with frequencies ν', ν'', with $\nu = \nu' + \nu''$. These two parametrically downshifted photons are in a quantum entangled state. As such these two photons can be absorbed by an atom to excite a state change in an atom with energy gap $h\nu$. If this experiment is conducted in a high-Q cavity the atom will re-emit a photon of frequency ν and the process is repeated. The one effect of this is that the uncertainty in the magnetic field **B** can become arbitrary while the uncertainty in the conjugate electric field **E** can go to zero: squeezed states. The quantum gravity vacuum does this to itself with virtual wormholes, which this leads to an uncertainty in defining what the vacuum state is. As such one can have negative energy fluctuations and wormholes. This then reduces the ZPE of the quantum gravity vacuum state, but at the cost of being unable to define it in a traditional way.

Wormholes require negative mass-energy, where negative energy is what causes the idea of macroscopic wormholes trouble. Classically if one has negative energy in a thin shell situated above where the Schwarzschild horizon should exist then a wormhole is possible. The Hawking-Penrose theorems indicate that if one has positive stress energy or if

$$T_{\mu\nu}U^\mu U_\nu \geq 0, \tag{7.1}$$

then momentum-energy will focus (gravity gravitates!) and the result is geodesic incompleteness at a singularity [1]. This is the strong energy condition. There are other conditions such as the dominant energy condition, where the above condition is maintained on a spacial surface $T_{ij}U^iU^j \geq 0$ $(i,j = 1, \ldots, 3)$ and $T_{\mu\nu}U^\mu$ is time-like. The weak energy condition is where $T^{00} \geq 0$. In all these conditions it appears that gravity focuses light rays into caustics and there is geodesic incompleteness.

In the Kip Thorne wormhole the weak energy condition is violated [2][3]. This means that in a region with negative energy geodesics defocus or "antigravitate." For the wormhole these geodesics have nowhere else to go but somewhere else in the universe.

This is where things go awry. All matter fields ultimately have a quantum mechanical origin. As such an expectation of a momentum-energy based on the Hamiltonian for these fields is derived. So the Einstein field equation in a semiclassical form looks like

$$R_{\mu\nu} - (1/2)Rg_{\mu\nu} = 8\pi(G/c^4)\langle |T_{\mu\nu}| \rangle \tag{7.2}$$

Consider a Hamiltonian for a free field

$$H = -\sum_k (\omega(k)a(k)^\dagger a(k)) \tag{7.3}$$

with a negative energy. The momentum-energy is not properly bounded below, where consequently total chaos results. Fields with negative energy spectra are not bounded below. Quantum mechanics is normally formulated with a lower bound, the vacuum or ZPE, with nice properties of L^p spaces such as Frechet spaces, Banach spaces and Hilbert spaces. So a quantum field with a negative energy spectrum would have a vacuum that is inherently unstable. Fields not bounded below as a source of spacetime curvature results in huge spactime fluctuations that are impossible to reconcile physically.

The quantum interest conjecture of Ford and Roman [4] illustrates that acquiring negative energy is impossible without also acquiring an equal or larger amount of positive energy. That the positive energy is in general larger is due to the work that must be done. This suggests that acquiring negative energy in sufficient quantities to construct a wormhole is practically impossible, as are other such solutions [5]. However, near the Planck scale quantum fluctuations are such that a black hole can be measured to have a negative energy or be a wormhole. Indeed it appears that as one approaches the Planck scale negative mass black holes, or wormholes, appear to predominate. A bit further from the Planck scale there is a domain where black holes and wormholes "blink' between each other. This then suggests that the quantum gravity vacuum has an energy expectation that is difficult to understand. It is argued here that the ZPE for gravity is impossible to define in the standard way, as gravity and quantum mechanics are ultimately aspects of a general theory. In effect gravity and quantum mechanics may be transformed into each other near the Planck scale. There is then a new uncertainty which permits physics to remove the divergences that are so prevalent in current theories of quantum gravity.

7.2: Differential Structure of Fields in Spacetime

To start an examination of the differential structure of general relativity is needed. This examination will lead to the action principle for gravity, which is the classical limit of a quantum theory of gravity. All classical field theories, whether Yang-Mills, σ-models, or general relativity, exist as the limit $\hbar \to 0$ of quantum field theories or topological field theories with quantum eigen-numbers that are topological invariants of a manifold.

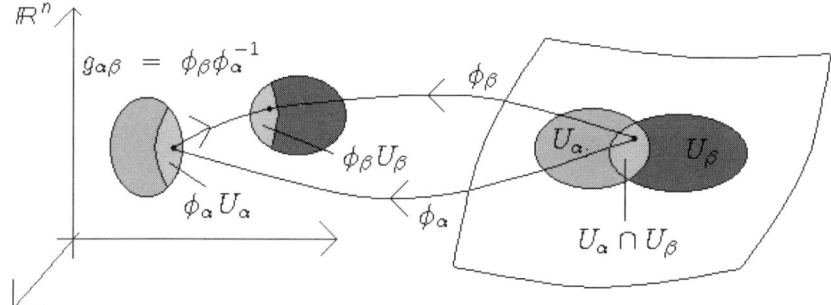

Atlas construction of a manifold.
Figure 7.1

Let $\mathcal{M}$ be the base manifold in four dimensions so the differential structure is determined by a set of symmetries described by a fibre bundle. Let U_α, where $\alpha = 1, 2, ..., n$, be an atlas of charts on the $\mathcal{M}$. The transitions from one chart to another is given by $g_{\alpha\beta} : U_\beta \to U_\alpha$, where these determine the transition functions between sections on the principal bundle over these charts. The transformation is a change of frame, where if that transformation involves a group $\mathcal{G}$ with dimension k the fibre is the product $\mathcal{M} \times V$. Here V is the vector space of the elements of $\mathcal{G}$. This is the projection $\pi(p, v) = p$ for $p \in \mathcal{M}$ and $v \in V$. This projection $\pi^{-1}U$, for $U \subset \mathcal{M}$, is isomorphic to $E = U \times V$. Given the atlas of charts the fibre bundle on the entire manifold is defined. $\mathcal{G}$ defines a transformation between sections of the bundle so that their overlaps define connection coefficients. The transform between one section to another is given by

$$s_\alpha = g_{\alpha\beta} s_\beta = e^{iX_{\alpha\beta}} s_\beta. \tag{7.4}$$

From now on the chart indices are suppressed to indicate sections. The notation s, s' for the two charts with $gs = s'$ is used. Now let the differential operator d act on s'

$$ds' = (gds + (dg)s). \tag{7.5}$$

This is written according to a total operator that acts on s as

$$ds' = g(d + ig^{-1}dg)s. \tag{7.6}$$

The action of g on $(d + \omega)s$ which equals $(d + \omega')s'$ gives

$$(d + \omega')s' = g(d + \omega)s$$
$$= g(d + \omega)g^{-1}gs = (d + g\omega g^{-1} + gdg^{-1})s'. \tag{7.7a}$$

This is a fundamental definition for how a connection transforms:

$$\omega' = g\omega g^{-1} + gdg^{-1}. \tag{7.7b}$$

The group element g are defined by algebraic generators so that $g = e^{iX}$. For the transformation sufficiently small so that $e^{iX} \simeq 1 + iX$,

$$A's' = ((1 + iX)A(1 - iX) - idX)s' = (\omega + i[X, \omega] - idX)s'. \tag{7.8}$$

For ω a constant or flat connection a pure connection term $(dg)g^{-1} = idX$ defines the connection in a frame with $g\omega g^{-1} = 0$.

The physical fields are derived from the action of a differential on this connection one-form

$$d\omega' = d(g\omega g^{-1} + (dg)g^{-1}). \tag{7.9}$$

From this it is easily seen that

$$d\omega' = g(d\omega + \omega \wedge \omega)g^{-1}. \tag{7.10}$$

The physical fields transform homogeneously under local gauge transformations

The concept of the fibre bundle may be generalized to include prolongation structures (Bäcklund transformations etc) with the inclusion of a jet structure. This general fibre bundle is defined as $(Y, \pi, \mathcal{M})$. This leads to an extension of sections according to jet structures, which permit connection coefficients and vector fields to include pseudopotentials and Pfaffians. Define $(J^k(Y), \pi_k, \mathcal{M})$ as the k-jet bundle [6] that consists of equivalence classes of sections where this equivalency is define according to

$$s_\alpha \sim s_\beta, \text{ if } D^j s_\alpha(x_1, \ldots, x_4) = D^j s_\beta(x_1, \ldots, x_4), \tag{7.11}$$

where D^j refers to a set of differentials up to k, $j = 1, \ldots, k$ with respect to the coordinates on $\mathcal{M}$ with $j = j_1 + \ldots + j_4$

$$D^j s_\alpha = \frac{\partial^j s_\alpha}{\partial^{j_1} \ldots \partial^{j_4}}(x_1, \ldots, x_4). \tag{7.12}$$

This jet construction permits a Finsler-like construction of geometry. Coordinates are defined in the following manner

$$\{x^\mu\} \in \mathcal{M}, \ \{x^\mu, y^i\} \in E, \ i = 1, \ldots, dim(\mathcal{G}), \tag{7.13}$$

and the coordinates on the jet

$$\{x^\mu, y^i, y^i_\mu, y^i_{\mu_1 \mu_2}, \ldots y^i_{\mu_1 \ldots \mu_k}\} \in J^k(Y). \tag{7.14}$$

A field theory on the jet bundle has an action given by the Lagrangian as [7]

$$S = \int d^4 x \mathcal{L}(j^k \phi), \tag{7.15}$$

where the field ϕ is identified as a section of $(Y, \pi, \mathcal{M})$ and the Lagrangian is a function of the k-jet extension of the field $j^k \phi$. This extension means that the Lagrangian is a function over all elements of the k-jet,

$$S = \int d^4 x \mathcal{L}\big(x, \ \phi(x), \ \partial_\mu \phi(x), \ \ldots, \ \partial_{\mu_1} \ldots \partial_{\mu_k} \phi(x)\big). \tag{7.16}$$

The action is a functional defined on the entire space of smooth sections of $(Y, \pi, \mathcal{M})$, which may be demonstrated to be a smooth function on $\sec Y$.

The action is an invariant if $\frac{d\mathcal{L}}{dt} = 0$, which may be written according to the Lie derivative

$$\frac{d\mathcal{L}}{dt} = \mathcal{L}_{j^k v} \mathcal{L} = 0, \tag{7.17}$$

where v is a vector field on Y and $j^k v$ is the jet prolongation of v on $J^k(Y)$. From this action the Euler-Lagrange equation of motion are

$$\frac{\delta \mathcal{L}}{\delta y^i}(j^k \phi) = \frac{\partial \mathcal{L}}{\partial \phi^i} - \frac{\partial}{\partial x^\mu} \frac{\partial \mathcal{L}}{\partial (\partial_\mu \phi^i)} - \ldots - \frac{\partial^k}{\partial x^{\mu_1} \ldots \partial x^{\mu_k}} \frac{\partial \mathcal{L}}{\partial (\partial_{\mu_1} \ldots \partial_{\mu_k} \phi^i)}. \tag{7.18}$$

For $k = 2$ this is the standard Euler-Lagrange equation of motion, where the stationary condition or invariance of the action obtains with the left hand side of this equation vanishes.

This jet bundle formalism is applicable to the formalism of "space plus time" formalism of general relativity. A spacial surface $\Sigma^{(3)}$ and a real line $\mathbf{R}$ with the embedding diffeomorphism $\Phi : \Sigma^{(3)} \times \mathbf{R} \rightarrow \mathcal{M}$ define spacetime. The Cauchy data on an initial spacial slice is given in the embedding $\Phi(\Sigma^{(3)}, 0) = \Sigma^{(3)}$. The foliation of spacial surfaces is given by $\Phi(\Sigma^{(3)}, t)$ and contain the information for the diffeomorphisms of the Einstein field equation. Here the curve is the parameter t which defines the vector field $t_\mu = \partial_\mu t$. Given the tangent bundle on $\mathcal{M}$, $T^*\mathcal{M}$ and the tangent bundle on Y as a vertical bundle in the Finslerian sense, the splitting of spacetime can be see on $J^1(Y)$ the projection $\pi_1 : J^1 \rightarrow Y$ maps to an affine bundle on Y, with the remaining vector bundle $V^*Y \otimes_Y T^*\mathcal{M}$. The splitting of spacetime splits the tangent bundle on $\mathcal{M}$ as well as $T^*\mathcal{M} = T^*\mathbf{R} \oplus T^*\Sigma^{(3)}$. The bundle on Y is then

$$V^*Y \otimes_y T^*\mathcal{M} = V^*Y \otimes_y \left(T^*\Sigma^{(3)} \oplus \mathbf{R}\right) = RV^*Y \oplus \left(V^*Y \otimes T^*\Sigma^{(3)}\right) \qquad (7.19)$$

Consequently the jet bundle correspondingly splits as $J = J^t \oplus_y J^\Sigma$ corresponding to differentials in time and space. The elements of the split jet bundle then define equivalence classes of sections on each subjet according to equation 7.18. Each subjet J^t and J^Σ are bundles on Y with the associated vector bundles V^*Y and $V^*Y \otimes_y T^*\Sigma^{(3)}$. The fibre bundle on $\Sigma^{(3)}$ is restricted to the bundles Y, J^Σ and J_t^Σ, where the latter is the jet on $\Sigma^{(3)}$ at a time t. For all t J^Σ is obtained by the diffeomorphisms that "push" $\Sigma^{(3)}$ through the foliation of spacetime [8].

Lagrangian dynamics is written in configuration space that consists of n-tuplets of points for particles. In general relativity these points consist of various $\Sigma^{(3)}$. If this space is defined as $\mathcal{C}$ then at $t = 0$ the boundary $\partial \mathcal{C}$ defines all the Cauchy data for spacetime. Further, at any t these Cauchy data are mapped to another $T\mathcal{C}$. The Euler-Lagrange equation is then a system of second order differential equations with Cauchy data contained in the sections of J^Σ. Hence the Cauchy data ϕ are elements of the set of sections $\phi \in Sec(J^\Sigma)$. This space of sections has a Frechet point-set open topology and defines the affine bundle by the projection $Sec(J_t^\Sigma) \rightarrow J^\Sigma$.

The connections are identified by the bundle morphism $\mu : T^*M \rightarrow TY$. The bundle TY splits into horizontal and vertical bundles $T_HY \oplus T_VY$, with a connection $\omega \in TY$. The connection then acts on vectors v_x, u with $\omega(v_x, u) \in T_uY$. For any section ϕ of J_t^Σ the vertical bundle is defined as $T_x\bar{\phi}\partial_t - \omega(\partial_t, \bar{\phi})$. This gives the time vector field on $\mathcal{M}$ which generates the spacetime foliation. It is then apparent that $SecY^\Sigma$ is the configuration space $\mathcal{C}$ and that $SecT_VY^\Sigma$ is the Cauchy data $T\mathcal{C}$. The Lagrangian then defines a map from the configuration space to the reals, where that real number is the action. The section J^Σ is determined by the data $\phi. \partial_t, \partial_1\phi, \ldots, \partial_4\phi$ that enters into the Lagrangian with the action

$$S[\phi] = \int d^4x \big(\mathcal{L}(\phi. \partial_t, \partial_1\phi, \ldots, \partial_4\phi)\big). \qquad (7.20)$$

The variation of the action defines the Euler-Lagrange equation as equation 7.18. This then illustrates how the Cauchy data enters the equations of dynamics.

It is evident that $SecY = C^\infty(\mathbf{R}, SecY^\Sigma)$, since $C^\infty(\mathbf{R} \times \Sigma^{(3)}, Y) = C^\infty(\mathbf{R}, C^\infty(\Sigma^{(3)}, Y))$, where $SecY \subset C^\infty(\mathbf{R}, C^\infty(\Sigma^{(3)}, Y))$. Then for $\phi \in SecY$

the curve $c(\phi) : \mathbf{R} \to Sec Y^\Sigma$ defines the parallel transport $Y(0, x) \to Y(t, x)$ as $c(\phi)(x) = \mathcal{P}_{(t,x)}\phi(t, x)$. Hence the parallel transport on $\phi \to c(\phi)$ is a bijective map.

A Legendre transformation casts the Lagrangian formalism into the Hamiltonian formalism. This maps the dynamics from configuration space to phase space of conjugate variables. In general relativity these dual variables are the metric for $\Sigma^{(3)}$ and a momentum conjugate, where these satisfy a Poisson bracket relationship. With conjugate variables this dualism is mapped into complementary operators of quantum mechanics. This dualism has interpretations in vector spaces and Hilbert or Banach spaces with dual bases of vectors. As a result the dual observables have certain point-set topological properties, where for general relativity these have potentially subtle implications. Further, these properties carry over to a quantum description of gravitation.

The symplectic form Ω is a map on the space of vectors $V = C \oplus C'$ such that for $a_i \in C$ and $b_i \in C'$

$$\Omega : V \times V \to \mathbf{R}. \quad \Omega\big((a_1, b_1), (a_2, b_2)\big) = \langle b_2, a_1 \rangle - \langle b_1, a_2 \rangle, \tag{7.21}$$

where the norm is over a Banach space identified as V. This product is also an antisymmetric product. This map is zero if and only if the two norms vanish, so for $\forall b_1, b_2 \neq 0$ this implies $a_1 = a_2 = 0$, so the map is injective. This defines the symplectic form as nondegenerate. The space C and its dual C' then satisfy a norm $\langle\,,\,\rangle : C \times C' \to \mathbf{R}$ that is injective, which implies that C, C' are Hausdorff. Of principal importance is that if C is composed of a sequence of subsets C_α then each C_α contains elements such that a sequence of norms is convergent. The duality operator $*$ that acts as $*C \to C'$ is an isomorphism.

The Hamiltonian is a C^∞ function $H : V \to \mathbf{R}$, where the range of reals defines the energy of the system. This gives the Hamiltonian vector field x_H defined by $dH \in V$ by

$$x_H = \Omega dH. \tag{7.22}$$

Poisson bracket structures in $C^\infty(V)$ are defined by their action of x_H on any vector $v \in V$

$$x_H \odot v = \Omega v dH = \Omega^{ij} \nabla_i v \nabla_j H := \{v, H\}_{PB}, \tag{7.23}$$

Here the indices on the symplectic form are written explicitly and are antisymmetric. The differentials depend on variables that are dual to each other in C, C'. The variables in these spaces are the conjugate variables q, p, which leads to the Poisson brackets $\{q_i, p_j\} = \delta_{ij}$. In general this bracket structure does not close and the vectors v and x_H do not commute. Further, as the symplectic form is smooth, but not necessarily continuous, the Poisson brackets in general do not have a continuous dependency on v and x_H.

The Lagrangian formalism works within configuration space $\sigma = Sec Y^\Sigma$, with the tangent space $T_\phi \sigma = Sec \phi^* T_V Y^\Sigma$. The cotangent bundle is found in the duality between T_V and T_V^*. The cotangent bundle is $T'_\phi \sigma$, where the space of vectors the symplectic form acts on is $V = T_\phi \sigma \oplus T'_\phi \sigma$. The inner product of dual vector fields is

$$Sec\phi^* T_V Y^\Sigma \times Sec\phi^* T_V^* Y^\Sigma \to \mathbf{R}, \tag{7.24}$$

which is defined on the whole $\Sigma^{(3)}$ by

$$(\xi, \pi) = \int_{\Sigma^{(3)}} d\Sigma^{(3)} \langle \xi(x), \pi(x) \rangle. \tag{7.25}$$

From here the symplectic form $\Omega : V \times V \to \mathbf{R}$ is defined by the inner product in equation 7.21

$$\Omega\bigl((\xi_1, \pi_1), (\xi_2, \pi_2)\bigr) = \int_{\Sigma^{(3)}} d\Sigma^{(3)} \bigl(\langle \xi_1(x), \pi_2(x) \rangle - \langle \xi_2(x), \pi_1(x) \rangle\bigr). \tag{7.26}$$

This symplectic form is nondegenerate and smooth, but not in general continuous.

For $f : \sigma \to \mathbf{R}$ there is the differential $d_\phi f \in T^*_\phi \sigma$. This enters an inner product with the tangent a curve γ, $\gamma(0) = \phi$ as

$$\langle d_\phi f, c'(0) \rangle = \int_{\Sigma^{(3)}} d\Sigma^{(3)} \langle d_\phi F, \gamma'(0) \rangle, \tag{7.27}$$

for $F : Y \to \mathbf{R}$. By equation 7.21 and 7.26 this defines the Poisson bracket between two classical variables A, B as

$$\{A, B\}_{PB} = \Omega^{ij} \nabla_i A \nabla_j B = \int_{\Sigma^{(3)}} d\Sigma^{(3)} \left(\frac{\partial A}{\partial \pi^i} \frac{\partial B}{\partial \phi^i} - \frac{\partial A}{\partial \phi^i} \frac{\partial B}{\partial \pi^i} \right). \tag{7.28}$$

The differential of ϕ along a Hamiltonian vector field v_H is given by the Poisson bracket $d\phi(v_H) = \{\phi, H\}_{PB}$. If $d\phi(v_H) = 0$ ϕ is a constant of the motion with respect to H and ϕ is in involution with H.

Poisson brackets and the symplectic form act on functions ϕ^a and π^a in $T^* = C^\infty(\Sigma^{(3)}, C) \times C^\infty(\Sigma^{(3)}, C')$. To give the fundamental bracket

$$\{\phi^a(x), \pi^b(x')\}_{PB} = \delta^{ab} \delta^3(x - x')$$

$$\{\phi^a(x), \phi^b(x')\}_{PB} = \{\pi^a(x), \pi^b(x')\}_{PB} = 0. \tag{7.29}$$

The presence of the delta function shows that for classical physics these functions are smeared out around a point with a test function. Such a test function will provide a domain of compact support for the functions in the bracket structure, which converts the Dirac delta function into a Gaussian function. The relationship between the Lagrangian and Hamiltonian formalism is the Legrendre transformation $T\sigma \to T^*\sigma$, where the Lagrangian and Hamiltonian are related by

$$H(\phi, \pi) = \int_{\Sigma^{(3)}} \bigl(\pi \partial_0 \phi - \mathcal{L}(\phi)\bigr) d\Sigma^{(3)}. \tag{7.30}$$

7.3: Nonlocalizability of Energy and Noncommutative Geometry

With this formal differential geometry consider the nature of observables in general relativity, with extensions to the quantum mechanical domain. This permits a generalization this theory. This opens the way to cast quantum gravity as a generalization of symplectic and Kähler structure.

In general relativity there is a complicating aspect to this formalism. All of the above machinery is set up so that energy-momentum of the field is defined on an infinitesimal region $d\Sigma^{(3)}$. Yet this leads to some conundrums. Identify the four momentum **p** in these infinitesimal regions by equating this with the stress energy tensor

$$\mathbf{p} = \mathbf{e}_\mu T^\mu{}_\nu d\Sigma^\nu. \tag{7.31}$$

The symmetry of the momentum-energy tensor $\nabla_\mu T^\mu{}_\nu = 0$ insures continuity of momentum-energy [9], but **p** on a spacial surface $\sigma^{(3)}$ gives $d\mathbf{p}(v_H)$ equal to

$$d\mathbf{p}(v_H) = d\mathbf{e}_\mu(v_H) T^\mu{}_\nu d\Sigma^\nu. \tag{7.32}$$

The right hand side is then expressed according to connections $d\mathbf{e}_\mu = \omega_\mu{}^\nu \mathbf{e}_\nu$ and as well as according to the Poisson bracket $d\mathbf{e}_\mu = \{\mathbf{e}_\mu, H\}$, which leads to the troubling equation

$$\omega_\mu{}^\nu \mathbf{e}_\nu = \{\mathbf{e}_\mu, NH_{ADM}\}_{PB}. \tag{7.33}$$

This equation is troubling because this connection coefficient on the right hand side is a matter of a coordinate condition. A coordinate condition may give connection coefficients $\omega^\nu{}_\mu = \Gamma^\nu{}_{\mu\rho} dx^\rho$ so that $\Gamma^\nu{}_{\mu\rho} = 0$, or the gravity field disappears. Such a choice tracks energy-momentum on a geodesic flows with zero connections through this $d\Sigma^{(3)}$ volume element with $d\mathbf{e}_\mu = 0$. Yet another choice of connection will give an answer $d\mathbf{p} \neq 0$. For the above Poisson bracket defined with the symplectic form Ω^{ij} then

$$\{\mathbf{e}_\mu, NH_{ADM}\}_{PB} = \Omega^{\alpha\beta} \nabla_\alpha \mathbf{e}_\mu N \nabla_\beta H, \tag{7.34}$$

where ∇_α are covariant differentials and under projections P_i^α exist on $\Sigma^{(3)}$. For any nonzero spacial curvature $R^{(3)}$ on $\Sigma^{(3)}$ that enters the ADM Hamiltonian, this quantity is contained in $T^*\mathcal{M} \otimes V$. This is then a frame dependent Hamiltonian vector field, which is a completely different structure than what pertains in standard classical mechanics.

Define a bundle morphism $\mu : T^*M \to TY$, such that TY is split into horizontal and vertical bundles $T_H \oplus T_V$. This determines connections $\omega \in TY$. Then on any section **e** of J_t^Σ, Hamiltonian vector fields are defined as $(T_x \bar{\mathbf{e}} \partial_t - \omega(\partial_t, \mathbf{e})) dH$ contained in $T_V Y \times V$. The configurations space is $Sec Y^\Sigma$, but now the Cauchy data TC is contained in $Sec T_V Y^\Sigma \times V$. Physically this implies that the momenta "flaps in the breeze" by coordinate choice, but one must make a choice.

This nonlocalizability of energy and momentum has a remarkable similarity to nonlocality in quantum mechanics. The inability to localize energy and momentum is due

252 *Quantum Fluctuations of Spacetime*

to a connection coefficient involved with the definition of the conservation of momentum in a small region. This connection coefficient is completely arbitrary, where energy and momentum an analyst attempts to locate in a small region is involved with energy and momentum anywhere. Consequently it can slip between your fingers and show up instantaneously almost anywhere else. The Christoffel connections are purely fictional quantities and can gremlin-like play havoc with attempts to localize energy and momentum. As such one is not able to perform a proper integration between two regions $U \subset \Sigma^{(3)}$ and $U' \subset \Sigma^{(3)'}$ in a foliation $\int_U^{U'} \mathbf{p} = \int_{Vol^4 \subset \mathcal{M}} d\mathbf{p}$ and state that energy and momentum flowed through this region in a manner that satisfies a continuity equation. The appearance of this presumed continuity equation would depend entirely on the coordinate condition imposed on the problem. Hence no continuity can exist.

It has to be pointed out that the one exception to this is occurs if one chooses a large enough region of integration so that the boundaries of U, U' are in an asymptotically flat region. Here as an approximation one can arrive at an estimate of the gravitational energy in a system. This would be the case if one were detecting weak gravity waves far removed from their source. In this case one has an anchor that the connection coefficients can be naturally pinned to, the Hubble frame, so that in an asymptotic region the errors induced by arbitrary connections is small. Similarly this approximation can also hold if the gravity fields are weak, where the ultimate case would be a Newtonian gravity field.

A consequence of this nonlocalizability of momentum and energy leads to the following interesting observation. The commutator between two momenta vectors leads to the result

$$[\mathbf{p}(v_H), \mathbf{p}(v'_H)] = [\mathbf{e}_\mu, \mathbf{e}_\nu] T^\mu{}_\sigma T^\nu{}_\rho d\Sigma^\sigma \wedge d\Sigma^\rho. \tag{7.35}$$

If $\mathbf{e}$ is a tangent vector, then in an anholonomic basis $[\mathbf{e}_\mu, \mathbf{e}_\nu] = c^\alpha{}_{\mu\nu} \mathbf{e}_\alpha$, where $c^\alpha{}_{\mu\nu}$ are commutation coefficients. This requires that commutators of momentum vectors satisfy

$$[\mathbf{p}(v_H), \mathbf{p}(v_H)] = c^\alpha{}_{\mu\nu} \mathbf{e}_\alpha T^\mu{}_\sigma T^\nu{}_\rho d\Sigma^\sigma \wedge d\Sigma^\rho. \tag{7.36}$$

The question is how to interpret this result. Consider the following commutator,

$$[\mathbf{p}(v_H), \mathbf{e}]_\nu = c_{\mu\nu}{}^\alpha \mathbf{e}_\alpha T^\mu{}_\sigma d\Sigma^\sigma, \tag{7.37}$$

where from physical thinking this corresponds to an angular momentum quantity. This commutator may be written as

$$[\mathbf{p}(v_H), \mathbf{e}]_\nu = \frac{1}{2} c_{\mu\nu\alpha} \mathbf{e}^{[\alpha} T^{\mu]}{}_\sigma d\Sigma^\sigma = \frac{1}{2} c_{\mu\nu\alpha} \mathbf{J}^{\alpha\mu}{}_\sigma d\Sigma^\sigma, \tag{7.38}$$

so that $\mathbf{J}^{\alpha\mu}{}_\sigma d\Sigma^\sigma = d\mathbf{J}^{\alpha\mu}$ and an infinitesimal angular momentum vector is recovered. For a quantum mechanical system, where the infinitesimal angular momentum approaches $\hbar$, this angular momentum tensor is $d\mathbf{J}^{\alpha\mu} = \hbar \mathbf{I}^{\alpha\mu}$, where the tensor on the right hand side has unit entries. This may be written according to the basis vectors $\mathbf{e}^\beta$ as $\mathbf{I}^{\alpha\mu} = \epsilon^{\alpha\mu}{}_\beta \mathbf{e}^\beta$. The contraction of the commutator coefficient with $\mathbf{I}$ then leaves $\epsilon^{\alpha\mu}{}_\beta c_{\alpha\mu\nu} \mathbf{e}^\beta = c_{\beta\nu} \mathbf{e}^\beta$. This is a vector with an antisymmetric coefficient that under normalization is a unit pseudocomplex matrix and quantum mechanically as $i = \sqrt{-1}$. Substitute the classical variables with quantum operators for $\hat{\mathbf{p}} = \mathbf{e}_\mu \hat{T}^\mu_\nu d\Sigma^\nu$ so that

$$[\hat{\mathbf{e}},\, \hat{\mathbf{p}}]_\nu = \frac{i}{2}\hbar \mathbf{n}_\nu. \tag{7.39a}$$

This is the standard quantum mechanical commutator, for $\mathbf{n}_\nu$ a unit vector for the spin direction normal to the plane spanned by $\mathbf{e}$ and $\mathbf{p}$. In equations 2.117 and 2.121 the commutator between complementary operators contains a curvature term due to quantum fluctuations. This term is included and generalized this commutator as

$$[\hat{\mathbf{e}}_\mu,\, \hat{\mathbf{p}}_\nu] = \frac{i}{2}\hbar\left(c_{\mu\nu}{}^\alpha \mathbf{n}_\alpha + \mathcal{R}_{\mu\nu\alpha\beta}\mathbf{e}^\alpha \mathbf{e}^\beta\right). \tag{7.39b}$$

The matrix $c_{\mu\nu}{}^\alpha$ now contains nonholonomic connections as well as symplectic information from the Poisson bracket of the classical variables. This has effectively "mixed" general relativity with quantum mechanics. Similarly the commutator between momentum operators corresponding to spacetime is

$$[\hat{\mathbf{p}}_\mu,\, \hat{\mathbf{p}}_\nu] = \frac{i}{2}\hbar c_{\mu\nu}{}^\alpha \mathbf{e}^\rho \left(T^\sigma{}_\rho d\Sigma^\rho\right)_\alpha = \frac{i}{2}\hbar e_{\mu\nu}^{-1\,\alpha}\hat{\mathbf{p}}_\alpha, \tag{7.40}$$

where $e_{\mu\nu}^{-1\,\alpha}$ has units of m^{-1}. This is a nonlocalizable quantity and an operator as well. In the case that $R_{\mu\nu} = 0$ $e_{\mu\nu}^{-1\,\alpha} = 0$ so this commutator vanishes. When spacetime is flat the standard result of quantum mechanics that $[\hat{\mathbf{p}}_\mu,\, \hat{\mathbf{p}}_\nu] = 0$ is recovered. This is a situation where the commutation of both $\hat{\mathbf{e}}$ and $\hat{\mathbf{p}}$ operators with themselves results in nonlocalizable operators, but are physically well behaved for commutators between each other.

Quantum gravity likely implies a noncommutative geometry of operators. This noncommutative geometry above contains elements of quantum mechanics (the Heisenberg algebra) and algebra of general relativity. However, it also contains an additional commutator between momentum operators that are not localizable. It then appears that the above algebra needs to be expanded to let the momentum operators be covariant, where equation 7.39b further demands this. From this we then demand that the commutator in equation 7.40 be extended to

$$[\hat{\mathbf{P}}_\mu,\, \hat{\mathbf{P}}_\nu] = \frac{i}{2}\hbar\left(e_{\mu\nu}^{-1\,\alpha}\hat{\mathbf{P}}_\alpha + \mathcal{R}_{\mu\nu\alpha\beta}\mathbf{n}^\alpha \mathbf{n}^\beta\right). \tag{7.41}$$

In this way the quantum gravity theory contains the gauged-like information of fluctuation of spacetime and general relativity. The term $c^{\beta\alpha\sigma}\mathcal{R}_{\mu\nu\alpha\beta}\mathbf{e}_\sigma$ is the number of gravitational quanta that pass through a region bounded by the momentum loop defined by the commutator. This is analogous to a magnetic flux that passes through a current loop that contains some integer number of magnetic quanta $B_0 = \sqrt{\hbar V \epsilon_0 / 2\omega}$.

254 Quantum Fluctuations of Spacetime

There is yet a further problem. In this case of nonholonomic coordinates $[\mathbf{e}_\mu, \mathbf{e}_\nu] = c_{\mu\nu}{}^\alpha \mathbf{e}_\alpha$, which results in a striking asymmetry between the action of the position commutators and the momentum commutators. Nature cannot be this unbalanced near or at the Planck scale of energy! There must then be an associated curvature for the basic commutation between a set of dual basis elements for a curvature in momentum space. Now impose the condition that

$$[\hat{\mathbf{e}}_\mu, \hat{\mathbf{e}}_\nu] = c_{\mu\nu}{}^\alpha \hat{\mathbf{e}}_\alpha + \frac{i}{2}\hbar S_{\mu\nu\alpha\beta}\hat{\mathbf{p}}_\alpha\hat{\mathbf{p}}_\beta.$$

$$= c_{\mu\nu}{}^\alpha \hat{\mathbf{e}}_\alpha + \frac{i}{2}\hbar c_{\alpha\beta}{}^\sigma S_{\mu\nu\alpha\beta}\hat{\mathbf{p}}_\sigma + O(\hbar^2). \tag{7.42}$$

Here $S_{\mu\nu\alpha\beta}$ is interpreted as a conjugate curvature on the momentum-energy manifold, which for weak linear gravity waves or gravitons is the Fourier transform curvature. This is now a completely symmetric theory of bracket structures. This general formalism is a noncommutative geometry of operators. Below is a discussion of the symplectic structure of fields and its quantization. Once this theory is laid out the geometry of these noncommutative operators can be determined. Physically this will be a complete fusion of gravitation and quantum mechanics into a larger system where both are low energy aspects of this system.

7.4: Quantization of Fields in Spacetime and Noncommutative Geometry

The symplectic form in equation 7.21 upon quantization turns Poisson brackets into commutators of operators. The symplectic form is a bilinear map from phase space $\mathcal{P}$ to the reals $\Omega: \mathcal{P} \times \mathcal{P} \rightarrow \mathbf{R}$, so that two points $x_1, x_2 \in \mathcal{P}$ are mapped to a real number. This further defines

$$\{\Omega(x_1, (\ )), \Omega(x_2, (\ ))\} = -\Omega(x_1, x_2). \tag{7.43a}$$

This is also seen according to vector bundles in equation 7.21, where this vector bundle appears to have relationships to Hilbert space as well. Given complementary variables that enter the Poisson bracket these have a quantum analogue:

$$[\hat{\Omega}(z_1, (\ )), \hat{\Omega}(z_2, (\ ))] = -\Omega(z_1, z_2). \tag{7.43b}$$

This is a bilinear map $\hat{\Omega}: V^\mathcal{C} \times V^\mathcal{C} \rightarrow \mathcal{C}$, that complexifies the basis elements z of $V^\mathcal{C}$. Elements in the complexified configuration spaces $C^\mathcal{C}$ the inner product analogue of equation 7.21 are

$$\hat{\Omega}((z_1, (\ )), (z_2, (\ ))) = |z_1\rangle - |z_2\rangle = \hat{z}_1|\ \rangle - \hat{z}_2|\ \rangle, \tag{7.44}$$

where $\hat{z}_i$ are operator valued functions. This is a map from our vector bundle space of solution to the Hilbert space $V \rightarrow \mathcal{H}$, where in the complexified $V^\mathcal{C}$ there are elements mapped to the Hilbert space $\mathcal{H}$ and its dual $\mathcal{H}^*$. This determines positive and negative frequency (particle-antiparticle) components, with $z_i = \mathcal{F}x_i + (\mathcal{F}x_i)^*$ then equation 7.44 with $z_1 = z_2$ is

$$\hat{\Omega}((z_1,(\ )),(z_1,(\ ))) = i\mathcal{F}x(t)\hat{a}|\ \rangle - i(\mathcal{F}x(t))^*\hat{a}^\dagger|\ \rangle,$$

or

$$\hat{\Omega}(z,(\ )) = i\mathcal{F}x(t)\hat{a} - i(\mathcal{F}x(t))^*\hat{a}^\dagger. \qquad (7.45)$$

Here the fields are set as Cauchy data on $\Sigma^{(3)}$ at $t = 0$. The multiplication of the right hand side on equation 7.45 by its complex conjugate is $\hat{\Omega}(z, z)$. Therefore the overlap of states may be written according to classical solutions

$$\Omega(z_1, z_2) = \Omega((\mathcal{F}x_1)^*, \mathcal{F}x_2) + \Omega(\mathcal{F}x_1, (\mathcal{F}x_2)^*) = 2\mathcal{R}e\Omega((\mathcal{F}x_1)^*, \mathcal{F}x_2), \qquad (7.46)$$

where by equation 7.45 and the inner product defined in equation 7.21 is equal to $2\mathcal{I}m\langle\ \mathcal{F}z_1|\mathcal{F}z_2\ \rangle$. On the real valued space V an inner product defined as $r(x_1, x_2)$ is the real part of the inner product over Hilbert space. Correspondingly $\langle \mathcal{F}z_1|\mathcal{F}z_2\rangle$ by equation 7.45 is $\mathcal{I}m\Omega((\mathcal{F}z_1)^*|\mathcal{F}z_2)$. This leads to the definition of the inner product in Hilbert space as

$$\langle \mathcal{F}x_1|\mathcal{F}x_2\rangle = r(x_1, x_2) - \frac{i}{2}\Omega(x_1, x_2). \qquad (7.47)$$

The characterization of the Hilbert space is entirely due to the choice in $r(x_1, x_2)$. Physically equation 7.47 shows that the phase in the overlap is determined by the degree to which the classical variables are not in involution in phase space. By the Schwartz inequality it is apparent that $r(x_1, x_1)r(x_2, x_2) \leq 1/4(\Omega(x_1, x_2))^2$. Thus there is an upper bound in Hilbert space, where for some $x_2 \neq 0$ it is evident that

$$r(x_1, x_2) = \frac{1}{4}\frac{(\Omega(x_1, x_2))}{r(x_2, x_2)}\bigg|_{x_2(max)}. \qquad (7.48)$$

As an example consider the momentum in equation 7.31 as due to the Klein-Gordon equation, $\Delta\psi + m^2\psi = 0$, with $\psi|\ \rangle = |\psi\rangle$. The Lagrangian is $\mathcal{L} = \frac{1}{2}|\nabla_\mu\psi|^2 + m^2|\psi|^2$ and the Hamiltonian is $\mathcal{H} = \frac{1}{2}(|\pi|^2 + |\nabla\psi|^2 + m^2|\psi|^2)$, for $n_i - i[\hat{\Pi}, \psi(r,t)] - \dot{\psi}(r_i)$. Here ∇_μ is a covariant spacetime operator. The stress energy tensor is then

$$T_{\mu\nu} = \nabla_\mu\bar{\psi}\nabla_\nu\psi - \frac{1}{2}g_{\mu\nu}(\nabla^\alpha\bar{\psi}\nabla_\alpha\psi + m^2\bar{\psi}\psi). \qquad (7.49)$$

The evaluation of equation 7.33 with the above Hamiltonian gives

$$\omega_i{}^\nu\psi_\nu = \mathbf{p}_i = \int d^3r e^{i\mathbf{k}\cdot\mathbf{r}}\hat{\pi}_i(r,t). \qquad (7.50)$$

This may be reinserted into equation 7.32 to give the following

$$d\mathbf{p} = i[e_\mu, \hat{H}_{ADM}]T^\mu{}_\nu d\Sigma^\nu = \int d^3r e^{i\mathbf{k}\cdot\mathbf{r}}d\hat{\pi}_i(r,t). \qquad (7.51)$$

Define **p** according to solutions of the Klein-Gordon equation $\mathbf{p} = \langle\psi_1|\hat{\mathbf{p}}|\psi_2\rangle$. Equation 7.51 is then a differential equation

$$\frac{d}{d\lambda}\langle\psi_1|\hat{\mathbf{p}}|\psi_2\rangle = i\langle\psi_1|[\mathbf{e}_\mu, \hat{H}_{ADM}T^\mu{}_\nu|\psi_2\rangle, \qquad (7.52)$$

where λ is an arbitrary timelike parameter. The equation for the evolution of $\hat{p}$ by equation 7.33 indicates that a continuity of momentum through a region can not be established.

Given the inner product in equation 7.47 the meaning of the time derivative of the inner product is examined. Start with the differential of the symplectic form

$$d\Omega_{\alpha\beta}(z_1, z_2) = \{\Omega(z_1,(\;)), H\}_{PB} = \Omega^{\mu\nu}(z_1,(\;))\nabla_\mu\Omega_{\alpha\beta}(z_2,(\;))\nabla_\nu H. \qquad (7.53)$$

In component form is is also

$$\frac{1}{2}\nabla_\mu[\Omega^{\mu\nu}(x_1,(\;)), \Omega_{\alpha\nu}(x_2,(\;))]\nabla_\beta H = -\frac{1}{2}\nabla_\mu\Omega^\mu{}_\alpha(z_1, z_2)\nabla_\beta H. \qquad (7.54)$$

The square brackets are used, where implicitly by the inner product definition of the symplectic form there is a Poisson bracket in this step. Quantum mechanically in the last step the $1/2$ is then replaced by an $i/2$. Equation 7.48 is then applied to obtain

$$\frac{d}{d\lambda}\langle\psi_1|\mathbf{p}|\psi_2\rangle = \frac{1}{4}\frac{\nabla_\mu\Omega^\mu{}_\alpha(x_1, x_2)\nabla_\beta H}{r(x_2, x_2)} + \frac{i}{4}\nabla_\mu\Omega^\mu{}_\alpha(x_1, x_2)\nabla_\beta H, \qquad (7.55)$$

where $r(x_1, x_2)$ is normalized to unity. With the use of equation 7.45 $\frac{d}{d\lambda}\langle\psi_1|\psi_2\rangle$ is then

$$\frac{d}{d\lambda}\langle\psi_1|\mathbf{p}|\psi_2\rangle = \frac{1}{2}(a^\dagger a + aa^\dagger - a^2\mathcal{F}^2 - a^{\dagger 2}\mathcal{F}^{*2})x(t). \qquad (7.56)$$

For $\mathcal{F} = i\phi e^{i\theta}$ the off diagonal term becomes $a^2(\mathcal{F})^2 - a^{\dagger 2}(\mathcal{F}^*)^2$, where the imaginary part of ϕ is a form of the squeezing operator. The Wigner function contains a term e^{2ipx} and gives a measure of the phase space of a system. Similarly the squeezed state operator changes the form of the measure of a system with the effect of collapsing the uncertainty in one variable and expanding it with the conjugate variable. This is a direct link between the uncertainty in locating mass-energy and its quantum mechanical interpretation as associated with the squeezed state of the vacuum state.

Now transformation principles are examined. Quantum mechanics transforms Hilbert space basis elements by unitary operators $U^\dagger = U^{-1}$. Here such transformations exist, but are more complex due to the squeeze state operator implicit in this. The squeeze state operator is

$$S(|z|, \theta) = exp\left(\frac{1}{2}|z|(a^2 e^{-2i\theta} - a^{\dagger 2}e^{2i\theta}\right). \qquad (7.57)$$

The action of $S(|z|, \theta)$ on the annihilation operator gives

$$S(|z|, \theta)aS^\dagger(|z|, \theta) = acosh(|z|) + a^\dagger e^{2i\theta} sinh(|z|)., \tag{7.58}$$

The corresponding transformed creation operator may be similarly found. This squeezed state operator is then a transformation on $\mathcal{H}$. The uncertainty in the position and momentum variables is

$$\langle(\Delta \mathbf{e})^2\rangle = \frac{1}{2}(cosh2|z| - cos(2\theta)sinh(2|z|))$$

$$\langle(\Delta \mathbf{p})^2\rangle = \frac{1}{2}(cosh2|z| + cos(2\theta)sinh(2|z|)) \tag{7.59}$$

$$\langle(\Delta \mathbf{e}\Delta \mathbf{p})\rangle = -\frac{1}{2}sin(2\theta)sinh(2|z|).$$

Now apply these transformations to the Wheeler-DeWitt equation for a cosmology of radius r,

$$\frac{d^2\Psi(r)}{dr^2} + \omega^2(r)\Psi(r) = 0 \tag{7.60}$$

The analyst is free to choose any basis for $\mathcal{H}$. Define two representations for the wave function in two bases as

$$\Psi(r) = a\phi(r) + b^\dagger \phi^\dagger(r), \quad \Psi(r) = c\chi(r) + d^\dagger \chi^\dagger(r). \tag{7.61}$$

The transformation in equation 7.58 defines the transformation according to the following operators

$$T(\alpha) = exp\big(-i\alpha(r)(c^\dagger c + d^\dagger d)\big)$$

$$S(r, \theta) = exp\Big(\frac{1}{2}|z(r)|(cde^{-2i\theta(r)} - c^\dagger d^\dagger e^{2i\theta(r)})\Big), \tag{7.62}$$

with $a = T^\dagger S^\dagger cST$ and $b^\dagger = T^\dagger S^\dagger d^\dagger ST$ as

$$a = ce^{-i\alpha}cosh(|z|) - d^\dagger e^{-i(\alpha - 2\theta)}sinh(|z|)$$

$$b^\dagger = d^\dagger e^{i\alpha}cosh(|z|) - ce^{i(\alpha - 2\theta)}sinh(|z|). \tag{7.63}$$

Define $U = ST$ and consider the transformation $U^\dagger aU$. There exist the two Hilbert spaces $\mathcal{H}_1$ and $\mathcal{H}_2$ and the process where by the operators a and $b^\dagger$ are scattered into c and $d^\dagger$. There are then the following operators with their maps

$$A : \mathcal{H}_2 \to \mathcal{H}_1, \; B : \mathcal{H}_2 \to \bar{\mathcal{H}}_1$$

$$C : \mathcal{H}_1 \to \mathcal{H}_2, \; D : \mathcal{H}_1 \to \bar{\mathcal{H}}_2. \tag{7.64}$$

The transformation of the operator a and b are

$$Ua(\psi)U^{-1} = c(\bar{C}\psi) - d^\dagger(\bar{D}\psi)$$

and
$$U b^\dagger(\psi) U^{-1} = d^\dagger(\bar{C}\psi) - c(\bar{D}\psi), \tag{7.65}$$
so the inner product for $\psi, \phi \in \mathcal{H}_2$ gives
$$\langle \psi | \phi \rangle_{\mathcal{H}_2} = -i\Omega(\bar{\psi}, \phi) = -i\Omega(\psi \mathcal{F} + \psi\bar{\psi}, \psi\mathcal{F} + \psi\bar{\psi})$$
$$= \langle A\psi | A\psi \rangle_{\mathcal{H}_1} - \langle B\psi | B\psi \rangle_{\bar{\mathcal{H}}_1} = cosh^2(|z|) - sinh^2(|z|) = 1. \tag{7.66}$$
Compute the inner product with $\psi \in \mathcal{H}_2$ and $\phi \in \bar{\mathcal{H}}_2$
$$\langle \psi | \bar{\phi} \rangle_{\mathcal{H}_2} = \langle A\psi | \bar{B}\psi \rangle_{\mathcal{H}_1} - \langle B^\dagger \psi | \bar{A}\psi \rangle_{\bar{\mathcal{H}}_1} = i sin(2\theta) sinh(|z|) cosh(|z|). \tag{7.67}$$
From the inner product $\langle \psi | A \phi \rangle_{\mathcal{H}_1}$ $A^\dagger = C$. This gives an algebra
$$AA^\dagger + BB^\dagger = 1, \quad A^\dagger \bar{B} - B^\dagger \bar{A} = i sin(2\theta) sinh(|z|) cosh(|z|)$$
$$CC^\dagger + DD^\dagger = 1, \quad C^\dagger \bar{D} - D^\dagger \bar{C} = i sin(2\theta) sinh(|z|) cosh(|z|) \tag{7.68}$$
$$A^\dagger = C, \quad \bar{B}^\dagger = -D.$$

Under an average over the squeeze angle phase term this algebra recovers the Bogoliubov algebra. In this approximation quantum information about coherence is lost, with an entropy associated with the system.

The entropy of quantum gravity is found from the uncertainties in equation 7.59 as
$$S = \frac{1}{2} ln \big(\langle (\Delta \mathbf{e})^2 \rangle \langle (\Delta \mathbf{p})^2 \rangle - \langle \Delta \mathbf{e} \Delta \mathbf{p} \rangle^2 \big). \tag{7.69}$$

This may be coarse grained for an average over the squeezing angle with $\langle (\Delta \mathbf{e})^2 \rangle = \langle (\Delta \mathbf{p})^2 \rangle = (1/2) cosh(2|z|)$. This leads to an entropy
$$S = ln cosh(2|z|) \simeq 2|z|. \tag{7.70}$$

For the transmission amplitude $T^2 = c^\dagger c / a^\dagger a$ this is related to the squeezing amplitude by $T^2 = sech^2(|z|)$. Now integrate the Wheeler-DeWitt equation in the classically permitted region $r \geq \Lambda^{-1/2}$ to obtain the squeezing amplitude given by the tunnelling barrier [10]
$$|z| = ln 2 + \frac{1}{3\Lambda} = ln 2 + \frac{3}{16 G^2 \epsilon}, \tag{7.71}$$

where ϵ is the vacuum energy. It now appears that the algebra of quantum gravity in a coarse grained setting is involved with the vacuum energy value.

7.5: Connections to One-Loop Quantum Gravity

The result of equation 7.70 is a curious one. The entropy in the coarse grained view is
$$S = ln(a^\dagger a) - ln(c^\dagger c). \tag{7.72}$$
This in conjuction with equations 7.59 so that
$$S = ln(a^\dagger a) - ln(c^\dagger c). \tag{7.73}$$
This squeezing of the vacuum state can be also seen in equation 4.99. For the replacement $a \rightarrow a exp(-\pi H/g)$ this entropy takes the form
$$S = ln\left(\sum_n e^{-2\pi E_n/g}\right) + ln(a^\dagger a) - ln(c^\dagger c), \tag{7.74}$$
with $g \propto T$. This is the standard form of the entropy. For a partition function $Z = \sum_m e^{-E_n/kT}$ the entropy is then $S = k\frac{\partial}{\partial T}(T ln Z)$. The partition function is then
$$Z = \sum_n e^{E_n/kT + a^\dagger a/kc^\dagger c}, \tag{7.75}$$
so the energy in the partition function contains $a^\dagger a/c^\dagger c$.

Consider the action for this system, not to confuse the use of the symbol S for entropy above, as [11]
$$S = -\frac{1}{2}Tr ln(\Delta + m^2), \tag{7.76}$$
where this has been written to consider a grand sum over all modes that are scattered through the barrier. This action is for the sum of all one-loop diagrams coupled to an arbitrary number of tree level external gravitons. This action is seen to be equal to $\zeta(0)$

Zeta function realizations may be employed to regularize this function. The action is
$$S = \int_0^\infty \frac{ds}{s}\left(e^{-sa^\dagger a} - e^{-sc^\dagger c}\right), \tag{7.77a}$$
and equivalently
$$S = S_0 + \int_0^\infty \frac{ds}{s}\left(e^{-s\Delta + m^2} - e^{-s\Delta_0 + m^2}\right), \tag{7.77b}$$
This is a computation of the correction due to one loop fluctuations between two metrics. Define the integral according to its eigenvalues $a^\dagger a|n\rangle = \lambda_n|n\rangle$ and represent the integral as
$$Tr(e^{-sa^\dagger a} - e^{-sc^\dagger c}) = \sum_n exp(-s\delta\lambda_n) \tag{7.78}$$

where the sum comes from the trace and the δ comes from the change in eigenvalues due to the squeezed state tunnelling. A Taylor expansion of the exponential and expand this function is

$$\zeta(0) = \int_0^\infty (b_0 s^{-2} + b_1 s^{-1} + b_2 s^0 + \ldots \tag{7.79}$$

where the DeWitt result illustrates that the b_i functions are

$$b_0 = \frac{1}{(4\pi)^2}$$

$$b_1 = \frac{1}{6(4\pi)^2} R$$

$$b_2 = \frac{1}{180(4\pi)^2} \Big(R^{\alpha\beta\mu\nu} R_{\alpha\beta\mu\nu} - R^{\alpha\beta} R_{\alpha\beta} + 30R^2 + 6\Delta R \Big). \tag{7.80}$$

The total divergence ΔR is dropped from the Lagrangian. This gives the unregularized action of the form

$$S = S_0 + \frac{\kappa^4}{2} Tr \int_0^\infty d^4x \delta \sqrt{-g} + \kappa^2 Tr \frac{1}{6(4\pi)^2} \int_0^\infty d^4x (\sqrt{-g} R - \sqrt{-g_0} R_0) +$$

$$\frac{1}{180(4\pi)^2} ln(\kappa^2/m^2) Tr \int_0^\infty d^4x \delta \Big(\sqrt{-g} (R^{\alpha\beta\mu\nu} R_{\alpha\beta\mu\nu} - R^{\alpha\beta} R_{\alpha\beta} + 30R^2 - \Lambda) \Big). \tag{7.81}$$

The first terms are standard gravitational actions, where the first gives geodesic motion and the second is the action for spacetime. The third term is the action for gravitational self-coupling. The constant κ^{-2} is a lower cut off and set to $12G/c^4$. In this term the cosmological constant has been introduced. The properly regularized version of this with the cutoff is then

$$S = S_0 + \Big(\frac{\kappa^4}{2} - m^2\kappa^2 + \frac{m^4}{2} ln(\kappa^2/m^2) \Big) Tr \int_0^\infty d^4x \delta \sqrt{-g} +$$

$$(\kappa^2 - m^2 ln(\kappa^2/m^2)) \kappa^2 Tr \frac{1}{6(4\pi)^2} \int_0^\infty d^4x (\sqrt{-g} R - \sqrt{-g_0} R_0) +$$

$$\int_0^\infty d^4x \delta \Big(\sqrt{-g} (-\frac{1}{16\pi G} R + a(\kappa) R^{\alpha\beta\mu\nu} R_{\alpha\beta\mu\nu} - a(\kappa) R^{\alpha\beta} R_{\alpha\beta} + 30a(\kappa) R^2 - \Lambda) \Big), \tag{7.82}$$

where the one loop corrections in the self-coupling action are

$$\Lambda = \Lambda_0 - \frac{1}{32\pi^2} \Big(\frac{\kappa^4}{2} - m^2\kappa^2 + \frac{m^2}{2} ln(\kappa^2/m^2) \Big)$$

$$\frac{1}{G} = \frac{1}{G_0} - \frac{1}{2\pi} \frac{1}{180(4\pi)^2} (\kappa^2 - m^2 ln(\kappa^2/m^2)) \tag{7.83}$$

$$a(\kappa) = \frac{1}{180(4\pi)^2} + \frac{1}{32\pi^2}ln(\kappa^2/m^2)$$

The most relevant quantity is the correction on the cosmological constant. If $\Lambda_0 \ll \Lambda$ the observed cosmological constant is due to the correction term at the one-loop Feynman diagram level. In fact if the uncorrected terms are set to either zero or unity (e.g. $G_0 = 1$) then all physical quantities are entirely due to quantum gravity at the one loop level and its associated corrections. This is the program Andre Sakharov advocated [12].

This has a correspondence to equation 7.71. For classical probability one may define entropy as

$$S = -k \int d^3x d^3p P(x, p) ln P(x, p). \tag{7.84}$$

Quantum mechanically the analog of equation 7.84 stems from decoherence. For the density matrix $\hat{\rho} = |\psi\rangle\langle\psi|$ with $|\psi\rangle = \sum_n a_n |n\rangle$, then the density matrix is of the form

$$\hat{\rho} = \sum_{mn} \rho_{mn} |m\rangle\langle n|, \quad \rho_{mn} = a_m^* a_n. \tag{7.85}$$

The off diagonal terms determine nonlocal correlations or coherence between states. With a fully coherent system the fine grained entropy is

$$S = -kTr(\hat{\rho} ln \hat{\rho}). \tag{7.86}$$

The above averaging of the squeeze induced phase destroys coherence involved with momentum operators. In the standard approach this is equivalent to letting $\rho_{mn} = |a_m|^2 \delta_{mn}$ with the coarse grained entropy,

$$S = -k \sum_m |a_m|^2 ln|a_m|^2. \tag{7.87}$$

The change in entropy ΔS is given by the off diagonal terms,

$$\Delta S = k \sum_{m \neq n} \rho_{mn} ln \rho_{mn} > 0. \tag{7.88}$$

In the Hilbert space $\mathcal{H}$ a sub matrix of the density matrix is over the noninvolutory nature of the momenta, where the phase averaging means that the strict coherence between two observables is "spread out" amongst all superpositions or phase configurations. This implies that coherences between momenta states are spread out amongst a set of states Ω and the phase space volume the system occupies increases. If the coherent phase configuration between momenta are equiprobable $\sim 1/\Omega$ the entropy increase in equation 7.88 can be represented by

$$\Delta S = -k \sum_{states} \left(\frac{1}{\Omega} ln \frac{1}{\Omega}\right). \tag{7.89}$$

This requires that the state space involved with the noninvolutory structure of momenta be sufficiently large as to give rise to the entropy required to define the cut-off in one-loop gravity, but not too large.

For a large squeezing parameter the cosmological entropy is

$$\Delta S \simeq 2|z| = \frac{3}{8G^2 E_v}, \quad (7.90)$$

where G and E_v are given by equation 7.78. For large κ then $E_v \propto \kappa^2$. Consequently the vacuum energy and the cosmological constant are due to the entropy in the coarse grained view of the tunnelling of the cosmology from the squeeze induced barrier.

7.6: The NonCommutative Algebra

A noncommutative algebra has been found by appealing to symplectic geometry. It gives rise to one-loop gravity under phase averaging that restores involution within the position and momentum variables. Based on equations 7.38, 7.39 and 7.42 the operators $z^a{}_\mu = (\mathbf{e}^1{}_\mu, \mathbf{p}^2{}_\mu)$ satisfy the algebra,

$$[\hat{z}^a_\mu, \hat{z}^b_\nu] = \frac{i}{2}\hbar\left((C^{abc})_{\mu\nu}{}^\alpha \hat{z}^c_\alpha + (C^{abc})^{\alpha\beta}{}_\sigma \mathcal{R}_{\alpha\mu\beta\nu}\hat{\mathbf{z}}^{\sigma c}\right). \quad (7.91a)$$

By construction $(C^{abc})_{\mu\nu}{}^\alpha z^c{}_\alpha$ is a unit symplectic matrix for $a = 1, 2$ $b = 2, 1$ and $c = 3$ and z^3 equal to unity. The transformation matrices on the state space $\mathcal{H}$ obeys the corresponding algebra,

$$AA^\dagger + BB^\dagger = 1, \quad A^\dagger\bar{B} - B^\dagger\bar{A} = isin(2i\theta)sinh(|z|)cosh(|z|)$$

$$CC^\dagger + DD^\dagger = 1, \quad C^\dagger\bar{D} - D^\dagger\bar{C} = isin(2i\theta)sinh(|z|)cosh(|z|) \quad (7.91b)$$

$$A^\dagger = C, \quad \bar{B}^\dagger = -D.$$

This algebra is designated as the B^* algebra, where under coherence the B^* algebra becomes the Bogoliubov algebra. Equation 7.91 should be regarded as a representation of this algebra as derived by symplectic methods. Below a more general form that this algebra is found without appealing to symplectic geometry.

This is an algebra that satisfies $\{\hat{\Omega}(z_1, (\ )), \hat{\Omega}(z_2, (\ ))\} = i\hat{\Omega}(z_1, z_2)$, but where on a region $U \subset \Sigma^{(3)}$ momentum-energy is not localized by $\int_U \hat{\Omega}(z_1, z_2)$. Ω is a map on the space of vectors $V = C \oplus C^*$, such that for $z_i, \in C$ and $z_j \in C^*$

$$\Omega : V \times V \rightarrow \mathbf{R}. \ \Omega((z_1, z_1'), (z_2, z_2')) = \langle z_2', z_1\rangle - \langle z_1', z_2\rangle, \quad (7.92)$$

but where the Hilbert space norm V is nonlocalizable. The norm $\langle\ |\ \rangle : C \times C \rightarrow \mathbf{R}$ is not localizable. This indicates the ability to determine the vacuum energy-momentum $\langle 0|\mathbf{P}|0\rangle$ is not possible. The vacuum energy is not in general definable. The above phase structure of the B^* algebra illustrates this in the language of quantum logic with a weaker orthomodularity condition on quantum propositions. This is explored in section 7.7 and 7.8 as a way of constructing the B^* algebra in a more general manner.

The Cauchy data on a spacial surface $\Sigma^{(3)}$ is defined by z_μ, $\mu \in \{1, \ldots, 4\}$. The embedding diffeomorphism $\Phi : \Sigma^{(3)} \times \mathbf{R} \to \mathcal{M}$ is then given by the dynamics of the B^* algebra. The Cauchy data on an initial spacial slice is given in the embedding $\Phi(\Sigma^{(3)}, 0) = \Sigma^{(3)}$. The foliation of spacial surfaces $\Phi(\Sigma^{(3)}, t)$ is governed by the B^* algebra, as diffeomorphisms of the of the space of variables z_μ. The evolution is parameterized by t, which defines the vector field $t_\mu = \partial_\mu t$ as an element of B^*. The B^* algebra defines a tangent bundle on Y as a vertical bundle according to nonlocalizable variables. The splitting of spacetime on $J^1(Y)$ is given by the projection $\pi_1 : J^1 \to Y$, with a remaining vector bundle $V^*Y \otimes_Y T^*\mathcal{M}$. This splits the tangent bundle on $\mathcal{M}$ as well as $T^*\mathcal{M} = T^*R \oplus T^*\Sigma^{(3)}$. The bundle on Y is then formally the same as equation 7.19

$$V^*Y \otimes_y T^*\mathcal{M} = V^*Y \otimes_y \left(T^*\Sigma^{(3)} \oplus R\right) = V^*Y \oplus \left(V^*Y \otimes T^*\Sigma^{(3)}\right). \quad (7.93)$$

The jet bundle splits as $J = J^t \oplus_y J^\Sigma$ corresponding to differentials in time and space. Each subjet J^t and J^Σ are bundles on Y with the associated vector bundles V^*Y and $V^*Y \otimes_y T^*\Sigma^{(3)}$. Hence there exist diffeomorphisms that "push" $\Sigma^{(3)}$, or the spacial variables in the set z_μ, through the quantum foliation of spacetime. However, it is impossible to identify in a local region the conjugate variables that define the bundle V^*Y under a Legendre transformation.

$Sec Y^\Sigma$ is the configuration space C and $Sec T_V Y^\Sigma$ is the Cauchy data TC. Yet Y cannot be locally specified. The Lagrangian, a map from the configuration space to the reals, determines the action. The section J^Σ is determined by the data $\phi, \partial_t, \partial_1\phi, \ldots, \partial_4\phi$ that enters the Lagrangian with the action $S[\phi] = \int d^4x\bigl(\mathcal{L}(\phi.\,\partial_t,\,\partial_i\phi)\bigr)$. Nonlocality of the energy-momentum infers that this Cauchy data cannot be completely specified. Neither can the ϕ be specified that are the required data input into the Lagrangian. For the Hamiltonian $H(\phi, \pi) = \int_{\Sigma^{(3)}} \bigl(\pi\partial_0\phi - \mathcal{L}(\phi)\bigr)d\Sigma^{(3)}$ there is insufficient data to determine $\langle 0|H|0\rangle$ or $\langle 0|[a, a^\dagger]|0\rangle$. This indicates the zero point energy is uncertain or indeterminate. Physically this suggests that at this scale wormholes, geons and closed time like curves corresponding to the vacuum state are prevalent and prevent the establishment of any data. Since the Hamiltonian is the generator of time the inability to determine the nature of the Hamiltonian means that time is as well undefined. The B^* algebra involves the structure of the vacuum where neither mass-energy nor time are uniquely defined.

Equation 7.91a may be extended into a graded algebra. Consider this direct sum of the algebra of momentum operators $L_{B^*} \subset B^*$ with the algebra L_Q, $L = L_{B^*} \oplus B_Q$. For the general product $\odot$ so that $\odot : L \times L \to L$ there are the following properties for any element $x_i \in L$:

$$x_1 \odot x_2 \in L \ closure \quad (7.94)$$

$$x_1 \odot (x_2 + x_3) = x_1 \odot x_2 + x_1 \odot x_3 \ linearity$$

$$x_1 \odot x_2 = -x_2 \odot x_1 \; antisymmetry$$

$$x_1 \odot (x_2 \odot x_3) + x_2 \odot (x_3 \odot x_1) + x_3 \odot (x_1 \odot x_2) = 0 \; Jacobi \; identiy$$

The elements of L are broken up into those in L_{B^*} and those in L_Q as

$$L_{B^*} = span\{p_\mu\}, \; \mu = 1, \ldots, dim(L_{B^*}) = 4$$

$$L_Q = span\{Q_a\}, \; a = 1, \ldots, dim(L_Q), \tag{7.95}$$

and where the product $\odot$ further obeys the rules

$$L_{B^*} \odot L_{B^*} \subset L_{B^*}, \; L_{B^*} \odot B_Q \subset L_Q, \; L_Q \odot L_Q \subset L_{B^*}. \tag{7.96}$$

Associated with this product is a degree $d \in \{0, 1\}$ for any $x \in L$ according to

$$d(p_\mu) = 0, \; \forall \; z_\mu \in L_{B^*}$$

$$d(Q_a) = 1, \; \forall \; Q_a \in L_Q. \tag{7.97}$$

The product $\odot$ is then defined according to

$$x_i \odot x_j = x_i x_j - (-1)^{d(x_1)d(x_2)} x_j x_i. \tag{7.98}$$

Hence for p_μ and Q_a, with $d(z_\mu) = 0$ and $d(Q_a) = 1$ then

$$p_\mu \odot Q_a = p_\mu Q_a - (-1)^0 Q_a p_\mu = [p_\mu, Q_a], \tag{7.99}$$

and

$$Q_a \odot Q_b = Q_a Q_b - (-1)^1 Q_b Q_a = \{Q_a, Q_b\}. \tag{7.100}$$

The commutators between elements of L_{B^*} satisfy equation 8.91a. The anticommutator $\{Q_a, \bar{Q}_b\}$ is is supersymmetric

$$\{Q_a, \bar{Q}_b\} = -i\sigma^\mu{}_{ab} p_\mu, \tag{7.101}$$

with $\{Q_a, Q_b\} = \{\bar{Q}_a, \bar{Q}_b\} = 0$. The commutator of this with p_ν is then

$$[\{Q_a, \bar{Q}_b\}, p_\nu] = -i\sigma^\mu{}_{ab}[p_\mu, p_\nu] =$$

$$= -\frac{1}{2}\sigma^\mu{}_{ab}\hbar\left((C^{22c})_{\mu\nu}{}^\alpha p_\alpha + (C^{22c})^{\alpha\beta}{}_\sigma R_{\alpha\mu\beta\nu} z^{\sigma c}\right). \tag{7.102}$$

Further, the angular momentum tensor defined in equation 7.38 satisfies

$$[Q_a, J^{\mu\nu}] = i(\sigma_2^{\mu\nu})_a{}^b Q_b. \tag{7.103}$$

The noncommutative B^* algebra determines a form of the horizon algebra. The equation 4.126 is replaced with equation 7.91. The supersymmetrization just illustrated may then be performed to find the moduli space for the supersymmetric theory. Below the B^* algebra is advanced according to a space with a structure more general than symplectic geometry. From there the supersymmetrization may be introduced according to a graded version of a general exterior calculus.

7.7: General Relativity and Quantum Mechanics as Subsets of a Single Theory

It appears that quantum mechanics and general relativity are two aspects of one theory according to the B^* algebra. In the parlance of quantum logic a set of propositions with a correspondence to eigenvectors in a Hilbert space are non-Boolean. This non-Boolean structure is due to the non-commensurability of conjugate observables. In the B^* algebra there exists a non-commensurability of all observables as a measure of flux quanta of curvature evaluated around a loop. It would then appear that there is a deep relationship between the non-Boolean nature of quantum mechanics (logic) and general relativity. Further, this relationship puts quantum mechanics and general relativity on the same "footing" in the B^* algebra. Hence there exists a functor between gravity and quantum mechanics based on an equivalence between two orthomodular lattices.

Consider the measurements of p, q, where boundary conditions or Cauchy data cannot be established simultaneously. There are two possible outcomes for these variables p^+, p^- and q^+, q^-. For *not* the complementary operation *not* $p^\pm = p^\mp$. The incommesurability of complementary variables and the Cauchy data for these two observable means the two propositions p and q are disjoint and obey

$$0 = p^+ \cap q^+ = p^+ \cap q^-, \qquad (7.104)$$

where $p^+ \neq 0$. Since $p^\pm$ and $q^\pm$ correspond to quantum complementary variables they contain the same state space information. This means that

$$p^+ \neq (p^+ \cap q^+) \cup (p^+ \cap q^-), \qquad (7.105)$$

where this weak orthomodularity which is contrary to what is expected from boolean logic [13]. This is the essence of quantum logic: Quantum propositions involving incommensurate observables do not obey the distributive property of Boolean logic [13].

The distributive law is replaced with a weaker law. For $p^+ \subseteq q^+$, with the complementary law $p^- \subseteq q^-$, the following is evident

$$q^+ = p^+ \cup (q^+ \cap p^-) \qquad (7.106)$$

The inclusiveness of p^+ in q^+ is a measure of how compatible a simultaneous measurement is between the two observables. Complete inclusion means that an experiment may be performed that specifies both completely. Partial inclusion means that complete information on q^+ is obtained at some loss of information on p^+. this might correspond to a measurement of the spin in the $\hat{z}$ direction simultaneously with a spin measurement in another direction with a component in the $\hat{z}$ direction.

There exists a similar situation in general relativity. A black hole is a spacetime with the structure $\mathbf{R}^4/(\mathbf{B}^3\times\mathbf{R})$ in the exterior region. If the black hole has a negative ADM mass then it is a wormhole with acausal structure due to closed timelike curves. Due to either horizon structures or nontrivial topology equation 7.106 obtains. The horizon algebra is an $SO(2,1)$ algebra that is isomorphic to the angular momentum algebra in quantum mechanics. It implies an incompatibility in measuring fields along certain directions due to spacetime curvature. Hence general relativity has embedded in its structure a logic that is equivalent to equations 7.105 and 7.106. In a topologically nontrivial spacetime, such as a wormhole closed timelike curves will prevent the establishment of Cauchy data which results in a complete uncertainty in measuring fields in the spacetime. The B^* algebra is then a covering space that embeds the equivalent logic of quantum mechanics and general relativity.

In the tunnelling of the universe out of the vacuum state quantum correlations between coordinate directions are lost. The decoherence of orthogonal position and momentum coordinates in the B^* algebra results in the one-loop quantum gravity that describes the extremely early universe. In this unravelling of the vacuum structure there is the emergence of what is called time. The cut off in one-loop quantum gravity is then a "wall" that defines reality as understood according to spacetime and momentum-energy. This is a curious situation where as we probe deeper into the foundations of physics increasingly the universe melts away into nothingness. Conversely as we observe the universe at ever larger scales structure emerges that identified as tangible "reality." There is a similarity between the logic of general relativity and quantum mechanics. As one approaches the Planck scale these two logics merge into coherent whole, but where our ability to identify space, time and energy collapse into complete uncertainty.

There is an alternative route to give consideration of the equivalence of quantum mechanics and general relativity. This route is an "artistic" approach to this that might lead to a new form of mathematics. It has happened that artistic works have had an impact on mathematical and scientific thinking. The Italian renaissance was such a period in history where artistic developments stimulated intellectual and scientific progress.

In the moduli space there is the "blow up" of singularities corresponding to points on the horizon. These blow ups consist of "cones" in projective geometries

$$C^3/S^1 = CP^2 \times [0,1)/(x,0). \tag{7.107}$$

This leads to the existence of two Virasoro algebras with gauge theoretic context. There is a way to look at the quantization of these spaces that leads to a coordinate noncommutative algebra. Let us assume there exist two quantum entangled black holes has a set of such blow up cones $\{\mathcal{C}_{one}\}$ in their respective moduli spaces. Consider subspaces of these cones $X \subset \mathcal{C}_{one}$ defined by intersection of projective cones. Now consider the total space of these cones and maps between these subspaces. These maps involve a mixing of coordinate directions, for the "blow-up points" have different coordinate positions. So there is a mixing of projective directions. This posits that coordinates in one direction in a blow up are mixed with different coordinate directions in another. There then exists a map M

$$M : X \to *X', \tag{7.108}$$

where $*$ is a duality operator. This mapping can be "pictorially" represented by a painting by Picasso. In these paintings the profile and frontal representation of a person are depicted on the same scape. This is illustrated in Picasso's painting *Girl Reading at a Table* (1934). The projective rays in the moduli space cones are quantum correlated. So the projective rays exhibit a phase change, or a change in perspective. This is illustrated by the intersection of two sets of projective rays on a plane. In the accompanying diagram the two projections are presented by the two perspectives of the face of the girl. This illustrates a quantum correlation entirely due to a coordinates incommensurate under simultaneous measurement. Position and momentum operators amongst themselves assume a structure similar to angular momentum operators, or equivalently exhibit properties seen in the horizon algebra. This advances the prospect that spaces formed from moduli space "blow-ups" in a quantum mechanical setting form what might be termed a Picasso space.

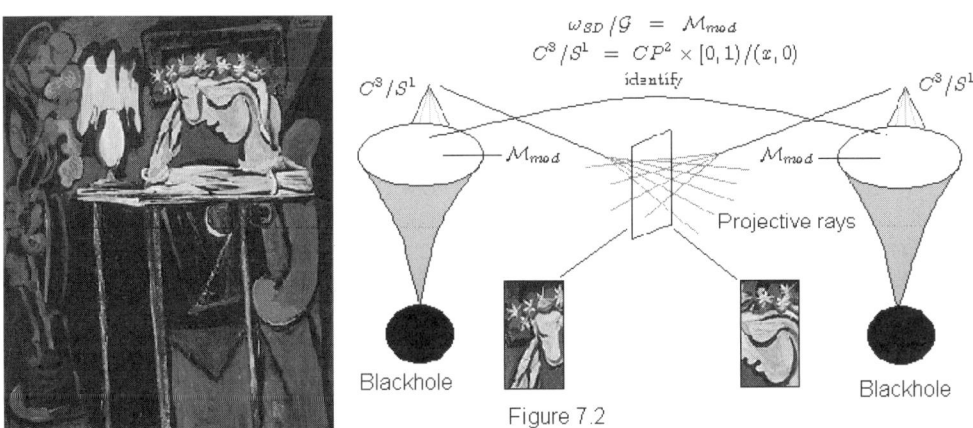

Figure 7.2

Artistic analogue of moduli for two entangled black holes.

This geometry implies "meets" and "joins" between different projective rays, which are points in projective geometry. The mixing of different directions in the moduli space has the structure of the B^* algebra if the meets and joins satisfy an orthomodular lattice with a nondistributive structure. The moduli space concerns gauge equivalent self-dual connections. In the context of spacetime consider these to be a choice of space-like surface. Yet if there are quantum meets and joins between elements of two moduli spaces this infers a noncommutative algebra. This gives a connection between this "Picasso" space concept and noncommutative algebra. Physically this is an entanglement between "cones" in moduli spaces. Such an entanglement suggests that gauge equivalent connections on two black holes will nonlocally correlate different gauge-like coefficients and mix how vectors are parallel translated at different regions of spacetime. The above projective structure shows that the general mathematics for the B^* algebra should have similarities to the Grassmann geometry of space formed from subspaces of a manifold.

7.8: A Proposed General Mathematics for the B^* Algebra

The above construction of the B^* algebra was based on symplectic manifold structure. This is still an inconvenient language in which to describe this structure as illustrated in the discussion on centered around equations 7.91-7.93. The symplectic structure is employed, but where there is no Cauchy data prescription. The noncommutative B^* algebra can be generalized to mathematical structures beyond sympletic geometry. It is then advanced that the B^* algebra is based on Giuseppe Peano's geometry [14]. Peano geometry is a vector space of dimension n with a multilinear form $[x_1, x_2, \ldots, x_n]$ that is antisymmetric. This multilinear form obeys

$$[x_1, \ldots, x_i, \ldots, x_i, x_n] = 0, \; \forall \, i \in [1, \ldots, n]$$
$$[x, \ldots, ax_i + bx_i, \ldots, x_n] = [x, \ldots, ax_i, \ldots, x_n] + [x, \ldots, bx_i, \ldots, x_n] \quad (7.109)$$
$$\exists \, \{e_1, \ldots, e_n\} \; such \; that \; [e_i, \ldots, e_n] \neq 0.$$

The first of these is antisymmetrization, the second is linearity and the final is completeness. This algebra applied in three dimensions would indicate $[x_1, x_2, x_3] \neq 0$. The interchange of x_1 and x_2 changes the sign of the bracket and thus defines an oriented parallelogram of two dimensions within a three dimensional simplex. Thus the antisymmetry of the x_3 with the parallelogram illustrates geometrically the different orientations of a plane. There is a similar bracket structure between the momentum variables $[p_1, p_2, p_3] \neq 0$, so the multilinear bracket for position and momentum is $[x_1, x_2, x_3, p_1, p_2, p_3] \neq 0$.

For V a vector space of elements the bracket structure defines the field F that the algebra $G(V)$ acts on. This field has the following property that for any $x_i = \sum_{ij} x_{ij} e_j$ that

$$[x_1, \ldots, x_n] = det(x_{ij}). \quad (7.110)$$

If the V defines the solution to a differential equation for elements entered into the bracket equation 7.110 defines the Wronskian. The field F is then determined by elements $x_i \in V$, $i \in [1, \ldots, k]$ $k \leq n$. The algebra $G(V)$ is determined by all possible brackets with maximum length n. $A(v)$ is an associative algebra over F generated by V. So for $k \leq n$ let $W_i(V)$ represent brackets of length i, where all possible brackets of size $k \leq n$ define

$$W(V) = \oplus_{i=1}^{n} W_i(V), \quad (7.111)$$

so the algebra is $G(V) = A(V)/W(V)$. Hence elements of $G(V)$ are defined by a map $\mu : A(V) \to G(V)$ given by

$$\mu(x_1, \ldots, x_n) = x_1 \vee x_2 \vee \ldots \vee x_n. \quad (7.112)$$

In the algebra $G(V)$ is then defined by the join operator $\vee$. Similarly define subspaces of $G(V)$ for each $G_k(V)$ as $G(V) = G_0(V) \oplus G_1(V) \oplus \ldots \oplus G_n(V)$, where $G_0 = F$ and $G_1(V) = V$ By the linearity on F there exists linearity on $G(V)$ with respect to the join operation. Further, for elements $x \in G_i(V)$ and $y \in G_j(V)$ the join satisfies

$$x \vee y = (-1)^{i+j} y \vee x. \tag{7.113}$$

The join then is an algebra of exterior multiplication.

Consider the two elements A and B formed by the join operation

$$A = x_1 \vee z_2 \vee \ldots \vee x_j, \quad B = y_1 \vee y_2 \vee \ldots \vee x_k. \tag{7.114}$$

The bracket structure for these two is

$$[A, B] = [x_1, x_2, \ldots, x_j, y_1, y_2, \ldots, y_k], \tag{7.115}$$

which in general is

$$[A, B] = \text{sign}(A, B)[B, A]. \tag{7.116}$$

Here $\text{sign}(A, B) = \pm$ for $A \vee B = \pm 1$, where in the case of duality where $j = n - i$ this is

$$[A, B] = (-1)^{i(n-i)}[B, A]. \tag{7.117}$$

Converse to the join operation is the meet operation. The meet product for A and B, which results in $C_2 \in G_{|i-j|}$ for $i + j \geq n$, is defined as

$$A \wedge B = \sum_{C_1, C_2} \text{sign}(C_1, C_2)[C_1, B]C_2. \tag{7.118}$$

It is not difficult to show that the meet operation is linear and are factorizable algebras. Define $I \in G_n(V)$ as an identity element for $A \in G_k(v)$ and $B \in G_l(v)$ with $k + l = n$ under the join operation

$$A \wedge B = (A \wedge B) \vee I = [A, B]I. \tag{7.119}$$

Elements with the meet and join operation in conjunction have nondistributive properties. For $i + j + k = n$ with $A \in G_i(v)$, $B \in G_j(v)$ and $C \in G_k(v)$ it may be shown that

$$A \wedge (B \vee C) = [A, B, C] = (A \vee B) \wedge C. \tag{7.120a}$$

For $C \in G_n(V)$ so that $C = D \vee E$, with $D \in G_j(V)$ and $E \in G_k(V)$ for $i + j = n$ it is also possible to demonstrate that

$$A \vee (B \wedge C) = \sum_{C_1, C_2} \text{sign}(C_1, C_2)[C_1, B] A \vee C_2. \tag{7.120b}$$

Similarly calculate

$$(A \vee B) \wedge (A \vee C) = \sum_{C_1, C_2} \sum_{D_1, D_2} sign(C_1, C_2) sign(D_1, D_2)[C_1, B][D_1, C]C_2 \wedge D_2. \tag{7.120c}$$

Equations 7.120b and 7.120c are equal only for $\sum_{D_1, D_2} sign(D_1, D_2)[D_1, C]D_2 = -A$. Hence the distributive property does not in general hold.

This is a general exterior calculus where the operation $\vee$ assumes the role of $\wedge$ in the theory of differential forms. Here $\wedge$ projects subspaces onto one another. Along with the logical operations $\vee$ and $\wedge$ there exists the logical complementation. For a subspace of dimension p in the total space of dimension n the complementation is the orthogonal space of dimension $n - p$. This complementation is then the Hodge dual star operator. As a result the complementation of $A \vee B$, for $A \in G_j(V)$ and $B \in G_k(V)$, exists in the orthogonal complement of $G_j(V) \otimes G_k(V)$ with dimension $n - j - k$. Thus $*(A \vee B) = (*A) \wedge (*B)$. Similarly for $I \in G_n(V)$ then $*I = 1 \in G_0(V)$.

Now consider the action of differential operators $d \in G_1(V)$. It is obvious for $A \in G_j(V)$ that

$$d \vee A = \sum_{a=1}^{n} B_a \vee \partial_a A \in G_{j+1}(V), \tag{7.121}$$

where the complement of this is

$$*(d \vee A) = \sum_{a=1}^{n} (*B_a) \wedge \delta_a * A \in G_{n-(j+1)}(V), \tag{7.122}$$

where $\delta_a = *\partial_a*$. It is simple to observe that $d \vee d \vee A = 0$. For $ker d: G_j(V) \to G_{j+1}(V)$ and $im d: G_{j-1}(V) \to G_j(V)$ then $ker d/im d$ defines a cohomology.

Connection coefficients are defined on the space $\omega = \omega^a B_a \in G_1(V)$ by the covariant operator $\mathcal{D} = d + \omega$, with $\omega = \omega_a B^s$, so that

$$\mathcal{D} \vee A = \sum_{a=1}^{n} \left(B_a \vee \partial_a A + B_a \omega^a \vee A \right) \in G_{j+1}(V). \tag{7.123}$$

The curvature is then calculated from $\mathcal{D} \vee \mathcal{D} \vee A \in G_{j+2}(V)$ as

$$\mathcal{D} \vee \mathcal{D} \vee A = \sum_{a,b=1}^{n} \left(\partial_b \omega_a + \omega_b \omega_a \right) B^b \vee B^a \vee A = \mathcal{R} \vee A. \tag{7.124}$$

This defines the curvature form $\mathcal{R} \in G_2(V)$. The curvature may be expressed according to the generalized bracket structures $[a_1, a_2, \ldots, a_j]$ $j \leq n$. The action of $\mathcal{D}$ on the dual $*(\mathcal{D} \vee A)$ is

$$\mathcal{D} \vee *(\mathcal{D} \vee A) = \mathcal{D} \vee (*\mathcal{D} \wedge *A) = \mathcal{D} \vee \sum_{a=1}^{n} (*B^a) \wedge \left(\delta_a * A + \omega_a^* * A \right)$$

$$= \sum_{a,b=1}^{n} (\partial_b \omega_a^* + \omega_b \omega_a^*) B^b \vee (*B^a \wedge *A), \qquad (7.125)$$

where $\omega_a^* = *\omega_a *$. Consider the gauge choice of flat connections where $\partial_b \omega_a^* = 0$. for $2 + k = n$ with $\mathcal{D} \in G_1(v)$ and $A \in G_k(v)$ equation 7.125 is

$$\mathcal{D} \vee (*\mathcal{D} \wedge *A) = \sum_{C_1, C_2} sign(C_1, C_2)[C_1, \mathcal{D}] * \mathcal{D} \vee C_2 =$$

$$\sum_{C_1, C_2} \omega_a^* \omega_b sign(C_1, C_2)(B^a \vee C_1) \wedge (*B^b \vee C_2). \qquad (7.126)$$

By the nondistributivity of this lattice given by equations 7.120 then

$$(\mathcal{D} \vee *\mathcal{D}) \vee A = \sum_{B^a, C_1} \sum_{B^b, C_2} sign(B^a, C_1) sign(B^b, C_2) \omega_a^* \omega_b (B^a \vee *B^b) \wedge (C_1 \vee C_2),$$

$$(7.127)$$

which is not in general equal to eqn. 7.126.

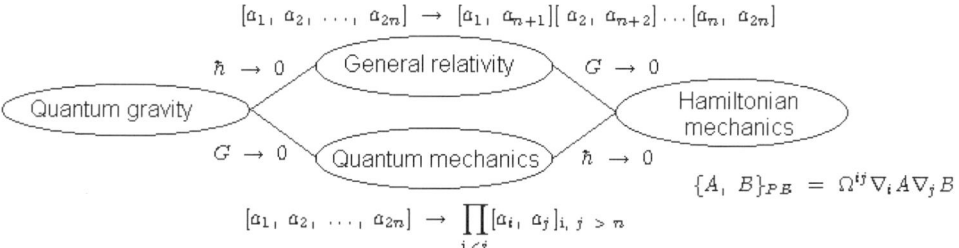

The limits between quantum gravity, quantum mechanics and general relativity.
Figure 7.3

This mathematics may be used to consider the Hodge star operator as an intertwiner between the position and momentum variables. If $G_2(V)$ is over a vector space with $j = 1, \ldots, n$ as position variables and $j = n+1, \ldots, 2n$ as momentum variables, then in general the commutators amongst all variables are not zero. This mathematics then has the type of structure that could be used to describe the B^* algebra in general. If commutativity is imposed on a_i, $i = 1, \ldots, n$ the position variables and a_i, $i = n+1, \ldots, 2n$ the momentum variables this is a Kähler geometry of quantum mechanics (geometric quantization). This may be seen if the general bracket structure assumes the form

$$[a_1, a_2, \ldots, a_{2n}] \rightarrow ([a_1, a_{n+1}][a_2, a_{n+2}] \ldots [a_n, a_{2n}]). \qquad (7.128a)$$

If there is then lattice distributivity with respect to ∨ and ∧ this defines symplectic geometry of Hamiltonian mechanics. Similarly if commutativity is removed on a_i, $i = n+1, \ldots, 2n$, or equivalently retain lattice nondistributivity amongst these variables, this arrives as general relativity.

$$[a_1, a_2, \ldots, a_{2n}] \to \prod_{i<j} [a_i, a_j]_{i, j > n}. \qquad (7.128b)$$

From there the subsequent imposition of distributivity on the entire lattice Hamiltonian mechanics is recovered. According to units equation 7.128a involves first letting the Planck scale approach zero by $G \to 0$ to recover quantum mechanics and then letting $\hbar \to 0$ to recover Hamiltonian mechanics. The other approach, equation 7.128b, is to let $\hbar \to 0$ first to recover general relativity, then take the limit $G \to 0$ to recover Hamiltonian mechanics.

These results are connected to orthomodular lattices. The most elementary lattice is one with propositions $a_i, -a_i$ so that $a_i \wedge a_j = \phi$, $i, j \in [1, \ldots, b]$, where $\wedge$ is the meet operation. Such a lattice appears as

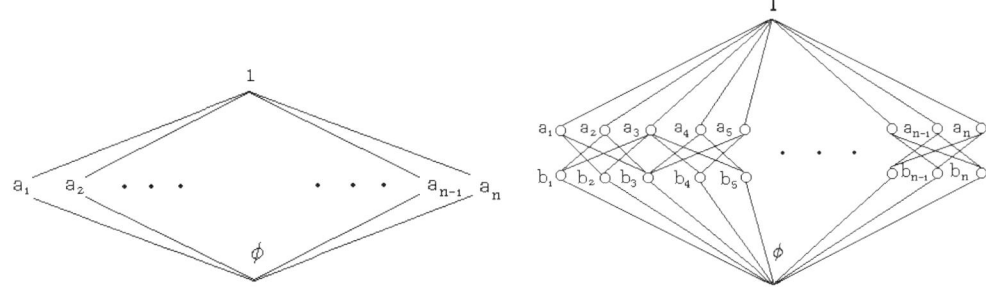

A completely nondistributive lattice.
Figure 7.4

A lattice with Boolean sublattices.
Figure 7.5

This lattice is one with propositions $a_i, -a_i$ with $a_i \wedge a_j = \phi$, $i, j \in [1, \ldots, b]$. If a_i is true or if a_j is true for $i, j \leq b$ the join is true. At the bottom the meet of these two is false. This lattice is then a general nondistributive lattice with ∨ and ∧ are generalizations of "or" and "and". . Assume that this is a sublattice with the top at the join replaced by the proposition A and the bottom replaced by the proposition A', so that "all true" and "all false" are replaced by more restricted quantum propositions. Then $(a_1 \vee a_2) \wedge a_n = a_n$ and $(a_1 \wedge a_n) \vee (a_2 \wedge a_n) = A'$. Hence this lattice is nondistributive. This sublattice, within a completely nondistributive lattice, will then obtain for equations 7.126 and 7.127 in general. Under phase averaging there will be a splitting of the propositions into two classes of quantum proposition a_i and b_i that under meets and joins in a lattice are nondistributive, but that contains a Boolean (distributive) sublattice. Such a lattice will be of the form as seen in Figure 7.5 above.

Here it is evident that the sublattice a_1, a_2, a_3, b_1, b_2, b_3 and a_3, a_4, a_5, b_3, b_4, b_5, as well as presumably a repeated sequence of them up to the right hand side, are boolean sublattices. In this case phase averaging has resulted in an increased "distiguishability" between quantum propositions. Just as in the lattice in figure 7.4 this could also be a sublattice in a larger lattice of propositions.

Another underlying aspect of geometry comes down to Cayley's numbers. If one has a sphere S^m, of dimension m, then for $m = 1, 2, 4, 8$ the sphere is a projective space over $\mathbf{R}, \mathbf{C}, \mathcal{Q}, \mathcal{O}$, the real numbers, complex numbers, quaternions and octonians. In these spaces the norm of vector x and y is a probability

$$p(x|y) = tr(|x\rangle\langle x||y\rangle\langle y|) = (|x\rangle\langle y|)^2, \qquad (7.129)$$

and so these norms are real valued and can be assigned to a probability.

The B^* algebra can be represented with a vector space $V = (e_1, \ldots, e_{2n})$, one imposes a general bracket structure $[e_1, \ldots, e_{2n}]$ that satisfies

$$[e_1, ..., e_i, e_{i+1} e_{2n}] = -[e_1, ..., e_{i+1}, e_i, e_{2n}]. \qquad (7.130)$$

This is by default a system over $\mathbf{C}$. For each e_i there is a form B_i, where the join operation $B_i \vee B_j$ is analogous to the wedge product between two one-forms. In order to extend this to Q, the B_i are Clifford valued (Cliff-forms).

There is a connection to computability theory since the join $\vee$ and meet $\wedge$ operations are logically analogous to "or" and "and," and that an equation with meets and joins elements on $(e_1, \ldots, e_{2n})$ is a logical proposition, or an algorithm. With Cliff-forms then calculations are spin calculations, with an added spin "up vs. down" logic to the orthomodular lattice of calculations. This means that an added spin bundle is imposed on the basis elements. The bundle is defined by the map

$$Spin(4) \rightarrow SO(4), \qquad (7.131)$$

where $Spin(4)$ is the Lie algebra for the Clifford algebra. Given the vector space $V = (e_1, \ldots, e_{2n})$ replace G_{2n} with G_{2n}^{cl}, $dim G_{2n}^{cl} = 2^{2n}$ so that

$$e_i \cdot e_j + e_j \cdot e_i = 0, i \neq j, \ e_i \cdot e_i = -2\langle e_i \cdot e_i \rangle, \qquad (7.132)$$

where the first condition infers the bracket relation $[e_i, e_j] = -2e_j \cdot e_i$. This imposes a spin structure on the interpretation of the bracket structure of the quantum gravi-logic. In general the algebra G_{2n}^{cl} is

$$G_{2n}^{cl} \simeq \otimes_{i=1}^{2n-1} \mathcal{Q}. \qquad (7.133)$$

This is simply an introductory view of this structure. These developments are beyond the state of progress at this time. However, the introduction of a spin structure into this orthomodular construction, where B^* algebra is required for extensions to spinorial quantum gravity.

This represents a fusion of geometry with the theory of algorithms and quantum computing. Possibly in some sense the laws of physics may be due to a sort of quantum cellular automata theory, where physical law is due to a vast network of elementary computing devices consisting of quantum propositions. Nature just eliminates the device and retain the quantum propositions. How these propositions define a calculation is how they are networked together. For Cliff-forms this would be a quantum computing spin network. Computing with quantum spin as a factor is a possibility in the future for computing machinery. It could be an example of how some trends in technology are converging with concepts on the foundations of physics.

The situation for octonions becomes very strange. Octonions involve a nonassociative algebra. What this means for quantum logical types of constructions is uncertain. Octonions are involved with $N = 8$ supersymmetry (32 supersymmetries) and may enter this scheme. Quantum gravity with octonions is developed in the next chapter.

7.9: Concluding Words

With the introduction of quantum logic and the tie between general relativity and quantum mechanics, in particular with the apparent "disappearance" of reality at the Planck scale and approaching the end of what can be known. The inability to define $\langle 0|H|0\rangle$ means that time is indefinable and a similar result occurs for momentum and the meaning of space. The concept of quantum logical or quantum-gravi-logical propositions, along with orthomodular lattices of propositions which describe all states, leads to a generalized incommensurability between observables. This suggests that the sort of modified exterior calculus in the above section is required. The complete structure of this exterior calculus and its underlying B^* algebra is a subject that requires considerable exploration.

The B^* algebra indicates that the measure of an operator $\mu(\mathcal{O}) = Tr\mathcal{O}$ is not generally obtainable in a one to one fashion associated with linear states. The meet and joins in the system infer that the quantum gravity vacuum is ultimately a set of logical, or quantum-gravito logical, propositions or algorithms. Kurt Gödel illustrated that any axiomatic system inherently contains propositions that are true but unprovable. Alan-Turing demonstrated a similar result with algorithms for general computing systems. The prospect exists that the truth nature of some quantum propositions may not be due to any logically consistent reason, but due to complete randomness. This orthomodular lattice structure, based on an exterior calculus, is a system of propositions will likely involve quantum-gravi-logical statements that are true but self-referentially unprovable. Such a statement must be true, for its negation clearly leads to a logical contradiction. The above B^* algebra then indicates a potential route towards the immeasurability of space under all circumstances. There is then the implication that physical laws involve "immeasurable" measures. Such situations do occur in problems with the axiom of choice and the ability to define universally a measure. Chapter 9 discusses the issue of whether physical law itself has any basis in "provability" in a standard axiomatic or logico-algebraic context. This then might lead to a form of "Pre-Geometry" of the form advocated by John Wheeler.

The nature of quantum spacetime fluctuations appears to require a core theory on the nature of quantum gravity according to a general commutative algebraic construction. The "meet" and "join" construction is a generalization of the mathametical constructions most often used in differential geometry. There also exist commutators of $[\hat{p}_\mu, \hat{p}_\nu]$ that are non-vanishing, as well as the same for position operators. It is possible that careful experimentation with interferometers could detect small phase shifts in the frequency of a photon due to this affect.

References

[1] S. W. Hawking, G. R. F. Ellis, "The Large Scale Structure of Spacetime," Cambridge Pub. Cambridge (1973).

[2] M. Morris, K. Thorne, *Am. J. Phys.* **56**, 395 (1988).

[3] M. Morris, K. Thorne, U. Yurtsever, *Phys. Rev. Lett.* **61**, 1446 (1988).

[4] L. H. Ford, T. A. Roman, *Phys.Rev.* **D60** (1999) 104018

[5] M. J. Pfenning, L.H. Ford, *Class.Quant.Grav.* **14** (1997) 1743-1751.

[6] M. C. Abbati, R. Cirelli, S. DeSantis, E. Ruffini, *J.Geom. and Phys.* **17** (1995).

[7] M. J. Gotay, J. M. Nester, *Ann. Inst. Henri Poincaré* Vol XXXII (1980).

[8] M. J. Gotay, *Diff. Geom. Appl.* **1** (1991) 375-390.

[9] C. Misner, K. Thorne, J. A. Wheeler, "Gravitation," Freeman pub., San Francisco (1973).

[10] A. Feinstein, M. Sebastian, *Found. Phys. Lett.* **13**, 2 (2000).

[11] M. Visser, *Mod.Phys.Lett.* **A17** (2002) 977-992.

[12] A. D. Sakharov, *Sov. Phys. Dokl.* **12** (1968) 1040 [*Dokl. Akad. Nauk Ser. Fiz.* **177** (1968) 70]. Reprinted in *Gen. Rel. Grav.* **32** (2000) 365-367.

[13] E. G. Beltrametti, G. Cassinelli. *The Logic of Quantum Mechanics*, Vol. 15 *Encyclopedia of Mathematics and its Applications*. Addison-Wesley Publishing Company, (1981).

[14] G. Peano, "Calcolo geometrico secondo l'Ausdehnungslehre di H. Grassmann," Fratelli Bocca Editori, (1888).

8: Octonionic Quantum Gravity

In the last chapter an extension of physics to a generalized bracket structure was presented. A possible approach to this is the octonions [1]. The Wronskian in equation 7.110 for three elements in a nonassociative bracket structure defines the octonions. The octonions are a potential extension of quantum field theory to describe nonlinear field theory without perturbative methods. It is further a potential path for string theory and quantum gravity. Here physical reasoning is given for why such structures may underlie string theory beyond the string length $\sqrt{8\pi}L_p$. The physics lie with connections between the holographic theory and nonassociative quantum gravity. Currently string theory lacks a number of predictive features, which may reflect a theoretical incompleteness due to the existence of some underlying structure. It is possible that octonionic quantum gravity is that structure.

This chapter starts with the mathematics of octonions. This is line with how students are taught quantum mechanics, where often they first learn linear algebra before they are taught quantum mechanics which employs the techniques of linear algebra. The latter part of the chapter then illustrates the physical motivation for octonionic quantum gravity. These arguments in part have their basis with Ahlawalia's observation that quantum fluctuations of the spacetime metric leads to a breakdown of any locality of between coordinates in the definition of a propagator. Quantum gravity then likely extends the concept of nonlocality to a noncommutative geometry. The octonions are then illustrated as a likely operator method for such a description.

8.1: Algebra and Basis Elements

The octonions $\mathcal{O}$ are a nonassociative and noncommutative set of basis elements of a normed division algebra. The complexes, quaternions and octonions are formed from a vector space with dimensions 2, 4, 8 by [1]

$$\begin{aligned}\mathbf{R}^2 &\to \mathcal{C} \\ \mathbf{R}^4 &\to \mathcal{H} \\ \mathbf{R}^8 &\to \mathcal{O},\end{aligned} \quad (8.1)$$

and define normed division algebras. Any octonion may then be written as $o = \sum_{i=1}^{8} c_i e_i$ [2]. The basis elements e_i, $i = 1, \ldots 8$ are then various square roots of -1 with $e_i^2 = -1$ The anticommutivity of octonions $e_i e_j = -e_j e_i$ leads to the general product rule $e_i e_j = \epsilon_{ijk} e_k$ so that a triplet product may be associated [3]

$$\begin{aligned}(e_i e_j)e_k &= \epsilon_{ijl} e_l e_k = \delta_{ik} e_j - \delta_{jk} e_i \\ e_i(e_j e_k) &= \epsilon_{jkl} e_i e_l = \delta_{ik} e_j - \delta_{ij} e_k,\end{aligned} \quad (8.2)$$

with the result that

$$(e_i e_j)e_k - e_i(e_j e_k) = \delta_{ij} e_k - \delta_{jk} e_i. \quad (8.3)$$

In a suitable basis this multiplication rule produces the Fano plane. This results in the trilateral associative product or bracket $[a,\ b,\ c] = (ab)c - a(bc)$

Any octonion o may be written as the linear combination

$$o = c_1 1 + \sum_{i=1}^{7} c_i e_i, \qquad (8.4)$$

so that octonion conjugation may be defined according to $\overline{ab} = \bar{b}\bar{a}$. This defines the inner product on this basis mapped from $\mathcal{R}^8$ as the Jordan product [1]

$$\langle a,\ b \rangle = \sum_i a_i b_j = \frac{1}{2}(ab + b\bar{a}) = \frac{1}{2}(ba + \bar{a}b) \qquad (8.5)$$

Octonions are constructed in the manner that complexes and quaternions are built. The complex numbers are formed by pairs of reals as

$$z = a + be_1 = \begin{pmatrix} a \\ b \end{pmatrix} \qquad (8.6)$$

These elements must transform by some matrix elements. The first element is the identity 2×2 matrix. The other is a matrix that mimics the multiplication by $i = \sqrt{-1}$ that is

$$\begin{pmatrix} 0 & -1 \\ 1 & 0 \end{pmatrix} \begin{pmatrix} a \\ b \end{pmatrix} = \begin{pmatrix} -b \\ a \end{pmatrix}, \qquad (8.7a)$$

and the other describes complex conjugation as

$$\begin{pmatrix} 1 & 0 \\ 0 & -1 \end{pmatrix} \begin{pmatrix} a \\ b \end{pmatrix} = \begin{pmatrix} a \\ -b \end{pmatrix}. \qquad (8.7b)$$

Finally a third matrix exists that reflects the combination of these two operations by

$$\begin{pmatrix} 0 & 1 \\ 1 & 0 \end{pmatrix} \begin{pmatrix} a \\ b \end{pmatrix} = \begin{pmatrix} b \\ a \end{pmatrix}. \qquad (8.7c)$$

Interestingly these define the Pauli matrices σ_x, σ_y, σ_z

In the same manner quaternions are formed from pairs of complex numbers. A quaternion is defined by basis elements mapped from $\mathcal{R}^4$ by

$$q = \sum_i q_i e_i = \begin{pmatrix} q_1 \\ q_2 \\ q_3 \\ q_4 \end{pmatrix}. \qquad (8.8)$$

The matrices that describe the conjugation and multiplication by $\sqrt{-1}$ are then the matrices

$$\gamma_i = \begin{pmatrix} \sigma_i & 0_{2\times 2} \\ 0_{2\times 2} & \sigma_i \end{pmatrix}, \; i = 1, \ldots 3$$

$$\gamma_4 = \begin{pmatrix} 1_{2\times 2} & 0_{2\times 2} \\ 0_{2\times 2} & 1_{2\times 2} \end{pmatrix}, \quad (8.9)$$

which are the Dirac matrices used in the theory of spin 1/2 particles. The theory of quaternions then has the group structure $SU(2)_R$ and $SU(2)_L$, which gives the adjoint action on a principle bundle. There is further the $SO(4) \sim SU(2)_L \times SU(2)_R$ group structure for quaternions. Finally with $\gamma_5 = -\gamma_1\gamma_2\gamma_3\gamma_4$ the group $spin(2,3)$ group structure for $\mathcal{H}$.

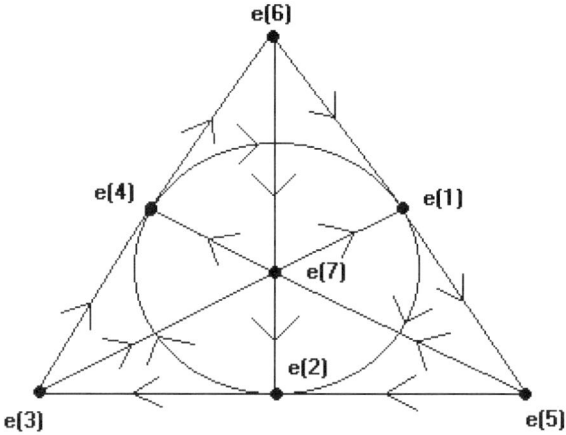

The Fano plane for the octonions.
Figure 8.1

With octonions there is again a pairing with quaternions. The elements of $\mathcal{O}$ are then formed by $\mathcal{H}|_p\mathcal{H}$ and defined by pairs of eight matrices. This will lead to two copies of $\mathcal{H}$ where the nontrivial elements of each are paired up with the unit matrix $\gamma_4 = 1_{4\times 4}$. With one of these matrices being the unit matrix, there are then 7 of these elements that are the "square root" of the negative unit matrix.

The octonionic algebra has a relationship to the $SO(8)$ algebra. The octonions are strictly a normed division algebra that obey no algebraic constraint, such as commutativity and associativity. The basis for $\mathcal{O}$ is determined by the eight elements $\mathbf{1}$, e_i, $i = 1, \ldots, 7$ that satisfy the relationship

$$e_i e_j = v_{ijk} e_k - \delta_{ij}, \quad (8.10)$$

where v_{ijk} is a totally antisymmetric term with

$$v_{ijk} = +1 \; for \; ijk = 124;\; 346;\; 615;\; 523;\; 371;\; 574;\; 672: \quad (8.11)$$

This may be seen with the Fano plane. This 3×3 matrix is dual to the 4×4 matrix defined by the Levi-Civita matrix in 7 dimensions by

$$v_{ijkl} = \frac{1}{3!}\epsilon_{ijklabc}v^{abc},\tag{8.12}$$

with

$$v_{ijkl} = +1 \; for \; ijkl = 7461; \; 1527; \; 3247; \; 6123; \; 2743; \; 3467; \; 3567, \tag{8.13}$$

where the last one is found by equation 8.12 and $v_{3567} = \frac{1}{2}v_{3541}v_{67}{}^{41}$ generally shown below in equation 8.14. These satisfy the relations

$$v^{ijk}v_{abk} = \delta^i_a\delta^j_b - \delta^i_b\delta^j_a - v^{ij}{}_{ab}$$

$$v^{ijkl}v_{abcl} = 4(\delta^i_a\delta^j_b - \delta^i_b\delta^j_a) - 2v^{ij}{}_{ab} \tag{8.14}$$

$$v^{ijk}v_{abjk} = -4v^i{}_{ab}.$$

These relations are similar to those of the quaternions (Pauli and Dirac matrices), yet since octonions are nonassociative they do not obey the Jacobi identity. These matrices then define the Cayley four-form [4]

$$\Upsilon = \mathbf{e}^1 \wedge \mathbf{e}^2 \wedge \mathbf{e}^4 \wedge \mathbf{e}^8 + \mathbf{e}^3 \wedge \mathbf{e}^4 \wedge \mathbf{e}^6 \wedge \mathbf{e}^8 +$$

$$\mathbf{e}^6 \wedge \mathbf{e}^1 \wedge \mathbf{e}^5 \wedge \mathbf{e}^8 + \mathbf{e}^5 \wedge \mathbf{e}^2 \wedge \mathbf{e}^3 \wedge \mathbf{e}^8 + \mathbf{e}^3 \wedge \mathbf{e}^7 \wedge \mathbf{e}^1 \wedge \mathbf{e}^8 + \mathbf{e}^5 \wedge \mathbf{e}^7 \wedge \mathbf{e}^4 \wedge \mathbf{e}^8$$

$$\mathbf{e}^6 \wedge \mathbf{e}^7 \wedge \mathbf{e}^2 \wedge \mathbf{e}^8 + \mathbf{e}^7 \wedge \mathbf{e}^4 \wedge \mathbf{e}^6 \wedge \mathbf{e}^1 + \mathbf{e}^1 \wedge \mathbf{e}^5 \wedge \mathbf{e}^2 \wedge \mathbf{e}^7 + \mathbf{e}^3 \wedge \mathbf{e}^2 \wedge \mathbf{e}^4 \wedge \mathbf{e}^7 + \tag{8.15}$$

$$\mathbf{e}^6 \wedge \mathbf{e}^1 \wedge \mathbf{e}^2 \wedge \mathbf{e}^3 + \mathbf{e}^2 \wedge \mathbf{e}^7 \wedge \mathbf{e}^4 \wedge \mathbf{e}^3 + \mathbf{e}^3 \wedge \mathbf{e}^4 \wedge \mathbf{e}^6 \wedge \mathbf{e}^7 + \mathbf{e}^3 \wedge \mathbf{e}^5 \wedge \mathbf{e}^6 \wedge \mathbf{e}^7$$

For Latin indices ranging from 1, ..., 7 and Greek indices ranging from 1, ..., 8, it is evident that the v_{ijk} and v_{ijkl} matrices have an $SO(8)$ representation as

$$\Upsilon_{\alpha\beta\gamma 8} = v_{ijk}, \; \Upsilon_{ijkl} = v_{ijkl}. \tag{8.16}$$

The duality principle is expressed according to the Hodge star operator on the form $*$: $\Upsilon \to \Upsilon$, which in components is

$$\Upsilon_{\alpha\beta\gamma\delta} = \frac{1}{4!}\epsilon_{\alpha\beta\gamma\delta\alpha'\beta'\gamma'\delta'}\Upsilon^{\alpha'\beta'\gamma'\delta'}. \tag{8.17}$$

This defines three different **35** representations of $SO(8)$ **35**$_0$ and **35**$_\pm$ that are related to each other by the triality of the associator. Equation 8.14 in this generalized setting leads to the following [5]

$$\Upsilon_{\alpha\beta\gamma\delta}\Upsilon^{\alpha'\beta'\gamma'\delta} = (\delta^{\alpha'}_\alpha\delta^{\beta'}_\beta - \delta^{\alpha'}_\beta\delta^{\beta'}_\alpha)\delta^{\gamma'}_\gamma + (\delta^{\gamma'}_\alpha\delta^{\alpha'}_\beta - \delta^{\alpha'}_\alpha\delta^{\gamma'}_\beta)\delta^{\beta'}_\gamma + (\delta^{\beta'}_\alpha\delta^{\gamma'}_\beta - \delta^{\gamma'}_\alpha\delta^{\beta'}_\beta)\delta^{\alpha'}_\gamma$$

$$+ \Upsilon_{\alpha\beta}{}^{(\alpha'\beta'}\delta_{\gamma}^{\gamma')} + \Upsilon_{\gamma\ \alpha}{}^{(\alpha'\beta'}\delta_{\beta}^{\gamma')} + \Upsilon_{\beta\gamma}{}^{(\alpha'\beta'}\delta_{\alpha}^{\gamma')}, \tag{8.18}$$

where the parentheses indicate a cyclic permutation of the indices enclosed. These elements define the octonion basis according to

$$(\gamma_i)_{ij} = iv_{ijk}, \quad (\gamma_i)_{b8} = i\delta_{ij}, \quad \{\gamma_i, \gamma_j\} = 2\delta_{ij}, \tag{8.19}$$

where $\gamma_{ij} = 1/2[\gamma^i, \gamma^j]$ are the generators of the $SO(7) \subset SO(8)$ generators, just as $\sigma^{ij} = 1/2[\sigma^i, \sigma^j]$ are the generators of $SO(4)$. Now pair the Dirac matrices [5],

$$\Gamma_i = \begin{pmatrix} 0 & i\gamma_i \\ -i\gamma_i & 0 \end{pmatrix} \tag{8.20a}$$

with Γ_8

$$\Gamma_8 = \begin{pmatrix} 0 & 1 \\ 1 & 0 \end{pmatrix} \tag{8.20b}$$

as $\Gamma_\alpha = (\Gamma_i, \Gamma_8)$ [5]. The decomposition of the group representations is then $\mathbf{8}_0 = 7 + 1$ and $\mathbf{8}_\pm = 8$ for the vector and chiral representations of $SO(8)$. The chiral representation of the group demands the existence of the chiral matrix

$$\Gamma^9 = \begin{pmatrix} 1 & 0 \\ 0 & 1 \end{pmatrix} \tag{8.21a}$$

which is used to define dual representations of the commutator $\Gamma^{ij} = 1/2[\Gamma^i, \Gamma^j]$ as

$$\Gamma^{ij} = v^{ijk}\Gamma^9\Gamma_{8k}, \quad v_{ijkl}\Gamma^{ij} = -(4 + 2\Gamma^9)\Gamma_{kl}. \tag{8.21b}$$

The following combination of these elements will leave a right or left handed spinor $\xi_\pm$ invariant

$$\Sigma_\pm^{\alpha\beta} = \Gamma^{\alpha\beta} \pm \frac{1}{6}\Upsilon^{\alpha\beta\gamma\delta}\Upsilon_{\gamma\delta}, \tag{8.22}$$

which is a stability group by $\Sigma_\pm^{\alpha\beta}\xi_\pm = 0$. This defines an additional $SO(7)$ stability group on $\xi_\pm$ by

$$\mathbf{8}_0 = 8, \quad \mathbf{8}_\pm = 7 + 1, \quad \mathbf{8}_\mp = 8. \tag{8.23}$$

The subgroup $SO(7) \subset SO(8)$ reflect the Hopf fibration

$$S^7 \hookrightarrow S^{15} \hookrightarrow S^8. \tag{8.24}$$

Given the action of $SO(8)$ over $SO(7)$ is determined by a line bundle. Transition functions $g_{ab}U_a \cap U_b \to SO(8)$ defines

$$c^{1,1} = \frac{i}{2\pi}d\,log(g_{ab}), \tag{8.25}$$

which is closed under d and defines the first Chern class $c_1(E)$ in a Cech-de Rham cohomology complex. This further defines the higher Chern classes

$$c_i(E) = det\left(1 + \frac{i}{2\pi}K_i\right), \tag{8.26}$$

with $K_i = d\theta_i + \frac{1}{2}\theta_i^2$ and $\theta_i = \sum_{i=1}^{n} g_{ij}^{-1} dg_{ij}$ where n is the number of charts this function is defined over. From this it should then be possible to define instantons, which physically would correspond to quantum black holes.

8.2: Triality and Right/Left Eigenvectors Over $\mathcal{O}$

The octonions are a nonassociative set of basis elements $\{1, \mathbf{e}_i\}$, $i = 1, \ldots, 7$ defined on the Fano plane. For the exception of the elements with the indicial triples 124, 346, 615, 523, 371, 574, 672 they obey the nonassociative rule

$$(\mathbf{e}_i\mathbf{e}_j)\mathbf{e}_k - \mathbf{e}_i(\mathbf{e}_j\mathbf{e}_k) = v_{ijkl}\mathbf{e}_l, \tag{8.27}$$

with $(\mathbf{e}_i\mathbf{e}_j)\mathbf{e}_k = -\mathbf{e}_i(\mathbf{e}_j\mathbf{e}_k)$. There is a context sensitivity associated with multiplication that depends upon whether the ( ) is on the left or the right. This structure is demonstrated to give rise to right and left eigenvectors for the elements $\mathbf{e}_4$, $\mathbf{e}_7$, which are $SL(2, \mathcal{H})$ eigenoperators.

Define these right and left vectors as u_r and u_l according to

$$\mathbf{e}_I u_l = \lambda_l u_l, \quad u_r \mathbf{e}_I = u_r \lambda_r. \tag{8.28}$$

Here the capital index is restricted to $I = 4, 7$ for the eigenvalue. There is an implicit sum over different eigenvalues here, but are ignored for brevity.

A standard rule is that eigenvalues are invariant under unitary transformations. Thus for $\mathbf{e}_I \to U\mathbf{e}_I U^{-1}$ the norm $\overline{\mathbf{e}_I u_l} \mathbf{e}_I u_l$ gives the eigenvalue λ^2. The transformation matrices are defined according to the other basis elements for $i = 1, 2, 3, 5, 6$ by

$$U = exp(ig_i\mathbf{e}_i), \quad U^{-1} = \bar{U} - exp(-ig_i\mathbf{e}_i), \tag{8.29}$$

where $\bar{\mathbf{e}}_i = -\mathbf{e}_i$ and $\bar{g}_i = -g_i$. The standard expansion of these unitary matrices is then

$$U = 1 + ig_i\mathbf{e}_i - \frac{1}{2}g_i g_j \mathbf{e}_i \mathbf{e}_j + \ldots \tag{8.30}$$

It is evident that right and left vectors transform according to

$$U\mathbf{e}_I u_l = (U\mathbf{e}_I)U^{-1}Uu_l, \quad u_r \mathbf{e}_I U^{-1} = u_r U^{-1}U(\mathbf{e}_I U^{-1}), \tag{8.31}$$

This defines the left and right transformations of the octonion eigenbasis elements according to

$$left: \mathbf{e}_I \to (U\mathbf{e}_I)U^{-1}, \quad right: \mathbf{e}_I \to U(\mathbf{e}_I U^{-1}), \tag{8.32}$$

which differ only by how the matrices are associated. This gives the expansions for these transformations as

$$(U\mathbf{e}_I)U^{-1} = \mathbf{e}_I + ig_j[\mathbf{e}_j, \mathbf{e}_I] - \frac{1}{2}g_jg_k((\mathbf{e}_j\mathbf{e}_k)\mathbf{e}_I + \mathbf{e}_I(\mathbf{e}_j\mathbf{e}_k)) + g_jg_k\mathbf{e}_j(\mathbf{e}_I\mathbf{e}_k) \quad (8.33a)$$

$$U(\mathbf{e}_IU^{-1}) = \mathbf{e}_I + ig_j[\mathbf{e}_j, \mathbf{e}_I] - \frac{1}{2}g_jg_k((\mathbf{e}_j\mathbf{e}_k)\mathbf{e}_I + \mathbf{e}_I(\mathbf{e}_j\mathbf{e}_k)) + g_jg_k(\mathbf{e}_j\mathbf{e}_I)\mathbf{e}_k. \quad (8.33b)$$

By use of commutation and associativity $(\mathbf{e}_j\mathbf{e}_k)\mathbf{e}_I + \mathbf{e}_I(\mathbf{e}_j\mathbf{e}_k) = 0$ is found, where

$$(U\mathbf{e}_I)U^{-1} = \mathbf{e}_I + ig_j[\mathbf{e}_j, \mathbf{e}_I] + g_jg_k\mathbf{e}_j(\mathbf{e}_I\mathbf{e}_k) \quad (8.34a)$$

$$U(\mathbf{e}_IU^{-1}) = \mathbf{e}_I + ig_j[\mathbf{e}_j, \mathbf{e}_I] + g_jg_k(\mathbf{e}_j\mathbf{e}_I)\mathbf{e}_k \quad (8.34b)$$

Now compute $(\overline{\mathbf{e}_Iu_l})'(\mathbf{e}_Iu_l)'$ and $(\overline{\mathbf{e}_Iu_r})'(\mathbf{e}_Iu_r)'$ with

$$(\overline{\mathbf{e}_Iu_l})'(\mathbf{e}_Iu_l)' = -\mathbf{e}_I\mathbf{e}_I + ig_jg_k[\mathbf{e}_I, \mathbf{e}_k][\mathbf{e}_j, \mathbf{e}_I] + \frac{1}{2}g_jg_k(\mathbf{e}_I(\mathbf{e}_j\mathbf{e}_k) + (\mathbf{e}_j\mathbf{e}_k)\mathbf{e}_I) = -1, \quad (8.35)$$

where the appropriate terms are found to vanish. Hence the sum over the square of the eigenvalues is normalized to unity.

This contrasts with the projection of $\overline{\mathbf{e}_Iu_r}'$ onto $\mathbf{e}_Iu_l'$

$$\overline{\mathbf{e}_Iu_r}'\mathbf{e}_Iu_l' = 1 + 2ig_j[\mathbf{e}_j, \mathbf{e}_I] \mapsto exp(2ig_j[\mathbf{e}_j, \mathbf{e}_I]). \quad (8.36)$$

This demonstrates that the left and right eigenstates of the Cartan matrices for $\mathcal{O}$ transform differently. With the associative subalgebras defined by triplets with indices 124, 346, 615, 523, 371, 574, 672 define the seven sphere S^7 embedded in the 8-sphere in the Hopf fibration. This means that the octonions are determined by the two copies of $SO(8)$ decomposed as $SO(8) \sim S^7 \times SO(7)$

8.3: Quantum Mechanical Interpretation

Physically this has an interesting interpretation. This system is similar to the neutral K-meson or B-meson problem. The octonions correspond to a system that does not have a single set of eigenvalues. Given any state of the system $|\psi\rangle$ the left and right portions of the octonions project out the two sets of states

$$a_l = \langle u_l|\psi\rangle, \ a_r = \langle u_r|\psi\rangle. \quad (8.37a)$$

If these amplitudes obey a Schrödinger equation then

$$i\dot{a}_{l,r} = \lambda a_{l,r}. \quad (8.37b)$$

A dynamical equation for left and right states implies that a left and right state will involves a component of a right and left state. This is analogous to the Schrödinger equation that rotates a states by a phase that moves it through the complex plane. Thus our equation 8.12 will contain terms $V_{lr} = \langle u_l|V|u_r\rangle$ and $V_{rl} = \langle u_r|V|u_l\rangle$ and so

$$i\dot{a}_{l,r} = \lambda a_{l,r} + V_{rl}a_{r,l}, \qquad (8.38)$$

which leads to a "beating" due to the coupling of states without simultaneous eigenstates. This is similar to the $K - \bar{K}$ meson problem [6]

8.4: Eigenvalues, Clifford Algebras and Quaternions

The octonions $\mathcal{O}$ then have independent right and left representations as quaternions $\mathcal{H}$. The octonions then have a left $\mathcal{H}$ vector space and a right $\mathcal{H}$ vector space. Assume for the left $\mathcal{H}$ vector space there are the bases $\{u_1, u_2, \ldots, u_m\}$ and $\{v_1, v_2, \ldots, v_n\}$ related to each other by $U : \{u\} \to \{v\}$,

$$v_i = \sum_{r=1}^{n} U_{ij}u_j, \qquad (8.39a)$$

and for $X = \sum_i x_i v_i$ this vector is transformed into

$$\sum_{i=1}^{m} x_i v_i = \sum_{i=1}^{m}\sum_{j=1}^{n} x_i U_{ij} u_j. \qquad (8.39b)$$

If $m \neq n$ then one basis has eigenvalues λ_i, $i = 1, \ldots, m$ and the other λ_i, $i = 1, \ldots, n$. However, this contradicts the requirement that eigenvalues are distinct. Assume that $n > m$ and $\mathbf{e}_I$ has eigenvalues λ_i, $i = 1, \ldots, m$. This means that equation 8.38b gives

$$A\sum_{i=1}^{m} x_i v_i = \sum_{i=1}^{m} x_i \lambda_i v_i = \sum_{i=1}^{m}\sum_{j=1}^{n} x_i A U_{ij} u_j$$

$$- \sum_{i,k=1}^{m}\sum_{j=1}^{n} x_i U_{ij}\lambda_k U_{kj}^{-1} v_k. \qquad (8.40a)$$

Since j ranges over a larger set than i, k for some set of $j > i$,

$$\sum_{i,k=1}^{m}\sum_{j>i}^{n} x_i U_{ij}\lambda_k U_{kj}^{-1} u_j = \sum_{i=1}^{n} x_i \lambda_i u_i, \qquad (8.40b)$$

there is a set of eigenvalues where $\sum_j U_{ij}\lambda_k U_{kj}^{-1} = \lambda_i$. Hence the eigenvalues are not distinct. Hence the basis set $\{v_i, \ldots v_n\}$ is not linearly independent and there are then a set of $m - n$ of these vectors that are linearly dependent. Thus the transformation on this basis has a square matrix representation. The above result follows similarly for right valued eigenvectors.

It is further obvious that for distinct eigenvalues their respective eigenvectors must be orthogonal. Since $\langle \overline{e_I u}, e_I v \rangle = \lambda \langle \bar{u}, v \rangle \lambda'$ one obtains $\lambda \propto \langle \bar{u}, v \rangle \lambda'$ for two nonorthogonal vectors with eigenvalues that are not distinct. Therefore the eigenvectors must be distinct.

The octonions then satisfy two Jordan canonical form for left and right eigenvectors. For the eigenvalues $\lambda_{l,r}$ there are two forms for $\lambda_{l,r} \in \mathcal{H}_{l,r}$

$$J_\lambda^l = \{u_l : u_l(A - I\lambda) = 0\}$$
$$J_\lambda^r = \{u_r : (A - I\lambda)u_r = 0\} \quad (8.41)$$

Each of these defines an **R** valued vectors space invariant under all transformations

$$J^l[\lambda] = \sum_U J_{(Ue_I)U^{-1}}, \quad J^r[\lambda] = \sum_U J_{U(e_I U^{-1})} \quad (8.42)$$

that act as centralizers

$$C(\lambda) = \{u \in \mathcal{H}, \lambda u = u\lambda\} \quad (8.43)$$

for sums over triplets of basis indices 124, 346, 615, 523, 371, 574, 672 that obtain in these transformations with $U \sim exp(ig_i e_i)$.

The octonions reflect the final Cayley number 1, 2, 4, 8 for the normed division algebras **R**, $\mathcal{C}$, $\mathcal{H}$, $\mathcal{O}$ respectively. This defines spheres with the Cayley numbered dimensions with the Hopf fibrations [1][7]:

$$\begin{aligned} \mathbf{R} &\simeq S^1, \ S^0 \hookrightarrow S^1 \to S^1 \\ \mathcal{C} &\simeq S^2, \ S^1 \hookrightarrow S^3 \to S^2 \\ \mathcal{H} &\simeq S^4, \ S^3 \hookrightarrow S^7 \to S^4 \\ \mathcal{O} &\simeq S^8, \ S^7 \hookrightarrow S^{15} \to S^8 \end{aligned} \quad (8.44)$$

The triality condition for octonions is a map $m : V_1 \times V_2 \times V_3 \to \mathbf{R}$, which gives the zero element for an associative algebra. This type of algebra is generically an $SO(8)$ algebra, with the double cover $spin(8)$ for spinor representations. $Spin(n)$ exists within the Clifford algebra $Cliff_0(n) \subseteq Cliff(n)$. The Clifford algebras for $Cliff(n)$, $n = 1, 2, 4, 8$ are then

$$m_n : V^n \times S_+^n \times S_-^n \to \mathbf{R} \quad (8.45)$$

which determine the elements in equation 8.20. Hence there then exist for the octonions the right and left handed groups. For $S_\pm^8$ the group $SO_\pm(8) \sim S^7 \times spin_\pm(7)$.

The pairing of quaternions in $\mathcal{O} = \mathcal{H} \oplus \mathcal{H}$ may easily be seen with the expansion $\mathbf{o} = \sum_{i=0}^7 o_i \mathbf{e}_i$. Since $\mathbf{e}_5 = \mathbf{e}_2 \mathbf{e}_3$, $\mathbf{e}_6 = -\mathbf{e}_4 \mathbf{e}_3$, $\mathbf{e}_7 = \mathbf{e}_2 \mathbf{e}_3$ this expansion may be written as

$$\sum_{i=0}^7 a_i o_i = a_0 o_0 + a_1 o_1 + a_2 o_2 + a_4 o_4 + (a_3 + a_5 o_2 - a_6 o_4 + a_7 o_2)\mathbf{e}_3, \quad (8.46)$$

8.5: Holographic Principle from the Quantum Theory of Spin

The holographic principle states that fields in a volume of spacetime are determined by fields on the boundary of that space [8][9]. As such measured fluctuations in a length δL are related to the Planck scale according to $(\delta L/L)^3 \simeq (L_p/L)^2$, $L_p = \sqrt{G\hbar/c^3}$ [10]. This principle may be derived from basic notions of spin in quantum mechanics.

Start with the spins in 3-space $\{S_x, S_y, S_z\}$ that are represented by Pauli matrices. The standard equations of motion are

$$\frac{\delta S_i}{\epsilon t} = \frac{i}{\hbar}\omega \epsilon_{ijk} S_j S_k, \tag{8.47}$$

where $\epsilon t \sim \sqrt{G\hbar/c^5} = t_p$. $\delta S/\epsilon t$ is a functional derivative $\lim_{\epsilon \to 0}(S(t+\epsilon t) - \langle S \rangle(t))$ that defines $\delta S = \langle \epsilon t, \delta S/\epsilon t \rangle$, as a physical fluctuation. The frequency of the system is $\omega = (n + \frac{1}{2})\omega_0$, for $\omega_0 = 2\pi/t$, where the periodicity of the system is $t \gg t_p$. The square of the spin fluctuation is

$$\left(\frac{\delta S}{S}\right)^2 = 4\pi^2(n + 1/2)^2 \left(\frac{t_p}{t}\right)^2, \tag{8.48}$$

and is a measure of the fluctuation on the energy surface embedded in phase space. The fluctuation in the volume $\delta V_s \sim (\delta S)^3$ is

$$\delta V_s = -i[\delta S', \delta S''] \cdot \delta S$$

$$= \hbar \delta S''' \cdot \delta S \simeq \hbar \left(\frac{t_p}{t}\right)^2 S^2. \tag{8.49}$$

With $V_s = \hbar^3 \omega^3 t^3$ and $\delta V_s = \hbar^3 \omega^3 (\delta t)^3$ equation 8.49 is

$$\omega^3(\delta t)^3 \simeq 4\pi^2 \left(\frac{S}{\hbar}\right)^2 \left(\frac{t_p}{t}\right)^2. \tag{8.50a}$$

Since $S = (n + \frac{1}{2})\hbar$ this leads to the final result

$$\left(\frac{\delta t}{t}\right)^3 \simeq \frac{(n + \frac{1}{2})^2}{2\pi} \left(\frac{t_p}{t}\right)^2, \tag{8.50b}$$

which is similar to the result for time fluctuations due to the holographic principle.

It is of physical interest that the above used absolutely nothing involving general relativity, except a fundamental cutoff in the scale of spacetime. This is given by the $U(1)$ fibration $\pi^{-1}\mathcal{C}^1 \to \mathcal{C}P^1$. Yet the same result may be derived using general relativity. The above suggests that general relativity and quantum mechanics are identical when it comes to the holographic principle. The $SU(2)$ spinor algebra of general relativity, reduced to the $U(1)$ algebra for horizons, [11] with respect to the diffeomorphism constraint is an event horizon the induces a censorship, which has its analogue in decoherence that acts as a quantum censorship [12]

An analysis of metric fluctuations may be conducted with the Dirac operator. The relationship between the Dirac field and general relativity is easily seen with the "Diracology" $tr(\gamma_\mu \gamma_\nu) = 4g_{\mu\nu}$, for the Dirac operator $i\partial_\mu(\gamma^\mu \psi)$, where γ^μ has a position dependent representation of the Dirac matrices $\gamma_\mu = \gamma_\mu(x)$. The Dirac operator is then expanded as

$$\partial_\mu(\gamma^\mu \psi) = \gamma^\mu \partial_\mu \psi + \Gamma^\mu{}_{\nu\mu}\gamma^\nu \psi, \tag{8.51}$$

where $\Gamma^\mu{}_{\nu\mu}$ is the Christoffel symbol for a spacetime connection. This is consistent with the covariant constancy of the metric $\mathcal{D}_\sigma g^{\mu\nu} = 0$.

The holographic fluctuation of the metric is $\delta g_{\mu\nu} \sim (L_p/L)^{2/3} g_{\mu\nu}$, with $4\delta g_{\mu\nu} = (\delta\gamma_\mu)\gamma_\nu + \gamma_\mu \delta\gamma_\nu$. For the Dirac field there is the conserved current $J_\mu = i\bar{\psi}\gamma_\mu\psi$, where spacetime fluctuations are such that $\delta J_\mu = 0$, or

$$(\delta\bar{\psi})\gamma_\mu \psi + \bar{\psi}\gamma_\mu \delta\psi + \bar{\psi}(\delta\gamma_\mu)\psi = 0. \tag{8.52}$$

Fluctuations of the field are coupled to metric fluctuations and the spinor field is dependent inhomogenously upon local metric fluctuations. This coupling appears questionable, where an additional gauge-like term is required to cancel this inhomogeneity.

Now write the fluctuation according to $\delta = \delta x^\mu (\delta/\delta x^\mu)$ so equation 8.52 becomes a functional differential equation. This functional derivative is not according to a covariant derivative with a geodesic connection. A quantum spacetime fluctuation induces a motion of a test mass that deviates from the standard geodesic equation of motion. The connection is then going to be torsional, or equivalently it will contain a teleparallel connection. This requires a torsional construction of this functional derivative according to connection terms with spinor content so continuity is preserved and ψ fluctuates in a homogeneous manner.

The Wheeler-DeWitt equation is a quantum form of the Hamiltonian constraint equation $H = 0$ in the ADM space plus time formalism of general relativity. The Wheeler DeWitt equation is $\hat{H}\Psi = 0$, where Ψ is the wave functional for a spacial surface and the Hamiltonian operator is

$$\hat{H} = G_{ijkl}\hat{\pi}^{ij}\hat{\pi}^{kl} - \sqrt{g}R^{(3)}, \tag{8.53}$$

and $\hat{\pi}^{ij} = -i\delta/\delta g_{ij}$ are momentum operators conjugate to spacial metric elements. Classically an additional momentum Hamiltonian constraint $H^i = -2\partial_j \pi^{ij} = 0$ is required. ADM relativity treats spacetime as foliated by spacial surfaces that are "pushed" forward by lapse functions. Consider the spacetime manifold $\mathcal{M}$ decomposed into $\Sigma^{(3)} \times \mathbf{R}$, where $\Sigma_t^{(3)}$ is a spacelike region and $\mathbf{R}$ is a real interval [13]. On $\mathcal{M}$ the metric $g_{\mu\nu}$ that is covariantly constant with respect to the operator ∇_μ. The spacetime $\mathcal{M}$ is foliated into spacelike slices $\Sigma_t^{(3)}$, where the parameter is a defined along **R** and defines the vector field $t_\mu = \partial_\mu t$. Thus for some $\Sigma_t^{(3)}$ there exists the future directed normal vector n_μ with the Gauss fundamental forms

$$h_{\mu\nu} = g_{\mu\nu} + n_\mu n_\nu, \; K_{\mu\nu} = -\frac{1}{2}\mathcal{L}_n h_{\mu\nu}. \tag{8.54}$$

Given a projector operator $P_i{}^\mu$ its action on $h_{\mu\nu}$ gives $h_{i\nu}$. $P_i{}^\mu$ projects $h_{\mu\nu}$ from $\mathcal{M}$ to $\Sigma_t^{(3)}$ according to $P_i{}^\mu P_j{}^\nu h_{\mu\nu} = h_{ij}$. This defines Gauss fundamental forms on $\Sigma_t^{(3)}$. The lapse and shift functions for the foliation are then $N = -t^\mu n_\mu$ and $N^i = h^i{}_\mu t^\mu$ with $t^\mu = Nn^\mu + N^\mu$.

The connection term for the fluctuation induced functional derivative should then contain spinor variables in addition to the spatial momentum π^{ij}. This invites the ansatz that fluctuations require a connection term $\delta \to \delta + \Omega$ of the form

$$\Omega = \Gamma_{ij}\pi^{ij} - i\sqrt{g}[\gamma^\mu, \gamma^\nu]\nabla_\mu\psi_\nu, \tag{8.55}$$

such that ψ_j is a Majorana spinor variable, ∇_μ is a covariant derivative and

$$\Gamma_{ij} = \frac{1}{2}(\gamma_i\psi_j + \gamma_j\psi_i). \tag{8.56}$$

This is a gauge term associated with a functional derivative of the Dirac matrix for $\delta\gamma_\mu \to \delta\gamma_\mu + \Omega\gamma_\mu$ and $\delta J_\mu = 0$.

This particular choice for the connection is motivated by the fact that it constructs a spinor version of the Hamiltonian constraint equation by anticommutation

$$\{\Omega(x), \Omega(x')\} = \delta(x - x')\gamma^\mu \hat{\mathcal{H}}_\mu^{spin}, \tag{8.57}$$

where for $\mu = 0$ this is the Hamiltonian operator and $\mu > 0$ gives the Hamiltonian momentum constraint operator. The ordinary Hamiltonian operators are modified by $\hat{\mathcal{H}}_\mu^{spin} = \psi_j \hat{\mathcal{H}}_\mu \psi_j$. These operators vanish when they act upon the wave functional of the universe Ψ. For $|g_{ij}(x)\rangle$ a state vector for the metric the overlap of two nearby metric states is

$$\langle g_{ij}(x)|g_{kl}(x')\rangle = \delta_{ik}\delta_{jl} + \langle g_{ij}^0|\gamma^\mu\hat{\mathcal{H}}_\mu^{spin}|g_{jl}^0\rangle \tag{8.58}$$

This is analogous to the variation $x + \delta x = x + \dot{x}\delta t$ with the variation in action $\delta S = p\delta x - H\delta t$. So for x and x' close the overlap of state vectors gives $\langle g(x)|g(x')\rangle = 1 + \langle g(x)|\gamma^\mu\mathcal{H}_\mu^{spin}|g(x')\rangle$. From completeness sums this will construct a general path integral for the transition between any widely separated metric configurations,

$$\langle g_{ij}|g'_{ij}\rangle = \int \mathcal{D}[g_{ij}]e^{iS/\hbar}. \tag{8.59}$$

The fluctuation operator Ω defines a Hamiltonian that acts upon the cosmological wave function as $\mathcal{H}_\mu\Psi = 0$. That it is derived by equation 8.57 illustrates that Ω is a supersymmetric generator. The Hamiltonian defined by Q and $\bar{Q}$ annules the cosmological wave function, which is probably a state function of a total "nothingness," or the vacuum.

At this stage a spinorial Hamiltonian is derived and $\hat{H} = \int dx \hat{\mathcal{H}}_0 = \langle \hat{H} \rangle + \delta \hat{H}$, where the expected value is zero. The fluctuation satisfies $\hat{H}\Psi = \delta\hat{H}\Psi = 0$. The fluctuation of the metric is shown below to have a topological content. Return to the non-spinor framework for torsion. The topological content is identified with $i\sqrt{g}[\gamma^\mu, \gamma^\nu]\nabla_\mu \psi_\nu$ in equation 8.55, with a cohomology that is identified with the topology of the Hopf fibration.

A torsional connection is one where there is exists an acceleration that deviates the motion of a particle from a geodesic. Accelerations in spacetime are $\mathbf{a} = d\mathbf{u}/d\tau$, with $\mathbf{u}^2 = -1$ and $\mathbf{a} \cdot \mathbf{u} = 0$. Assume that an observer in a rest frame observes this accelerated system so at that instant $\mathbf{u} = \mathbf{n}$. The accelerated frame is seen according to the covariant constancy of the metric $h_{\mu\nu}$

$$\frac{Dh_{\mu\nu}}{d\tau} = 0 = \frac{Dn_\mu}{d\tau} n_\nu, \tag{8.60}$$

where the second fundamental form may be expressed with the Lie derivative as,

$$\mathcal{L}h_{\mu\nu} = n_{\mu;\nu} + n_{\nu;\mu} + n_\mu a_\nu + n_\nu a_\mu. \tag{8.61}$$

The acceleration is then

$$a^\mu = n^\nu \nabla_\nu n^\mu, \tag{8.62}$$

with ∇_ν being the covariant derivative. The covariant derivative of the normal is

$$n_{\mu;\nu} = \frac{1}{2} n_{(\mu;\nu)} + \frac{1}{2} n_{[\mu;\nu]} = -K_{\mu\nu} - n_\mu a_\nu + \frac{1}{2} n_{[\mu;\nu]}. \tag{8.63}$$

$n_{[\mu;\nu]}$ is the antisymmetric $\nabla_{[\mu}\psi_{\nu]}$ term in equation 8.63. For an accelerated particle the normal n^μ pertains to a classical spacial manifold, with $t_\mu = \partial_\mu t$ and $n_{[\mu;\nu]} = 0$ so the fluctuation vanishes classically.

Let the normal n_μ pertains to a manifold $\Sigma^{(3)}$ within a foliation of spacial manifolds that is one "path" within a quantum superposition of foliations and spacial manifolds. This foliation is defined by a path integral for the transition from g_{ij} to g'_{ij}. This foliation in the spacial metric indicates the existence of quantum fluctuations in the spacial manifold that have their classical analogues in variations. The variation in the action is defined between the volumes $\Sigma^{(3)}$ and $\Sigma'^{(3)}$. This variation is a bubble in spacetime $\delta \mathbf{n} \cdot \mathbf{\Sigma}^{(3)}$, where $\mathbf{\Sigma}^{(3)}$ is a vector dual to the volume form for $\Sigma^{(3)}$. The standard action

$$S = \int d^4x (-g)^{-1/2} R = \int d^4x (-g)^{-1/2} \left((TrK)^2 - Tr(K^2) + R^{(3)} \right) \tag{8.64}$$

is derived from

$$S = \int d^4x (-g)^{-1/2} \left(n^\mu{}_{[;\nu} n^\nu{}_{;\mu]} + R^{(3)} \right), \tag{8.65}$$

where $n^\mu{}_{[;\nu} n^\nu{}_{;\mu]}$ is the kinetic energy component of the Lagrangian. The variation in the action results in

$$g^{\rho\nu} n_{\sigma\nu\rho} - n^\nu{}_{;\nu\sigma} + \frac{1}{2} R^{(3)}{}_{;\sigma} = 0, \tag{8.66}$$

and
$$K_{\mu\nu;\rho} - g_{\mu\nu}K_\sigma{}^\sigma{}_{;\rho} + (n_\mu a_\nu)_{;\rho} - \frac{1}{2}n_{[\mu;\nu];\rho} = 0. \tag{8.67}$$

The antisymmetric term $n_{[\mu;\nu]} = \nabla_{[\mu}\psi_{\nu]}$ are torsional metric fluctuations. Consider a fluctuation in the scalar $t = t_0 + \delta t$, where δt may be complex valued in general. The time vector $t^\mu = t_0{}^\mu + \delta t^\mu$, where δt^μ is a fluctuation in the vector field, is physically motivated by above discussions of quantum fluctuations. As $t^\mu = Nn^\mu + N^\mu$ then $\delta n^\mu = \frac{1}{N}\delta t^\mu$ and $\delta n^\mu = n^\mu{}_{;\nu}\delta x^\nu$. It is evident that $dt = t_\mu dx^\mu$, but with fluctuations $t_\mu dx^\mu = (t_{0\mu} + \delta t_\mu)dx^\mu$. Since the fluctuation δt_μ is not determined by $K^i{}_j$ it is not the result of the application of a differential operator on t. By the definition of n^μ the same applies. There is a cocyle associated with both t^μ and n^μ. This fluctuation is a loop that takes t_0 to $t = t_0 + \delta t$ and back, as parameterized according to an additional variable such as a scalar field. A loop integral $\oint dt$ is by Stokes' law $\int t_{\mu;\nu}dx^\mu \wedge dx^\nu$. This evaluates the antisymmetric tensor $n_{[\mu;\nu]} = \nabla_{[\mu}\psi_{\nu]}$ on a two-chain. Since quantum fluctuations vanish under expectations $\langle \delta t \rangle = 0$, expectations of this antisymmetric term in turn vanishes. However, quadratic terms in the fluctuation and this antisymmetric tensor are delta function correlated which indicate Markovian statistics. Markovian statistics give $\langle (\delta t)^2 \rangle \neq 0$. t^μ is dual to Σ^3 and so $\oint dt$ is defined by $\mathcal{H}^3(M, \mathbf{R}) = Z$. This result has a connection to the Hopf fibration and Bott periodicity.

The above analysis illustrates a cohomology associated with fluctuations, as well as the holographic principle. This is related to the antisymmetric term $i\sqrt{g}[\gamma^\mu, \gamma^\nu]\nabla_{[\mu}\psi_{\nu]}$ with a $U(1)$ content. The theory for quaternionic quantum gravity is $SL(2, \mathcal{Q})$ and the $U(1)$ fibration defines a $PSL(2, \mathcal{Q})$ projective space. This construction is parallel to the structure for the nonrelativistic spin $1/2$ theory above.

The nonrelativistic spin $1/2$ theory of fluctuations (equation 8.47) is defined by the Hopf fibration

$$S^1 \hookrightarrow S^3 \to S^2, \tag{8.68}$$

which defines the holographic fluctuations. The two-sphere is the energy surface and the three sphere is the the phase space. The Lagrangian is a constraint that defines this fibration. In general relativity the $U(1)$ fiber corresponds to fields on the event horizon that define a Hilbert space $\mathcal{H}_e$ [11]. The fields in the volume removed from the event horizon obey $SO(3,1) \times Z_2 \simeq SL(2,\mathcal{C})$ or $SU(2) \times SU(1,1)$, where these define the Hilbert space $\mathcal{H}_v$. The $SU(2)$ valued fields are reduced to $U(1)$ valued fields on an event horizon. The total Hilbert space for quantum gravity would then be $\mathcal{H}_v \otimes \mathcal{H}_e$. The $SU(2)/Z_2 \simeq SO(3)$ valued fields are constrained to exist on a two sphere with a bundle connection defined by the Hopf fibration. As demonstrated above spin systems generically have a similar structure to general relativity, which means the holographic principle obtains from quantum mechanics with no references to spacetime physics.

The spinorial theory of spacetime fluctuation has the same basic structure, but where the fluctuations are written according to connection terms. The spinorial fluctuations of spacetime is also an example of the Hopf fibration [1][7]

$$S^3 \hookrightarrow S^7 \to S^4 \tag{8.69}$$

The holographic principle is inherited by quaternions through the Hopf fibration by

$$S^1 \hookrightarrow S^3 \to S^2$$
$$\downarrow$$
$$S^3 \hookrightarrow S^7 \to S^4 \tag{8.70}$$

The quaternionic projective map by the Hopf construction inherits holographic fluctuations with small but measurable effects on a large scale.

The fluctuation in the time vector further is an indication of a fundamental property of the Hopf fibration, the Bott periodicity. R. Bott computed the the homotopy groups of the topological groups $O(n)$ for $n \to \infty$ [1]. This resulted in a periodicity $\pi_{n+8}(O(\infty)) = \pi_n(O(\infty))$ with specific values [1]:

$$\pi_0(O(\infty)) = Z_2, \ \pi_1(O(\infty)) = Z_2, \ \pi_2(O(\infty)) = 0, \ \pi_3(O(\infty)) = Z$$

$$\pi_4(O(\infty)) = 0, \ \pi_5(O(\infty)) = 0, \ \pi_6(O(\infty)) = 0, \ \pi_7(O(\infty)) = Z. \tag{8.71}$$

The division algebras on the reals, complexes, quaternions and octonions are formed from vector spaces with dimensions 1, 2, 4, 8 and Cayley numbers by [1],

$$\mathbf{R} \to \mathbf{R}, \ \mathbf{R}^2 \to \mathcal{C}, \ \mathbf{R}^4 \to \mathcal{H}, \ \mathbf{R}^8 \to \mathcal{O}, \tag{8.72}$$

where $\pi_3(O(\infty)) = Z$ indicates the quaternions [1]. The map $S^3 \to O(\infty)$ defines the orthogonal transformations of the division algebra for the quaternions. The homotopy group determines the topology in that dimension. The cohomological nature of the time vector is given by $H^1(\mathcal{M}, \mathbf{R}) \sim \pi_1(\mathcal{M})/commutators$ with the dual $\pi_3(\mathcal{M}) \to \pi^3(\mathcal{M}) \to \pi_3(O(\infty))$.

The above model of quantum gravity is quaterionic, which shares problems inherent with quantum gravity models that have been advanced. Essentially this type of model in a strong coupling limit exhibits difficulties with divergences. This has been the case with string theories at the Hagedorn temperature as well. The above construction indicates this theory for quantum metric fluctuations is purely topological. That this type of theory has topological information similar by Bott periodicity suggests this is only an approximate theory that emerges from a lower level in the Hopf fibration with octonions. In such an approach there is nonassociativity between field operators, which gives a new level of uncertainty in physics. Within this extended uncertainty the divergences that obtain in current theories are absorbed. This would have parallels with the how quantum uncertainty absorbed the infinities in classical mechanical attempts to understand atomic physics.

The octonions $\mathcal{O}$ are a nonassociative and noncommutative set of basis elements of a normed division algebra. Any octonion may then be written as $o = \sum_{i=1}^{8} o_i e_i$. The basis elements e_i, $i = 1, \ldots 8$ are then various square roots of -1 with $e_i^2 = -1$. The anticommutivity of octonions $e_i e_j = -e_j e_i$ leads to the general product rule $e_i e_j = v_{ijk} e_k - \delta_{ij}$ so that a triplet product may be differently associated

$$(e_i e_j) e_k = g_{ik} e_j - g_{jk} e_i - \delta_{ij} e_k$$
$$e_i (e_j e_k) = g_{ik} e_j - g_{ij} e_k - \delta_{jk} e_i, \quad (8.73)$$

where v_{ijk} are to be determined and g_{ij} is the metric in seven dimensions determined by $v_{ijm} v_{klm} = (g_{ik} g_{jl} - g_{il} g_{jk})$. This gives the result

$$(e_i e_j) e_k - e_i (e_j e_k) = (g_{ij} - \delta_{ij}) e_k - (g_{jk} - \delta_{jk}) e_i. \quad (8.74)$$

This multiplication rule produces the Fano plane. This defines the triality associative product or bracket $[a, b, c] = (ab)c - a(bc)$

Any octonion o may be written as the linear combination

$$o = o_0 \mathbf{1} + \sum_{i=1}^{7} o_i e_i. \quad (8.75)$$

Octonion conjugation is defined according to $\overline{ab} = \bar{b}\bar{a}$, with an inner product on this basis on $\mathcal{R}^8$

$$\langle a, b \rangle = \sum_i a_i b_j = \frac{1}{2}(ab + ba) = \frac{1}{2}(\overline{b\bar{a}} + \overline{\bar{a}b}) \quad (8.76)$$

Octonions are a pairing of quaternions. The elements of $\mathcal{O}$ are then formed by $\mathcal{Q} \oplus_p \mathcal{Q}$ and defined 16 elements. These elements are not matrices, since matrices are associative. This will lead to two copies of $\mathcal{Q}$ where the nontrivial elements of each are paired up with the unit matrix $\gamma_4 = 1_{4 \times 4}$. The octonions have 16 elements as are pairs of identical matrices. There are 8 independent matrices. One of these matrices is the unit matrix, where the seven remaining elements are the "square root" of the negative unit matrix.

An octonionic quantum gravity would possess the octonionic projective map on $SL(2, \mathcal{O})$ and will be defined by the Hopf fibration

$$S^7 \hookrightarrow S^{15} \to S^8. \quad (8.77)$$

The holographic fluctuations will obtain in the octonionic domain as they do in the quaternionic domain. This mean that the initial discussion on how quantum theory and gravitation are similar by their results on length fluctuations is fully realized when quaternionic quantum theory and spinorial gravitation are "democratically" embedded into a single system.

292 *Quantum Fluctuations of Spacetime*

The Cayley numbers 1, 2, 4, 8 define the sets **R**, $\mathcal{C}$, $\mathcal{Q}$, $\mathcal{O}$ as division algebras where for $xy = 0$ then either $x = 0$ or $y = 0$. The reals are well ordered, commutative and associative. $\mathcal{C}$ loses the well ordered property, $\mathcal{Q}$ loses commutativity and $\mathcal{O}$ loses associativity. The octonions are a nonassociative set of basis elements $\{1, e_i\}$, $i = 1, \ldots, 7$ defined on the Fano plane. For a choice of multiplication table, elements with the indicial triples 124, 346, 615, 523, 371, 574, 672, are are the nonzero octonion cyclic elements. In general octonions obey the nonassociative rule

$$(e_i e_j)e_k - e_i(e_j e_k) = v_{ijkl} e_l. \tag{8.78}$$

The associator is evaluated as $(e_i e_j)e_k = -e_i(e_j e_k)$ for associative triplets of basis elements e_i. Here v_{ijkl} is a structure constant for the multiplication of octonions. Hence there is a context sensitivity associated with multiplication dependent upon whether the ( ) is on the left or the right. This structure is demonstrated to give rise to right and left eigenvectors for the elements e_4, e_7, which are $SL(2, \mathcal{Q})$ eigen-operators. A Galois code emerges from the octonions from this that has deep algebraic content for graphic and combinatoric representations. [14]

The discussion on ADM relativity focused on the torsional aspects of quantum fluctuations. The nonassociativity of octonions is a measure of the torsional nature of in the geometry of the Hopf fibration. The seven associative subgroups identified below indicate quaternions that have zero nonassociative torsion, but associators across elements of these triplets exhibit torsion where geodesic description according to commutators fails. The Hopf fibration in fact indicates maps from S^0, S^1, S^3 and S^7 respectively to S^1, S^2, S^4 and S^8 define holonomies of a line bundle and are a S^1 with a $U(1)$ Lie algebra. There are maps

$$\mathbf{R}P^1 \to S^1, \ \mathcal{C}P^1 \to S^2, \ \mathcal{Q}P^1 \to S^4, \ \mathcal{O}P^1 \to S^8 \tag{8.79}$$

for which there is an infinite set of geodesics on each S^n restricted to $n = 1, 2, 4, 8$.

8.6: Associative and Co-Associative SubAlgebra and Spacetime

The first part of this section is largely due to R. Betts.

Let ϵ^{ijk} denote the usual permutation depending on how i, j, k permute. Thus to illustrate allow $ijk = 615$ to be an even permutation equal to $+1$ as "6" moves before "1" in an even number of steps in $(1, 2, 3, 4, 5, 6, 7)$. Since it takes an index such as 7 an odd number of steps to move before 1 $\epsilon^{715} = -1$

Within the commutative algebra $*e_i \wedge e_j \wedge e_k = \epsilon_{ijk}{}^l * e_l$, be equivalent to $(*e)_{ijk} = (e_{ijk}{}^l)e_l$. For each commutative algebra there exist seven Hodge star operators:

$$(*e)_{124} = (e_{124}{}^l)e_l \to (e^{1244})e_4 = (e^{12})e_4 = e_4,$$

for contraction in the upper and lower indices "4,"

$$(*e)_{346} = (e_{346}{}^l)e_l \to (e_{346}{}^6)e_6 = (e_{34})e_6 = e_6,$$

$$(*e)_{615} = (e_{615}{}^l)e_l \to (e_{615}{}^5)e_5 = (e_{61})e_5 = e_5,$$
$$(*e)_{523} = (e_{523}{}^l)e_l \to (e_{523}{}^3)e_3 = (e_{52})e_3 = e_3, \qquad (8.80)$$
$$(*e)_{371} = (e_{371}{}^l)e_l \to (e_{371}{}^1)e_1 = (e_{37})e_1 = e_1,$$
$$(*e)_{457} = (e_{457}{}^1)e_l \to (e_{745}{}^7)e_7 = (e_{45})e_7 = e_7,$$
$$(*e)_{672} = (e_{6721})e_l \to (e_{672}{}^2)e_2 = (e_{67})e_2 = e_2.$$

Now compare those seven results using the Hodge star operators with the seven octonionic products:

$$(e_1)(e_2) = e_4, \ (e_3)(e_4) = e_6, \ (e_6)(e_1) = e_5, \ (e_5)(e_2) = e_3,$$
$$(e_3)(e_7) = e_1, \ (e_4)(e_5) = e_7, \ (e_6)(e_7) = e_2. \qquad (8.81)$$

The Hodge star operator on the elements of the seven algebras the seven octonion products lead to the same result. To show that this is no accident, prove the following Proposition:

Proposition: Let $\{1, e_1, e_2, e_3, e_4, e_5, e_6, e_7\}$, where $(e_i)^2 = -1$, $i = 1, 2, \ldots, 7$, be the set of octonionic unit vectors for the algebra of octonions $\mathcal{O}$. Furthermore, let $\{i, j, k\}$, or (124), (346), (615), (523), (371), (574), (672) be seven 3-cycles on (ijk) standing for seven triplets $\{e_i, e_j, e_k\}$ from the seven unit octonionic vectors which satisfy the triality "associator" condition:

$$[e_i, e_j, e_k] = (e_i e_j)e_k - e_i(e_j e_k) = 0. \qquad (8.82)$$

For each of these seven triplets an endomorphism from $\mathcal{O}$ to $\mathcal{O}$ exists. The octonion product is mapped by the Hodge star operator.

First the following definition is needed:

DEFINITION: An algebraic structure is a nonempty set E with a binary operation $*$, for the elements of E.

Consequently the set e_i, e_j, e_k as defined in the Proposition with the operation $*$ constitutes an algebraic structure.

Proof: First define, exclusively for these seven triplets, an endomorphism map

$$\phi : \mathcal{O} \to \mathcal{O} \qquad (8.83)$$

from the algebra of octonions $\mathcal{O}$ to $\mathcal{O}$:

$$(e_i) * (e_j) \to \phi(e_i)\phi(e_j) = (*e)_{ijk}. \qquad (8.84)$$

where $(*e)_{ijk}$ denotes the Hodge star operator. Suppose for each of the seven triples $\{e_i, e_j, e_k\}$ satisfying the triality associator condition, $(e_i)(e_j) = (e_k)$, where the $\{i, j, k\}$

are the seven 3-cycles as specified in the Proposition. The map ϕ is well-defined by the following:

$$(e_i) * (e_j) \;\to\; \phi(e_i)\phi(e_j) = (*e)_{ijk} = (e_{ijk}{}^l)e_l \;\to\; (e_{ijk}{}^k)e_k = (e_{ij})e_k = e_k,$$

$$(e_i * e_j) = e_k \;\to\; \phi(e_i * e_j) = \phi(e_k) \tag{8.85}$$

$$= *((*e)_{ijk}) = (-1)^{7-1}e_k = (-1)^6 e_k = e_k,$$

where the end result is from the property of the double Hodge star operator ** is used. Hence the map ϕ is well-defined and the product in the octonionic algebra is carried over to a Hodge star operator, for these specific seven triplets from $\mathcal{O}$.

Physics has been proposed as a more fundamental basis for quantum field theory [15]. Commutative quantum field theory is defined within the seven associative 3-cycles. Consider the consequences of these seven subalgebras. There exist seven triplets denoted by these seven 3-cycles and all which satisfy the associator condition can also denote seven 3-spheres S^3 with vectors $\{e_i,\ e_j,\ e_k\}$ which can exist on a four dimensional manifold, meaning for computation of the Hodge Star Operator with $n = 4$ and $p = 1$ 3-forms are converted into 1-forms on the 4-manifold. For $\mathbf{P}^1 \times S^7$ these subalgebras are $\mathbf{P}^1 \times S^3$. With the projective real this includes a unit element with the three associative octonionic elements. This is a perfect model for spacetime of general relativity. In eight dimensions these constitute different realizations of a four dimensional subspace of S^8. An associative three-form is then defined as

$$\Psi(x,\ y,\ z) = \langle x,\ yz \rangle \tag{8.86}$$

The remaining components are then nonassociative elements of S^8. This nonassociativity means that the structure constant for any algebraic structure is not constant so that for $[e_i,\ e_j] = 2e_i e_j = t_{ijk}e_k$ so that the associator gives

$$(e_i e_j)e_k - e_i(e_j e_k) = t_{ijl}e_l \mathbf{e_k} - e_i t_{jkl}e_l \tag{8.87}$$

and there is a commutator $[e_i,\ t_{ijk}] \neq 0$. This means that there is a teleparallel connection or torsion associated with nonassociativity. To define the nonassociative half of S^8 according to an associative algebra Define the four-form

$$\Phi(w,\ x,\ y,\ z,) = \frac{1}{2}(\langle w,\ x(\bar{y}z)\rangle - \langle w,\ (\bar{z}y)\rangle), \tag{8.88}$$

over the coassociative algebra dual to the associative subspace [16]. In the methods often used in physics the full eight dimensional space will have the two dual operators $\Psi \pm i\Phi$ as the twistor description of the spacetime.

The spacetime is then $\Psi = \mathcal{R}e\mathcal{C}^4$, which has the group action $U(4)$. This is decomposed into $U(4) = spin(6) \times U(1)$. The 15 dimensional $spin(6)$ is a conformal group that contains the $spin(5) \sim SO(3,2)$ de Sitter group, which defines spacetime and gravitation. The remaining 5 parameters from the $spin(6)$ are a diliton factor and four conformal factors. Physically this has apparent connections to the problem of AdS and conformal field connection.

The coassociative space, for $S^8 \sim spin(8)$ is then given by $spin(8)/U(4)$, which is not a gauge group description. However, this 12 dimensional space may represent the weights of the gauge group. A choice of the octonions as a basis for this space is

$$\{1,\ i,\ j,\ ,k, (\pm 1,\ \pm i,\ \pm j,\ \pm)\}, \tag{8.89}$$

which is the sum of the 8 quaternions $\{\pm 1,\ \pm i,\ \pm j,\ \pm\}$, the 3 quaternions $\{i,\ j,\ ,k\}$ and the unit term 1. This suggests that the co-associative space gives the weights for $SU(3) \times SU(2) \times U(1)$[16].

8.7: Polyhedra and Jordan Algebraic Representations of Octonions

The octonions are a basis of elements that define

$$\sum_i^7 (e_i)^2 = -1 \tag{8.90}$$

There are $7! = 2 \times 3 \times 4 \times 5 \times 6 \times 7 = 5040$ permutation changes, with $2^7 = 128$ with possible sign changes. These contain the set of possible multiplication tables for octonions. The subalgebras are then a subset of these permutations. The set of $128 \times 7!$ over determine the multiplication tables, since the $2^3 = 8$ changes of e_1, e_2, and e_4 do not give a different multiplication. The number of different multiplications due to sign changes and permutations is then:

$$\frac{128 \times 2 \times 3 \times 4 \times 5 \times 6 \times 7}{8 \times 2 \times 2 \times 2 \times 3 \times 7} = 16 \times 5 \times 6 = 480 \tag{8.91}$$

The 480 multiplications consist of two sets of 240 multiplications, where products in one is in reverse order in the other set. The two sets of 240 multiplications are called sets of opposite multiplications. These correspond to multiplication orders $c_i c_j$ and $e_j e_i$.

$PSL(2,\ 7)$ is the 168-element simple group that is the central quotient group of $SL(2,\ 7)$. $SL(2,\ 7)$ is the 336-element group of discrete rotations. It consists of 2×2 matrices with determinant 1 with entries that define the finite group Z_7. Z_7 is the cyclic group of 7 letters, or the integers $mod\ 7$. Z_7 is a represented by the vertices of a heptagon and $PSL(2, 7)$ is the group of the vertices of the heptagon, which represents the Fano plane diagram.

Consider the Fano plane diagram "blown up" into an octahedron. One exterior triangle face will have the e_1, e_2, e_4 quaternionic cycle on it, the others will contain the e_2, e_3 and e_5, e_3, e_4, e_6 and e_1, e_5, e_6 nodes. The e_7 nodes sits at the center of the octahedron. These four triangles on the exterior correspond to lie algebraic endomorphisms. The three interior edgelinks connected to the e_7 node are also Lie algebraic endomorphisms as well.

296 *Quantum Fluctuations of Spacetime*

There are 8 tetradedra in this solid, 18 edgelinks and 7 nodes. The 3 − 0, 2 − 1 dual polytope will then contain 8 nodes as the 0-form is dual to the 3-form in 3 dimensions This is a cube without a central node. Similarly 4 triangles for the lie algebraic endemorphism have dual edgelinks. Similarly the edgelinks in the octahedron will have dual faces on the cube. If the cube is tesselated by tetrahedra each edgelink will attach to another by a unique tetrahedron. By considering internal tetrahedra one arrives at a dual cube with opposite faces that stand for the internal lie endomorphism. A second dual is the 2 − 0, 1 − 1 duality that again produces a cube, but where the triangular faces of the octahedron are associated with vertices or nodes. The interior triplet groups form edgelinks that connect these nodes. There is a similar octahedral representation for the co-associative algebras as well.

This suffers from an obvious problem in that the 7 triplets are not geometrically equivalent. There are either 4 faces and 3 edgelinks or where these roles are switched. However, recall this is a projection from $spin(8)$ onto $spin(7)$. For $spin(8)$ there is a similar diagram to the bottom cube, but now in four dimensions has become a tesseract. There exist 8 of these triplets, where 4 are on separate vertices of the two cubes in the tesseract!

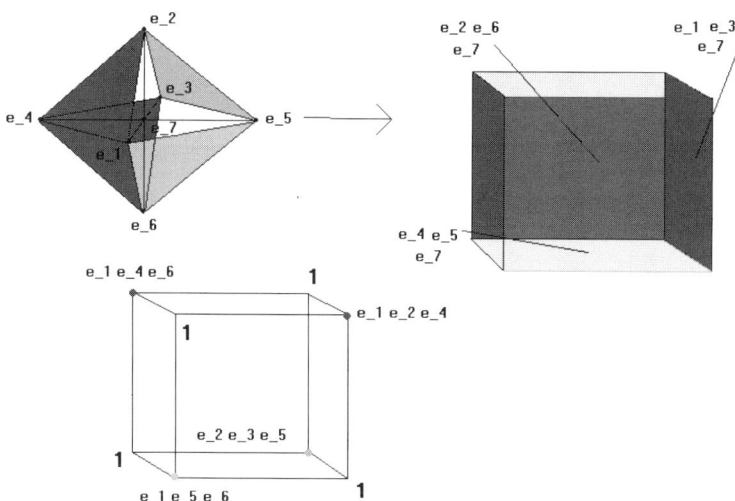

Octahedral representation of the Fano plane and its dual.
Figure 8.2

The 480 products are determined by $Z_2^7 \times (S^7/Z_2)^3 \times PSL(2, 7)$. Each of the 480 products has an internal symmetry group $Z_2^3 \times PSL(2, 7)$ so that $PSL(2, 7)$ is a symmetry group of the octonion product. $PSL(2, 7)$ is double covered by 336-element $SL(2, 7)$. $SL(2, 7) = Z_2 \times PSL(2, 7)$ is also a symmetry group of a given octonion product. The 168-element simple group $PSL(2, 7)$ contains the 24-element octahedral $\{4, 3, 2\}$ subgroup. $SL(2, 7)$ has a 48-element binary octahedral $\{4, 3, 2\}$ subgroup. The binary octahedral $\{4, 3, 2\}$ group is a subgroup of order 7 in $SL(2, 7)$. For this octonion product, there are 7 associative 3-dimensional quaternionic triangles and correspond to the 7 coset spaces of $SL(2,\ 7)/\{4,3,2\}$.

Each of these tripets is then a local representation of the Jordan group $J^3(V)$ with a matrix of the form
$$\begin{pmatrix} 0 & a & b \\ 0 & 0 & a^\dagger \\ 0 & 0 & 0 \end{pmatrix} \quad (8.92)$$
the a and $a^\dagger$ are the annihilation and creation operators and their commutator at each node determines the b. These are elements of $S^7 \times P^1$, where the b is uniquely determined by a commutator of the a and $a^\dagger$. However in S^8 this does not completely hold in general
$$[e_i,\ e_j] = 2e_i e_j = T_{ijk} e_k \quad (8.93a)$$
and
$$e_i(e_j e_k) = e_i T_{jkl} e_l, \quad (8.93b)$$
but the structure constant is outside of the tesseract nodes depends on the e_i elements. This is the source of nonassociativity, where this nonassociative part is then dual to an identical co-associative part. As such in general on the S^7 there is the fibration of $G_2 \to S^7$, where the S^8 has this exceptional algebraic structure [17].

At this point now construct the 3×3 Octonion matrices. These are are $9 \times 8 = 72$-dimensional.
$$\begin{pmatrix} e_1 & e_4 & e_5 \\ e_7 & e_2 & e_6 \\ e_8 & e_9 & e_3 \end{pmatrix}. \quad (8.94)$$
This general matrix is over determined with an e_8 and e_9. These matrices correspond to the vertices of the E_6 root vector polytope. The Jordan matrices are then 3×3 Hermitian Octonion matrices are $3 \times 8 + 3 \times 1 = 27$-dimensional [17].
$$\begin{pmatrix} Re(e_1) & e_4 & e_5 \\ e_4^* & Re(e_2) & e_6 \\ e_5^* & e_6^* & Re(e_3) \end{pmatrix}. \quad (8.95)$$
They form the exceptional Jordan algebra $J^3(\mathcal{O})$. The 26-dimensional traceless subalgebra $J^3(\mathcal{O})$ can represent the 26-dimensional bosonic String. The anti-Hermitian matrices defines the gauge theory as 3×3 Octonion matrices in $3 \times 8 + 3 \times 7 = 45$-dimensional.
$$\begin{pmatrix} Im(e_1) & e_4 & e_5 \\ -e_4^* & Im(e_2) & e_6 \\ -e_5^* & -e_6^* & Im(e_3) \end{pmatrix}. \quad (8.96)$$
These must be restricted to $45 - 7 = 38$ dimensional tracelss matrices of to represent a Lie algebra. This is accomplished with the addition of the 14-dimensional octonion derivation algebra G_2:
$$\begin{pmatrix} Im(e_1) & e_4 & e_5 \\ -e_4^* & Im(e_2) & e_6 \\ -e_5^* & -e_6^* & 0 \end{pmatrix} \times G_2. \quad (8.97)$$

The resulting Lie algebra is the $38 + 14 = 52$-dimensional F_4. The 26-dim traceless part of $J^3(\mathcal{O})$ can be combined with F_4

$$\begin{pmatrix} Im(e_1) & e_4 & e_5 \\ -e_4^* & Im(e_2) & e_6 \\ -e_5^* & -e_6^* & 0 \end{pmatrix} \times G_2 \times \begin{pmatrix} Re(e_1) & e_4 & e_5 \\ e_4^* & Re(e_2) & e_6 \\ e_5^* & e_6^* & 0 \end{pmatrix} \qquad (8.98)$$

to make the $26 + 52 = 78$ dimensional Lie algebra E_6.

Now consider the case with three independent off diagonal elements. The $J^3(\mathcal{C})$ is a matrix of the form

$$\begin{pmatrix} 1 & 2 & 2 \\ 0 & 1 & 2 \\ 0 & 0 & 1 \end{pmatrix}, \qquad (8.99)$$

where the antisymmetric matrix for the Lie Algebra $L^3(\mathcal{C})$ is of the same form. This corresponds to $U(3) = SU(3) \times U(1)$. The "2" on the $1-2$ and $2-1$ entries correspond to a and $a^\dagger$, and the 3 is from $[a, a^\dagger]$. If the $U(1)$ is removed the J^3 and L^3 matrices must be traceless. For octonions the Jordan algebra is

$$\begin{pmatrix} 1 & 8 & 8 \\ 0 & * & 8 \\ 0 & 0 & 1 \end{pmatrix}, \qquad (8.100a)$$

with the antisymmetric matrix

$$\begin{pmatrix} 7 & 8 & 8 \\ 0 & * & 8 \\ 0 & 0 & 7 \end{pmatrix} + 14. \qquad (8.100b)$$

The $*$ represents a term required to give $tr(J^3) = tr(L^3) = 0$. Thus $J^3(\mathcal{O})$ is a 26 dimensional algebra and $L^3(\mathcal{O})$ is 38 dimensional. There is a double covering relation between the J^3 elements as the conformal group and the gauge theory as L^3 elements. Therefore an additional a 14 dimensional automorphism is attached toto L^3, which is the exceptional group G_4. This is extended to $\mathcal{O} \times \mathcal{C}$ which as the J^3 group

$$\begin{pmatrix} 2 & 16 & 16 \\ 0 & * & 16 \\ 0 & 0 & 2 \end{pmatrix} \text{ AND } \begin{pmatrix} 8 & 16 & 16 \\ 0 & * & 16 \\ 0 & 0 & 8 \end{pmatrix} + 14, \qquad (8.101)$$

which requires the addition of a 14-dim group, G_4 to give a 78 dimensional group that corresponds to the 34 dimensional $L^3(\mathcal{C} \times \mathcal{O})$. This is due to the addition of one dimension from $\mathbf{R} \to \mathcal{C}$. Further, these extensions contain a $2 \times 26 + 52$ dimensional object in 130 dimensions, but the J^3 must be 133 dimensional. Hence an additional 3 dimensions must be added. Thus the $L^3(\mathcal{H} \times \mathcal{O})$ is

$$\begin{pmatrix} 10 & 32 & 32 \\ 0 & * & 32 \\ 0 & 0 & 10 \end{pmatrix} + 14 + 3. \qquad (8.102)$$

The $L^3(\mathcal{H} \times \mathcal{O})$ requires the addition of a 17 dimensional group that is G_4 plus $SU(3)$. Finally, for $\mathcal{O} \times \mathcal{O}$ there is $7 \times 26 + 52 = 234$ and an additional **14** for G_4 is required. This then gives

$$\begin{pmatrix} 14 & 64 & 64 \\ 0 & * & 64 \\ 0 & 0 & 14 \end{pmatrix} + \mathbf{14} + \mathbf{14}, \tag{8.103}$$

for $L^3(\mathcal{O} \times \mathcal{O})$. These are the F_4 for $\mathcal{C} \times \mathcal{O}$, the E_6 for $\mathcal{H} \times \mathcal{O}$ and E_8 for $\mathcal{O} \times \mathcal{O}$.

Finally there is an obvious pattern with all of this extended to the exceptional group. There are in each family of elementary particles two leptons (neutrino plus the electron) and two quarks. However, the quarks have 3 colors. This makes a total of 8 types of particles. Further, there are three types of particle families, eg, the triplet description with $J^3(V)$. This appears related to the problem of families of elementary particles.

8.8: Nonassociative Fields and Topological Supersymmetry

Octonionic quantum field theory appears to naturally embed supersymmetry. The octonions $\mathcal{O}$ is E_8, which is one half of the heterotic string theory. A more physical approach to supersymmetric octonionic field theory will necessarily involve the physics of fermions and bosonization of spin fields. The extension of a boson field to a supersymmetric pair $\xi \cdot \psi$, for ξ a supergenerator, is similar to the coupling of two fermion fields into a boson field. A connection between the octonions and physics may then be established within a toy model.

The Dirac algebra is a 32-part algebra produced with the combination of the four-vector units $(i, \mathbf{i}, \mathbf{j}, \mathbf{k})$ with the unit quaternions $(1, e_1, e_2, e_3)$. Physically the two sets denote spacetime and mass-charge parameters. These enter into the dynamics of a relativistic particle with the relativistic momentum-energy invariant interval

$$m^2 = E^2 - p^2. \tag{8.104}$$

This may then be factorized according to the quaternion elements with the additional phase term $e^{-i(et - \mathbf{p} \cdot \mathbf{x})}$, so this interval assumes the form

$$(\pm e_3 E \pm i e_1 \mathbf{p} + i e_2 m)(\pm e_3 E \pm i e_1 \mathbf{p} + i e_2 m) e^{-i(et - \mathbf{p} \cdot \mathbf{x})} = 0. \tag{8.105}$$

By the standard quantization rule $E \to i\partial/\partial t$ and $\mathbf{p} \to -i\nabla$ the energy momentum interval is reproduced as

$$\left(\pm i e_3 \frac{\partial}{\partial t} \pm \mathbf{e}_1 \nabla + i e_2 m\right)(\pm e_3 E \pm i e_1 \mathbf{p} + i e_2 m) e^{-i(et - \mathbf{p} \cdot \mathbf{x})} = 0. \tag{8.106}$$

The wave function is then a quaternionic state vector

$$\psi = (\pm e_3 E \pm i e_1 \mathbf{p} + i e_2 m) e^{-i(et - \mathbf{p} \cdot \mathbf{x})} = 0, \tag{8.107}$$

which obeys the invariant interval is the quantum wave equation

$$\left(\pm i\mathbf{e}_3\frac{\partial}{\partial t} \pm \mathbf{e}_1\nabla + i\mathbf{e}_2 m\right)\psi = 0. \tag{8.108}$$

This illustrates how the Dirac operator and the quaternionic state vector are related to each other by the standard quantization rule [18]. The Pauli exclusion principle states that $\psi^2 = 0$, which is identical in form to the coboundary condition $d^2 = 0$. Further, the operator formed from the square of the Dirac operator acts upon a state vector to give zero. For $\mathcal{D}$ the Dirac operator $\mathcal{D}\psi = 0$. Similarly the square of the Dirac operator $\mathcal{D}^2 = \psi^2 = 0$ illustrate that $\mathcal{D}$ contains the same cohomology, "boundary of a boundary = 0," information as does the state vector.

The associator in field theory $[a, b, c] = (ab)c - a(bc)$ in the algebra of octonions is generally nonzero [1]. Consider the associated transformed fermion $b'_k = U(b_k U^{-1})$. Since the Dirac operator and the quaternionic state vector ψ are topologically equivalent ψ acts as the generator of transformations of the fermion operators b_k, $b_k^\dagger$. Define the transformation matrix as $U = exp(i\epsilon(\bar{\xi}\psi + \xi\bar{\psi}))$, with $UU^{-1} = 1$, for ξ a Grassmannian variable and ϵ a parameter. Then to $O(\epsilon^2)$ the transformed Dirac operator b'_k is

$$b'_k = U(b_k U^{-1}) = b_k + i\epsilon\big((\bar{\xi}\psi + \xi\bar{\psi})b_k - b_k(\bar{\xi}\psi + \xi\bar{\psi})\big) - \epsilon^2(\bar{\xi}\psi + \xi\bar{\psi})(b_k(\bar{\xi}\psi + \xi\bar{\psi})). \tag{8.109}$$

The fermion field is

$$\psi = \sum_k \big(u(k)b_k + v(k)b_k^\dagger\big), \tag{8.110}$$

where $u(k)$ and $v(k)$ are spinors. The nonassociative fields b_k $b_k^\dagger$ b_{-k} and $b_{-k}^\dagger$ determine an associative subaglebra e_1, e_2, e_4, one of the seven 3-cycles, with the multiplication table,

	e_1	e_2	b_k	e_4	$b_k^\dagger$	b_{-k}	$b_{-k}^\dagger$
e_1	-1	e_4	$b_{-k}^\dagger$	$-e_2$	b_{-k}	$-b_k^\dagger$	$-b_k$
e_2	$-e_4$	-1	$b_k^\dagger$	e_1	$-b_k$	$b_{-k}^\dagger$	b_{-k}
b_k	$-b_{-k}^\dagger$	$-b_k^\dagger$	0	b_{-k}	e_2	$-e_4$	e_1
e_4	e_2	$-e_1$	$-b_{-k}$	-1	$b_{-k}^\dagger$	b_k	$-b_k^\dagger$
$b_k^\dagger$	$-b_{-k}$	b_k	$-e_2$	$-b_{-k}^\dagger$	0	e_1	e_4
b_{-k}	$b_k^\dagger$	$-b_{-k}^\dagger$	e_4	$-b_k$	$-e_1$	0	e_2
$b_{-k}^\dagger$	b_k	b_{-k}	$-e_1$	$b_k^\dagger$	$-e_4$	$-e_2$	0

$$\tag{8.111}$$

$H = \frac{1}{2}\sum_k \epsilon(k) b_k^\dagger b_k$ is invariant since $(b'_k)^\dagger b'_k = b_k^\dagger b_k$. On the vacuum state terms $q \neq -k$ contribute an ignorable random phase, so H contains a term that interchanges the k and $-k$ states,

$$\frac{1}{2}\sum_k \epsilon(k)(b'_k)^\dagger b'_k = \frac{1}{2}\sum_{k,q} \epsilon(k) b_k^\dagger b_k + \sum_k \mathcal{V}(k)(b_k^\dagger b_{-k} + b_{-k}^\dagger b_k), \tag{8.112}$$

which has zero contribution to energy eigenstates.

Consider a canonical transformation of the fermion operators by

$$b_k = A\beta_k + B\beta^\dagger_{-k}, \quad b^\dagger_k = A^*\beta^\dagger_k + B^*\beta_{-k} \tag{8.113}$$

$$b_{-k} = A\beta_{-k} + B\beta^\dagger_k, \quad b^\dagger_{-k} = A^*\beta^\dagger_{-k} + B^*\beta_k,$$

where the transformed operators are nonassociative by equation 8.111 and $|A|^2 - |B|^2 = 1$. The resulting Hamiltonian is

$$H = \sum_k (\epsilon(k)(|A|^2\beta^\dagger_k\beta_k + |B|^2\beta_{-k}\beta^\dagger_{-k}) + \sum_k \epsilon(k)(A^*B\beta^\dagger_k\beta^\dagger_{-k} + AB^*\beta_{-k}\beta_k)$$

$$+ \sum_k V(k)(|A|^2(\beta^\dagger_k\beta_{-k} + \beta^\dagger_{-k}\beta_k) + |B|^2(\beta_{-k}\beta^\dagger_k + \beta_k\beta^\dagger_{-k})). \tag{8.114}$$

Each one of the terms in this Hamiltonian involve the quaternions e_1, e_2, e_4 which result from the multiplication table in equation 8.111. For $AB^* = -A^*B$ this is a case of the general B^* algebra in Chapter 7. The cross term in equation 8.114 is physically the same as equation 7.68 and recovers a squeezed state operator [19].

e_1, e_2, e_4 are defined by fermion operators according to

$$e_1 = b^\dagger_k b_{-k}, \quad e_1 = b_k b^\dagger_{-k}$$

$$e_2 = b_{-k} b^\dagger_{-k}, \quad e_2 = b_k b^\dagger_k \tag{8.115}$$

$$e_4 = b_{-k} b_k, \quad e_4 = b^\dagger_k b^\dagger_{-k},$$

with their conjugates. This degeneracy has no influence on the e_i level, but differ in their actions on the fermion basis $|\{0,1\},\{0,1\}\rangle$. There exist the maps between fermion states,

$$e_1 : |0, 1\rangle \rightarrow \pm|1, 0\rangle, \quad e_1 : |1, 0\rangle \rightarrow \pm|0, 1\rangle$$

$$e_2 : |\{0,1\},\{0,1\}\rangle \rightarrow \pm|\{0,1\},\{0,1\}\rangle \tag{8.116}$$

$$e_4 : |0, 0\rangle \rightarrow \pm|1, 1\rangle, \quad e_4 : |1, 1\rangle \rightarrow \pm|0, 0\rangle,$$

where the $\pm$ sign is a Z_2 covering by conjugation. e_1, e_2, e_4 give respectively the kinetic Hamiltonian, the potential terms and the squeezed state terms found in the third, first and second terms in equation 8.114. The latter is a squeezing operator used in quantum optics [19]. There are 6 independent operators, with a Z_2 phase. These 6 operators are then a projective realization of a seven dimensional space. Fields defined as $b^\dagger_k b^\dagger_{-k}$ and $b_k b_{-k}$ are analogues to Cooper paired states, or quark-antiquark pairs in a spin-1 meson[15].

The above construction may be made explicit if the six operators are defined as

$$e^1_1 = b_k b^\dagger_{-k}, \quad e^1_2 = z b^\dagger_k b^\dagger_{-k}, \quad e^1_4 = b_{-k} b^\dagger_{-k}$$

$$e_1^2 = b_k^\dagger b_{-k}, \ e_2^2 = z^* b_k b_{-k}, \ e_4^2 = b_k b_k^\dagger, \tag{8.117}$$

which defines the algebra

$$[e_1^1, e_2^1] = 2ze_4^1, \ [e_2^1, e_4^1] = 2ze_1^1, \ [e_4^1, e_1^1] = -\frac{2}{z}e_2^1$$

$$[e_1^2, e_2^2] = 2z^* e_4^1, \ [e_2^2, e_4^2] = 2z^* e_1^1, \ [e_4^2, e_1^2] = -\frac{2}{z}e_2^1$$

$$[e_1^1, e_2^2] = 2z^* e_4^2, \ [e_2^1, e_4^2] = -2ze_1^2, \ [e_4^1, e_1^2] = \frac{2}{z^*}e_2^2 \tag{8.118}$$

$$[e_1^2, e_2^1] = -2ze_4^2, \ [e_2^2, e_4^1] = z^* e_1^2, \ [e_4^2, e_1^1] = -\frac{2}{z^*}e_2^2,$$

where z is the squeezing parameter. The Moufang identity $(xy)(zx) = x(yz)x$ is used above. The first line of equation 8.118 is $SU(2)$ or $SU(1,1)$ depending upon the values of z and z^*. Now define $E_i^1 = e_i^1 \mathbf{1}_{2\times 2}$ and $E_i^2 = e_i^2 \sigma_i$ and this algebra is then a graded algebra, where the second line of equation 8.118 becomes anticommutators.

Define the operator

$$\mathcal{Q} = \gamma_0 \begin{pmatrix} b_k \\ b_{-k} \end{pmatrix}, \ \bar{\mathcal{Q}} = \gamma_0 \begin{pmatrix} b_k^\dagger \\ b_{-k}^\dagger \end{pmatrix}, \tag{8.119}$$

where the supersymmetry operators obtain as $b \to \mathcal{Q}$. The operators in equation 8.117 are not determined by $\mathcal{Q}^2$ or $\bar{\mathcal{Q}}^2$. Consider the different representations of these basis elements as components of the spinors. The action of $\mathcal{Q}$ on E_x, E_y, E_z is

$$\mathcal{Q}^T E_x^1 = \gamma_0 \begin{pmatrix} 0 \\ b_k(b_k b_{-k}^\dagger) \end{pmatrix}, \ \mathcal{Q}^T E_y^1 = iz\gamma_0 \begin{pmatrix} b_k(b_k^\dagger b_{-k}^\dagger) \\ b_k(b_k^\dagger b_{-k}^\dagger) \end{pmatrix},$$

$$\mathcal{Q}^T E_z^1 = \gamma_0 \begin{pmatrix} b_k(b_{-k} b_{-k}^\dagger) \\ 0 \end{pmatrix}, \tag{8.120}$$

so $E_i^1 \notin \ker \mathcal{Q}^T / \mathrm{im} \mathcal{Q}^T$. The same holds for E_i^2. This is a problem that needs to be corrected with the addition of a term that imposes a requirement these fields exist in $\ker \mathcal{Q}^T / \mathrm{im} \mathcal{Q}^T$. Just as gauge fields are imposed to prevent inhomogeneous transformation of observables they are imposed here to maintain the topological nature of observables. Gauge fields in the case of the E_y^1 components must emerge as $b_k^\dagger b_{-k}^\dagger \to \beta_k^\dagger \beta_{-k}^\dagger + \mathcal{A}_y^1$, where the action of $\mathcal{Q}^T$ on $\mathcal{A}_y^1$ cancels out the action of $\mathcal{Q}^T$ on E_y^1. The $\beta_k, \beta_{-k}, H.C.$ are given by equation 8.113. This construction lets $\{e_1, e_2, e_4\} \in \ker \mathcal{Q}^T / \mathrm{im} \mathcal{Q}^T$, which defines a BRST cohomology [20]. Yang-Mills equations are then derived, for example

$$[b_k^\dagger b_{-k}^\dagger, b_k b_{-k}] = 0 = [\beta_k^\dagger \beta_{-k}^\dagger, \mathcal{A}_y^2] + [\mathcal{A}_y^1, \beta_k \beta_{-k}] + [\mathcal{A}_y^1, \mathcal{A}_y^2]$$

$$[b_k b_k^\dagger, b_k b_{-k}] = 0 = [\beta_k \beta_k^\dagger, \mathcal{A}_y^2] + [\mathcal{A}_z^2, \beta_{-k} \beta_k] + [\mathcal{A}_z^2, \mathcal{A}_y^2], \tag{8.121}$$

where Lie algebraic indices are implied. Thus bracket operators of the form $[\beta_k^\dagger \beta_{-k}^\dagger, \ ]$ are coboundary operators analogous to d, with $d^2 = 0$. It is evident from this requirement that $A_i^2 = -A_i^{1\dagger}$.

The supersymmetric cohomology defines $1/2$, 1, $3/2$ spin fields that span three, seven and four dimensions respectively. The total supersymmetric map is then determined by the Hopf fibration

$$S^3 \to S^7 \to S^4. \tag{8.122}$$

The $S^7 \sim SO(7)$ contains the six operators contained in e_1, e_2, e_4 are a projective subspace of this space. The Z_2 defines two copies of $SO(6)$ in $SO(7)$ along a vector direction in S^7. For the Z_2 as the signs of determinant the covering group for these fields is then $U(4) \sim spin(6) \times U(1)$. The Hopf fibration then has the group theoretic description

$$\left(SO(3) \simeq SU(2)/Z_2\right) \to SO(7) \to SO(4). \tag{8.123}$$

This sequence is given by the the Hopf fibration which connects the quaternions to the octonions [1][7]. Further these fields are annulled by the Hamiltonian

$$Q\bar{Q} \to H = 1/2(Q\bar{Q} + \bar{Q}Q). \tag{8.124}$$

This defines the system of quaternions on an associative subgroup in $\mathcal{O}$ chosen as above. There are also six other such systems, where the octonion variables assume different roles as e or products of b, $b^\dagger$ operators. The quaternions in S^7 exists are given by S_i^3 $i = 1,\ldots 7$, which gives seven distinct field theories. This leads to the prospect that these fields compose what is called dark matter. The gauge fields and associated fermion in these seven quaternion groups gravitate, but fields from one of the seven quaternions are detected. This may mean the other six unobserved fields compose dark matter. To "see" other fields from the other subgroups requires a transformation between multiplication tables that permute the nonfermionic e_i amongst the other octonion basis elements.

The horizontal Hopf fibration for octonions

$$S^7 \to S^{15} \to S^8 \tag{8.125}$$

is connected to quaternionic S^7. The $S^8 \sim \mathcal{C}^4$ contains topological spin-2 fields for gravitation. Products of nonassociative fermions determine associative quaternions $\{e_1, e_2, e_4, 1\}$, which by cohomology also determines the associative gauge connections. These connections define field strength tensors and bundle curvatures. Spacetime is then $\Psi = \mathcal{R}e\mathcal{C}^4$, with group action $U(4)$. This is decomposed into $U(4) = spin(6) \times U(1)$. The $spin(6)$ is a conformal group that embeds the $spin(5) \sim SO(3,2)$ deSitter group for gravitation. The remaining 5 parameters from the $spin(6)$ are a diliton factor and four conformal factors. [16]

The coassociative space, for $S^8 \sim spin(8)$ is $spin(8)/U(4)$, defined by the Jordan product

$$\Phi(w, x, y, z) = \frac{1}{2}(\langle w, x(\bar{y}z)\rangle - \langle z, z(\bar{y}x)\rangle) \tag{8.126}$$

is not a gauge group, but may represent the weights of the gauge group. A choice of a basis for $spin(8)/U(4)$ is

$$\{1,\ i,\ j,\ k, (\pm 1,\ \pm i,\ \pm j,\ \pm k)\}, \tag{8.127}$$

as the sum of the 8 quaternions $\{\pm 1,\ \pm i,\ \pm j,\ \pm k\}$, the 3 quaternions $\{i,\ j,\ ,k\}$ and the unit term 1. The co-associative space gives the weights for $SU(3) \times SU(2) \times U(1)$, or family structures [16]. Beyond the octonions are sedenions

$$S^{15} \rightarrow S^{31} \rightarrow S^{16} \tag{8.128}$$

that have the spin-3/2, 2 and 5/2 fields. The theory cuts off here, for S^{16} defines unphysical 5/2-spin fields as nonalgebraic sedenions.

The connection between the two S^7 spheres in the quaternionic and octonionic portions of the Hopf fibration give weight to the concept advanced in Chapter 2 that quantum gravity fluctuations posses a "mirror" gauge symmetry. If the structure of supersymmetry is extended across the whole Hopf fibration this suggests that the structure of gauge fields is tied to an underlying symmetry of gravitation, which as seen earlier is argued to be a mirror of these gauge symmetries in the structure of spacetime fluctuations.

A Galois number code is defined as [14]:

$$s_n = g^n \cdot mod p, \tag{8.129}$$

where, n is a positive integer and p is a prime. $g = 3$ is a primitive root of $p = 7$. If each of the numbers $g,\ g^2,\ \ldots g^{n-1}$ leaves a different remainder when divided by p. Then $s_n = 3^n \cdot mod 7$, the produces the Galois code [3, 2, 6, 4, 5, 1, 0] and the following $mod 2$ sequence $[-1, 0, 0, 0, -1, -1, 0]$. A permutation of this gives the diagonal entries of the octonion multiplication table. All possible permutations of this array produces tables for all seven quaternions. For the octonions a similar issue involves a more general choice of all eight elements as an octonion. This gives all possible 480 multiplication tables.

8.9: Bohr's Gravitization of the Quantum

At the Solvay conferences Neils Bohr and Albert Einstein debated the nature of quantum mechanics [21]. Einstein was convinced of reality and locality and argued staunchly for an incompleteness of quantum mechanics. Quantum theory could only be made complete if there are some hidden variables that underlay the probabilistic, nonlocal quirky aspects of quantum mechanics. At the 1930 Solvay conference Einstein proposed an interesting thought experiment. Einstein considered a device which consisted of a box with a door in one of its walls controlled by a clock. The box contains radiation, similar to a high-Q cavity in laser optics. The door opens for some brief period of time t, which is known to the experimenter. The loss of one photon with energy $E = \hbar \omega$ reduces the mass of the box-clock system by $m = E/c^2$, which is weighed. Einstein argued that knowledge of t and the change in weight provides an arbitrarily accurate measurement of both energy and time which may violate the Heisenberg uncertainty principle $\Delta E \Delta t \simeq \hbar$ [21][22].

Bohr retreated for the day to address this problem. He later realized that the weight of the device is made by the displacement of a scale in spacetime [21]. The clock's new position in the gravity field of the Earth, or any other mass, will change the clock rate by gravitational time dilation as measured from some distant point the experimenter is located. The temporal metric term for a spherical gravity field is $1 - 2GM/rc^2$, where a displacement by some δr means the change in the metric term is $\simeq (GM/c^2 r^2)\delta r$. Hence the clock's time intervals T is measured to change by a factor

$$T \rightarrow T\sqrt{(1 - 2GM/c^2)\delta r/r^2} \simeq T(1 - GM\delta r/r^2 c^2), \tag{8.130}$$

so the clock appears to tick slower. This changes the time span the clock keeps the door on the box open to release a photon. Assume that the uncertainty in the momentum is given by the $\Delta p \simeq \hbar/\Delta r < Tg\Delta m$, where $g = GM/r^2$. Similarly the uncertainty in time is found as $\Delta T = (Tg/c^2)\delta r$. From this $\Delta T > \hbar/\Delta mc^2$ is obtained and the Heisenberg uncertainty relation $\Delta T \Delta E > \hbar$. This demands a Fourier transformation between position and momentum, as well as time and energy.

Consider an example with the Schwarzschild metric terms. The metric change is then $\sim 1. \times 10^{-12} m^{-1} \delta r$, which for $\delta r = 10^{-3}$m is around 10^{-15}. Thus for a open door time interval of 10^{-2}sec, the time uncertainty is around $\Delta t \sim 10^{-17} sec$. The uncertainty in the energy is further $\hbar \Delta \omega$, where by Fourier reasoning $\Delta \omega \sim 10^{17}$. Hence the Heisenberg uncertainty is $\Delta E \Delta t \sim \hbar$.

This was one of a succession of defeats Albert Einstein endured by Niels Bohr due to his stance on the epistemology of nature as quantum theory. The growing side of physics in decades forth swayed completely against Einstein. The more current experimental situation on the nature of the quantum has clearly shown Einstein was indeed wrong on this, in particular due to Bell's theorem and experimental situations which have consistently tested the theorem in the affirmative. So far Bell's theorem appears to be correct experimentally to within errors of 10^{-3}. Albert Einstein was an early founder of quantum mechanics, but soon in his researches on the quantum found the consequences of the quantum world too bizzar for his Swiss watch view of the universe.

Bohr's answer to Einstein's Gedankin experiment has a unique feature to it. It argues for a quantum interpretation of the world by appealing to general relativity. Bohr's argument involves a small displacement of the system in a weak gravity field. Consider this clock system as very small and in the environment of a micro-black hole. The system could be arranged so that the emission of a quantum from the box could result in a large displacement of the system relative to the scale of the black hole. Let that displacement be $r_1 \rightarrow r_2$. As a result the observer will see the clock slow down by a factor of $T \rightarrow T\sqrt{(1 - 2M/r_2)/(1 - 2M/r_1)}$. It is interesting to ponder whether this suggests something about the tie between quantum mechanics and general relativity. In other words presume the clock is a black hole with some time interval as defined by the number of Planck masses which compose it.

The time interval for the open door is given by the radiation production rate of photons of some wavelength λ from a black hole due to Bekenstein's thermal result [23].

This was refined by Hawking's result on the quantum emission by black holes. This process has recently been found by Vafa to be consistent with superstring theory and M-theory according to strings that scatter off D-branes. Yet these programs have always had a one way thrust — quantization. With Bohr's argument there is a gravitization of the quanta. There is a complementary way of viewing quantum gravity: quantizing gravity and gravitizing the quantum.

Consider the clock in Einstein's box as a black hole with mass m. The quantum periodicity of this black hole is given by some multiple of Planck masses. For a black hole of integer number n of Planck masses the time it takes a photon to travel across the event horizon is $t \sim Gm/c^3 = nT_p$, which are considered as the time intervals of the clock. The uncertainty in time the door to the box remains open is

$$\Delta T \simeq Tg/c(\delta r - GM/c^2), \tag{9.131a}$$

as measured by a distant observer. Similarly the change in the energy is given by $E_2/E_1 = \sqrt{(1 - 2M/r_1)/(1 - 2M/r_2)}$, which gives an energy uncertainty of

$$\Delta E \simeq (\hbar/T_1)g/c^2(\delta r - GM/c^2)^{-1}. \tag{9.131b}$$

Consequently the Heisenberg uncertainty principle still holds $\Delta E \Delta T \simeq \hbar$. Thus general relativity beyond the Newtonian limit preserves the Heisenberg uncertainty principle. It is interesting to note in the Newtonian limit this leads to a spread of frequencies $\Delta \omega \simeq \sqrt{c^5/H\hbar}$, which is the Planck frequency.

The above illustrates that the uncertainty principle is given by a mutual uncertainty in coordinates normally regarded as commutative in quantum mechanics. The uncertainty in the mass of the box-clock is given by its change in position with respect to coordinates in a gravity field. Hence the uncertainty in mass-energy $\Delta M = \hbar \Delta \omega$ is given by a ΔR of the box, which has an uncertainty relationship to the time ΔT. There is a subtle example of this in quantum mechanics. The DeBroglie wave-particle duality is well known. Yet the wave aspects of quanta are deduced by experiments that ultimately require the measurement of particle positions obtained by the reduction of a state on a detector or measurement screen. Experiments in quantum systems never directly measure the wave aspect of the quanta, but only indirectly deduce it from the statistics obtained from many particle measurements.

Equations 8.131 illustrate that the transformations of spacetime are such that quantum mechanics still obtains. Spacetime according to a local symmetry preserves the quantum uncertainty. This suggests a subtle connection between quantum mechanics and general relativity. Coordinate transformation are such that quantum uncertainty is invariant. This leads to a physical ansatz that quantum mechanics and general relativity have a similar structure. This suggests that quantum mechanics has an underlying structure with parallel transport. Conversely spacetime parallel transport structure should share properties of quantum mechanics. In effect the bracket structures of the two theories are fused into a signel general algebraic structure. This is shown below to have a connection with octonions. From this the commutator structure of quantum mechanics is found to have a parallel transport structure near the Planck scale. This is a complementary way of examining the problem of quantum gravity.

This also appears to have connections with the holographic principle. The metric fluctuations can just as easily be considered as due to the change in the mass of the black hole clock, for once the black hole clock emits a quantum the metric is changed by conservation of momentum. The box is weighed according to the kick back or back reaction of the metric. Hence the metric term of consideration can just be due to the black hole clock in flat spacetime.

To model this consider the fluctuation of a spatial volume as due to virtual black holes. Consider the timelike volume form $\mathbf{V} = \mathbf{e}_1 \wedge \mathbf{e}_2 \wedge \mathbf{e}_3$ for a three dimensional cube The variation in the volume form may be written as

$$\delta \mathbf{V} = \delta \mathbf{S}_{12} \wedge \mathbf{e}_3 + \mathbf{e}_1 \wedge \delta \mathbf{S}_{23} + \delta \mathbf{S}_{31} \wedge \mathbf{e}_2, \tag{8.132}$$

where S_{ij} are elements of the two dimensional surface 2-form bounding the volume and $\delta \mathbf{S}ij = \delta \mathbf{e}_j \wedge \delta \mathbf{e}_i = \epsilon^2 \mathbf{e}_i \wedge \mathbf{e}_j$, for ϵ a small parameter. ϵ is cut off at the Planck length $L_p = \sqrt{G\hbar/c^3}$ so that $|\epsilon \mathbf{e}_i| = L_p$. This then gives the variation in the volume form in components

$$\delta V_0 = \epsilon_{ijk0} \delta S_{ij} e_k, \tag{8.133}$$

The volume fluctuation is then seen to be written as

$$\frac{\delta V}{V} \simeq \frac{\delta S}{S}. \tag{8.134}$$

The variation in the volume is then due entirely from the fluctuation in the surface area bounding that volume. This leads to the holographic result $(\delta L/L)^3 \simeq (L_p/L)^2$. This is accompanied by a fluctuation in the metric $\delta g_{\mu\nu} \simeq (L_p/L)^{2/3} g_{\mu\nu}$, which defines the gravity force on a particle, which is really the force required to prevent the particle from moving, in this volume $L\delta g \simeq \delta g_{00} c^2$. The uncertainty in the momentum of the particle is then assumed to be $\Delta p \simeq \hbar/\Delta r < T\Delta gm$ and $\Delta T = TL\delta g/c^2 \simeq T(L^2 L_p)^{1/3}/c^2 = (T^2 T_p)^{1/3}$. The relativistic mass of the particle is $M = m\gamma_{00}$, so $m\delta y = \delta M$ is the fluctuation in the mass due to spacetime foam. From which the Heisenberg uncertainty principle is then found.

Bohr's gravitization of the quantum further indicates that coordinates in spacetime are themselves not commutative. The measurement of a spatial distance and a time interval determine the standard Heisenberg uncertainty principle. It is physically necessary that x and t are on some scale incommensurate. Hence standard quantum mechanics applies on scale of $> 10^5$, but as one approaches the Planck scale commuting observables cease to be commensurate. Of course Bohr's argument involved a large mass measuring a small mass change. This reflects a disparity in scale, where realistically these effects only become apparent as physics approaches the Planck scale. Yet Bohr's argument is a mark of brilliance, for it indicates the underlying spacetime geometry is noncommutative and so previously commuting quantum operators are no longer commensurate. Gravity and noncommutative quantum geometry are embedded as a pair of quaternions in the octonions. This is the topic of the next section. The noncommutative underlying geometry of quantum mechanics will have on the large a small dispersive effect interpreted according to the holographic principle. This in principle can be detected [24]. These developments lead to connections with theories covered by nonassociative quantum gravity, such as string theory.

Bohr's observation suggests that noncommutativity between momentum and position operators is intertwined with noncommutativity between geometric coordinates. Let p_i and x_i obey $[x_i, p_j] = i\hbar\delta_{ij}$. To find a tie between this standard quantum commutator and noncommutative geometry consider the octonionic nonassociator

$$[p_i, x_j, x_k] = p_i \odot (x_j \odot x_k) - (p_i \odot x_j) \odot x_k, \tag{8.135}$$

with the octonionic product rule $x_i \odot x_j = \xi_{ijk} x_k$ and $p_i \odot p_j = \pi_{ijk} p_k$, where ξ_{ijk} and π_{ijk} are scale dependent coefficients. The commutator product means that $z_i \odot z_j = 1/2[z_i, z_j]$, $z \in \{x_i, p_i\}$. By basic algebra it is possible to demonstrate that

$$[p_i, x_j, x_k] = 1/2([x_j, p_i]x_k + p_i[x_j, x_k]). \tag{8.136}$$

Since the nonassociator reflects Planck scale physics set $[p_i, x_j, x_k] = \alpha L_p T_{ijk0} x^0$ for α a constant with magnitude $\alpha \sim 1 - 10$ and T_{ijk0} a structure constant. The standard quantum commutator is then

$$[x_j, p_i]x_k = 2\alpha L_p T_{ijk0} x^0 - p_i[x_j, x_k] \rightarrow i\hbar\delta_{ij}x_k. \tag{8.137}$$

This similarly obtains for associators of the form $[x_i, p_j, p_k]$ with the commutator structure

$$[p_j, x_i]p_k = 2\alpha^{-1} M_p c T_{ijk0} x^0 - x_i[p_k, p_k] \rightarrow -i\hbar\delta_{ij}p_k. \tag{8.138}$$

Here the arrow indicates large scale $\gg L_p$ behavior. Consider the situation for scales an order of magnitude above L_p or near the string length, where the nonassociative contribution becomes smaller. Let the structure factor $\xi_{ijk} = \beta L_p T_{ijk}$, for β a scaling factor $\beta = \beta(nL_p)$ and T_{ijk} a structure constant. It is evident that

$$[x_j, x_k] = \beta L_p T_{jkl} x_l. \tag{8.139}$$

By DeBroglie $\oint p_i dx_k = \hbar\delta_{ik}$. Consider the x_k as the result of the parallel translation of a vector around a loop. Since the commutator $[x_j, p_i]$ is a constant now express the middle term in equation 8.136 as

$$\oint [x_j, p_i]dx'_k \simeq -\oint p_i[x_j, dx'_k]$$

$$\simeq -\beta L_p T_{jkl} \oint p_i dx'_l = -\beta\hbar L_p T_{ijk}. \tag{8.140}$$

Equations 8.136 and 8.140 indicate that the commutator between the position and momentum operators is on a large scale

$$[x_j, \ p_i] \simeq -\beta(L_p/L)\hbar T_{ijk}x_k \rightarrow i\hbar\delta_{ij}. \tag{8.141}$$

This commutator result suggests something about quantum wave equations. Energy-momentum is conserved as an average around quantum fluctuations. Now perform the replacement $\beta(L_p/L) \rightarrow \beta$ for brevity of notation. It is clear that for $L \gg L_p$ this scaling factor approaches unity. This scaling factor is defined as $\beta = 1 + (E/E_p)^\gamma$. The second term is a dispersion in the wave evolution of a massless particle. Consider the square of the energy-momentum vector

$$(\langle E \rangle + \delta E)^2 - (\langle \mathbf{p} \rangle + \delta \mathbf{p})^2 = m^2, \tag{8.142}$$

where commutation relations imply

$$\delta \mathbf{p} = A\langle \mathbf{p} \rangle \left(\frac{p}{m_p c}\right)^\gamma, \ \delta E = B\langle E \rangle \left(\frac{E}{E_p}\right)^\gamma. \tag{8.143}$$

The invariant momentum-energy interval is then

$$\langle E \rangle^2 - \langle \mathbf{p} \rangle^2 + B\langle E \rangle^2 \left(\frac{E}{E_p}\right)^\gamma - A\langle \mathbf{p} \rangle^2 \left(\frac{p}{m_p c}\right)^\gamma = m^2. \tag{8.144}$$

For $E \gg m$ and $\langle E \rangle \simeq \langle \mathbf{p} \rangle$ there is a dispersion term $(B - A)(E/E_p)^\gamma$. The velocity of a massless particle or highly relativistic particle is then

$$v = \frac{\partial \langle E \rangle}{\partial \langle \mathbf{p} \rangle} = c\left(1 + (B - A)\frac{1 + \gamma}{2}\frac{E^\gamma}{E_p^\gamma}\right). \tag{8.145}$$

This gives the speed of light for $A = B$ for a massless particle. For $A \neq B$ the velocity of a massless particle fluctuates around c, which gives an energy dependent uncertainty spread in the arrival times of massless bosons of the same energy E. The fluctuation in time is $\delta t \sim |A - B|t(E/E_p)^\gamma = |A - B|t(t_p/t)^\gamma$. This recovers the holographic result if $|A - B| \sim 1$ and $\gamma = 2/3$. [24]

A cosmological version of the Wheeler delayed choice experiment has been proposed. The SETI (Search for Extra-Terrestrial Intelligence) program could provide a way for performing a delayed choice experiment on photons from a galaxy split by the Einstein lens effect. A photon will exist in an entangled state of states along the two paths. The problem is that the path difference for the two paths could be very large and the time of arrival for the two states very different. However, by making the band pass in the radio spectrum very short the Heisenberg uncertainty principle will result in a huge uncertainty in the time. In this way an interference pattern between radio wave photons on the two paths could in principle be detected. This is quantum mechanics experimentation on a cosmological scale! An overview of this may be found at [25]. If photons exhibit a dispersion due to the holographic principle, ultimately tied to octonionic quantum gravity, this dispersion will effect the "sharpness" in which the narrowing of the bandpass will give a quantum uncertainty in the time of arrival. There will be a frequency dependency that slightly changes the criticality in the bandpass. The reader is left to use equation 8.145 to determine this.

The automorphism, $A = G_2$, of the octonions $\mathcal{O}$ is the transformation which satisfies $A(xy) = A(x)A(y)$ [17]. G_2 is a simply connected and compact real Lie group of dimension 14. This group pertains to the basic real-octonion, and is the smallest of the five exceptional Lie groups. The discrete group $PSL(2, 7)$ is the automorphism group of the octonionic Fano plane. For higher level octonions, complex octonions, quaternion-octonions and octo-octonions higher exceptional groups are required. For complex-octonions the automorphism is F_4 of 52 dimensions. F_4 is equivalent to adding 16 short roots vectors to the four roots of the group $SO(9)$. This is also the symmetry of the 24-cell. The short roots define the quotient $F_4/SO(9)$ which defines the sequence [26]:

$$F_4/SO(9): 1 \to spin(9) \to F_{4\backslash 52} \to \mathcal{O}P^2, \tag{8.146}$$

where $4\backslash 52$ means the group is restricted to 36 of the 52 dimensions. The $\mathcal{O}P^2$ defines the Cayley plane. F_4 is the automorphism group of the Jordan algebra with the symmetric product $X \odot Y = (1/2)(AY + YX)$. The group F_4 acts transitively on $\mathcal{O}P^2$ with isotropy subgroup $spin(9)$. For the octo-octonions the group is the exceptional $E_8 \times E_8$ with heterotic structure.

This result is demonstrated in the next section based on quantum information theory of black holes. It is found that black holes conserve quantum information as an error correction code associated with sphere packing in four and eight dimensions. This is an underlying nonassociative quantum or octonionic structure to superstring theory. A second approach from the holographic principle illustrates how quantum gravity may be best viewed as a pair of quaternions, one for spinorial gravity and the other for spinor quantum fields. These two pairs define an octonionic structure [27]. This structure to quantum gravity appears to emerge from quantum gravity as the gravitization of the quantum. This may then be a dual approach to quantum gravity as the quantization of gravity.

8.10: Quantum Black Holes as Strings and Error Correction Algorithms

Tidal forces distend an extended body in a spherically symmetric gravity field. The Weyl tensor components are $W \simeq 2GM/(c^2r^3)$. Integrate W up to the event horizon $W \simeq c^4/4G^2M^2$ with time measured by an external observer during the lifetime of the black hole. This gives an action-like term. Similarly integrate W over the time it takes an observer to fall into the black hole and approach within a Planck unit of length of the singularity. The approximate results are comparatively close, $S \simeq 3M/4\hbar$ versus $S \simeq M/\hbar$, where the difference is likely due to incomplete knowledge of black hole states. A complementarity between probability and action [28] suggests information conservation. Vacuum polarizations may involve nascent cosmologies, where this rough calculation suggests that no information leaks out of our cosmology. This crude calculation suggests that quantum information is preserved in a black hole. This also agrees with the information bound predicted by the holographic principle [8].

Let a_k and b_k be the lowering operator for bosonic fields with wave number k on the flat spacetime vacuum $|0\rangle$. These are related to the operators A_k and B_k which annihilate the vacuum outside and inside the black hole, with a Minkowski state $|\ \rangle_m$. The flat space operators are related by the Bogoliubov transformation [29]

$$A_k = \alpha a_k - \beta b_{-k}, \quad B_k = \alpha b_k - \beta a_{-k}. \tag{8.147}$$

Here

$$\alpha^2 = cosh^2(g) = \frac{1}{1 - e^{k/T}}, \quad \beta^2 = sinh^2(g) = \frac{1}{e^{-k/T} - 1}, \tag{8.148}$$

for g the surface gravity at the horizon determined by the Killing vector field ξ^μ as $\nabla^\nu(\xi^\mu \xi_\mu) = -2g\xi^\nu$. Transformations $A_k = e^{iH} a_k e^{-iH}$, $B_k = e^{iH} b_k e^{-iH}$ are given by the Hamiltonian

$$H = ig\left(a_k^\dagger b_{-k}^\dagger - a_{-k} b_k + a_{-k}^\dagger b_k^\dagger - a_k b_{-k}\right). \tag{8.149}$$

This is a generator of a squeezed state operator and parametric down shifting of laser light [19].

Consider the scattering of a boson with momentum k by a black hole. The black hole adjusts its mass by $M \to M + \delta M$ if it absorbs the particle and $M + \delta M \to M$ if it emits the particle. When the black hole absorbs the particle there is the transition operator T_{in},

$$T_{in} a_k : |M\rangle_{bh} |n_k\rangle_m \to \sqrt{n}|M + \delta M\rangle_{bh}|n_k - 1\rangle_m, \tag{8.150}$$

with emission corresponding to

$$T_{out} a_k^\dagger : |M + \delta M\rangle_{bh}|n_k - 1\rangle_m \to \sqrt{n}|M\rangle_{bh}|n_k\rangle_m. \tag{8.151}$$

Fields that exit and enter a black hole are coherently correlated, with a scattering operator

$$e^{iH'} a_k e^{-iH'} = \sqrt{1 - P} e^{iH} a_k e^{-iH} + \sqrt{P} e^{iH} b_k e^{-iH}, \tag{8.152}$$

so that $H' = H + H_s$. P and $1 - P$ give the absorption and emission probabilities of a particle in and out of a black hole. Amplitudes for creation and destruction of a boson are

$$_{bh}\langle M + \delta M|_m \langle n_k - 1|T_{in} a_k|M\rangle_{bh}|n_k\rangle_m = {}_{bh}\langle M + \delta M|T_{in}|M\rangle_{bh}\ {}_m\langle n_k - 1|a_k|n_k\rangle_m$$

$$= (n_k)^{-1/2}{}_{bh}\langle M|T_{in}^\dagger T_{in}|M\rangle_{bh}\ {}_m\langle n_k|(\alpha a_k^\dagger - \beta b_{-k}^\dagger)a_k|n_k\rangle_m \tag{8.153a}$$

and

$$_{bh}\langle M|_m\langle n_k|T_{out}a_k^\dagger|M+\delta M\rangle_{bh}|n_k-1\rangle_m = {}_{bh}\langle M|T_{out}|M+\delta M\rangle_{bh}\, {}_m\langle n_k|a_k^\dagger|n_k-1\rangle_m$$

$$= (n_k)^{-1/2}{}_{bh}\langle M+\delta M|T_{out}T_{out}^\dagger|M+\delta M\rangle_{bh}\, {}_m\langle n_k-1|(\alpha a_k - \beta b_{-k}^\dagger)a_k^\dagger|n_k-1\rangle_m. \tag{8.153b}$$

A normal ordering of $a_k a_k^\dagger$ is used and matrix elements with $b_{-k}a_k^\dagger$ and $b_{-k}^\dagger a_k$ vanish. The probability for the absorption and emission of a particle are

$$w_{in} = \frac{2\pi}{\hbar}|\rho(M+\delta E)\langle M+\delta M|T_{in}|M\rangle|^2|\langle n_k-1|a_k|n_k\rangle|^2$$

$$w_{out} = \frac{2\pi}{\hbar}|\rho(M)\langle M|T_{in}|M+\delta M\rangle|^2|\langle n_k|a_k^\dagger n_k - 1\rangle|^2. \tag{8.154}$$

The Fermi golden rule gives that $w_{out}/w_{in} = e^{-8\pi M\delta M}$ [9]. With an evaluation of ${}_m\langle |a_k^\dagger a_k|\rangle_m$ this gives,

$$e^{-8\pi M\delta M} = \frac{\rho(M)}{\rho(M+\delta M)}\left(\frac{(1-P)n_k + \beta^2\{1+(1-P)n_k\}}{(1-P)(n_k-1) + \beta^2\{1+(1-P)(n_k-1)\}}\right)^2. \tag{8.155}$$

Let the ratio of the density of states be $\sim PQ(T)$ for $Q(T) = 1/(1+ze xp(-8\pi M\delta M))$ the partition function for degrees of freedom on the horizon, with $P \simeq exp(-8\pi M\delta M)$ the emission probability of a particle of energy δM. $z = e^{-4\pi M^2}$ is the fugacity of the system. Consider the emission of one particle by the black hole $n_k = 1$

$$1 + ze^{-8\pi M\delta M} = \left((1-P)\beta^{-2} + 2 - P\right)^2, \tag{8.156a}$$

so the absorption coefficient is then

$$P \simeq 1 - \frac{1}{2}e^{-2\pi M^2}. \tag{8.156b}$$

So for quantum unitarity P is nonzero in general.

Consider the state vector

$$|\chi\rangle = \frac{1}{\sqrt{n}}\sum_i |i\rangle_{in}|j\rangle_{out}, \tag{8.157}$$

where n is the dimension of the bosonic Hilbert space containing $|i\rangle_{in}$ and $|j\rangle_{out}$, in a Minkowski basis. Introduce the black hole state $|M\rangle$ as the ancillary state associated with the channel. A trace over the "in" and "out" states gives the two measurable configurations

$$|\psi\rangle_i = \sum_{j(out)}\langle j|\chi\rangle|M\rangle$$

$$|\psi\rangle_o = \sum_{i(in)} \langle i|\chi\rangle |M\rangle. \tag{8.158}$$

For an appropriate initial and final state description of the black hole, the quantum description is unitary [30].

Let Alice send a message to Bob as quantum bits. These quantum bits are sent through a black hole, a noisy communication channel. The message is received by Bob as black hole radiation. The black hole is a processor that converts Alice's message into a different form received by Bob. The state $|\phi\rangle_i = \langle M|\psi\rangle_i$ is the initially prepared state Alice sends to Bob and $|\psi\rangle_o$ is accessible to Alice and Bob. Both $|\psi\rangle_i$ and $|\psi\rangle_o$ are entangled states between the black hole state and the flat space boson states observed by Alice and Bob. A measurement of the "in" and "out" states are performed so these states are in an entanglement. Since $\sum_i |i\rangle_{in}$ are orthogonal the eigenstates of $|M\rangle$ are also orthogonal to $\sum_j |j\rangle_{out}$. Hence a subset of eigenstates of $|M\rangle$ are those of $\sum_i |i\rangle_{in}$. Further, given the scattering matrix $S = 1 - 2\pi i T$ the density matrix for the black hole $\hat{\rho}_{bh} = |M\rangle\langle M|$ evolves by unitarity

$$\hat{\rho}'_{bh} = S\hat{\rho}_{bh}S^\dagger. \tag{8.159}$$

This is still the case for a completely random scattering matrix. By unitarity of S the subspace of $|M\rangle$ that contains $\sum_i |i\rangle_{in}$ is preserved as orthogonal states to $\sum_j |j\rangle_{out}$. The black hole then teleports states from Alice to Bob. If the states of the black hole are unknown Alice still teleports quantum bits to Bob, but the subset of states in $|M\rangle$ equivalent to $\sum_i |i\rangle_{in}$ are indecipherable. Access to $|M\rangle$ is the key that permits Bob to reconstruct $\sum_i |i\rangle_{in}$ from the randomness due to S. This means that the quantum states of a black hole are an error correction code that preserves Q-bits.

It is then apparent that the black hole state must be a coherent state. For the operators a and $a^\dagger$ a coherent state $|\xi\rangle$ is defined

$$|\xi\rangle = exp(\xi a^\dagger)|0\rangle = \sum_{n=0}^\infty \frac{\xi^n}{\sqrt{n}}|n\rangle, \tag{8.160}$$

which are eigenstates of a

$$a|\xi\rangle = \xi|\xi\rangle, \quad e^{\gamma a}|\xi\rangle = e^{\gamma\xi}|\xi\rangle. \tag{8.161}$$

From coherent states a vertex operator may then be defined according to a sum over all modes [31]

$$V(k, x) = exp\left(k \cdot \sum_{n=0}^\infty \frac{1}{n} a_n^\dagger x^n\right) exp\left(k \cdot \sum_{n=0}^\infty \frac{1}{n} a_n x^{-n}\right). \tag{8.162}$$

Given the propagator propagator $\Delta = \int_0^\infty dt e^{-t\mathcal{H}} = \mathcal{H}^{-1}$, The ground state tachyon amplitude for the 26 dimensional bosonic string emerges [31]

$$\langle 0, k_1|V_0(k_2)\Delta V_0(k_3)|0, k_4\rangle = \int_0^1 dx\, x^{k_3 k_4 + 1}(1-x)^{k_2 k_3} = \frac{\Gamma(a)\Gamma(b)}{\Gamma(a+b)} \tag{8.163}$$

for $a = k_3k_4 + 1$ and $b = k_2k_3$. An elementary model for a black hole of coherent states is one composed of bosonic strings in 26 dimensions [31].

This indicates black holes are quantum computers that process Q-bits and teleport them through the universe. An essential aspect of a quantum computer is it has an error correcting capacity. The scattering matrix is completely random, so the set of Q-bits sent through the black hole quantum computer require an error correction algorithmic construction for teleportation.

To avoid transPlanckian physics the quantum state of spacetime must be completely specified by Planck 4-volumes with their centers separated by a Planck length. This is related to the Kepler Conjecture [32]. In the minimal sphere packing in four dimensions each 4-sphere is touched by 24 other 4-spheres [32]. Consider messages sent according to 24 states or frequencies, which define 24 letters. Alice sends a code on a noisy channel and Bob receives a corrupted version. The quantum black hole scattering matrix is completely random, so this channel is noisy. If more than two of the 24 frequencies are sufficiently close a noisy ambiguity can exist in the message Bob receives. Thus the signal must be composed of bits, or Q-bits, that each have a Planck length distance between them in the signal space of 24 dimensions. A minimal sphere packing is required. Non-overlapping packed spheres then separate out the unique quantum bits sent through a noisy channel. This is an error correction code [33].

The 24-cell reflects the minimal packing of Planck units of 4-volume. The 24-cell is the 4-dimensional convex regular polyhedron with Schläfli symbol {3, 4, 3} [34]. The 24 3-facets are octahedral. The number of vertices, segments, facets and three volumes are 24, 96, 96, 24, so the 24-cell is self-dual. Vertices, centered at the origin of 4-space, can be given as follows: The 8 unit length vertices from permutations of $(0, 0, 0, \pm 1)$, in addition there are 16 short vertices of the form $(\pm 1/2, \pm 1/2, \pm 1/2, \pm 1/2)$. The 16 vertices are the vertices of a tesseract, while a dual to the tesseract is formed by the other 8 vertices. This is equivalent to adding to the four roots of $B_4 = so(9)$ 16 short root vectors to form the algebra F_4, which is the symmetry of 24-cell.

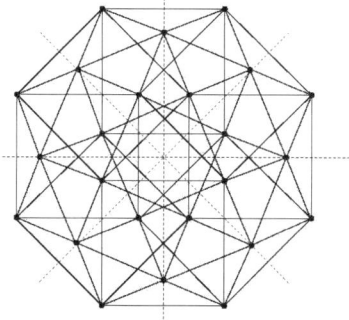

The 24-cell projected onto the x-y plane.
Figure 8.3

F_4 is one of the five exceptional simple Lie groups. F_4 has rank 4 and dimension 52 with fundamental representation in 26-dimensions. The F_4 Lie algebra obtains from 16 spinorial generators added to the 36-dimensional Lie algebra $so(9)$. The 24-cell has an enclosing form, or envelope of projection into three dimensions, as a cuboctahedron and a dual representation as the rhombic dodecahedron covering form. This reflects the symmetry of F_4 with Dynkin diagram O-O=O-O. This duality is a quantum analogue of the Buckminster Fuller jitterbug construction [35]. This duality between ingoing and outgoing states dual copies of a Hilbert space $\mathcal{H}_{in} \sim \mathcal{H}_{out}$ with a total Hilbert space $\mathcal{H} = \mathcal{H}_{in} \oplus \mathcal{H}_{out}$. This defines amplitudes $_{out}\langle\{q^i\}, \{y^i\}|\{p^i\}\{x^i\}\rangle_{in}$ holographically.

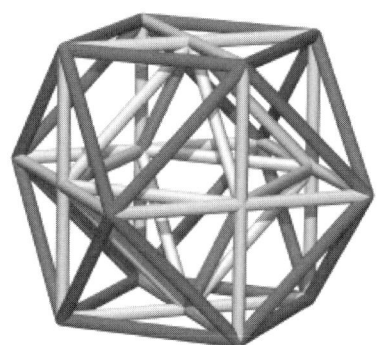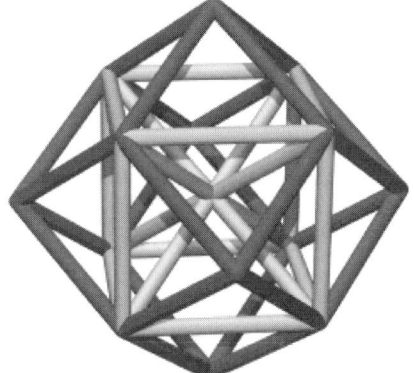

Cuboctohedral and rhombic dodecahedral representation of the 24-cell in three dimensions.
Figure 8.4

This structure defines a target map, or sigma model, from the spacetime to a higher dimension space of string states. The roots added to $so(9)$ is the quotient F_4/B_4 which defines the sequence

$$F_4/D_4 \cdot 1 \rightarrow spin(9) \rightarrow \Gamma_{4\backslash 52} \rightarrow \mathcal{O}P^2 \rightarrow 1, \tag{8.164}$$

where $4 \backslash 52$ means the group is restricted to 36 of the 52 dimensions. The $\mathcal{O}P^2$ is the 16-real dimensional Cayley plane. The group F_4 is the automorphism group of the Jordan algebra with the symmetric product $A \odot B = (1/2)(AB + BA)$. The group F_4 acts transitively on $\mathcal{O}P^2$ with isotropy subgroup $Spin(9)$. The octonions $\mathcal{O}$ have heterotic structure, where the sphere packing on the octonionic space of 8 dimensions is determined by the rank 8 E_8 lattice of 248 dimensions [36]. $E_8 \times E_8$ clockwise modes in 10 dimensions on a heterotic string dual to the counter-clockwise 10-dimensional modes reduced from the bosonic string in 26 dimensions. This is related to the 8-dimensional Hamming code. The sphere packing results in 8 dimensions are numerical and not theorem-proof oriented. However, bounds on errors indicate this is the case to within one part in 10^{27}. Thus error codes underlying unitarity of quantum gravity leads naturally to this sort of structure.

8.11: Quantum States for Hamming and Golay Error Correction Codes

The teleportation of quantum states through black holes implies the existence of an error correction code. For the 24 cell there is the correspondence with an error code. Richard Hamming solved a problem with the error detection that plagued computer programming in the post World War II period. A parity bit error, where a binary $0 \to 1$ or $1 \to 0$ by input or operating error can be detected by the Hamming distance. If the bit string 101111 is replaced by 111011 the Hamming distance measures the number of entries where the strings differ. In this case the Hamming distance is 2. Yet there is no knowledge of where the error occurs. If a dummy binary bit is appended to this binary string and the Hamming distance calculated this can be determined. If a bit error happens in transmission then the change in the Hamming distance can be used to back out where the error is.

For a code that is sent with m words separated by a "," with n letters the information content of that code is $log_2(m)/n$. For there being r relevant digits out of the total n digits an (n, r) code is defined. The number of words is then $m = 2^r$, which defines the information content of the message. Hamming then found an algorithm for determining the minimum information required to send a message. Such a minimizing code must determine the number of up to n possible errors. For $s = n - r$ then $2^s > n$. A multiplication of this by 2^r and a division by $n + 1$ gives the result $2^r > 2^n/n$. This is the basis for Hamming's linear error correction code [37].

A Hamming code word of m bits is generated by multiplying the data bits, of length n by an $m \times n$ generator matrix G using modulo-2 arithmetic. This multiplication's result is called the code word vector $(x_1, x_2, x_3, \ldots, x_m)$, which contains the original data bits plus the calculated parity bits. The generator matrix G used in constructing Hamming codes consists of I (the identity matrix) and a parity generation matrix A, so $G = [I : A]$. A seven length Hamming word may be generated by the matrix:

$$G = \begin{pmatrix} 1 & 0 & 0 & 0 & | & 1 & 1 & 1 \\ 0 & 1 & 0 & 0 & | & 0 & 1 & 1 \\ 0 & 0 & 1 & 0 & | & 1 & 0 & 1 \\ 0 & 0 & 0 & 1 & | & 1 & 1 & 0 \end{pmatrix} \quad (8.165)$$

The multiplication of a 4-bit vector $\mathbf{e} = (e_1, e_2, e_3, e_4)$ by G results in a 7-bit code word vector of the form $(e_1, e_2, e_3, e_4, p_1, p_2, p_3)$. The A partition of G is responsible for the generation of the actual parity bits. Each column in A represents one parity calculation computed on a subset of $\mathbf{e}$. The Hamming rule requires that $p = 3$ for a (7, 4) code, where A contains three columns to produce three parity bits.

The columns of A represent distinct parity bits and must be independent vectors. Thus (p_1, p_2, p_3) represents parity calculations on a three dimensional subset of $\mathbf{e}$. Error correcting or insuring valid information content of the received code word w requires its multiplication it by a parity check to form $\mathbf{p}$ the syndrome or parity check vector, as a partition of the matrix $H = [A|I]$

$$\begin{pmatrix} 1 & 0 & 1 & 1 & | & 1 & 0 & 0 \\ 1 & 1 & 0 & 1 & | & 0 & 1 & 0 \\ 1 & 1 & 1 & 0 & | & 0 & 0 & 1 \end{pmatrix} \times \begin{pmatrix} 1 \\ 0 \\ 0 \\ 1 \\ 0 \\ 0 \\ 1 \end{pmatrix} = \begin{pmatrix} 0 \\ 0 \\ 0 \end{pmatrix} \qquad (8.166)$$

If all elements of **p** are zero, the code word is received with the original information. If **p** contains non-zero elements there is a bit error. So long as there is only an error with one bit that error can be determined by computing parity checks to find a discrepancy.

The Hamming distance defines a ball within a space with a separation from other balls. Any overlap introduces an ambiguity or an error. For the case of the (7, 4) Hamming code there are eight vertices these balls surround. There are $2^r = 2^4 = 16$ words in each code. Thus there are $8 \times 16 = 2^7$ vertices. This number is equal to the number number of possible sign changes on the number of permutations of the octonion tables. This system of the (7, 4) Hamming code has been show to work for Q-bit error correction. For quantum bits $|0\rangle$ and $|1\rangle$ expanded in a basis for the (7, 4) Hamming code this system is shown to work to provide a quantum erasure procedure to correct for decoherent quantum noise in a quantum computer.

As an historical aside this is the origin of octal and 16 bit systems used in computers.

The (7, 4) Hamming code is then a subspace of the $GF(2)$ vector space [38]. The Dixon Galois code for the octonions is $GF(2^n)$ for $n = 3$[14]. For quantum computing the Galois field is $GF(4) = (0, 1, z, z^2)$ with $z = \frac{1}{2}(i\sqrt{3} - 1)$ with $z^2 = z^*$. $GF(4)$ is the Dynkin diagram for the Lie Algebra $D_4 = spin(8)$. As seen below general relativity has a Galois field representation of $GF(4)$ which further substantiates $[GR] \sim [QM]$. The octonionic $GF(8)$ is then the appropriate

A geometric method for the parallel transport of a vector in spacetime is the Schild's ladder. This is a ruler and compass construction. These ruler and compass constructions have a correspondence with Galois groups [39][40]. Given a set of points $\mathcal{P}$ any further construction is a sequence of points $p_1, \ldots, p_n$ that defines the set $\mathcal{P} \cup \{p_1, \ldots, p_n\}$. This very naturally defines a field extension, where it K_0 is the field for the points $\mathcal{P}$, then the subsequent constructions define extensions of the sort

$$K_0 \subseteq K_1 \subseteq \ldots \subseteq K_n. \qquad (8.167)$$

Ultimately these constructions are built from the intersection of lines, the intersections between lines and circles, and between circles. Hence these constructions involve two linear equations, a linear and quadratic equation, and two quadratic equation. Hence if the radius of the circle is $r \in K_i$ the simultaneous solution between these equations gives points $\{x\} \in K_i$ that correspond to these intersections as zeros of a quadratic polynomial.

It is then evident that the degree of the polynomials in K_0 is 2. Then given any point $r_i = (x_i, y_i)$ define a subset of $\mathcal{P}$ on $\mathbf{R} \times \mathbf{R}$ the degrees $[K_0(x), K_0]$ and $[K_0(y), K_0]$ are seen to be 2. $K(x)$ is a ring of polynomials. Since the polynomials involve the above intersections then

$$[K_i(x_{i+1}), K_i] \in \{1, 2\} \ and \ [K_i(y_{i+1}), K_i] \in \{1, 2\}. \tag{8.168}$$

This means that

$$[K_i(x_{i+1}, y_{i+1}), K_i] = [K_i(x_{i+1}, y_{i+1}), K_{i+1}(x_{i+2})][K_{i+1}(x_{i+2}), K_i], \tag{8.169}$$

which implies that $[K_{i+1}, K_i]$ is a power of 2, and by induction $[K_n, K_0]$ is also a power of 2. It is evident that

$$[K_n, K_0(x)][K_0(x), K_0] = [K_n, K_0], \tag{8.170a}$$

and so $[K_0(0), K]$ has a power of 2. This then proves that these ruler compass constructions are Galois field extensions. By extension this implies that the parallel transport of a vector in a finite element construction is codified in Galois field extensions

The above illustrates that the parallel translation of vectors is associated with some Galois field extension $K_0 \subseteq K_1 \ldots \subseteq K_n$. For $K_0 = Q$ then

$$Q, \ Q(\sqrt{2}) \subseteq Q(2^{1/4}) \subseteq Q(2^{1/8}) \subseteq \ldots \subseteq Q(2^{1/2^n}) \ldots \tag{8.170b}$$

are field extensions seen in equation 8.170a. This is then a formal algebraic description for a finite difference approach to parallel translation. This is obviously is an approximation, for if the surface is a sphere the geodesics are circles. A circle is described by the number π, which is transcendental. A transcendental number does not have a root and is not algebraic. Consequently this is a finite element and approximation to the standard description in differential geometry. The nature of limits as these constructions become infinitesimal is a subtle problem in mathematics. This formal algebraic system indicates that general relativity may be expressed according to a simple set of algebraic rules. The above rule for parallel transport may be applied to the transport by the vector $\mathbf{u}$ and by the vector $\mathbf{v}$ to define $[\nabla_\mathbf{u}, \nabla_\mathbf{v}]$ in the Riemannian curvature $R^\alpha{}_{\mu\nu\sigma} = \langle dx^\alpha, [\nabla_{\mathbf{u}^\mu}, \nabla_{\mathbf{v}^\nu}] e_\sigma \rangle$, for $\mathbf{e}$ a vector along the $\mathbf{V}$ above. From this the geodesic deviation equation is computed

$$\nabla_\mathbf{u} \nabla_\mathbf{u} \mathbf{V} + \mathbf{R}(\mathbf{u}, \mathbf{V}, \mathbf{u}) = 0, \tag{8.171}$$

which defines the curvature.

To understand this more fully we have that equation 8.167, with connection

$$\nabla_\mathbf{u} \mathbf{e} = \omega \wedge \mathbf{e} \tag{8.172}$$

is a differential equation that has a representation as a Galois field. The Galois field is then a field of functions contained in the coefficients of the geodesic equations $\Gamma^\alpha{}_{\mu\nu}$. This means that the Galois field structure is only apparent if the connection coefficients are "chosen" to be flat under a coordinate condition. Hence there is a set of fields K that are invariant under a set of automorphisms g of this field that fixes all vectors. This defines a moduli, or a space of solutions within a certain gauge. Given a set of solutions $\mathbf{V} = V_i \mathbf{e}^i$ the automorphisms then act as an algebraic group as $V_i = g_{ij} V_j$ and $\mathbf{e}^i = g^{ij} \mathbf{e}^j$ with the matrix $g_{ij} \in \mathcal{G}$ for $\mathcal{G}$ a $GL_n(C)$ group. The action of these matrices is to transform the vector as an element of a field extension. These matrix elements define the connection coefficients in equation 8.172. The geodesic equation may then be written as $\nabla \mathbf{V} = g(\mathbf{V})$

The curvature is then evaluated as the second order differential equation,

$$\nabla_\mathbf{u} \nabla_\mathbf{u} V = g \odot g' \mathbf{V}, \tag{8.173}$$

where $g \odot g'$ is an algebraic operation in $\mathcal{G}$. By closure of the algebra $g \odot g' = g'' \in \mathcal{G}$, and further defines the commutator of the connection. The structure of this commutator contains the information for the curvature. This then defines a difference extension of the Galois field, with the equation

$$\nabla^2 \mathbf{V} = g \mathbf{V}. \tag{8.174}$$

This implies that there is an association of g with an element of K.

Now extend this description to four dimensions. It is then incumbent upon us to find a Galois field that the elements $\phi(g)$ exist in, for $\phi : \mathcal{G} \to \mathbf{Gal}$ an automorphism. A field with elements p^n is defined as the Galois field $\mathbf{GF}(p^n)$[40]. The above groups g, g' and their produce $g \odot g'$ are associated with vectors in $\mathbf{GF}(p^n)$ for some p and n. This is a functorial equivalency [41].. For g, g' matrices associate these with the produce $u = \langle a|b \rangle \in K$, for $g \mapsto a$ and $g' \mapsto b$. Assign $\phi(u) = ax + b\bar{x}$ for x, $\bar{x}$ basis elements and their dual of $\mathbf{GF}(p^n)$. Here complex, or pseudocomplex, variables are used to give a symplectic structure to products of the Galois field basis elements. Hence the product between two elements $\phi(u)$ and $\phi(u')$ is

$$Tr\big(\phi(u) \cdot \bar{\phi}(u')\big) = a \cdot a' \langle x\bar{x} \rangle + b \cdot b' \langle \bar{x}x \rangle + a \cdot b' \langle x^2 \rangle + b \cdot a' \langle \bar{x}^2 \rangle. \tag{8.175}$$

The properties of the basis elements that produce a commutator are

$$x^2 = x + I, \; x^3 = I, \; \bar{x} = x^2, \tag{8.176}$$

where I is a symplectic matrix so that $\langle I \rangle = 0$. Hence the product $Tr\big(\phi(u) \cdot \bar{\phi}(u')\big)$ is then

$$Tr\big(\phi(u) \cdot \bar{\phi}(u')\big) = a \cdot b' \langle \bar{x} \rangle + b \cdot a' \langle x \rangle. \tag{8.177}$$

All that is left is to identify the trace products $\langle x \rangle$ and $\langle \bar{x} \rangle$ For the traces as identified as the element x or $\bar{x}$ with its conjugate associated with the symplectic matrix $\langle x \rangle = x + I \cdot \bar{x}$ the trace product in equations 8.175 and 8.177 are

$$Tr\big(\phi(u) \cdot \bar{\phi}(u')\big) = \big(a \cdot b' - b \cdot a'\big)\big(x + \bar{x}\big) \cdot I \tag{8.178}$$

thus for $u = u'$ this product defines a commutator, or bracket structure, between elements a, b associated with the matrices g and g'. The algebra for the basis elements in equation 8.175, with the assignment of the symplectic matrix with the trace it is evident that this Galois field has the basis $\{I, x, \bar{x}, x^2\}$ with $p = 4$ and $n = 1$ with the trace property $Tr\mathbf{BF}(4) = \mathbf{Z}_2$. If $a - \nabla u$, $b = \nabla v$ and $a' = u$, $b' = v$ in equation 8.177 is then $\nabla_u v - \nabla_v u$, which defines the Riemannian curvature. Similarly if a, a' and b, b' are the conjugate variables of ADM relativity g_{ij} and π^{ij} then the above bracket structure could represent the Poisson bracket between these variables.

The [24, 12, 8] extended binary Golay code $C24$ consists of 4096 binary words of length 24. The Hamming weight of a binary word is the number of letters that consist of "ones." The [24, 12, 8] Golay code contain 759 words of weight 8 and an equal number of weight 16. This is in addition to the zero word and the word consisting of 24. The remaining 2576 words are of weight 12. The weight 8 words and weight 12 words are called octads and dodecads respectively. The Hamming weight indicates a duality exist in $C24$

The set (1, 2, ..., 24) is a set of points which contain words in the code which is an indicator function of a subset of the points [42]. The five fold duality $1 - 24$, $8 - 16 - 8$, $12\ 16 - 8$, $24 - 1$ between entries of zeros and ones are a "5"-design which contain these subsets. The octads form a $5 - (24, 8, 1)$ design and the dodecades for a $5 - (24, 12, 1)$ design isomorphic to a Steiner system $S(5, 8, 24)$. The automorphisms on $C24$ are a set of permutations that send words to words is the Mathieu group M_{24}. M_{24} is a simple sporadic group. The sporadic groups are the 26 finite simple groups outside of the four infinite families of finite simple groups. M_{24} is an automorphism group of Steiner systems, such as the "5"-design. M_{24} is of order 244823040 with prime factorization $2^{10} \times 3^3 \times 5^3 \times 7 \times 13 \times 29$. The Mathieu groups are the most elementary of the sporadic groups which extend into a hierarchy to the monster group of order

$$808017424794512875886459904961710757005754368000000000, \qquad (8.179a)$$

with prime factorization

$$2^{46} \times 3^{20} \times 5^9 \times 11^2 \times 13^3 \times 17 \times 19 \times 23 \times 29 \times 31 \times 41 \times 47 \times 59 \times 71 \qquad (8.179b)$$

The Golay code is the unique code which projects onto the $[6, 3, 4] \sim GF(4)$ hexacode with Hamming-like parity conditions. Let a Golay word be encoded into a 4×6 array which indexes the rows by the elements of $GF(4) \sim (0, 1, z, z^2)$. Each column of this array will determine a linear combination of elements in $GF(4)$. An array of zeros and ones is a Golay codeword if the length 6 vector exist in the columns in the hexacode and the first row and all columns have the same bit parity. A generator for the hexacode is the matrix

$$G = \begin{pmatrix} 1 & 0 & 0 & 1 & z^2 & z \\ 0 & 1 & 0 & 1 & z & z^2 \\ 0 & 0 & 1 & 1 & 1 & 1 \end{pmatrix} \qquad (8.180)$$

This is the hexacode for error correction in quantum computing within the Galois code $GF(4)$.

The $GF(4)$ code gives the Dynkin diagram for the $D_4 \simeq spin(8)$ group of the quaternions. The 5-design is then a system of automorphisms over the $8 + 16 = 24$ unit quaternions. The quaternion algebra D_4 in the standard basis e_0, e_1, e_2, e_3 may be decomposed into a two parts

$$D_4^+ = \{e_i\} \cup \{\frac{1}{2}(\pm e_0 \pm e_1 \pm e_2 \pm e_3)\}$$

and

$$D_4^- = \{\frac{1}{2}(\pm e_i \pm e_j)\} \tag{8.181}$$

for $i, j = 0, 1, 2, 3$. D_4^+ forms the inner shell of the Λ_4 lattice $\Lambda_4 = D_4$ and is closed under multiplication and forms $spin(4)$. The set D_4^- is not closed. However, these two may be used in a pairing process to construct the octonons. For elements $x^\pm$, $y^\pm \in D_4^\pm$

$$\{\langle x^+|0\rangle \langle 0|y^+\rangle\} \cup \{\langle x^-|y^-\rangle\}, \tag{8.182}$$

for $x^- y^{-\dagger} = \pm e_i$ the two sets in 8.182 respectively define the $48 + 192 = 240$ elements of the inner shell of $E_8 \subset S^7$. The order of E_8 is 240, which are the elements of the unit octonion 7-sphere. This gives various fibration from $S^n : S^7 \to S^4$. The two groups $D_4^\pm$ in this pairing then give the E_8 lattice inner shell in the basis

$$E_8^+ = \{\pm e_i\} \cup \{\frac{1}{2}(\pm e_i \pm e_j \pm e_k \pm e_l)\}, \tag{8.183}$$

for $i, j, k, l \in \{0, \ldots, 7\}$ all different and the association condition $e_i(e_j(e_k e_l)) = 1$ [43]. Here the elements of D_4^+ construct the associative group and D_4^- define the coassociative structure in the octonions in equation 8.119 - 8.120.

The inner shell of the E_8 lattice describes fibrations $S^7 : S^{15} \to S^8$ Hopf fibrations. Again this decomposition defines $\begin{pmatrix} x \\ y \end{pmatrix} \in S^{15}$ for $x, y \in \mathcal{O}$. For $xx^\dagger + yy^\dagger = 1$. For these elements $x^\pm$, $y^\pm \in E_8^\pm$

$$\{\langle x^+|0\rangle \langle 0|y^+\rangle\} \cup \{\langle x^-|y^-\rangle\} \tag{8.184}$$

for $x^- y^{-\dagger} = \pm e_i$ and the two sets in 8.184 respectively define the $480 + 3840 = 4320$ elements in $S^{15} \in \mathcal{O}^2$, or the pairing of the octonions. However, there is no algebra for this and the sedenions are not a division algebra. The lattice system for these elements are Λ^{16}, where the E_8 is then a map $E_8 : S^{15} \to Z^9$ [43].

It is interesting to ponder whether this sort of theory can be extended by appealing to the hierarchy of the 26 sporadic groups up to the monster group. This may be a route to deeper structures to quantum gravity as an error correction quantum code could be examined beyond the octonions. Physically this may also give physics closer to the Planck scale. Of course at this point physics has crossed into a domain of incredible levels of abstraction. Whether any of these possible future theories could ever be tested is problematic. If future generations of physicist should pursue this track it is advisable that they keep some aspect of physical meaning to these theories, with some contact with possible predictions they present that might be detected

The next chapter discusses whether physics ends at the Planck scale and whether what obtains there is completely "Bohu Vatohu." Here there is offered the speculation that physical laws and structures emerge for no reason at all on a larger scale. Appealing to any sort of structure at the Planck scale becomes a lost enterprize. Planck units of volume are discussed as lawless entities similar in character to Liebniz's monads. These entities "mirror" each other in a way similar to the mythological Indra net. In this process they self-referentially sort out emergent structures to a cosmology.

8.12: Phase Structure of Black Holes

A black hole absorbs and emits quanta according to some rearrangement of its internal quantum state. The event horizon prevents an external observer from ascertaining how the internal state of the black hole is rearranged due to the absorption of a quantum particle. A black hole which adjusts its mass $M \rightarrow M + \delta M$ may be formed by the the absorption of $\delta M = \hbar \nu$ or by $\delta M = \hbar(\nu_1 + \nu_2)$, with $\nu = \nu_1 + \nu_2$. According to the external observer there is no manner in which the internal state of the black hole may be distinguished by these two cases of input scattering quanta. Similarly, for the emission of a quanta from a black hole there is no manner in which the internal configuration of the black hole is accessed. So the black hole quantum scattering experiment appears not determined by a unitary quantum process.

Recently it has been argued that black holes preserve quantum information. This may be argued the case for a few Planck mass quantum black hole, where the event horizon is blurred by quantum fluctuations, yet not so easily for a semi-classical or massive black hole. The mass of the black hole has its classical correlation with the area of the event horizon, where for black hole of a few Planck masses this relationship exhibits a quantum uncertainty. Thus the event horizon is no longer a sharp barrier between the exterior universe and the interior of the black hole. Here the entropy of the black hole is less a matter of classical laws of thermodynamics but of quantum statistical mechanics. As the black hole becomes smaller the entropy decreases. For such a quantum black hole quantum fluctuations involve quantum entanglements between exterior states with interior states that are increasingly accessible to the exterior observer. The fluctuating event horizon is less a strict barrier, but is an aspect of a quantum black hole as a noisy channel. Thus there is some phase transition point where as a quantum black hole absorbs mass from quantum fields it transitions into being a semi-classical object, where coarse graining of interior states is all an exterior observer is able to perform.

It is apparent that a classical black hole, such as an astrophysical black hole of several solar masses, is not going to be an efficient channel for the teleportation of quantum states. Of course in the end, after the black hole has evaporated in a time $T \simeq G^2 M^3/(\hbar c^4) \sim 10^{60}$ years, the states are teleported and converted into a highly encrypted form that appears highly random. Yet from a practical experimental point of view access to these quantum bits is impossible. Besides the obviously large time scales, an exact description of the state of such a black hole is intractably impossible, which requires a coarse graining of such states. Further, on such time scales issues of the final state of the entire universe may be relevant as well.

This poses an interesting problem. A quantum black hole of some small number of Planck units of mass may well teleport quantum states, where Alice and Bob have access to the black hole as a set of ancillary quantum states. One obvious difference is that such a small quantum black hole must be maintained by the input of quanta, something similar to a continuous scattering experiment which inputs energy to the black hole in much the same way a population inversion is maintained with a laser. A large astrophysical black hole is maintained over a very long period of time, outlasting everything else in the universe. Even a semiclassical black hole, such as one with a billion grams and a lifetime $\sim$ .1 seconds, exists on a time scale far larger than the periodicities of quanta that compose it, in particular the Planck time. This then suggests there is some critical value of a parameter, below which a quantum black hole will act as a realistic teleporter of states and above where the behavior qualitatively changes so that teleportation is no longer experimentally realistic. However, from a "bird's eye view" states are still teleported.

Consider an experiment where a quantum black hole of some small number $\sim 10 - 100$ of Planck units of mass is maintained. The quantum black hole is produced by the high energy collision of two particles, where subsequent particles are sent to scatter in and out of the black hole at a rate sufficient to keep the black hole from quantum evaporating away. This is analogous to the maintenance of a population inversion in a lasing medium of atoms. The density $\rho(M)$ is given by the operator $\rho^M_{mn}(t)$, for internal states m, n and a black hole of mass M. A coarse grained change of this matrix traced over the black hole mass is changed by the interaction with a scattering particle in a time τ according to [44]

$$\delta\rho_{mn}(t, \tau) = Tr_M\big(\rho^M_{mn}(t + \tau) - \rho^M_{mn}(t)\big), \tag{8.185}$$

where the change due to the input of N such particles over a time Δt is given by

$$\Delta\rho_{mn}(t, \tau) = N(\Delta t)\delta\rho_{mn}(t, \tau). \tag{8.186}$$

The time τ is such that the input particle is completely absorbed by the black hole. The rate of particle injections r is such that the black hole persists and $N(\Delta t) = r\Delta t$. This is obviously difficult to arrange, as black hole thermodynamics avoids equilibrium, but the experimenter must maintain a dynamic quasi-equilibrium by adjusting the particle injection rate. However, for simplicity it is assumed r is the exact critical value required and constant. Here $\tau < \Delta t$, so we form a coarse grained differential equation for the evolution of the density matrix

$$\frac{\Delta\rho_{mn}}{\Delta t}(t, \tau) = r\delta\rho_{mn}(t, \tau). \tag{8.187}$$

The averaged time change in the density matrix of the black hole per input scattering particle may be found according to the probability distribution $P(\tau)$ as

$$\delta\rho_{mn}(t) = \int_0^T d\tau P(\tau)\delta\rho_{mn}(t,\tau), \qquad (8.188)$$

for $\tau < T < \Delta t$. Consider a probability distribution for a spontaneous emission process $P(\tau) = \alpha exp(-\alpha\tau)$. This leads to the averaged coarse grained evolution rate for the density matrix

$$\frac{\Delta\rho_{mn}}{\Delta t}(t) = r\int_0^T d\tau \alpha e^{-\alpha\tau}\big(\rho_{mn}(t+\tau) - \rho_{mn}(t)\big). \qquad (8.189)$$

This coarse grained evolution equation must be coupled to the boson field emitted by the black hole. The problem now is to probe for a boundary between a pure quantum description and a semi-classical or thermodynamic description.

The radiation field of bosons measured by Bob is of the form

$$|\psi\rangle = \sum_n C_n(t)|n\rangle, \qquad (8.190a)$$

with the total state for the black hole plus the emitted field

$$|\psi_T\rangle = A_M|M\rangle|\psi(t)\rangle. \qquad (8.190b)$$

The time evolved total state vector is then of the form

$$|\psi_T(t+\delta t)\rangle = \sum_n \Big(C_n(t+\tau)A_M|M\rangle|n\rangle + C_{n-1}(t+\tau)A_{M+\delta M}|M+\delta M\rangle|n-1\rangle\Big). \qquad (8.191)$$

For the black hole initially in the $|M\rangle$ state the corresponding initial condition of the field is $C_n(t) = 0$, with the resonant condition for the production of bosons as

$$C_{n-1}(t+\tau) = C_{n-1}(t)cos(g\sqrt{n}\tau), \quad C_n(t+\tau) = -iC_{n-1}sin(g\sqrt{n}\tau), \qquad (8.192)$$

where g is a coupling constant between the black hole states and the boson field. This result is similar to that obtained for the interaction of a photon with an atom in a high-Q QED cavity or a laser [45]. The evolution of the density matrix $\rho_{mn}^M(t+\tau) = C_m^*(t+\tau)C_n(t+\tau)|A_M|^2$ is

$$\rho_{mn}^M(t+\tau) = \rho_{mn}(t)cos(g\tau\sqrt{m+1})cos(g\tau\sqrt{n+1})$$

$$\rho_{mn}^{M+\delta M}(t+\tau) = \rho_{m-1n-1}(t)sin(g\tau\sqrt{m})sin(g\tau\sqrt{n}), \qquad (8.193)$$

This gives the difference equation for the evolution of the density matrix as

$$\frac{d\rho_{mn}}{dt} \simeq \frac{\Delta\rho_{mn}}{\Delta t}(t) = -r\rho_{mn}(t)\left(1 - \alpha\int_0^T d\tau e^{-\alpha\tau}cos(g\tau\sqrt{m+1})cos(g\tau\sqrt{n+1})\right)$$

$$+ r\rho_{mn}(t)\alpha \int_0^T d\tau e^{-\alpha\tau} sin(g\tau\sqrt{m+1})sin(g\tau\sqrt{n+1}). \tag{8.194}$$

The evaluation of the integral in equation 8.194 is straight forward. The time evolution of the density matrix is written then as

$$\frac{d\rho_{mn}}{dt}(t) = -\left(\frac{\mathcal{F}(m,n)\mathcal{G}}{1+\mathcal{F}'(m,n)\mathcal{S}/\mathcal{G}}\right)\rho_{mn}^M + \left(\frac{\sqrt{mn}\mathcal{G}}{1+\mathcal{F}'(m-1,n-1)\mathcal{S}/\mathcal{G}}\right)\rho_{m-1,n-1}^{M+\delta M}, \tag{8.195}$$

where the various terms are the gain coefficient $\mathcal{G} = 2r(g/\alpha)^2$, the self-saturation coefficient $\mathcal{S} = 4\mathcal{G}(g/\alpha)^2$ and the dimensionless coefficients $\mathcal{F}$ and $\mathcal{F}'$ are

$$\mathcal{F}(m,n) = \frac{1}{2}(m+1+n+1) + \frac{(m-n)^2\mathcal{S}}{8\mathcal{G}}$$

$$\mathcal{F}'(m,n) = \frac{1}{2}(m+1+n+1) + \frac{(m-n)^2\mathcal{S}}{16\mathcal{G}}. \tag{8.196}$$

The black hole is then a medium with a certain gain. Of particular interest are the diagonal elements of the density matrix, which gives the probabilities for the occurrence of n-numbers of bosons. An expansion of the denominators on these functions gives the approximate dynamical equation for the diagonal entries

$$\frac{d\rho_{nn}}{dt}(t) \simeq -(\mathcal{G}-(n+1)\mathcal{S})(n+1)\rho_{nn}^M(t) + (\mathcal{G}-n\mathcal{S})n\rho_{n-1,n-1}^M. \tag{8.197}$$

Now evaluate the average boson number $\langle n(t)\rangle = Tr(n\rho)$, where the density matrix is over the Minkowski basis $|\ \rangle_M$. This requires that the evaluation of the boson operator on the flat space be evaluated according the Minkowski basis. The number operator is then evaluated as

$$_m\langle\ |n(t)|\ \rangle_m = (1-P)\langle n_k\rangle + \beta^2\{1+(1-P)\}\langle n_k\rangle. \tag{8.198}$$

With $\frac{d\langle n\rangle}{dt}(t) = \langle n\frac{d\rho_{nn}^M}{dt}\rangle$ evaluated on the Minkowski basis, the evolution in the expectation of the number operator for the acceleration $y \gg 0$ is then

$$\frac{d\langle n\rangle}{dt} \simeq [(2\mathcal{G}+\mathcal{S})(1-P) - 3\mathcal{S}]\alpha^2\langle n\rangle + 3\mathcal{S}(1-P)\alpha^2\langle n^2\rangle + (2\mathcal{G}+\mathcal{S})\alpha^2. \tag{8.199}$$

This differential equation is a Langevin process [46], where a corresponding Fokker-Planck equation may be arrived at. This is found with a coordinate representation of a coherent state [47]

$$|a\rangle = \sum_0^\infty \frac{a^n}{\sqrt{n!}} exp(-|a|^2/2)|n\rangle = exp(aa^\dagger - a^*a/2)|0\rangle. \tag{8.200}$$

The operator $\partial/\partial a$ on this state gives the relationship

$$a^\dagger = \left(\frac{\partial}{\partial a} + \frac{1}{2}a^*\right)|a\rangle. \tag{8.201}$$

The associated Fokker-Planck equation for an occupation probability $\mathcal{P}(a)$ in the operator a is then

$$\frac{\partial \mathcal{P}}{\partial t} = \frac{1}{2}\frac{\partial}{\partial a}\Big([(2\mathcal{G} + \mathcal{S})(1 - P) - 3\mathcal{S}]\alpha^2 a\mathcal{P} + 3\mathcal{S}(1 - P)\alpha^2|a|^2 a\mathcal{P} + (2\mathcal{G} + \mathcal{S})\alpha^2\Big)$$

$$+ C.C. + \mathcal{G}\frac{\partial^2 \mathcal{P}}{\partial a \partial a^*}. \quad (8.202)$$

The last term is the diffusion of the boson field and the first order term is the drift term. The drift term gives the amplification or attenuation of the boson field, which has a Landau-Ginsburg type of potential term. The second order term gives the spread of the field.

Introduce a probe term of the form $-\Phi\mathcal{P}$, where Φ represents the input of an external signal. The drift and diffusion coefficients when integrated by a and a^* gives the probability distribution for a as

$$\mathcal{P}(a) = N exp\Big([(2\mathcal{G} + \mathcal{S})(1 - P) - 3\mathcal{S}]\alpha^2|a|^2 + 6\mathcal{S}(1 - P)\alpha^2|a|^4 + C.C.$$

$$+ (2\mathcal{G} + \mathcal{S})\alpha^2(a + a^*)\Big), \quad (8.203)$$

where N is a normalization constant. This is a Landau-Ginsburg type of potential, which enters into a probability function that defines the partition function for quantum fields interacting with the black hole. The terms

$$\xi = (2\mathcal{G} + \mathcal{S})(1 - P), \quad \xi_T = 3\mathcal{S} \quad (8.204)$$

are respectively the primary variable, ultimately determined by the gravity on the black hole horizon, and the critical value of that parameter. The generator of the partition function is a thermodynamic potential written as

$$F(\xi) = (\xi - \xi_T)|a|^2 + 2(1 - P)\xi_T|a|^4 - \Phi(a + a^*), \quad (8.205)$$

for $P = exp(-F(\xi)\alpha^2)$. It is apparent that above the threshold the quantum modes have the same phase and energy, but below it modes have a spread of frequencies and uncorrelated phases. Thus for small gravity g the black hole is associated with quantum fields in the exterior region that have a completely random appearance in Hawking radiation. Figure 8.5 illustrates the potential function for the probe term set to zero.

This theory is then entirely analogous to the ferromagnet with a Curie temperature T_C. A classical-like field Φ applied to the black hole as an input signal, or states to be teleported, is analogous to the external field $\mathbf{H}$ applied to a ferromagnet as a temperature T. The input field skews the Mexican hat potential to induce a symmetry breaking. The boson field a defines a critical ordered parameter $(\Phi/\xi_T)^{1/3}$ analogous to the critical isotherm $M = (H/T \times const)^{1/3}$, where the zero field susceptibility is

$$\chi = \frac{\partial a}{\partial \xi}\Big|_{\Phi = 0}. \quad (8.206)$$

This is a phase transition for black holes from a purely quantum description to a semi-classical description. With ferromagnetism a magnetization remains frozen for $T < T_C$ and the magnetized spins remain fixed for a long time period. For $T > T_C$ the spins are easily changed by the input of an external field, where for $\mathbf{H} = 0$ the orientations of spins are random. In the case of the quantum black hole the structure of the internal states of the black hole are entangled with the boson field for $\xi > \xi_T$, which is a domain where quantum teleportation through black holes is readily accomplished. For $\xi < \xi_T$ the external field does not have strong coupling to these internal states, where quantum information about the interior becomes less accessible. This is then a phase transition between a pure state quantum mechanical description of a black hole and a semi-classical one. This would suggest that the domain where teleportation is convenient is on a scale where the boson field, presumably a Yang-Mills gauge field such as electromagnetism, is unified with gravitation. Below ξ_T information sent into a black hole becomes intractably difficult to extract. The information sent into a black hole still remains, but the freezing out of the black hole into a semi-classical description at a large scale makes teleportation simply too difficult in practice.

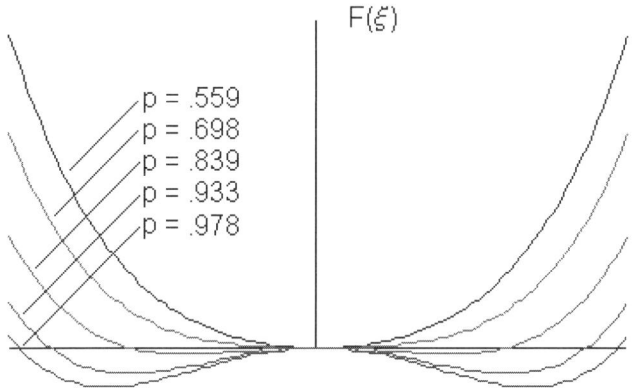

The potential function for various values
of the absorption coefficient.
Figure 8.5

Beyond the analogue with ferromagnetism are analogues with superconductivity and the Higgs field in elementary particles. In particular the analogue with the Higgs field is particular of interest. It is unlikely that this is the dual to the Higgs field in the TeV range of the $SU(3) \times SU(2) \times U(1)$ standard model This duality may reflect the inner gauge field in quantum gravity fluctuations. This phase transition may be responsible for the decoupling of gauge fields from quantum gravity or for the compactification of strings and Calabi-Yau spaces. Under special conditions the two Higgs fields may transform into each other so that the Higgs field of the standard model reflects some black hole physics.

It is also possible to extend this to cosmologies as well. For a cosmological constant Λ the magnitude of the Killing vector on the cosmological horizon is $|\xi| \sim \sqrt{\Lambda/3}r$. For the extremely early cosmos the value of $|\xi|$ is very large. This would lead to a similar phase transition between the extremely early quantum cosmology and the subsequent semi-classical cosmology, leading obviously to the current state of the observable universe. The field operators a_k and b_{-k} exist across the cosmological horizon, where a coherent description of this may be possible according to the B^* algebra or octonions for the extremely early nascent universe. This raises an interesting question concerning the "string landscape." The coupling of a cosmology with a Higgs type of field is given by the Lagrangian density

$$\mathcal{L} = \int d^4x \sqrt{g} \left(\frac{1}{8\pi} \phi R + |\nabla \phi|^2 - \frac{\mu}{2}(|\phi|^2 - \lambda^2 |\phi|^4) \right). \tag{8.207}$$

The phase transition suggests phenomenologically the Higgs field is determined in part by the structure of spacetime itself. Hence the landscape of cosmologies, if we are permitted to talk about this in the plural, is "crafted" in part by the evolution of the spacetime itself. If we are to think that cosmologies emerge from the vacuum state this means the process involves the guiding a cosmology by a scalar field is as much about the guiding of the scalar field by the spacetime cosmology. The process of by which cosmologies emerge from the vacuum state is likely considerably more complex than currently thought.

8.13: Higher Polyhedral Representations

From the 24-cell a polyhedral representation of the E_8 may be derived. The 24-cell has an underlying golden mean, which means the E_8 lattice embodies one of the most elementary structures known. The golden mean or ratio is defined by the construction of a rectangle that embeds another rectangle so its long side is the short side of the first and the process is continued. In the iterative process the ratio between the long and short sides of the two are equal. For a continued nesting of such rectangles this leads to the number

$$\phi = \cfrac{1}{1 + \cfrac{1}{1 + \cfrac{1}{1 + \cfrac{1}{1 + \cdots}}}}, \tag{8.208}$$

where algebraically ϕ satisfies the equation $\phi^2 - \phi - 1 = 0$ so that $\phi = (1 + \sqrt{5})/2$. ϕ is not transcendental and is algebraic in the sense of Galois theory, but it is a highly irrational number. ϕ further gives the ratio of the different sides of the pentagon and the ratio of adjacent number in the Fibonacci sequence 1 1 2 3 5 8 13 21 . . . .

The 24-cell has 24 vertices. The dual 24 cell also contains 96 edgelinks. Consider a point on each edgelink that divides each according to the golden ratio. This produces the $96 + 24 = 120$ vertices of the 600-cell. The 600-cell is dual to the 120-cell.

By one choice of 7 such constructions one is then able to construct the 240-vertex Witting polytope for the 8-dimensional E_8 lattice. Since the golden ratio contains the $\sqrt{5}$ this extends the D_4 lattice of the 24-cell in four dimensions, with rational lengths, to eight dimensions by the extension of the rational numbers to the irrational number $\sqrt{5}$. The 120-cell is then a half of the E_8 lattice.

The dual to the 120-cell is the 600-cell. The 600-cell is a 4-dimensional convex regular polytope with 600 facets. Its Schläfli symbol of the 600-cell is {3, 3, 5}. It is a 4-dimensional analogue of the icosahedron, constructed by the golden ratio on the edges of an octohedron. All of the 600 facets are tetrahedral, where 20 facets meet at each vertex. The number of vertices is 120 by duality of the 600-cell with the 120-cell.

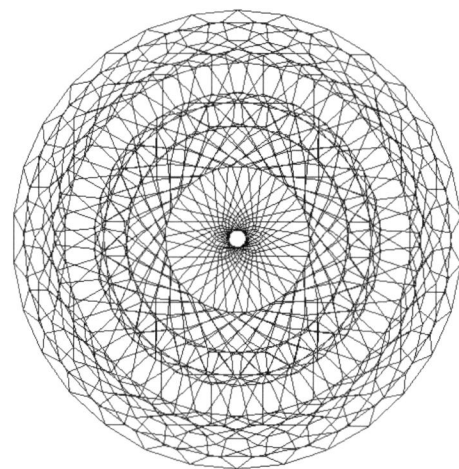

The 120-cell projected onto the x-y plane. The 600-cell projected onto the x-y axis.
Figure 8.6   Figure 8.7

The group structure is analyzed by the structure of its edges. The vertices of a 600-cell centered at the origin of four dimensional space can be given according to edgelinks of lengths $1/2$, 1 and $1/\Phi$ for $\phi = (1 + \sqrt{5})/2$. By permutations this is done as follows: 16 vertices of the form $(\pm 1/2, \pm 1/2 \pm 1/2, \pm 1/2)$, with 8 vertices obtained from $(0, 0, 0, \pm 1)$ and 96 vertices are obtained by taking even permutations of $(\pm 1, \pm \phi, \pm 1/\phi, 0)$. The first 16 vertices are the vertices of a tesseract, where these together with the next 8, form the vertices of the 24-cell. These then form the roots of the E_8 group, which is the Weyl group H_4 of order 14400. The vertices then define quaternions under quaternionic multiplication.

Octonions are infinitely reflexive and recursive. Their underlying structure with the golden mean is a path towards a Julia set constructions. G_2 preserves ϕ as the automorphism of the octonions. This leads to a recursive set of maps with an underlying golden mean. The fractional dimension of Julia sets and the Hausdorff dimension imply a regularization to this recursion. This sort of structure is seen in a number of ordinary situations, such as the recurrent Fibonacci sequence in a head of cauliflower.

330 *Quantum Fluctuations of Spacetime*

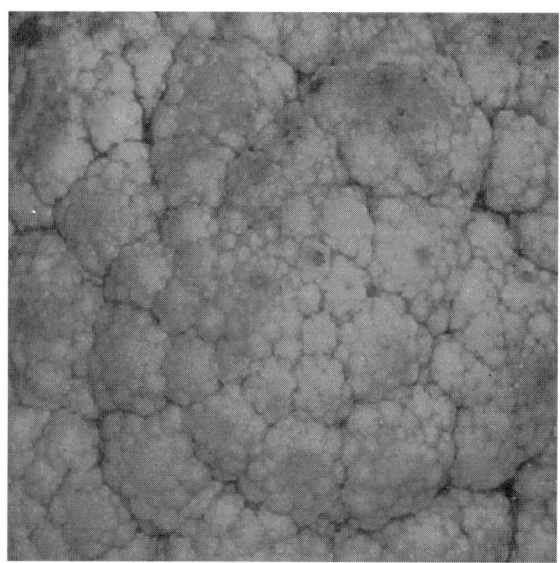

Recurring fibonacci sequence in cauliflower.
Figure 8.8

References

[1] J. Baez, "The Octonions", http://www.arxiv.org/abs/math.RA/0105155

[2] C. A. Manogue, *Octonionic Mobius Transformations*, Mod. Phys. Lett. **A14** (1999) 1243-1256.

[3] K. Abdel-Khalek, "Beyond Octonions," http://www.arxiv.org/PS_cache/math-ph/pdf/0002/0002023.pdf

[4] Y. Song, "Finite Invariance of Cayley Calibration Form," http://www.math.hmc.edu/seniorthesis/archives/2000/ysong/ysong-2000-thesis.pdf

[5] K. Abdel-Khalek *Unified Octonionic Representation of the 10-13 Dimensions Clifford Algebra* http://www.arxiv.org/PS_cache/hep-th/pdf/9704/9704049.pdf

[6] R. Feynman, "Feynman Lectures on Physics," bf 3 Addison Wesley Longman (1970).

[7] H. Hopf, "Uber die Abbildungen der dreidimendionalen Sphäre auf die Kugelfläche," *Mathematische Annalen*, **104** (5), (1931) 637-665.

[8] L. Susskind, *J. Math. Phys.* **36**, 11, (1995) 6377.

[9] G. 't Hooft, *Class. Quant. Grav.* **16**, (1999) 395-405.

[10] Y. J. Ng, "Spacetime Foam," *Int J. Mod. Phys.* **11** 1585 (2002).

[11] A. Ashtekar, J. Baez, K. Krasnov, *Adv. Theor. Math. Phys.* **4** (2000) 1-94.

[12] H. D. Zeh, "One the Interpretation of Measurement in Quantum Theory," *Found.*

Phys. **1**, 69 (1970), "What Is Achieved By Decoherence,"
http://www.arxiv.org/PS_cache/quant-ph/pdf/9610/9610014.pdf

[13] C. Miser, K. Thorne, J. Wheeler, *Gravitation*, Ch 21, Freeman Press, San Francisco (1973).

[14] G. Dixon, *Il. Nouv. Cim.* **105B**(1990), 349. and "Division algebras, Galois fields, quadratic residues,"
http://www.arxiv.org/PS_cache/hep-th/pdf/9302/9302113.pdf

[15] V. Dzhunushaliev, "Nonperturbative operator quantization of strongly nonlinear fields," *Found. Phys. Lett.* **16**, 1 (2003) 57-70.

[16] T. Smith, "Hermitian Jordan Triple Systems, the Standard Model plus Gravity, and $\alpha_E = 1/137.03608$,"http://www.arxiv.org/PS_cache/hep-th/pdf/9302/9302030.pdf

[17] L. Smolin, "The exceptional Jordan algebra and the matrix string,"
http://www.arxiv.org/PS_cache/hep-th/pdf/0104/0104050.pdf

[18] P. Rowlands, J. P. Cullerne, "The Dirac algebra and its physical interpretation,"
http://www.arxiv.org/PS_cache/quant-ph/pdf/0010/0010094.pdf

[19] R. Munoz-Tapia, *Am. J. Phys.* **61**, 11 (1993).

[20] C. Becchi, A. Rouet, R. Stora, *Ann. Phys.* **98** 287 (1976).

[21] A. D. Aczel, *Entanglement*, PLUME, Penguin Publishing, New York, NY (2003).

[22] N. Bohr, "Discussion With Einstein On the Epistemological Problems in Atomic Physics," *Albert Einstein: Philosopher-Scientist*, **1**, ed. P. A. Schilpp, Open Court Classics (1949).

[23] J. D. Bekenstein, "Generalized Second Law of Thermodynamics in black hole Physics," *Phys. Rev.* **D9**, 3292.

[24] Y. J. Ng, "Selected Topics in Planck-Scale Physics," *Mod.Phys.Lett.* **A18** (2003) 1073-1098.

[25] L. Doyle, "Quantum Astronomy,"
http://www.seti.org/site/apps/nl/content2.asp?c=ktJ2J0MMIsE&b=194993&ct=320725

[26] M. J. Duff, "M-theory on manifolds of G_2 holonomy: the first twenty years,"
http://www.arxiv.org/PS_cache/hep-th/pdf/0201/0201062.pdf

[27] L. B. Crowell, R. Betts, "Spacetime Holography and the Hopf fibration," *Found. Phys. Lett.*, **18**, 2, April 2005 (183).

[28] M. J. W. Hall, "Exact Uncertainty Approach In Quantum Mechanics and Quantum Mechanics and Quantum Gravity," plenary talk presented at the 4th Australian Conference on General Relativity and Gravitation (Melbourne, January 7-9, 2004); Proceedings to appear in GRG, http://www.arxiv.org/PS_cache/gr-qc/pdf/0408/0408098.pdf

[29] C. Adami, G. L. Ver Steeg, "black holes conserve information in curved-space quantum field theory".

http://www.arxiv.org/PS_cache/gr-qc/pdf/0407/0407090.pdf

[30] G. T. Horowitz, J. Maldacena, "The black hole Final State", *JHEP* 0402 (2004) 008.
http://www.arxiv.org/PS_cache/hep-th/pdf/0310/0310281.pdf

[31] M. B. Green, J. H. Schwarz, E. Witten, "Superstring Theory" **1**, Cambridge University Press, Cambridge UK, New York, NY (1988).

[32] T. C. Hales, "An Overview Of The Kepler Conjecture",
http://www.arxiv.org/PS_cache/math/pdf/9811/9811071.pdf

[33] O. R. Musin, "The Kissing Number In Four Dimensions,".
http://www.arxiv.org/PS_cache/math/pdf/0309/0309430.pdf

[34] H. M. S. Coxeter, "Regular Polytopes," Dover, London (1973).

[35] B. Fuller, "Synergetics", Macmillan Pub Co. (1982).

[36] H. Cohn and N. Elkies, "New upper bounds on sphere packing I," *Annals of Mathematics*, **157** (2003), 689-714,
http://www.arxiv.org/PS_cache/math/pdf/0110/0110009.pdf

[37] Ding, et al., "Chinese Remainder Theorem," World Scientific, Singapore (1996),

[38] J.H. Conway, N.J.A. Sloane, "Sphere packings, Lattices and Groups," Springer-Verlag, New York, (1988).

[39] F. Klein, *et. al., Famous Problems and Other Monographs*, Kegan Paul, London (1962).

[40] N. Jacobson, *Lectures on Abstract Algebra*, vol 3, Van Norstrand, Princeton (1964).

[41] R. Geroch, *Mathematical Physics*, University of Chicago, Chicago (1985).

[42] V. Pless, "Decoding the Golay codes," *IEEE Trans. Info. Theory* **32** (1986), 561-567.

[43] G. Dixon, "Octonions: E_8 Lattice to Λ_{16},".
http://xxx.lanl.gov/PS_cache/hep-th/pdf/9501/9501007.pdf

[44] W. H. Louisell, "Quantum Statistical Properties of Radiation in Pure and Applied Optics" John Wiley & Sons, New York (1973).

[45] M. Sargent, M. O. Scully, W. E. Lamb, "Laser Physics," Addison-Wesley Publishing Co., (1974).

[46] W. T. Coffey, Y. P. Kalmykov, J. T. Waldron, "The Langevin Equation" World Scientific, Singapore (1996).

[47] J. R. Klauder, "The Current State of Coherent States,"
http://www.arxiv.org/PS_cache/quant-ph/pdf/0110/0110108.pdf

9: Physical Law from No Law

Ultimately there is the question of "why physics?" This might appear to be a strange question for obviously physical reality exists and physicists have engaged in a multi-century endeavor to discover the fundamental principles that govern nature. This poses the question of why it is that physical laws exist as they do at all. Classically the idea has been there must exist some finite set of rules when properly applied explain everything. In fact some modern theoretical trends have portrayed themselves as a "theory of everything." Yet even if such an ultimate foundational principle were found we would still be confronted with the question of why those principles apply instead of something else. Wheeler points out that one can lay out mathematical formulae that describe physical processes in squares on a floor, but none of them "fly." Reality may be describable up to some level in its foundation, but that understanding fails to capture the existential property of nature: nature exists, exists as it does and not in some other form. It might then be that ultimately there are no absolute foundations in an axiomatic sense, but rather physical laws exist for purely accidental reasons and that there may be an infinite nesting of structures or principles that describe nature up to the Planck scale.

To start consider the nature of the chronology protection conjecture. This is a more accessible question than attempting to examine issues addressed in Chapter 8. Chronology protection is a principle by Stephen Hawking that nature is organized in such as way as to prevent closed timelike curves and time travel. This conjecture may be examined as if its truth were decided by the universe itself modeled as a quantum computer. The cosmological wave state, written according to a basis of states for spinor variables, is rotated in the evolution as bits in a universal quantum computer. Here it is demonstrated that the cosmic censorship hypothesis and chronology protection conjecture are not decidable propositions. Closed timelike curves define a recursively enumerable set, which in turn corresponds to a set of undecidable propositions in canonical quantum gravity. This means that these exist in quantum gravity as an ω-consistent set.

The phase structure of black holes appears to be the added requirement need for chronology protection. This involves an appeal to additional principles outside of canonical quantized gravity. By the same token the same issue probably exists with string theories and octonionic quantum gravity with error correction coding states. This discussion is meant to illustrate physical principle may exist by pure randomness. This randomness has its links to the formal structures used to describe them. This relationship between axiomatic incompleteness and the emergence of physical laws by random means suggests that mathematics is ultimately a model system.

Next the nature of space is examined in the light of an apparent logical paradox due to Banach and Tarski. This theorem illustrates that a set with a spatial measure may be decomposed into components by group rotations and then reassembled to duplicate that set. This would be a matter where a ball is disassembled and reassembled to give rise to two such balls. This is a result of some undecidable issues with the nature of a measure over a set or space, with connections with the Gödel-Turing theorems on the incompleteness of mathematics. This is applied as a method for duplicating simplices in a tessellated space or a CW complex. This then leads to topology changes. It is illustrated that this leads to a quantum orthomodular lattice that describes topology changing quantum fluctuations and states. Here it is illustrated that the foundations may in fact be due to the complete randomness of the truth value for propositions corresponding to physical states.

It is possible that this might have a bearing on the Anthropic Cosmological Principle (ACP). The ACP has gained some adherents in the last decade. Yet caution needs to be exercised, for the strong ACP does not appear to lead to anything that can be called a useful physical theory. The ACP may in fact be a true conjecture on the structure of the universe, but its truth may be unprovable and of a nature that it proves little or nothing else. A brief discussion is given on the problem of consciousness, where our near scientific ignorance on the nature of consciousness is such that no credible scientific conjecture, let alone a conclusion, on whether the existence of intelligent life is a requisite aspect for the existence of a cosmology.

9.1: Universal Computation in Quantum Cosmology and the Decidability of Cosmic Censorship

Mathematical logic has not traditionally been a major part of mathematical physics. Only recently has the interest in the physics of computation put logic as a mathematical formalism appropriate for physics. Currently these studies are involved with information theory and algorithmic complexity[1]. There is also the nascent study of quantum computers [2]. Yet here this is less a study of logic as a formalism for the foundation of the structure of reality. This may change with the subject of quantum gravity. The reason is that spacetime quantum foam is filled with virtual wormholes that shuffle the "time deck." A virtual wormhole or real wormhole, an exciton of the vacuum, in a spacetime of virtual fluctuations may be boosted so that the wormhole is a form of virtual or real time machine[3]. Therefore it is difficult to construct a time ordered product of fields required to describe a path integral. Yet, physics larger than the Planck scale is such that time ordered products are applicable without any sort of discrepancy or uncertainty in assigning time intervals. This raises the question of how it is that amongst the multiple possible shuffling of the "time deck" the universe is able to "decide" which time shuffling procedure is appropriate.

Wormholes with one opening boosted in a "send and return," or oscillatory manner act as a form of time machine[3]. Assume in the spacetime quantum foam that a virtual wormhole is created so that the two faces are created near each other. If during the time $\Delta t \simeq \hbar/\Delta E$ one face of a virtual wormhole is boosted and returns to the second face before the wormhole is absorbed into the vacuum, then the spacetime foam has a nonchronal region. This means fields within the spacetime foam can propagate through the time machine backward in time. The time ordering of a field path that propagates through the wormhole is fundamentally different if it is to the future of the Cauchy horizon.

For such a field path through the virtual time machine, the notion of time involves closed timelike loops. This means that a path integral of such a field has an ambiguity, for the time ordering of paths that do not wind through the wormhole are fundamentally different from those paths that do.

The wave functional defines a superposed foliation of spacelike manifolds of evolution that is described by the Wheeler-DeWitt equation [4]. This is canonically quantized according to conjugate observables π_{ij}, g^{ij}, the spacelike momentum and metric respectively, according to $\hat{\pi}_{ij} = -i\hbar\delta/\delta g^{ij}$. The Hamiltonian constraint in the ADM formalism of general relativity is an operator $\mathcal{H}$ that acts on the wave functional $\Psi[g]$ as $\mathcal{H}\Psi[g] = 0$, often called the wave function of the universe [4]. Introduce a scalar field ϕ onto the spacetime to extend the wave functional to $\Psi[g,\phi]$ and put the Wheeler-DeWitt equation in the form

$$-\left(G^{ijkl}\frac{\delta^2}{\delta g_{ij}\delta g_{kl}} + R^{(3)} - \frac{\partial^2}{\partial\phi^2} + \phi^2\right)\Psi[g,\phi] = 0. \tag{9.1}$$

G^{ijkl} is the supermetric for the minisuperspace of six dimensions. The scalar field is also a system of harmonic oscillators. The wave function is expanded into a set of modes of the field ϕ as $\Psi[g,\phi] = \sum_n \psi_n[g]a_n(\phi)$, with

$$\frac{1}{2}\left(\frac{d^2}{d\phi^2} - \phi^2\right)a_n(\phi) = (n + 1/2)a_n(\phi). \tag{9.2}$$

Here the scalar wave has been rescaled as $\phi \rightarrow \phi/(\sqrt{2}\pi\sigma R)$, where R is the radius of the cosmology and σ is a normalization factor. The eigenvalues may then be interpreted as due to the application of a time-like derivative on a basis of elements $\psi_n(t) = \psi_n^0 e^{-i(n+1/2)t}$,

$$\mathcal{H}\psi_n[g] = i\frac{\partial}{\partial t}\psi_n[g] = (n + 1/2 - E_0)\psi_n[g], \tag{9.3}$$

where the time derivative measures the numbers of Planck units. This scalar field then acts to define a time-like parameter so that the Wheeler-DeWitt equation becomes a Schrödinger equation. This is a curious element to the Wheeler-DeWitt equation, for without some auxiliary field it is simply a constraint equation with no sense of time. It is the introduction of an auxiliary field the permits one to consider the Wheeler-DeWitt equation as analogous to a Schrödinger equation that describes the evolution of a cosmological wave function.

The scalar field defines a time or a pseudotime as Planck units of length. The scalar field is also computed by Greene's functions and time ordered products of fields. Given the occurrence of the field at various times t_1, t_2, ..., t_n the Greene's function $\langle\phi_1, \phi_2, ...\phi_n\rangle$ may be constructed. Yet for the field amplitudes that loop through the virtual wormhole there are paths to the future of the Cauchy horizon and to the past. Since both of these fields converge to the Cauchy horizon there is an ordering reversal of those fields in the future region of the Cauchy horizon. Hence a path integral that computes the sum over all path histories will have a time ordering ambiguity. Paths that wind an infinite number of loops around the wormhole require an infinite number of time ordered fields to define a Greene's function. Such a path defines a caustic [3], where the winding through a wormhole is characterized by the first cohomology class of the spacetime [5]. The very early universe contained a large number of virtual wormholes in the quantum foam that played a significant role on the nascent cosmology. Hence a path integral for the cosmological wave function will in general be one that contains many such paths that are "time shuffled" in many different ways. A path integral for the early universe is one then that would likely suffer some serious time ordering ambiguities.

The path integral is effectively a machine that computes the interference between many possible paths that a quantum system follows. Each path is an element in a quantum ensemble of paths. The time ordering of this path is determined by the proper ordering of the fields ϕ in the computation of a Green's function. Hence the Wheeler-DeWitt equation is determined by time parameterizations of these fields, as a machine that computes the dynamics of each path. Each computation of a path is based on the input of a set of fields that are time ordered according to a geodesic winding of the path, where these fields interfer with themselves. The ordering of the scalar field is then an instruction set for the determination of a particular path, or set of paths that have the same topological winding.

The Wheeler-DeWitt equation for the wave functional $\psi[g]$

$$\mathcal{H}\psi[g] = i\frac{\partial}{\partial t}\psi[g], \tag{9.4}$$

then dictates that the evolution of the wave function is given by the unitary operator

$$U(t)\psi[g] = e^{-i\mathcal{H}t}\psi[g]. \tag{9.5}$$

Here t is a time-like variable defined by an auxiliary field. An ordered sequence of fields ϕ is then an instruction set, where this evolution equation can be regarded as the quantum mechanical description of a universal digital computer. If the spacelike variables are written according to Ashtekar's variables [6] $g_{ij} = \sigma_i^a e^a{}_j$, $\pi^{ij} = \sigma^{ai}\omega^{aj}$ the theory is described according to spinors. This suggests an equality between the evolution of these spinor variables and the quantum computer of Deutsch [7], where the rotation of a spin determines the evolution of a single bit. Bennett [8] demonstrated that the map from one state to another must be bijective to have a description as a pure quantum state. Thus, each state has a unique image and preimage. This is necessary in order to describe a quantum computer as the proper quantization of a Turing machine or VonNeumann machine. There must be a record of the computation performed within the machine, since logical operations can erase information. This requirement is made to preserve the value of $Tr|\psi\rangle\langle\psi|$ and maintain quantum coherence.

The Hamiltonian is then an operator that determines the phase of the wave function. With each time interval δt, determined uniquely by a particular path according to the ordering of ϕ, there is a change in the phase of the wave function. So each path is a sort of machine that changes its phase to compute the Greene's function, whose output is the final wave function. The fields ϕ_n are associated with the integer for their eigenvalue n. A particular time ordered product is a set of symbol strings for the computational input.

Now find the eigenvalues or equivalently the diagonalization of the generator of the evolution operator U. The wave function is expanded in a basis $|e^a{}_i\rangle$, where $e^a{}_i$ is the spinor triad for the spacelike metric. A local Lorentz transformation will then be a rotation of this variable $(ReR^\dagger)^a{}_j$ plus the determination of an appropriate lapse function. This rotation is of the form $(e^{i\theta}e)^a{}_j$. The corresponding transformation of the basis element is $e^{i\theta}|e\rangle$, where the spinor and coordinate indices have been suppressed. Consider the eigenvalues as the set of discrete rotations of the spinor. So according to this set of rotations the basis elements are $|e\rangle = \sum_m e^{i2\pi/m}|e_m\rangle$, where m are equally spaced eigenvalues. Rotations due to the evolution operator carry one basis element into another according to such rotations. This demands that the evolution operator will have eigenvalued generators of the form $i2\pi/m + n$, where for $r = nm/2\pi$ this gives the eigenvalued generators $exp(i2\pi(r+1)/m) = v_m{}^r$. The action of the evolution operator on the state $|e\rangle$ is then

$$U|e\rangle = \sum_{m=0}^{\infty} \frac{1}{\sqrt{m}} \sum_{r=1}^{m-1} v_m{}^r |e_m\rangle, \qquad (9.6)$$

which for each basis element e_m gives the entire set of independent rotations of the spinor. This evolution operator then describes the spacelike spinor as a spin element and as a bit in the Deutsch quantum computer.

A computer that halts is one that evolves through a finite number of states, where its final execution is to enter the halt state. A path integral for a computer is a superposition of computers that are started at an initial state and end at a final state. The entire wave function is a superposed collection of all possible such computers, which may be a set of algorithms that enters a universal quantum computer. For those paths given by a ordered products of fields ϕ which wind completely through the wormhole are ordered products infinite in length. In effect these correspond to quantum computers that fail to halt. Further, the path integral contains these quantum computers that approach the Cauchy horizon from the initial state and well as the final state. Computers that halt will have a finite set of spinor rotations whose Fourier transforms have a discrete set of frequencies. The computers that fail to halt induce a countably infinite number of spinor rotations and contribute to the Fourier transform a white noise type of spectrum. Those computers that do halt, within the superposition, will have a Fourier transform that returns the set of angles the spinor is rotated through. These angles will be a rational number times 2π. Those computers, within the superposition, that fail to halt contribute a set of frequencies that reflect a continued fraction expansion that approaches some irrational number within the interval $[0, 2\pi]$. Turing proved that no universal computing system is able to make a list of all possible halting machines, essentially through the Cantor diagonalization argument [9]. The cosmological path integral, considered as a universal computing system, is unable to formally list all those paths that halt from those that do not halt, which correspond to paths that infinitely wind through a multiply connected topology.

This leads to the question of the cosmological censorship hypothesis. Wormholes are spacetimes that produce a Cauchy horizon that lightlike paths asymptotically approach. This leads to a caustic that is a naked singularity in spacetime. Wormholes, as spacetime solutions, violate the cosmic censorship hypothesis. The cosmic censorship hypothesis then demands that all spacetime singularities are unable to propagate information to the future infinity $\mathcal{I}^+$. The related statement of Hawking is the Chronology Protection Conjecture that requires the laws of physics do not permit the existence of closed timelike curves [10]. For the Cosmological Censorship Hypothesis to be decided this would require a UTM that can solve the halting problem, which is impossible. So the Wheeler-DeWitt equation does not have the algorithmic capacity to prove cosmological censorship or chronology protection.

The ordered products of fields ϕ are a symbolic input, which defines the set of all time ordered products of scalar fields are $\cup_{m=1}^{\infty} \langle \prod_{n=1}^{m} \phi_n \rangle = \mathcal{S}$. For each one of these elements there is a path in the cosmological wave function that is a computer that accepts it. The universal language is $L_u = \{(M, \mathcal{S}) | M \text{ accepts } \mathcal{S}\}$. It is known that such a universal language is recursively enumerable [11]. A recursive set or algorithm is one who's complement is itself recursive, where a recursively enumerable set is one whose complement is not recursive or recursively enumerable. This leads to the existence of true self-referential statements that indicate their own unprovability. If these statements were false, the complement of these statements would themselves be provably true and thus recursive. The set of unprovable statements is then a recursively enumerable set. That their complements would be recursive is a contradiction. This would mean that the set L_u is not recursive. This is the essence of the Gödel theorem [11,12].

For L_u a language set of the cosmological wave function, this set of statements gives an unprovability of any decision process that can determine whether there is a consistent ordering procedure for the fields ϕ. Further, if this conclusion were false, then this means that L_u is recursive. Equivalently the existence of such a time ordering procedure contradicts the enumerable recursive property of this set. Hence this demonstration no complete time ordering procedure is ω-consistent leads to conclusion that the Cosmological Censorship Hypothesis is unprovably false.

This means a number of possible things. Either the cosmological wave function as described by the Wheeler-DeWitt equation is not physically complete, or that there exist limits to understood physical principles. In the first case this would mean that other general principles are required to understand quantum gravity. It appears likely such a general principle requires there is no general time ordering procedure for this scalar field. Physically this would mean that in general the universe is one that permits the existence of closed timelike curves; at least quantum mechanically. It is unlikely closed timelike curves are permitted on a very large scale. In the second case the truth or falsehood of the cosmological censorship hypothesis is undecidable. This is a matter determined by self-referential aspects of treating quantum cosmology as a universal computing system. In this situation the falsehood of the cosmic censorship hypothesis is a matter of ω-consistency. Some thought should make one see that the two situations are not that disparate; for appealing to new physical principles is a matter of invoking new postulates or axioms. This demand is imposed by the limitations of logic for a universal computing system. This also means that in quantum gravity the cosmic censorship and chronology protection are "self referentially false." Conversely, it may well be the case that these two principles obtain in classical gravitation.

It appears that mathematics is becoming more of a physical subject just as physics is becoming more of a mathematical subject. It is a situation with physics where various theories operate properly within certain domains of experience; where a bad theory is one with the null set for such a domain. Physicists are willing to take off the hat of gravitation and put on the hat of fluid dynamics or quantum theory. Mathematics, on the other hand, has a tradition, founded by Plato, of looking for some great and ultimate foundation of logical truth. Mathematics has in recent times become a subject where different styles of proof exist, each method appropriate for certain mathematics. Since Haken's proof of the four color problem computer proofs have become a growing industry. Ultimately the undeniable Gödel-Turing theorems hangs over the heads of all mathematics. Further, mathematics has to be ultimately regarded as a subject that involves computations done by physical objects such as neurons and brains. Physics, or at least attempts to understand the foundations, has become more of a mathematical pursuit. At some point it would not be surprising if the limitations of mathematics to appeal to logical unity should also begin to occur in physics as well. It also appears that mathematics is adopting a sort of empirical, or empiri-logical, nature to it. This is as computer use increases to understand the nature of fractals and various spaces and as different forms of mathematics pertain to various limited domains. Physics has always existed with this sort of limitation, but we may be faced with the prospect of a mathematical understanding of how it is that the physical world is compartmentalized into domains of experience.

For quantum gravity and cosmology it appears that the Cosmic Censorship and Chronology Protection principles may be self-referentially false as the physics runs into the limits of logic. Yet, classically one is tempted to invoke considerable skepticism about the existence of wormholes and time machines. Even with the mechanics offered by Thorne to prevent contradictory paradoxes in causality skepticism is still warranted as the physical world, even classically, is known to be stochastic and likely not constrained by this mechanics. So it is likely that the classical universe does indeed uphold the Cosmic Censorship and Chronology Protection principles. To argue for these principles meansthe large scale structure of spacetime is ultimately the result of a coarse graining of the quantum states in quantum gravity. This indicates that general relativity is a phase transition from quantum gravity. From this it appears that physics must appeal to additional principles.

9.2: PreGeometry and Mathematical Undecidability

Chapter 7 illustrated the nonlocalizability of energy-momentum and its potential implications for quantum gravity. A multilinear B^* algebra was advanced to describe the quantum mechanics of variables that are not localizable. Octonionic quantum gravity is a special case of the B^* system. As indicated in Chapter 8 the use of higher order sporadic algebras may surpass octionionic quantum gravity. This theory presents a way in which the entropy associated with a coarse grained approach to the squeeze state tunnelling from the vacuum results in decoherence of the cosmological wave function. This entropy becomes the natural cut-off in the energy at one-loop quantum gravity where coordinate and momentum variables are no longer noncommutative with themselves at lower energy. In the B^* theory the vacuum zero point energy $\langle 0|H|0\rangle$ is not definable, where as a consequence time is no longer definable as a consequence since the Hamiltonian is the generator of time. This infers that closed time-like curves in path integrals in the spacetime "foam" prevent the establishment of an appropriate Cauchy data set. As one views spacetime on a scale that approaches the Planck scale the degree to which one may specify any Cauchy data begins to breakdown. Ultimately the entanglement of multiply connected regions assumes a scale that overwhelms any ability to specify any sort of geometry. The B^* algebra may describe quantum gravity theories between L_p and $\sqrt{8\pi}L_p$ (the gravity scale), but in the limit that reality approaches the Planck length this description may begin to breakdown.

As scales approach the Planck length quantum gravity fields propagate upon themselves. When this is considered as a quantum computer the data stack for the algorithm is ultimately the algorithm itself. This leads to the problem of self-referential mathematics. This runs into the issues of the Turing-Gödel theorems of mathematical undecidablility. As physics approaches the Planck scale that the B^* algebra will "feedback" on itself and run into difficulties of mathematical incompleteness. For the 24-cell of Planck volumes a quantum gravity computer exists to perform an error correction for the teleportation of states. This exists on a scale of $\sim 2.2L_p$. For the 8-dimensional sphere packing an error correction permits quantum gravity teleportation of states within $1.2L_p$. Yet as the system approaches L_p itself one either may need an infinitely larger space, or equivalently that Planck unit states compute themselves. By a black hole duality between inside and outside states those states larger than a Planck volume might be determined by those at the Planck scale — or smaller. Hence physical law may ultimately emerge for completely random reasons. The randomness may well stem from the randomness of self-referential mathematical truth.

John Wheeler suggested that ultimately physics may be about pregeometry [13]. Pregeometry is the suggestion that geometry has an underlying basis in nongeometric structures similar to the monads of Liebniz. Wheeler proposed a PreGeometry with a basis in mathematical incompleteness or in the quantum entangled logic of a huge number (infinite) of mathematical propositions that correspond physically to virtual physical states. The calculus of propositions will then determine how these pregeometric objects link together quantum mechanically to give rise to a geometric, or quantum geometric, structure that on ever larger scales defines geometry. Geometry is then due to a "logic sieve" that filters appropriate propositions that are proper theorem-proofs corresponding to geometry. This all appears to be scale dependent: pregeometric objects that link to one another by geometric theorem-proofs define objects on a scale larger than pregeometry. Those pregeometric objects linked by rules that violate the proper theorems of geometry remain buried at the fine grained scale of pregeometry.

The following is a possible route to pregeometry based on paradoxical duplication and measure of geometric objects. This approach returns to basic commutative geometry, for the analogue with noncommutative B^* algebra remains an open question. This is offered as a way geometry may be based on a pregeometry derived from the paradoxical duplication of simplices that tessellate the manifold. These simplices are objects at or near the Planck length scale.

The theorem of Banach and Tarski describes paradoxical duplication of the measure of a volume that shares many of these qualitative features. Both describe operations on a volume that changes it into an entirely different volume. The theorem states that any volume, with $dim = 3$, may reconstruct any other volume for the initial volume disassembled into subsets derived from group rotations acting on these subsets and then reassembled into a different volume from the initial volume [14]. The group operations performed on the volume subsets are the generators, $\sigma, \tau \in \mathcal{G}$, of the familiar $SO(3)$ group. This group is volume preserving, so the discovery of these paradoxical rotations on a volume that reconstruct different volumes is surprising. A similar theorem also demonstrates that any given volume V may be similarly decomposed and reassembled to produce two copies of itself $V \rightarrow V + V$. These volumes before and after these operations are equidecomposible. The proof of the Banach-Tarski theorem rests on showing that the volume may be decomposed into subsets, where some of the subsets are immeasurable. On even deeper foundations these results rest on the truth of the axiom of choice. If the axiom of choice is true a measurable set may be decomposed into nonmeasureable subsets. These subsets may be rotated and reassembled into a measurable set that has a different measure than the initial set [14].

Another parallel has been drawn between the Banach-Tarski paradox and the physical world. The paper by Bruno Augenstein [15] makes the connection between particle physics and the paradoxical decomposition of a ball. A three dimensional ball is paradoxically decomposed into two equivalent balls using a minimum of five pieces taken from the ball. One equivalent ball is formed from three pieces and the other from two. The parallel is drawn that the pieces are quarks. The three piece ball is analogous to a baryon and the two piece ball is analogous to a meson. This connection drives one to ask if there are connections between quantum mechanics and the Banach-Tarski paradox. Quantum mechanics has the property of being a sort of virtual "mass-energy duplicator" by virtue of the Heisenberg uncertainty principle. The parallel between hadron physics and this paradoxical duplication of geometric objects leads one to ask if this might provide for a new foundation to the quantum principle.

The above two speculations make us speculate that the theories of general relativity and quantum mechanics are ultimately unified by appealing to a physics that uses the Banach-Tarski language of paradoxically equivalent spheres. The first goal is to introduce the Banach-Tarski paradox and its proof in a semiformal manner. Now consider the mathematics to examine the nature of geometrodynamic conservation principle stated as the boundary of a boundary is zero $d^2 = 0$ in the cohomological view or $\partial \partial = 0$ in homology. The mathematics becomes a language for modelling the duplication of geometrodynamic quantities as an elementary quantum phenomenon. The viewpoint has implications for cohomology as $d^2 = 0$ and the nature of topological spaces.

A sphere is an object whose surface contains a continuum of points. Hence the cardinality of the set of points on the sphere is identical to the cardinality of the set of real numbers. Assume there exists a bag containing a countably infinite number of dust particles, where the dust particles are infinitesimally small. From this bag of dust each dust mote is place on a unique point on the sphere. There is a list generated at each of these operations that gives the dust mote's ordinal number and its position on the sphere. This list then effectively enumerates a dense set of points that sit on the surface of the sphere. Each entry of the list contains the pair of reals $(m_1 m_2 \ldots, n_1 n_2 \ldots)$. This list may be compressed into a list of reals by constructing the number $(m_1 n_1 m_2 n_2 \ldots)$. This illustrates the curious nature of transfinite numbers. Given the cardinality of a countably infinite set, $\aleph_0$, adding this set to another countably infinite set does not change the cardinality of the set, $\aleph_0 = \aleph_0 + \aleph_0$ [16]. Another curious property emerges from our list. The sequence of diagonal entries in this list, e.g. the integers $s_1, s_2, \ldots,$, may be used to form the real $(s_1 - 1)(s_2 - 1) \ldots$ not contained in the list. Including this new real in our list results in a list with the same cardinality, $\aleph_0 + 1 = \aleph_0$. This process may be repeatedly executed to obtain a power set over a set with cardinality $\aleph_0$. This demonstrates that the cardinality of the continuum, $\mathcal{C}$, is greater that $\aleph_0$. George Cantor suggested that $\mathcal{C} = 2^{\aleph_0}$, which is consistent with Zermaelo-Fraenkel set theory, but independent of its axioms.

The duplication of spheres or balls is due to these curious results in the study of transfinite numbers. Any countably infinite list of points on the sphere is insufficient to enumerate the continuum of points existing on the sphere. It is possible to take this list and assign alternate entries to separate lists. The cardinality of each separate list is $\aleph_0$. This satisfies the statement for the addition of transfinite cardinal numbers. Alternatively for two disjoint lists of points on our sphere, the above process may be inverted by interweaving the entries of the two lists into a third list. By performing the duplication of lists this effectively duplicates the sphere.

Equidecomposible volumes A and B under the action of the group $\mathcal{G}$ are written as $A \sim_\mathcal{G} B$. Let the group $\mathcal{G}$ act on a volume or set X. Let $E \subset X$ with $E = A \cup B$. Then E is equidecomposible using r pieces if $A \sim_m E \sim_n B$ with $m + n = r$. Now let σ and τ be the generators of $\mathcal{G}$, e.g. each is a rotation about a linearly independent axis. If ρ is one of σ and τ operations define $W(\rho)$ to be any set of sequences of rotations, or matrix symbol strings, starting with ρ. Now decompose the group $\mathcal{G}$ according to $\mathcal{G} = \{1\} \cup \big(W(\sigma) \cup W(\sigma^{-1})\big) \cup \big(W(\tau) \cup W(\tau^{-1})\big)$. Note that $\mathcal{G} = W(\sigma) \cup \sigma W(\sigma^{-1})$ and $\mathcal{G} = W(\tau) \cup \tau W(\tau^{-1})$. Let the symbol string $\eta \in \mathcal{G} \setminus W(\sigma)$, then $\sigma^{-1} \eta^{-1} \in W(\sigma^{-1})$ or $\sigma \sigma^{-1} \eta^{-1} \in \sigma W(\sigma^{-1})$. Consequently this symbol string sits in both $\sigma W(\sigma^{-1})$ and $\mathcal{G} \setminus W(\sigma)$. Consequently the set E on which $\mathcal{G}$ acts is $\mathcal{G}$-paradoxical.

Each of these generators of $\mathcal{G}$ are the typical rotation matrices of the $SO(3)$ group,

$$\sigma^{\pm} = \begin{pmatrix} 1/3 & \mp 2\sqrt{2/3} & 0 \\ \pm 2\sqrt{2/3} & 1/3 & 0 \\ 0 & 0 & 1 \end{pmatrix}, \quad \tau^{\pm} = \begin{pmatrix} 1 & 0 & 0 \\ 0 & 1/3 & \mp 2\sqrt{2/3} \\ 0 & \pm 2\sqrt{2/3} & 1/3 \end{pmatrix}. \quad (9.7)$$

With the above matrices, $W(\sigma)$, $W(\tau)$ represent a sequences of operations that start with σ and τ respectively. Take a point, x, on our sphere. Each $\alpha \in W(\rho)$, for a given ρ as either σ or τ, will rotate x to a unique point on this sphere. Therefore each α is associated with a point on the sphere.

Now proceed to make a list of all $\alpha_i \in W(\rho)$. For each set $\{\alpha_i\}$ there exists a collection of points $\{y_i\}$ on the sphere. The list of symbols is isomorphic to a set that contains all these collections of points. By Cantor's diagonalization process that the list of symbols is countably infinite, but incomplete. Consequently each $W(\rho)$ is equivalent to a dense set of points on the sphere. There are four dense sets of points on the sphere. By the above paradoxical equivalence, $\mathcal{G} = W(\sigma) \cup \sigma W(\sigma^{-1})$ and $\mathcal{G} = W(\tau) \cup \tau W(\tau^{-1})$, two pairs of these dense sets reproduce two disjoint sphere.

The group $\mathcal{G}$ is composed as $\mathcal{G} = A \cup B$, where $\mathcal{G}$ is paradoxical using four pieces $A \sim_2 \mathcal{G} \sim_2 B$. Define M to be the set of points on a sphere S with respect to the group $\mathcal{G}$, with

$$A^* = \cup\{g(M) : g \in A\}, \ B^* = \cup\{g(M) : g \in B\}, \ A^*, B^* \subseteq S. \quad (9.8)$$

The only elements contained in A and B that fix a point on S is the identity element. All other elements contained in A and B do not fix a point on S. The lack of fixed points infers that for $y \in S$ there is a unique $g \in \mathcal{G}$ such that $y \in g(M)$, $A^* \cup B^* = S$ and $A^* \cap B^* = \phi$ [13]. $A^* \sim_2 S \sim_2 B^*$ is then a direct result from $A^* \sim_2 \mathcal{G} \sim_2 B^*$.

The paradoxical decomposition of the sphere rests on the fact that a countably infinite dense set on a continuum characterizes what is mathematically knowable about the space. Even though the dense set is of measure zero on the sphere, the continuum is "carried along for the ride" by the group structure over the dense set. The dense set is the outer skin of the sausage and the continuum is the meat contained by the skin. Without the skin the sausage has no form or structure and without the meat the sausage has no content or substance.

Quantum lattices are a construct involving the operations of join, meet and complementation, $(\vee, \wedge, -)$, between propositions pertaining to quantum states. These are generalizations of the Boolean operations when applied to specific propositions. The set theoretic constructs $\cup, \cap, -$ in quantum logic then differ from the standard set theoretic operations in that they in general obey the distributive property [17][18]:

$$X \cap (Y \cup Z) = (X \cap Y) \cup (X \cap Z), \ Boolean\ logic$$

$$X \cap (Y \cup Z) \neq (X \cap Y) \cup (X \cap Z), \text{ Quantum logic.} \tag{9.9}$$

The property of nondistributivity in quantum logic will produce a quantum lattice for a spin 1/2 system that is valid for Stern-Gerlach experiments. Quantum lattices may be constructed by the paradoxical equivalence of a sphere or ball. Below is the construction of a quantum lattice for a spin 1/2 particle from the paradoxical equivalence of a sphere in four pieces.

Extend this to the duplication of a three-ball $\mathbf{B}^{(3)} \subset \Sigma^{(3)}$. The matrices σ, τ are the generators of a group $\mathcal{G}$. For any symbol string that starts on the left with $\rho \in \{\sigma^{\pm}, \tau^{\pm}\}$, $W(\rho)$, the group $\mathcal{G}$ may be decomposed as:

$$\mathcal{G} = \{1\} \cup W(\sigma) \cup W(\sigma^{-1}) \cup W(\tau) \cup W(\tau^{-1})$$
$$\mathcal{G} = W(\sigma) \cup \sigma W(\sigma^{-1}) = W(\tau) \cup \tau W(\tau^{-1}). \tag{9.10}$$

For $\Sigma^{(3)}$ the spacial surface let x be any point on the ball $\mathbf{B}^{(3)}$. Each of the subsets $W(\rho)$ is a list of symbol strings that act on the point $x \in \mathbf{B}^{(3)}$, where for each $h \in W(\rho)$, $h : x \to x'$. As a result the list of symbol strings, $W(\rho)$, corresponds to a set of points $\{x_i\} \in \mathbf{B}^{(3)}$. By the diagonalization theorem of Cantor the cardinality of $\{x_i\}$, $C(\{x_i\}) = \mathcal{C}$, is greater than the cardinality of countably infinite sets $\aleph_0 < C(\{x_i\}) \leq 2^{\aleph_0}$.

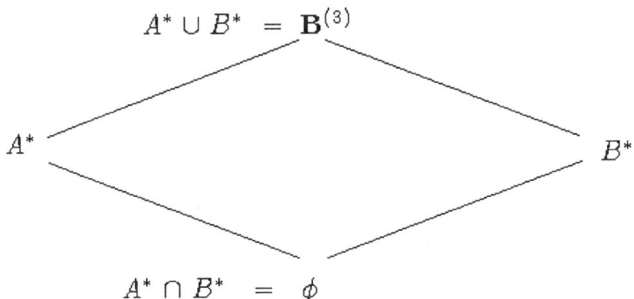

Elementary lattice for a basic duplication.
Figure 9.1

The set $\{x_i\}$ is the orbit on $\mathbf{B}^{(3)}$ of the subset $W(\rho)$ for some ρ. The above paradoxical equivalence of $\mathcal{G}$ by five pieces (split into two groups of two and three partitions at each step), $W(\sigma) \cup \sigma W(\sigma^{-1}) \sim_2 \mathcal{G} \sim_3 W(\tau) \cup \tau W(\tau^{-1})$, means that there exist two orbits A^*, B^*, that are sets of points on $\mathbf{B}^{(3)}$ that are reached from some $x. \mathbf{B}^{(3)}$ by the operations contained in $W(\sigma) \cup \sigma W(\sigma^{-1})$ and $W(\tau) \cup \tau W(\tau^{-1})$ respectively. The lack of fixed points in the $W(\rho)$, or that no nonidentity element for $\mathcal{G}$ fixes a point of the set, insures that $A^* \cup B^* = \mathbf{B}^{(3)}$ and $A^* \cap B^* = \phi$. This then duplicates the ball by paradoxical equivalence, $A^* \sim_2 \mathbf{B}^{(3)} \sim_3 B^*$. The set theoretic containment of A^* and B^* forms the lattice illustrated in figure 9.1.

Let $A^* = a^* \cup -a^*$ and $B^* = b^* \cup -b^*$, where a^*, $-a^*$ and b^*, $-b^*$ are the orbits of $W(\sigma) \cup \sigma W(\sigma^{-1})$ and $W(\tau) \cup \tau W(\tau^{-1})$ respectively. For any ρ $\rho' \in \{\sigma^{\pm 1}, \tau^{\pm 1}\}$ the following is true $W(\rho) \cup \rho'W(\rho'^{-1}) = W(\rho) \cup (G \setminus W(\rho')) = \mathbf{B}^{(3)}$, so $a^* \cup -a^* = \mathbf{B}^{(3)} = b^* \cup -b^*$. Since $A^* \cap B^* = \phi = (a^* \cup -a^*) \cap (b^* \cup b^*)$ it is evident that $a^* \cap b^* = a^* \cap -b^* = -a^* \cap b^* = -a^* \cap -b^* = \phi$. These relationships may be used to construct the lattice in figure 9.2. These set theoretic relationships give the quantum logical relationships on the quantum lattice, L, according to:

$$a^* \cup -a^* = A^* \sim_2 \mathbf{B}^{(3)} \sim_3 B^* = b^* \cup -b^*, \ \alpha \vee -\alpha = \bigvee,$$

$$a^* \cup b^* = -(-a^* \cap -b^*) = -\phi = \mathbf{B}^{(3)}, \ \alpha \vee \beta = \bigvee, \quad (9.11)$$

$$a^* \cap b^* = a^* \cap -b^* = \ldots = \phi, \ \alpha \wedge -\beta = \bigwedge, \ \alpha, \beta \in L.$$

for $\bigwedge$ = all false and $\bigvee$ = all true in the lattice L.

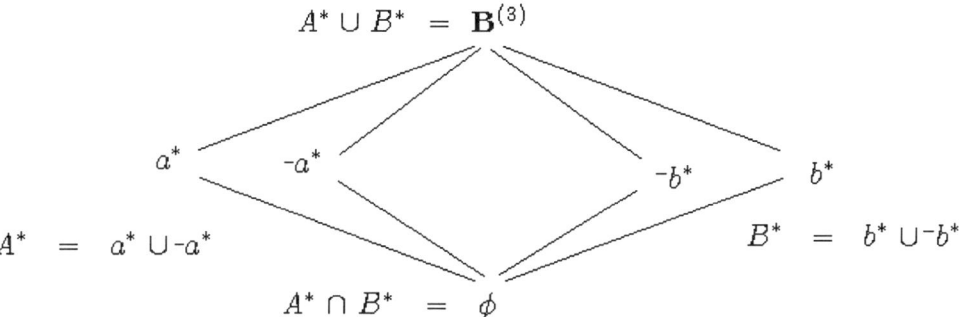

Lattice for a duplication of sphere with subsets.
Figure 9.2

This lattice generated by the paradoxical equivalency of a ball, $A^* \sim_2 \mathbf{B}^{(3)} \sim_3 B^*$, must have the nondistributivity property of a quantum lattice: $(a^* \cup -a^*) \cap b^* = b^*$, while $(a^* \cap b^*) \cup (-a^* \cap b^*) = \phi$, by equation 9.11. Other relationships derived in this fashion satisfy the criterion for being a quantum lattice that concerns a spin 1/2 system.

There are possible parallels between these peculiar reconstructions of three dimensional volumes and an underlying structure to differential geometry. The connection may be seen with general relativity. The curvature of spacetime governs the motion of particles along geodesics, and the masses contained in a region act to curve the manifold of spacetime. The possible parallels between these results in the foundations of mathematics and Einstein's gravitation are twice motivated: first by the analogy with the parallel transport of a volume, and next the need to understand physics, in particular quantum gravity, from a more fundamental anchor than geometry, called pregeometry.

The simplest avenue to start with is with the moment of rotation formalism of Elie Cartan [13 ch-15]. This is examined for moments of rotation and the paradoxical equivalence on a ball. Let $(1/2)\Delta y \Delta z \mathbf{e}_y \wedge \mathbf{e}_z$ be the triangular face of a tetrahedron. The rotation associated with this face is the quantity $e_\nu \wedge e_\mu R^{|\nu\mu|}{}_{yz} \Delta y \Delta z$, where $R^{|\nu\mu|}{}_{yz}$ is the

Riemann tensor with $\nu > \mu$. This quantity is associated with the parallel transport of a vector around the triangular face:

$$-\delta A^\mu = (1/2) R^\mu{}_{\mu yz}(x^\nu + \delta x^\nu)\Delta y \Delta z. \tag{9.12}$$

A rigid body remains at rest if the net force acting on the body is zero, i.e., $\sum_i \mathbf{F}(i) = 0$. However, the body may rotate under these forces. To state a condition of stationarity for this body $\sum_i (P^{(i)} - P) \wedge \mathbf{F}^{(i)} = 0$. The sum over the moments in classical mechanics has an analog within general relativity. Given a four dimensional region of spacetime, Ω, conservation principles demand that there be no creation of the source current $^*\mathbf{T}$ within this spacetime region. Let the four volume Ω have a boundary of three dimensions composed of five tetrahedra. Then an equivalent way to state this conservation principle is that $^*\mathbf{T}$ vanished on the three dimensional volume, $\partial \Omega$, bounding Ω

$$\int_\Omega d^*\mathbf{T} = \int_{\partial\Omega} {}^*\mathbf{T} = \sum_{tetrahedra} d P \wedge \mathbf{R} = 0, \tag{9.13}$$

where $\mathbf{R}$ is the (1, 1) tensor $\mathbf{e}_\nu \wedge \mathbf{e}_\mu R^{|\nu\mu|}{}_{\alpha\beta} dx^\alpha \wedge dx^\beta$, and $d\mathbf{P} = e_\sigma dx^\sigma$

The region Ω is bounded by five tetrahedral blocks. Each solid tetrahedron is topologically equivalent to a three dimensional ball. This construct defines an elementary volume element (EVE). Examine one EVE as the ball $\mathbf{B}^{(3)}$ that is paradoxical using five pieces $A^* \sim_2 \mathbf{B}^{(3)} \sim_3 B^*$. Again A^* and B^* are the orbits of the paradoxically equivalent elements of the group $\mathcal{G}$ with $A^* \cup B^* = \mathbf{B}^{(3)}$ and $A^* \cap B^* = \phi$. The set $\{x_i i = 1, \ldots, 4\}$ are the vertices of the tetrahedron, $T_{\mathbf{B}^{(3)}}$, that are shared by $\mathbf{B}^{(3)}$. Assume that there exists a small 3-disk $D_\epsilon(x_i)$, an ϵ neighborhood, around each x_i on S. The dense set of points in each $D_\epsilon(x_i)$ are given by a countably infinite set of rotations acting on a point $x \in \mathbf{B}^{(3)}$. The list of rotations are a subset $\omega \in \mathcal{G}$ and may be split according to those strings contained in $W(\sigma) \cup \sigma W(\sigma^{-1})$ and those in $W(\tau) \cup \tau W(\tau^{-1})$. This is then the paradoxical decomposition of $D_\epsilon(x_i)$ into the disjoint sets $D_\epsilon(x_i)|_A$ and $D_\epsilon(x_i)|_B$, according to:

$$D_\epsilon(x_i)|_A \subset A^* : D_\epsilon(x_i)|_A = \cup\{g(D_\epsilon(x_i)) : g \in A\}$$

$$D_\epsilon(x_i)|_B \subset B^* : D_\epsilon(x_i)|_B = \cup\{g(D_\epsilon(x_i)) : g \in B\}. \tag{9.14}$$

It is then apparent that there is an equivalence in five pieces according to $D_\epsilon(x_i)|_A \sim_2 D_\epsilon(x_i) \sim_3 D_\epsilon(x_i)|_B$, with $D_\epsilon(x_i)|_A \cup D_\epsilon(x_i)|_B = D_\epsilon(x_i)$. The moments of rotation are a "current" that travels flux tubes that have vertices contained in the disks $D_\epsilon(x_i)$. The edgelinks of the tetrahedra carry a current of stress-energy, where the vetrices are current junctions. Each point $x \in D_\epsilon(x_i)$ then carries an infinitesimal current. The duplication of these ϵ-disks carries with it a duplication of the current passing through a vertex. It is then apparent that the tetrahedron in A^* and B^*, along with their associated moments of rotation, are paradoxically equivalent under $^*T_{A^*} \sim_2 {}^*T_{\mathbf{B}^{(3)}} \sim_3 {}^*T_{B^*}$.

The duplication of a tetrahedron bounding Ω also duplicates the moment of rotation associated with this tetrahedron in $\partial\Omega$. The duplication of one tetrahedron in Ω, $\partial\Omega \to 2\partial\Omega$, introduces a curious anomaly:

$$0 = \int_\Omega d^*\mathbf{T} = \int_{\partial\Omega} {}^*\mathbf{T} \to \int_{\partial\Omega'} {}^*\mathbf{T}, \qquad (9.15a)$$

which duplicates the tetrahedron $tet \to tet'$ with

$$0 = \int_{tetrahedra} \mathrm{d}P \wedge \mathbf{R} \to \mathrm{d}P \wedge \mathbf{R} + \mathrm{d}P \wedge \mathbf{R}|_{duplicated\ tet} = \mathrm{d}P \wedge \mathbf{R}|_{duplicated\ tet}. \quad (9.15b)$$

The duplication of one tetrahedron in $\partial\Omega$ results in the violation of the Bianchi identity that obtains from $\mathbf{d}^*\mathbf{T} = 0$.

The above anomaly is identified with a quantum fluctuation, where $\langle {}^*\delta\mathbf{T}\rangle = 0$ is imposed by the lattices in figure 9.1 and 9.2. The Heisenberg uncertainty principle allows for the temporary violation of energy conservation. This fluctuation in the source, ${}^*\mathbf{T}$, on $\partial\Omega$, and equivalently the creation of the source $d^*\mathbf{T} \neq 0$ in Ω, are described by the lattice in figure 9.3.

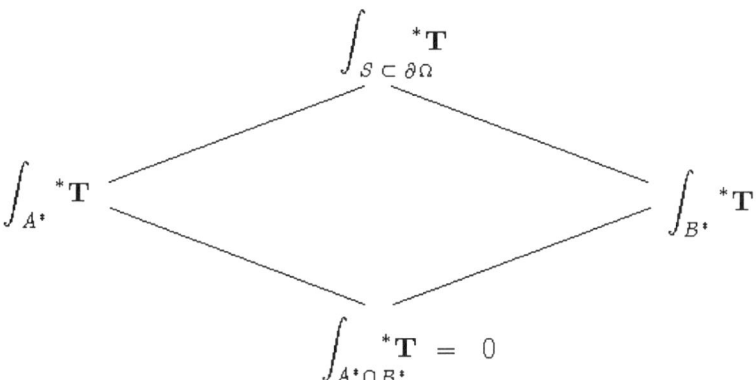

Duplication of moments of rotation.
Figure 9.3

This lattice is isomorphic to the lattice in figure 9.1. The duplicated moments of rotation T_{A^*}, and T_{B^*} are decomposed into T_{a^*}, T_{-a^*} and T_{b^*}, T_{-b^*} according to the orbits given by $W(\sigma) \cup \sigma W(\sigma^{-1})$ and $W(\tau) \cup \tau W(\tau^{-1})$ respectively. So the paradoxical equivalence of tetrahedra is determined by the paradoxical equivalence of the same group $\mathcal{G}$ that gave the quantum lattice in figure 9.2. There is then a lattice for the creation of $\mathrm{d}P \wedge \mathbf{R}$ that decomposes into a lattice isomorphic to the lattice in figure 9.3.

The tie between the creation of moments of rotation and the paradoxical equivalence of balls produces a lattice that is identical to the lattice for a spin $1/2$ system. A complex system of interacting 1/2 spins obeys a Hamiltonian with a spin coupling term $\sum_{i\neq j} \sigma_i \sigma_j$. Similarly all the EVEs are adjectent to one another on the spacelike surface $\Sigma^{(3)}$ so every point on $\Sigma^{(3)}$ is covered by at least on EVE. How these EVEs intersect one another gives the coupling term necessary for the dynamics of a spacelike manifold. A four dimensional manifold is split into a spacelike manifold of three dimensions that evolves through the fourth dimension called time. This formalism by Arnowit, Deser, and Misner (ADM) reproduces all the transform properties of general relativity, but regards three dimensional spacelike hypersurfaces as the dynamical entity that evolves through time [13 ch-21]. An initial spacelike surface must be specified, and from there the dynamical machine runs. An obvious extrapolation is quantum gravity. A wave function acts as a functional over various spacelike surfaces that designates a probability amplitude to each spacelike manifold. The quantum geometrodynamic machine must be fed an initial spacelike manifold, or its momentum conjugate manifold, and then the wave function evolves through a Hilbert space of spacelike manifolds [19].

Take one of these spacelike surfaces and grid it up into tetrahedra that are sufficiently small to represent a locally flat region of the manifold. The spacelike surface is then classically propagated by geometrodynamics to a subsequent spacelike manifold. How the curvature of the spacelike manifold evolves determines the curvature of the subsequent spacelike manifold and the curvature of spacetime between the two surfaces. How these unit blocks, that grid the initial spacelike surface, dilate and rotate to construct the final spacelike governs the change in the curvature of the initial and final spacelike manifold hypersurface. The entire dynamics may be thought in terms of how the elementary volume elements (EVEs) of a manifold reconstruct to give subsequent EVEs on the time developed spacelike manifold. This suggests the evolution of topologically changing quantum spacetime manifolds according the paradoxical equivalences of EVEs.

Given a spacelike manifold gridded into EVEs, what do these paradoxically equivalent EVEs generate? They will produce a collection of EVEs where $\partial\partial = 0$ is in general violated. Since these duplicated EVEs are disconnected from manifold structure then even with $\langle |\delta \mathbf{T}| \rangle = 0$ this has no connection to physics. A manifold may be buried with duplicated EVEs by paradoxical equivalence of EVEs. Consider a spacelike manifold obeying $\partial\partial = 0$. The manifold is propagated into a time developed manifold by geometrodynamics. Now include paradoxical equivalence of EVEs on this manifold to examine the quantum geometrodynamic evolution of this manifold. There is then a collection of lattices for the quantum time development of the spacelike manifold. There are collections of EVEs that are constructed by including EVEs contained in each level of every lattice. Any collection of EVEs, obtained from one level of each lattice, that includes paradoxically duplicated EVEs (from a lower level of a lattice) are such that $\partial\partial = 0$ is determined as an average.

9.3: Recovering Geometry from PreGeometry

There is an unsatisfactory situation that exists. Disconnected EVE has no geometric meaning with respect to a manifold. Given an initial spacelike surface this is a manifold where $\partial\partial = 0$ obtains. This manifold evolves into a collection of EVEs, where most of this collection has no manifold structure. The classically time developed manifold sits within a Hilbert space of states where moments of inertia are being violently created and destroyed. A space as understood by ordinary geometry can become "buried" in a dust of duplicated EVEs, and at a point the manifold is a measure zero aspect of the set of entangled states of the system. In fact it is possible to remove the manifold or space entirely and just have duplicated EVEs. So a mathematics of duplicating volume elements, where $\partial\partial \neq 0$ are deviations from geometric "sense" due to fluctuations. There must be a manner in which all these duplicating volume elements couple to one another to give rise to geometry that on a large scale is familiar.

The history of geometry is one of abstracting new concepts that, when found to be self consistent, give rise to mathematics that generalizes geometry. This extends from Pythagoras' theorem, the first axiomatic or proof theoretic concept of geometry, with modern notions of spaces. Topology is the study of spaces according to their connectedness on chains, manifold surgery which produces complex spaces from simple spaces and how these spaces are decomposed into simplices [21]. These decompositions are studied with sequences that provide algebraic techniques for classifying spaces. A vast leap of geometric understanding has occurred in 25 centuries.

Topology examines spaces according to how they decompose into cells within complexes, where as these simplices or cells may be paradoxically duplicated quirky results, such as the Banach-Tarski paradox, creep into geometry and topology. Spaces due to simplex duplication that violate $\partial\partial = 0$ may only recover well defined geometry if $\langle |\delta^* \mathbf{T}| \rangle = 0$. Hence a space with $\partial^2 = 0$ on average should still exhibit fluctuations due to the replication of EVEs. What would be comforting is the reappearance of geometry with $d^2 = 0$ on a scale larger than the EVEs.

Let there exist two EVEs, with the 3-balls $\mathbf{B}_1^{(3)}$ and $\mathbf{B}_2^{(3)}$, where each defines a spacetime simplex. These balls are then decomposed into orbits by the groups $\mathcal{G}_1$ and $\mathcal{G}_2$. The groups are decomposed according to:

$$\mathcal{G}_1 = \{1\} \cup W_1(\sigma) \cup W_1(\sigma^{-1}) \cup W_1(\tau) \cup W_1(\tau^{-1}) = \mathcal{G}_1(\sigma) \cup \mathcal{G}(\tau)$$
$$\mathcal{G}_2 = \{1\} \cup W_2(\sigma) \cup W_2(\sigma^{-1}) \cup W_2(\mu) \cup W_2(\mu^{-1}) = \mathcal{G}_2(\sigma) \cup \mathcal{G}(\mu). \quad (9.16)$$

There are then paradoxical equivalencies of the lattices:

$$\mathbf{B}_1^{(3)} = A_1^* \cup B_1^*, \ \phi = A_1^* \cap B_1^*$$
$$\mathbf{B}_2^{(3)} = A_2^* \cup B_2^*, \ \phi = A_2^* \cap B_2^* \quad (9.17)$$

If the rotations σ, τ, μ are along orthogonal axis then $\mu = [\sigma, \tau]$. This means that $W(\mu)$ is the set of all symbol strings that begin with σ and τ. The balls are paradoxically equivalent according to the following:

$$A_1^* \sim_2 \mathbf{B}_1^{(3)} \sim_2 B_1^* \qquad A_2^* \sim_2 \mathbf{B}_2^{(3)} \sim_2 B_2^* \quad (9.18)$$

350 *Quantum Fluctuations of Spacetime*

$$A_1^* = \cup\{\mathcal{G}_1(M_1)\} : \mathcal{G}_1 \in W_1(\sigma) \cup \sigma W(\sigma), \; A_2^* = \cup\{\mathcal{G}_2(M_2)\} : \mathcal{G}_2 \in W(\tau) \cup \tau W(\tau)$$
$$B_1^* = \cup\{\mathcal{G}_1(M_1)\} : \mathcal{G}_1 \in W_1(\sigma) \cup \sigma W(\sigma), \; B_2^* = \cup\{\mathcal{G}_2(M_2)\} : \mathcal{G}_2 \in W_1(\tau) \cup \tau W(\tau)$$

It is immediate that all symbol strings in $W_1(\sigma) \cup \sigma W_1(\sigma^{-1})$ have corresponding symbol strings in $W_2(\sigma) \cup \sigma W_2(\sigma^{-1})$. Then any matrix string $s \in W_1(\sigma) \cup \sigma W_1(\sigma^{-1})$ that computes a point $x \in A_1^*$, there is a corresponding string $s' \in W_2(\sigma) \cup \sigma W_2(\sigma^{-1})$ so that ${}^*\mathbf{T}|_{x \in A_1} = {}^*\mathbf{T}|_{x' \in A_2} + \delta^*\mathbf{T}$. This is a semi-isomorphism of the EVEs A_1^* and A_2^*, and connects the fluctuation $\delta^*\mathbf{T}$ to the manifold. Let the matrix τ be obtained from the rotation by σ on μ, $\tau = \sigma\mu\sigma^{-1}, \tau^{-1} = \sigma\mu^{-1}\sigma^{-1}$. Let the symbol strings $\gamma \in W_2(\tau)$, and $\eta \in G \backslash W_2(\tau^{-1})$. Then $\gamma \in \sigma W_2(\sigma)$ and $\eta \in \sigma W_2(\sigma^{-1})$. These strings are paradoxically duplicated in $W_2(\mu) \cup \mu W_2(\mu^{-1})$. Then any two such equivalent strings that give point x, x' in B_1^* and B_2^* will result in a semi-isomorphism ${}^*\mathbf{T}|_{x \in B_1} = {}^*\mathbf{T}|_{x' \in B_2} + \delta^*\mathbf{T}$. $\langle |\delta^*\mathbf{T}| \rangle = 0$ gives stress-energy junction condition and conservation on the vertex ϵ-disks by summing all the infinitesimal stress-energies. The interweaving of computable symbol strings assure us the conservation principle. There is a subtle question concerning symbol strings that are not computable, or that do not compute a point in a duplicated EVE, but still exist in the generic list $W(\rho) \cup \rho W(\rho^{-1})$.

The balls $\mathbf{B}_1^{(3)}$ and $\mathbf{B}_2^{(3)}$ each decompose into the two balls, where the pairs A_1^*, A_2^* and B_1^*, B_2^* have isomorphic orbits. If the stress energy associated with $\mathbf{B}_1^{(3)}$ and $\mathbf{B}_2^{(3)}$ have connected moments of inertia, then the two EVEs are semi-isomorphic. This gives us the first condition necessary for $\partial\partial = 0$. The union of the two balls A_1^* and A_2^* acts so their stress-energies subtract,

$$\int_{A_1^* \cup A_2^*} {}^*\delta\mathbf{T} = \int_{A_1^*} {}^*\mathbf{T} - \int_{A_2^*} {}^*\mathbf{T}. \qquad (9.19)$$

Let the balls $\mathbf{B}_1^{(3)}$ and $\mathbf{B}_2^{(3)}$ be separated so that $\mathbf{B}_1^{(3)} \cap \mathbf{B}_2^{(3)} = \phi$, and so $B_1^* \cap B_2^* = 0$. This infers that $\int_{B_1^* \cap B_2^*} {}^*\mathbf{T} = 0$. Now the stress energies sum to $\delta^*\mathbf{T}$ at every step on the lattice. Let the center of $\mathbf{B}_1^{(3)}$ and $\mathbf{B}_2^{(3)}$ sit on the point x_1 and x_2 on the spacelike manifold. Let $\mathbf{B}_1^{(3)}$ and $\mathbf{B}_2^{(3)}$ be the simplices Ω_1 and Ω_2 centered at x_1 and x_2 then the resulting lattice recovers $d^2 = 0$:

$$\int_{\Omega_1} d^*\mathbf{T} - \int_{\Omega_2} d^*\mathbf{T} = \int_{\partial\Omega_1} {}^*\mathbf{T} - \int_{\partial\Omega_2} {}^*\mathbf{T} = 0$$

$$(duplication) = \int_{\partial\Omega_1} {}^*\mathbf{T} - \int_{\partial\Omega_2} {}^*\mathbf{T} = \mathbf{T}|_{A_1^*} - {}^*\mathbf{T}|_{A_2^*} = \delta^*\mathbf{T}|_{A_1^* \cup A_2^*}. \qquad (9.20)$$

Since $\langle |\delta^*\mathbf{T}|_{A_1^* \cup A_2^*}| \rangle = 0$ the expectation values of the lattice obeys $d^2 = 0$, and this recovers this requirement on the manifold with average zero creation of moments of rotation. Further, this has duplicated EVEs that "connect" to one another and some geometry can now emerge. The pairs of stress-energy $\mathbf{T}_{A^{1*}}$, $\mathbf{T}_{A^{2*}}$, and $\mathbf{T}_{B^{1*}}$, $\mathbf{T}_{B^{2*}}$ are decomposed into $\mathbf{T}_{a^{1*}}$, $\mathbf{T}_{-a^{1*}}$, $\mathbf{T}_{a^{2*}}$, $\mathbf{T}_{-a^{2*}}$ and $\mathbf{T}_{b^{1*}}$, $\mathbf{T}_{-b^{1*}}$, $\mathbf{T}_{b^{2*}}$, $\mathbf{T}_{-b^{2*}}$ respectively according to the orbits given by the subsets of $\mathcal{G}_1$ and $\mathcal{G}_2$. This gives a full quantum lattice for the coupled simplices that is identical to the lattice for two coupled 1/2 spin particles. As differential forms are integrated on chains, boundaries and cocyles they are a measure on them. As a result there are ultimately these paradoxical aspects of an unmeasurable measure lurking behind this formalism. Yet out of the paradoxical confusion ultimately emerges geometry. The process is presented by the lattice in figure 9.4.

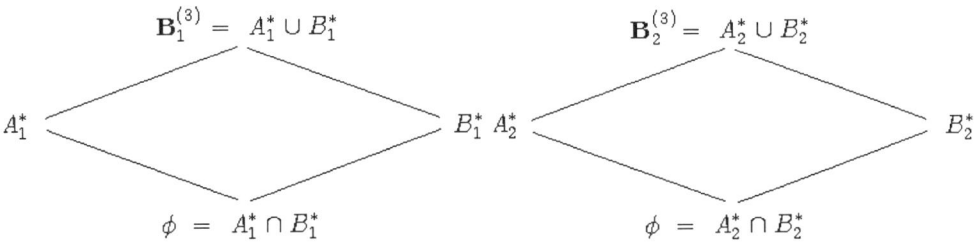

Duplication of two equivalent spheres.
Figure 9.4

The recovery of geometry with duplicated EVEs has curious consequences. Given $\partial\Omega_1$ and $\partial\Omega_2$ as the boundary of Ω_1 and Ω_2 on the manifold, this duplication of tetrahedra attaches handles to the manifold. If the spacelike manifold, containing $\partial\Omega_1$ and $\partial\Omega_2$, has the topology of a ball, then after the paradoxical equivalencies $\partial\Omega_{A^1} \sim_2 \mathbf{B}^{(3)} \sim_3 \partial\Omega_{A^2}$ with $\int_{A^1} {}^*\mathbf{T} - \int_{A^2} {}^*\mathbf{T} = \int_{A_1 \cup A_2} \delta^*\mathbf{T}$ the spacelike manifold has the topology $\Sigma^{(3)}/\mathbf{B}_1^{(3)} \oplus \mathbf{B}_2^{(3)}$, with the two balls identified point by point $\mathbf{B}_1^{(3)} \sim \mathbf{B}_2^{(3)}$.

The physical interpretation appears to be the following. With the uncoupled lattice in figure 9.4 there exists the quantum fluctuation of a moment of rotation. With the coupled lattice in figure 9.5 there is the subtraction of moments of rotation in region Ω_1 and the addition of moments of rotation in region Ω_2 in the spacetime manifold by quantal action. This is given by the semi-isomorphism of EVEs duplicated on the boundaries of the two spacetime simplices Ω_1 and Ω_2. The obvious physical interpretation is quantum tunnelling of mass-energy by the formation of a handle or worm hole throat that bridges the region Ω_1 and Ω_2. The lattice in figure 9.5 reflects this process. The EVEs B_1^* and B_2^* exist in $\partial\Omega_1$ and $\partial\Omega_2$ respectively. The two regions are connected by the handle given by $A_1^* \cup A_2^*$, so that the stress-energy fluctuation is associated with the handle.

These manifolds, generated by paradoxical rotations, represent probable pathways for quantum geometrodynamics. The wave function starts off as an initial hypersurface, or its momentum conjugate surface. From there the quantum wave function may evolve in Hilbert space to include other hypersurfaces. By the above formalism this wave function evolves as a superposition of topologically nonequivalent manifolds governed by the above

352 Quantum Fluctuations of Spacetime

extremal principle. If this quantum wave function evolves so $TrK \rightarrow L_p^{-1}$ on some regions of spacelike hypersurfaces this is catastrophe of gravitational collapse. The topology of the manifold has changed, where the $TrK \sim L_p^{-1}$ condition may only be one path in a superposition of possible geopmetric dynamical evolutions.

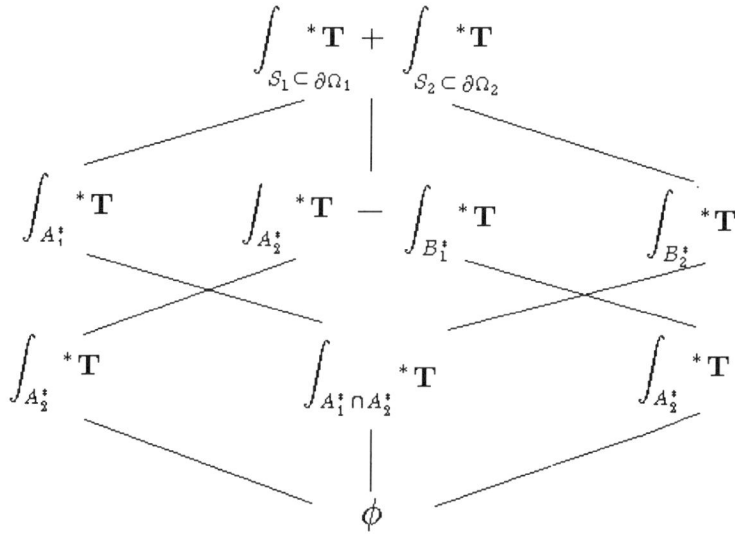

Duplication of moments of rotation for two entangled spheres.
Figure 9.5

There is then a fog of unconnected EVEs undulate by duplication and annihilation. Within this fog exits EVEs duplicated in a coupled manner that preserve $d^2 = 0$. It is here that geometry comes into being and to an end. The curvature of the manifold drives the tiling size on the manifold within the Regge calculus [20]. The tiling of the manifold becomes rapidly smaller nearer to the singularity. As the curvature approaches $1/L_p^2$, where L_p is the Planck length, each Regge tetrahedra approach the size of an EVE. On this scale the Regge tetrahedra fail to provide a manifold structure by paradoxical equivalence, or where the manifold is a measure zero over all EVEs. Only those EVEs that are paradoxically equivalent in a coupled lattice produce a manifold. Those EVEs that couple to form manifolds contribute their moments of rotation to quantum geometrodynamics. The undisciplined EVEs, those that are disconnected from a manifold, contribute to the quantum vacuum a part that may be subtracted away.

This lattice may be very complex. The very early universe was likely to have been a superposition of spacelike manifolds, where most of these manifolds contained many handles or worm holes. A single very large lattice embodied all the quantal propositions that embodied the superposition of manifolds of the early universe. As one pushes the early universe back to the singularity of its origin the universe, as a manifold, evaporates into unconnected EVEs. Manifold structure and geometrodynamics are lost in the dust. This presents a picture of how physical reality percolated out of nothingness, or an approximation to nothingness.

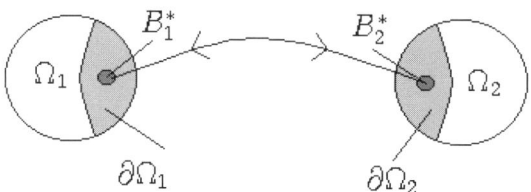

Formation of a handle or wormhole throat that bridges the regions Ω_1 and Ω_2.
Figure 9.6

The EVEs we have discussed come very close to being the elements of pregeometry [MTW]. Indeed they come close to the definition of Leibniz monads or the Buddhist allegory of "Indra's net." Here these elements have no absolute intrinsic structure, but have structure only through their relationships with other such elements. The problem of gravitational collapse brings the need for pregeometry to the forefront. When matter falls through an event horizon a region is later encountered where the curvature diverges. This singularity is the region where geometry ends and all information about manifolds is mixed into entanglements the bury geometric information. Within the context of pregeometry a singularity is a region packed with EVEs that have no geometric relation to one another. This is similar to the problem of Marczewski on how to pack measure zero sets into arbitrarily small volumes. The singularity may be thought of as all conceivable EVEs decomposed into nonmeasurable sets and packed into this small region. Where a manifold contains a singularity its geometry ends. This infers that a manifold has EVEs packed nicely together to form a geometric structure. This obtains for very large curvatures $R > L_p^2$ right up to the singularity. However, at the singularity where $R \sim L_p^2$ the EVEs break free of the constraint of geometry.

What is physically significant to an observer are those EVEs that are duplicated in a coupled lattice. The introduction of a singularity into a time developed spacelike manifold has the affect of rendering this manifold topologically inequivalent to the initial spacelike manifold. Regions that enclose a handle or singularity cannot be deformed smoothly into regions that do not enclose the handle or singularity. The condition on how EVEs form a connected order determines the topology of the surface. For a well behaved geometric surface the EVEs will all pack in so that the face of every EVE connects to another. The prescription for how to connect EVEs determines the topology of a manifold.. If this manifold is a spacelike surface that evolves into a subsequent manifold with a singularity, then the connecting order of these subsequent EVEs will be changed to give manifolds with different topologies. The quantum principle dictates that black holes, singularities, and worm holes should be continuously created and destroyed in the quantum vacuum. Thus a volume of spacetime has on the Planck scale, $10^{-33}cm$, an undulation of spacetime curvature that creates and destroys singularities. This requires that a wave function over spacelike hypersurfaces includes pregeometric EVEs that form no manifold. It is likely connected EVEs are a measure zero set in the entire collection of EVEs on which the wave functional acts. There is then a bewildering collection of EVEs, where only a diminutive set of them form any manifold structure. Only at a scale larger than the Planck length does geometry emerge from the fog of unconnected pregeometric elements. This is manifested by the subtraction of disobedient EVEs from the vacuum, and regarding the geometric residue as having physical significance. On an even larger scale will the interferences between each of the probable spacelike manifolds collapse to reveal the classical spacelike manifold of geometrodynamic evolution.

Ultimately the nature of quantum gravity may transcend the constraints of geometry. Gravitational collapse and the resulting singularities smash geometry into oblivion. The computing machinery of geometrodynamics suffers a major system failure. The geometric theory of gravity laid down by Einstein fails utterly at this point of collapse. This may lead beyond the bounds of space and time into the realm of pregeometry. What is the natural prescription for assigning a functional, or measure, over a manifold in our collections of manifolds generated by paradoxical rotations? The laws of geometrodynamics generates a classical time developed spacelike manifold. Within a variational formalism of general relativity, an evolutionary sequence of spacelike manifolds is the extremal path of geometrodynamic evolution. Extremizing the action integral [13 ch 21],

$$\delta S = \int \{(\pi \cdot \delta g)^{(3)}_{base} - H_{ADM}\delta(trK)\}dV, \qquad (9.21)$$

governs the evolution of the spacelike manifold, but fails near L_p.

At a point p on a spacelike surface, the parallel transport of a normal vector away from p, $p \rightarrow p + \delta p$, obeys $\delta \mathbf{n} = K(\delta p)$. This defines the extrinsic curvature tensor, $K^\mu{}_\nu$, at every point of the spacelike hypersurface. The trace of this quantity defines the fractional change in a volume element per unit of proper time. Proper time is measured by the outward normal at every point on the manifold. TrK is often called the York time [13 ch 21]. The change in the volume of a region on the manifold, whether by classical geometrodynamics or by the paradoxical duplication of EVEs, will change the value of TrK for the region of the next manifold. The change in the volume measure of over a CLEVE will enter the differential geometry of manifolds through this route. The measure of distance between the initial classical spacial manifold and the final manifold is given by the difference between TrK on the initial and final manifolds. As $TrK \rightarrow L_p^{-1}$ at some point x on a manifold, a singularity develops on the manifold. The collapse of a star into a black hole is the astrophysical example of gravitational collapse. If this occurs on all points of the spacelike manifold, then this is a "big crunch" or the complete collapse and disintegration of the manifold. This far more complete gravitational collapse has obvious cosmological implications, and is where geometry dissolves into disconnected EVEs.

For the problem above the undecidable statement of interest is the axiom of choice. The axiom of choice enters the problem of measure theory in the following way: let $S \subset \mathbf{R}^n$ be a subspace, then the outer measure obeys

$$\mu^*(A) = \mu^*(A \cap S) + \mu^*(A \cap -S). \tag{9.22}$$

Either this statement is true or false. The statement contains the clause $S \subseteq \mathbf{R}^n$. If $2^\mathbf{R}$ is well defined, then this clause is not decidable by the diagonalization argument. Adding the axiom of choice well orders S, and allows one to make the above conjecture. This leads to these paradoxical duplications. Negating the axiom of choice produces some difficulties in constructing an adequate set theory [22].

If Kurt Gödel's theorem is used in a description of the physics, it is likely to enter the description of measurement where the observer is described as a part of the system. This is tantamount to a "self-measurement". The standard approach to quantum measurement has been to regard the reduction of states, represented by the operator $\mathcal{R}$, as distinct from the Schrödinger-unitary operator U. The operator $\mathcal{R}$ is a projection operator in the states space, and as such is not an unitary evolution operator. This distinction, $\mathcal{R} \neq \mathcal{R}^{-1}$, is often attributed to the observer's role as a classical system. The reduction of states is the amplification of a quantum amplitude by a classical system. Here the reduction of states changes the objective status of the system. Indeed the Bohr interpretation is to give ψ no objective status until the act of an observation. The observer's objective status does not however change, other than the acquired knowledge of the system. Quantum measurement that arose with the famous Einstein-Podolsky-Rosen paradox results in a decoherence of a quantum system. This results in a coarse grained view of the energy surface of the system, or phase averaging, that reduces the quantum coherence of probabilities to a classical probability. It is possible this self-referential process is what ultimately lies behind the decoherence of noncommutative position and momentum operators in the B^* algebra. Self-referential statements involving the B^* algebra and its exterior calculus would lead to propositions that are "beyond" provability of B^*. The ensuing "chaos" could mean that on a large scale phase information is not tractibly computable.

On another speculative vein this tends to imply that existence is analogous to a vast (infinite?) set of cellular automata. Quantum gravi-logical propositions can be considered as produced by some logic engine, or a network thereof, and that the process of massively paralleled computing, or quantum computing, that lead to emergent structures. Under phase averaging one such structure is likely to be one-loop quantum gravity. Yet on a fine grained scale this also involves propositions that are true for utterly no reason what so ever. This leads to the possibility that the emergence of structures in cosmologies that tunnel out of the vacuum state is similar to the emergence of structure in cellular automata.

9.4: Gödel's Theorem in Quantum Mechanics

The process of quantum measurement is a physical analogue to the Gödel numbering of a system and a recursive feedback of those numbers within the system. The system measures itself, the apparatus and experimenter are not completely distinguished from each other. This has a certain philosophical appeal, but is difficult to apply. However, in recent years developments have emerged that make this a plausible development. Chaitan extended Turing's Halting Problem into a theory of the Halting Probability [23]. Curiously this makes the issue of undecidability "worse" than the situation advanced by Gödel and Turing. Turing illustrated that there cannot exist an algorithm that is able to determine if all algorithms will halt or not. In effect this algorithm has to determine if it can halt itself. Now consider a succession of pure coin tosses that generates a random string of $0's$ and $1's$. In principle a set of all possible such coin toss games will generate all binary symbol strings of algorithm. It is possible to define a probability measure Ω over any such symbol string, where that probability is a number in $[0, 1]$ on whether that algorithm will halt.

This halting probability Ω is maximally uncertain. A general binary symbol string will have no structure what so ever. The longer the symbol string is the probability it has any structure diminishes. One may see this as completely random strings run through a data compression program will not compress. Yet a theory, or axiomatic system is a sort of compression algorithm: enter an abstract proposition into it, turn the crank and if this proposition satisfies the axiomatic system it compresses into the axioms of the system. Yet all completely random N-bit length algorithms, for N large that do not compress require $\sim < 2^N$ axioms of length N to describe them. For all binary symbol strings of length N for N very large, the number of nonrandom strings becomes a vanishingly small measure over this set.

For N finite one may look at the set of all algorithms by brute force to determine if they halt. If one does it contributes a probability $1/2^N$ and so in a finite sense one has

$$\Omega = \sum_{p \text{ halts}} 2^{-p}. \tag{9.23}$$

However, this only works in a finite situation where one examines each binary string one by one. However, for N very large this becomes an intractable process, and equation 9.23 only gives a "weak" estimate for Ω. Equation 9.23 is only an estimate and there is no procedure that indicates how this is converging to the real value for N very large or infinite. As a result Ω is maximally uncomputable.

The implications for quantum mechanics are clear. Quantum mechanics is nonlocal. Assume an atomic decay with zero initial spin results in two spin particles (spin up and down). The two quantum states are in an entanglement. This "quantum two-coin toss" has a curious element to it. Let the two coins end up at opposite ends of the room so the two detectors measure the two spins at the same time. One coin will be heads and the other tails by conservation of angular momentum. This obtains even if the two detectors make the measurements at the same time, or within a time interval shorter than what it takes a light ray to travel across the room. The total quantum wave function is reduced to the two outcomes without a causal signal transmitted between the two detector points. If one detector finds the outcome "heads" then instantly or on an interval faster than light the other coin is automatically tails: a purely acausal process. This curious consequence of quantum mechanics, along with many others, has been experimentally tested. Quantum mechanics is then the ultimate perfect coin toss. Now assume that there must be some superluminal "inner causal" structure that underlies these measurements. If such hidden variables obey a dynamical principle for these wave function reductions on measurement, then a vast set of experiments with 1 and then 2 and ... N outcomes, where for each $n \leq N$ all possible outcomes are recorded, will generate binary strings that encode this dynamics. Instead most of these are completely random. As quantum measurement is a perfect coin toss there is no way to compute an Ω for the binary symbol strings generated. This means that there is no simple set of axioms, or physical postulates, that act as "data compressor" to reduce these data into any type of dynamical principle. Wave function measurement is both nonlocal and completely random, where no algorithmic principle for a dynamics can ever by found.

All this leads to some curious aspects to the theory of quantum mechanics. For each N there are 2^N possible symbol strings. Consequently the set of all possible binary symbol strings is a power set over the number of bits these strings contain. Hence for $N \to \infty$ the cardinality of each experiment is $\aleph_0$, but the cardinality of the set of possible algorithms is $\aleph_1 = 2^{\aleph_0}$. This is the heart of the Continuum Hypothesis [24]. The continuum hypothesis is well known to be consistent and independent from the formal system of Zermelo-Fraenkel set theory with the Axiom of Choice [24]. This structure is regarded as giving the foundations of Mathematics. Define the set of experiments as Γ, the set of outcomes as Γ^∞ and Ω is the probability distribution for this set.

Given a state $|\psi\rangle$ the outcome of a measurement is modelled according to a projector $\hat{P}_a$ so that

$$P(a_i) = \langle \psi | \hat{P}_{a_i} | \psi \rangle. \tag{9.24}$$

For the outcome of a measurement a the distribution of possible outcomes is given by the operator

$$A = \sum_i a_i \hat{P}_{a_i}, \tag{9.25}$$

so that the expectation for outcomes is

$$\epsilon(A) = \sum_i a_i P(a_i) = \sum_i a_i \langle \psi | \hat{P}_{a_i} | \psi \rangle \tag{9.26}$$

Regard the limit of the summation as an integration, then the measure for this integration is over Γ. Now consider the Bohm approach to quantum mechanics. This is equivalent to stating that a probability for the outcome a_i is

$$P(a_i) = \int_{\Sigma_i} P(a) \tag{9.27}$$

where Σ_i is a sample space of all outcomes projected onto the ith state. This means that

$$\epsilon(A) = \sum_i a_i P(a_i) = \sum_i a_i \int_{\Sigma_i} P(a). \tag{9.28}$$

If the Bohm interpretation of quantum mechanics implies some internal causality this would indicate that the integration measure in equation 9.28 is over Γ^∞, which is also equivalent to Γ. This violates the continuum hypothesis $\aleph_1 = 2^{\aleph_0}$ where $\aleph_1 \neq \aleph_0$. If the continuum hypothesis is false in quantum mechanics in Bohm's quantum theory the probability measure is indefinable by the halting probability. Yet as seen in chapter 2 there is little reason to assume that Bohm's quantum mechanics is any different from the more standard quantum mechanics. So evidently the Bohm probability measure cannot be defined over Γ^∞, but rather over $\Gamma \neq \Gamma^\infty$. The Bohm theory does not capture any type of hidden variable causality principle, where in fact any such theory that purports to do so fails due to the indefinability of the halting probability Ω.

Another way to see this is that quantum mechanics involves complex valued wave functions, but where the measured quantities are real valued. What is determined experimentally is completely real valued, but the expectation of those real valued eigenvalues are computed with the use the complex valued structure of quantum waves. However, to expect that real valued measurements can deduce some "inner causality" is tantamount to stating that the axioms of real numbers are able to tell us everything about complex numbers. This is obviously impossible: the axiomatic structure of real numbers say little about imaginary numbers for $arg(z) \neq 0$! The mathematics of real numbers pumps out "dada poetry" on the nature of imaginary numbers.

There are some interesting results in the foundations of mathematics in connection to quantum mechanics and the role of paradoxical equivalent replication of sets. Cohen showed that the Continuum Hypothesis infers the Axiom of Choice. Hence if the Axiom of Choice is false then the Continuum Hypothesis is false. As yet there is no proof of the Continuum Hypothesis, but it has been demonstrated to be consistent with the Zermelo-Fraenkel Set theory construction, but independent of it. It is then curious that the above concept of "no Universal Turing Machines" to determine the halting of all possible strings obtained from all possible quantum experiments infers $\aleph_0 < S(exp) \leq 2^{\aleph_0}$. The continuum hypothesis would infer that the lack of algorithmic constructions to describe wave function measurement is "hardened" by the requirement of the successor to $\aleph_0$ as $\aleph_1$. Without $\aleph_1 = 2^{\aleph_0}$ there is an ambiguity to what is meant by the cardinality of the set of sets of experimental outputs as larger than the total number of experiments. Further, as shown after equation 9.11, $(a^* \cup -a^*) \cap b^* \neq (a^* \cap b^*) \cup (-a^* \cap b^*)$ was derived using the Axiom of Choice. So without the Axiom of Choice, and by implication the falsehood of the Continuum Hypothesis, this derivation is not possible. Hence from a physical point of view it is most natural to regard the Continuum Hypothesis as a true statement in mathematics.

It is often said that the theory of quantum mechanics is as much epistemological as it is ontological. This assessment is true, for the uncomputability of Ω infers that there is nothing of value in hidden variable theory. The wave function can only give us expectations for the measurement of observables that are real valued. The same occurs with the B^* algebra and quantum gravity. It cannot tell us anything dynamical, beyond phase information associated with commutators of the form $[\hat{p}_\mu, \hat{p}_\nu]$. All it can do is to "set the stage" for the understanding of lower energy physics such as induced quantum gravity at the one loop level, or to give estimates for the tunnelling of a cosmology out of the vacuum. It predicts nothing about its own dynamics, for without definable concepts of time, space, energy and momentum there are no dynamics. This can only give estimates under phase averaging that removes noncommutativity between independent position as well as independent momentum operators. It will provide a tool to derive consistent theories of quantum gravity, string, and loop space gravity which might, if we are sufficiently clever, lead to detectable physics. Ultimately this is one motivation for these reviews on some of the current work on string theories and extra large dimensions. This suggests that the foundations of physics are becoming ever more epistemological and less concrete according to a naive objectivism. This is something that we might actually expect: nature at the foundations should become stranger rather than more in line with our older conceptions of things

The connections between Gödel's theorem and quantum mechanics and the paradoxical replication of geometric elements indicates that physical structure may emerge for reasons not deducible. In ordinary quantum mechanics a localized wave packet will spread and adjust itself to the boundary conditions imposed on it. The eigen-numbers associated with that wave packet emerge from the dynamics of the wave as it adjusts to its environment. In a similar way the laws of physics may similarly emerge. Completely random units of Planck mass will through their mutual entanglements with each other will enter into some pattern or structure for reasons established by random fluctuations. The randomness has connections to the halting probability and its ultimate randomness. What is regarded as physical law, as well as possibly mathematical truth, may well be the result of the particular accident that obtained with the quantum tunnelling process that lead to the observable universe.

9.5: The Anthropic Cosmological Principle? — Maybe

The concept of a self-referential cosmology has been boldly, if not rashly, advanced. This then leads directly into the question of whether conscious observership is an aspect of this process whereby the universe "observes itself." The validity of that proposition is contingent on a number of things. The first is that we have a reasonable understanding of consciousness. Currently there is a near paucity of scientific understanding of consciousness. The second is that a future scientific understanding of consciousness would involve the reduction of quantum states and decoherence under a measurement. Various ideas have been advanced for a theory of "quantum consciousness," but these remain unsatisfactory and raise far more questions than the putative answers they suggest. Thirdly it would likely require that consciousness strongly couples reality on a vast range of scales. This could involve the scale of the observable universe to the scale of the cell and bio-molecules and ultimately down to or near the Planck scale itself. Such a complete correlation of all these scales tends to go against the trend whereby physics on hugely different scales only perturb one another in very subtle ways. Note the tiny phase measured in detecting quantum gravity fluctuations with atom-laser physics discussed in Chapter 3. Penrose suggests that consciousness is tied to the reduction of states and quantum fluctuations involving Planck masses. Currently this proposition has not achieved much support either theoretically or experimentally. Wheeler further uses ideas of delayed choice in quantum experiments to advance the proposition that our observations of the universe effectively "bring the universe into being" throughout its history. A brief discussion on these ideas is presented here with the conclusion that little if nothing can be concluded on these matters.

Henry Stapp advances various notions of quantum consciousness. Stapp proposes the idea that the acausal nature of quantum mechanics in measurements is an aspect of consciousness. These ideas suffer from one apparent problem analogous to the Maxwell demon argument for reversing the flow of entropy [LBNL- 46870]. It appears that consciousness is a sort of demon that is able to prevent its own wave function "collapse" as a sort of quantum meta-observer. Yet this tends to put the ultimate observer in a status that is outside physics. In effect the human mind is a wave function that is somehow immune from the phase averaging and decoherence seen in other quantum systems. This despite the fact that the brain is a warm system with plenty of thermal noise to cause decoherence. Stapp further makes the problematic step of delving into the paranormal domain to argue that mental quantum waves are involved with ESP and telekinetic effects. He references Dean Radin in a July 1995 letter to Physics Today. Radin is a paranormal researcher who has been found to have made the classic "file drawer effect" error in his analysis of data where he argues for the reality of paranormal influences.

Penrose's argument for quantum consciousness has one aspect to it that is better than Stapp's, but suffers from another serious problem. Penrose argues that neurons involve a collapse mechanism associated with microtubulins within them. This has the advantage of admitting that the brain is a sort of "Zeno machine" that continually collapses wave functions, which avoids the quantum mysticism of Stapp. However, Penrose argues that the collapse involves fluctuations with Δm equal to the Planck mass. The argument is that the gravitational effects therein will result in a sort of orchestrated collapse mechanism. However, such extreme fluctuations are most often treated as above the renormalization cut-off in lower energy physics which have no measurable influence on such physics. Without some sort of coherence enhancement mechanism, such as squeezed states, this mechanism appears unlikely. Further, in order for this enhancement mechanism to exist in the brain this would probably require a "quantum brain," which gets one back into the problem with Stapp's ideas.

It also has to be mentioned that there is no agreement within the molecular biological and neurophysiology communities on these theories. It also appears that invoking microtubulins as the source of quantum consciousness is little different from Descarte's argument that the pineal gland in the brain was the "seat of the soul." Further there is no biophysical data that suggests that microtubulins are transmission lines for quantum waves. It is known that they are involved with cell shape and structure, are the spinal bodies attached to kinetichores during mitosis and are pathways for kinesin molecules that migrate along them. It is also an observation that most physicists who argue for quantum consciousness have a serious lack of knowledge of biology on the molecular and physiological scales. On the other side there is a complete absence of experimental information on the molecular biology of microtubulins to support any idea of their being quantum transducers or transmission lines.

Wheeler has argued that the universe is a self-excited circuit that requires observership internal to it. The argument harkens back to the idea of wave function collapse and observership. Wheeler's argument employs the delayed choice experiment in quantum mechanics, where a measurement beyond a slit to determine if a photon traversed that slit has the affect of changing the wave function at a time before it reached the slit. This is suggested as a way in which human observership reduces the wave function of the universe to bring the past into existence. Our observations of the distant universe are then argued to be a delayed choice quantum collapse that brings the universe into being. Wheeler further invokes the Gödel's theorem as an underlying mechanism for this self-excited circuit. There is a curious similarity to the capacity for consciousness to invoke the sense of "I" and Gödel's theorem, where as argued above the spontaneous tunnelling of the universe out of the vacuum may involve quantum states that encode their own quantum information.

It certainly cannot be argued that our observations of the universe have brought it from being a pure state wave function into a mixed state in decoherence. The entropy change associated with such would be $\Delta S \simeq 10^{100} Gev/^{o}K$, and it appears difficult to image how our two kilogram brains are able to generate that much entropy! Such an idea appears to be cosmological extension of the Cartesian theater, where the being of the universe is center on the brain and its capacity for mind. That consciousness might play some very subtle role in this matter is difficult to ascertain. Yet, it is likely that the universe existed in essentially the same state currently observed long before we came into existence. From cosmic ray tracks found in rocks, and moon rocks, it is likely that state reduction occurred then, rather than creating a situation where these rocks were in a quantum superposition of a "tracked" and "untracked" state before observed under a microscope.

The matter of the hard Anthropic Cosmological Principle will be impossible to address until we have reasonable scientific theory of consciousness with some experimental data to support it. The hard Anthropic Cosmological Principle argues that the only sort of universe that may exist is one that brings about intelligent life or consciousness. Daniel Dennett has offered an approach to the problem of consciousness in *Consciousness Explained* with a viewpoint that avoids the Descartes theater problem of an infinite regression of "ghosts" involved with observership and consciousness [25]. His heterophenomenology and multiple drafts approach offer up probably the best architecture for addressing the consciousness problem. Consciousness might involve some renormalized version of Gödel's theorem or that consciousness involves some biophysical process we have no concept of at all. The process behind consciousness might well be as outside our mindscape as was quantum mechanics to 19th century physicists pondering the problem of the stability of matter. The ideas offered by Dennett might be the best tools available for grabbing some thread of a principle that today we simply have no conception of. As yet too little is known about consciousness to support the hard ACP.

Barrow and Tipler in their book *The Anthropic Cosmological Principle* [26] discuss mostly the weak Anthropic Cosmological Principle, that argues that the laws of physics must be of such a nature to have brought us into existence. This is analogous to geology and paleo-ontology, where the Earth must have had certain geological conditions and a course of biological evolution to have given rise to Homo sapiens. This is not so much a principle as it is an operating guide line. It is also the logical opposite of the hard Anthropic Cosmological Principle, and so its support by experimental data cannot be used to argue for the hard Anthropic Cosmological Principle.

This is obviously a brief review of quantum consciousness and Anthropic Cosmological Principle. As Stapp writes papers at a great rate and I am not up on his more recent efforts along these lines. My review of his efforts is highly distilled, but captures what I think are the most troubling aspects of his work. Wheeler's writings on this subject have a compelling element to them, as he is a very good writer, but there are several questions that can be easily raised. Wheeler may be right in some very subtle sense, but at this time physics is inadequate and our understanding of consciousness is minimal. It is then difficult to judge whether there is any merit to them. Similar objections may also be raised with other ideas including Penrose's. There are other more fringe physics ideas, from those of Brian Josephson to variants of "hidden variable" theories. These ideas simply cannot be taken very seriously.

So currently it can only be concluded that the hard Anthropic Cosmological Principle is undetermined. Until there is some type of consciousness theory, some connection between consciousness and quantum mechanics, and further some type of experimental tests of these we are simply unable to determine if consciousness is some required aspect of physical existence. If this is found to be in the affirmative in the future this could be the ultimate principle of physics. However, it has to be pointed out that a science of consciousness involves consciousness knowing itself. This could run into issues of the Gödel theorem, so that we might never really know consciousness in its ultimate depth: solving the hard problem. We might at best be able to address the soft problem. Whether a scientific resolution of the soft problem can at all aid in looking at these more cosmological questions is impossible to determine at this time.

References

[1] W. H, Zurek, *Phys. Rev A.* **40**, 4731-4751 (1989).

[2] P. Benioff, *J. Stat. Phys.* **22** 563–591 (1980); *Phys. Rev. Lett.* **48**, 1581-1585 (1982).

[3] M. S. Morris, K. S. Thorne, U. Yurtsever, *Phys. Rev. Let.*, **61**, 1446-1449 (1988).

[4] J. B. Hartle, S. W. Hawking, *Phys. Rev. D*, **28** 2960-2974 (1984).

[5] V. P. Frolov, I. G. Novikov, *Phys Rev. D.*, **42**, 1057-1065 (1990).

[6] A. Ashtekar, *phys. Rev. Let.*,**57**, 2244-2247 (1986).

[7] D. Deutsch, *Proc. Roy. Soc. Lond.* **A 400**, 97–117 (1985); *Proc. Roy. Soc. Lond.* **A 425**, 73–90 (1989).

[8] C. H. Bennett, *Int. J. Theor. Phys.* **21**, 905-940 (1982).

[9] A. M. Turing, *Proc. London Math. Soc.*, **2**, 42, 230-265 (1936).

[10] S. W. Hawking, *Phys. Rev. D*, **46**, 603-611 (1992).

[11] G. S. Boolos, R. C. Jeffery, "Computability and Logic," p 180, Cambridge University Press, Cambridge, London (1980).

[12] K. Gödel, On Formally Undecidable Propositions in ... , *Manstshefte fur Mathematik und Physik*, **38**, 173-198 (1931), english translation in *From Frege to Gödel*, ed J. Heigenoort, Harvard, (1967).

[13] C. W. Misner, K. S. Thorne, J. A. Wheeler, *Gravitation*, Freeman pub. (1973).

[14] S. Wagon, *The Banach Tarski Paradox*, Cambridge (1985)

[15] B. Augenstein, *Int. J. Th. Phys.* **23** 12 (1984).

[16] G. Cantor, *Transfinite Numbers*, (1895), Dover edition (1955).

[17] E. G. Beltrametti, G. Cassinelli. *The Logic of Quantum Mechanics*, **15** "Encyclopedia of Mathematics and its Applications," Addison-Wesley Publishing Company, (1981).

[18] R. Hughes, *Scientific American* **245**, 4 (1981).

[19] J. B. Hartle, S. W. Hawking, *Phys. Rev.* **D 31** 1777 (1983).

[20] T. Regge, *Nuovo Cimento*, **19**, 558-571 (1961).

[21] W. S. Massey, "Singular Homology Theory,"Springer Verlag (1980).

[22] Lerman, *Degrees of Unsolvability*, Springer Verlag (1983).

[23] G. Chaitan, *Complexity* **5**, No. 5 (May /June 2000), pp. 12-21.

[24] P. J. Cohen, *Set Theory and the Continuum Hypothesis*, Benjamin Cumming Publication, Reading Mass. (1966).

[25] D. C. Dennett, Consciousness Explained, Little, Brown, (1991) 1st ed.

[26] J. Barrow, F. J. Tipler, "The Anthropic Cosmological Principle," Oxford Paperbacks (1989).

Index

120-cell, 328–329
24-cell, 19, 310–315, 328–329, 340
   cuboctoheral representation, 315
   rhombic dodecahedral representation, 315
600-cell, 328–329

Accelerated expansion of the universe, 41, 128–135
Accelerated reference frame, 11, 36–38, 87, 111, 288
Acceleration
   Equivalence to gravity "force", 85–87
   Spacetime fluctuations, 7, 11, 36–49, 53–64, 99–110
   Teleparallel connections, 6, 21, 36–38, 52, 217, 286, 294
   Unruh effect, 6, 11, 17, 21, 86–87, 110–114, 116, 146, 188, 243
Action
   Dp-brane, 178, 196–200
      Born-Infeld action, 178, 199
   Particle action
      Euclidean, 168, 221, 225
      Gauge field action, 39, 178–180, 199, 209–210, 213
      Gauge fixing condition, 165–166
   In curved spacetime
   Spacetime action
      ADM action, 34, 38, 211, 354
      Induced quantum gravity, 3, 250 261,358
      Torsional action, 34–39, 201, 288–289
   String action (string world sheet), 142, 200–202, 220
   bc action, 162–165
   $\beta\gamma$ action, 162–165
   Chern-Simons action, 199, 209, 223–224
   Faddeev–Popov ghost, 162, 165–166
   Gauge fixing condition, 165–166

   Gravitational, 34, 210, 260
   Polyakov, 8, 81, 169, 179, 183
AdS/CFT correspondence, 9, 134–135, 294–295, 303, 321
Anomaly
   Ghost number, 163
   Gravitational, 166
   Virasoro, 175
Antisymmetric field tensor
   Gauge field, 39–40, 168, 177–180, 195–200, 209, 213, 302
   Torsional field, 37–39, 42, 154, 168, 289
Atomic physics. 7, 22, 85, 87–92, 101, 114
   Atom-photon interaction, 7, 14–16, 86–92, 101, 114
   Detection of quantum gravity fluctuation, 7, 86–92, 101, 114
   Laser physics, 7, 16–17, 86–87, 93–97, 99, 102–110, 243, 311, 323–324
Atiyah-Singer index, 13, 186, 190, 231–232
Axiom of choice, 274, 341–342, 354–358

B^* Algebra, 258, 268–274, 276, 339
Banach-Tarski "paradox", 341–353
   Duplication of balls, spheres and elements of a CW complex, 341–349
   Pregeometry, 339–353
   Quantum gravity fluctuation, 350–353
   Randomness of physical law, 333–353
bc, 162 165
$\beta\gamma$, 162–165
Bell's theorem, 10, 24, 65, 69–73, 76–81, 305
Berry phase, 16, 32, 87, 91–95, 97
Betti number, 155–158, 169, 235
Black holes
   Black branes, 198
   Classical and semiclassical black holes, 171, 193, 322–323

Coherent states, 149–150, 310–315, 325
D2-branes and D5-branes, 135, 182–183, 193, 197–200, 205-206, 210–211
Hawking radiation, 106, 205–206, 236, 326
Higgs field, 8, 167, 183, 193–198, 202, 206–211, 213–217, 327–328
Holographic principle, 4, 16, 18, 75, 86, 99–109, 125, 134, 139, 160–167, 224–225, 276, 285–290, 307–311, 315
Horizon algebra, 8, 19, 139–141, 154, 160–170, 173, 181, 182, 198, 205–207, 233, 264, 267
Phase structure, 193, 322–328
Singularity, 171–173, 212, 310
Teleportation of states through, 312–316, 322–323, 327
Vacuum structure, 55–57, 211–212, 312–316,
Zero mass black hole, 211–212
Bogoliubov algebra, 48, 63, 133, 147, 156, 181–182, 187–188, 257–258, 262, 311
Bohm's quantum mechanics, 5, 10, 21, 23–33, 35, 39
  Quantum gravity, 35, 39
Born-Infeld action, 178, 199
Bose-Einstein condensate, 17, 86–87, 97–98, 110–114, 153
Bosonic string, 19, 200–201, 297, 313–315
BRST quantization, 18, 52, 162–166, 183, 302
  String field theory, 52

Calabi-Yau compactification, 53, 140, 194, 208, 224, 237–238, 327
Cauchy horizon, 6, 49–58, 63–65, 334–335, 337–338
Central charge, 162–165, 175, 182
  bc, 162–165
  $\beta\gamma$, 162–165
  Virasoro, 175
  Chan-Paton factors, 5, 19, 139, 170, 180, 211
Charge, 12, 40, 49, 299
  BRST, 165–166

Dp-brane charge, 182, 198–200, 205–206, 210
Dyon charge, 179–180, 196
Magnetic monopole, 179, 196–200, 206, 211
Topological charge, 171, 184, 223
Virasoro, 175–176
Chiral gauge field, 134–135, 195, 205, 213, 234, 237, 280
Clifford algebra, 183, 273, 283–284
  Cliff-form, 273
Closed string, 148–149, 237
Closed quantum state, 52, 162, 165, 302
Coherent states of black holes, 311–315, 325
Cohomology
  BRST, 18, 52, 162–166, 183, 302
  Dolbeault, 238
  Elementary volume elements (EVE), 346–349
  Floer, 155–158
  Octonionic, 302–303
  Quantum cohomology, 158, 170
Coleman-Weinberg potential, 214
Compactification, 8, 53, 140, 194–195, 208, 224, 237–238, 327
Conformal field theory, 134–135, 167, 182, 189, 201, 220–222, 225–226, 294–298, 303
  Conformal Killing equation, 167
Cosmic delayed choice experiment, 75, 309
Cosmology, 1–2, 6, 12–13, 34–35, 41, 50–51, 64, 121–135
  Constant
  DeSitter, 35, 64, 131–135, 184, 225
  Robertson-Walker, 121–130
  "sum over all = 0" cosmology, 125–135
  quantum cosmology, 13, 63–64, 133–134, 154, 160, 184, 225–226, 257–261 328, 335–358
  tunnelling amplitude, 13, 63–64, 133–134, 257–258
CP symmetry, 183, 194, 202, 204, 228,
CPT symmetry, 204
Cubic Schrödinger equation, 116

Index 367

Delayed choice experiment, 75, 76–81, 309
Density of states, 140, 149, 220, 312
DeSitter cosmology, 35, 64, 131–135, 184, 225
  AdS/CFT correspondence, 9, 134–135, 294–295, 303, 321
Detection of quantum gravity fluctuations, 4–7, 14–18, 85–125
  Atom-photon interactions, 7, 14–16, 86–92, 101, 114
  Berry phase, 16, 32, 87, 91–95, 97
  Fluctuations of metric and lengths, 4–12, 35, 43–44, 99–109, 131
  Holographic principle, 4, 16, 18, 75, 86, 99–109, 125, 134
  Interferometry, 5, 16, 98–110
  LIGO detection of early universe, 5, 16, 119–125
  Renormalization of $\hbar$ and Planck length, 3, 6–8, 41, 50, 87, 117, 121, 125
Dilaton, 150–151, 180, 220
Dimensional reduction
  Calabi-Yau space, 53, 140, 194, 208, 224, 237–238, 327
  Dp-brane, 206–211, 237–242,
  Extra large dimensions, 193–200, 213–218, 232–233
Dirichlet boundary condition, 139, 169, 199, 234
Donalson theorem, 19, 141, 167–170, 179
  Donaldson polynomial, 167, 170
Dp-branes
  Action, 178, 196–200
  Black hole, 135, 182–183, 193, 197–200, 205–206, 210–211
  Horizon algebra, 170–173, 182, 198, 205–211
  Strings, 198, 205–211, 237–238
Duality Dyon, 196, 206
  Magnetic monopole, 179, 196, 206
  S-duality, 205
  T-duality, 140, 178–180, 199, 202, 205–206, 234–237

Einstein field equation, 39, 46–47, 50, 56, 62–63, 121, 126, 157, 171, 211, 244, 248
Electromagnetic field, 7, 15, 50, 88–95, 103–105
  QED, 88–95, 103–105
Electroweak interaction, 4, 213, 215
Elementary volume element, 346–349
Euclidean action, 168, 221, 225
Euclidean metric and path integral, 43, 75, 167, 169, 173–174, 180, 184, 188, 221, 236
Euler index, 75, 155–158, 166–167
Event horizon
  Black hole, 1–5, 18–19, 86, 100, 106, 111, 139–141, 147–149, 170–174, 181–182, 188, 236, 198, 206, 306, 322
  Cosmic censorship, 286, 334–339
  Cosmological, 128–133, 198
  D2-brane, 181–182, 236
  Fluctuations, 21, 42
  Holographic principle, 139–141
  Horizon algebra, 160–167, 210, 224–225
  Particle horizon, 5, 11, 21, 42, 86, 111, 113–114, 145
  Rotating reference frame, 145, 147–149
  Split SL(2, C) bundle, 289, 171, 184
Exceptional Lie algebra
  E_6, 297–299
  E_8, 8, 19, 169, 234, 299, 310, 315, 321, 328–329
  $E_8 X E_8$, 310
  E_8 Lattice, 315, 321, 328–329
  F_4, 298–299, 310, 314–315
  G_2, 239–242, 297–298, 310, 329
Expectation values
  Lack of in B*, 243–244, 251–258
  Vacuum expectation values (vev), 127, 213, 228, 236
Extra large dimensions, 193–200, 213–218, 232–233

Faddeev-Popov ghost field, 162, 165–166
Field strength, 39–40, 168, 177–180, 195–200, 209, 213, 302
Finsler geometry, 6, 41–42, 51–52, 247–248

Flat spacetime,   14, 51, 88, 157, 188, 238, 311
  Asymptotic regions,   157, 211–212, 252
Fokker-Planck equation,   25–26, 96, 325–326
Floer cohomology,   155–158
Friedman-Robertson-Walker cosmology,   121–130

Galois codes,   292, 304, 317, 328
  Galois equivalence GF(4) of gravity and GF(4) for quantum information,   317, 321
  GF(8) and octonions,   317
  Quantum mechanics,   318–320
Gauge fields
  Chiral gauge boson,   134–135, 195, 205, 213, 234, 237, 280
  Mirror gauge field,   40–41, 193, 214, 242, 304
  Moduli space,   8, 19, 141, 167–174, 176, 180, 182–191
  Octonionic,   302–304
  Supersymmetric,   159, 207–210, 232–233
  Vector gauge boson,   195–196
  Virasoro realizations,   176–178
    Weighted projective space,   174–175
Gauge transformation,   178, 185–187, 195, 246,
Gaussian second fundamental form,   34, 51, 129–130, 287
Generalized uncertainty principle,   6, 12, 21, 47–48, 114, 156–157, 243–245, 251–258
Ghosts:see Faddeev-Popov ghosts
Gödel-Turing theorem's,   80, 274, 334–339, 354–359
  Banach-Tarski "paradox",   341–353
  Gödel numbering,   230, 355–358
  Incompleteness in cosmic quantum computer,   334–339
  Turing machine,   80, 336, 358
Goldstone boson,   167, 197
Graded Lie algebra,   159, 207
Grand unification,   123, 179, 294–295
Grassmann spaces,   73–75, 268–272

Grassmann variable,   162–164, 200, 300
Gravitational coupling,   3, 194–200, 206, 260, 290
  4 and 5 dimensions,   194–200, 206, 210–213
  acceleration,   17, 113–114, 147–149, 157, 217, 289, 325
  Dp-brane,   198–200, 202, 206, 210–213
  Fluctuations in an atom,   7, 37–41, 43, 100
  Gravity force,   7, 17, 85–86, 113, 307
  Teleparallel connections,   38, 217, 286–289
  Unruh effect,   17, 86, 108–112, 147–149
Gravitization of the quantum,   304–309

Hagedorn temperature,   8, 140, 149, 207, 237, 290
  Fourth law of thermodynamics,   140
Hamiltonian
  ADM Hamiltonian constraint,   35, 43–44, 57, 130, 251, 255–256, 286, 335, 354
  Coupling between atomic states and linearized graviton,   15, 88–91
  Quantum fluctuations,   37–39
  String,   142
  Vector field,   249–251
  Wheeler DeWitt,   35, 43, 57, 286, 335, 354
Han-Nambu quark model,   40
Heterotic string,   40, 234–238, 242, 299, 315
Higgs field
  Black holes,   8, 167, 183, 193–194, 207, 211–212, 214–215, 327–328
  Electroweak theory,   193, 213–214
  Goldstone boson,   167, 197
  Gravity,   206–207, 212, 327–328
Holographic principle
  Experimental tests,   4, 7, 99–110
  Fluctuations of length and metric,   7, 11, 36–49, 53–64, 99–110, 114, 130–131
  Hopf fibration,   285, 289–290
  Horizon algebra,   8, 139–141, 160–167, 198, 285–290

Octonions, 290
Holomorphic fields, 177–180, 238
Holonomy, 32, 183–184, 187–188
    Detecting quantum gravity, 16, 87, 92–97
    SU(3) subgroup of G2, 238–239
Hopf fipration, 284–285, 289–292
    Octonions, 284, 290–292, 303–304
Horizontal bundle, 51–52
Horizon algebra, 8, 139–141, 160–167, 198, 285–290

Icosahedron, 329
Induced quantum gravity, 3, 259–261, 358
    One loop quantum gravity, 259–260
Inflation, 17, 41, 64, 119–129, 200,
Instanton, 134, 167–168, 183–184, 188, 199, 206, 223–225, 233–237, 281
Interaction
    Atom-graviton, 7, 14–16, 86–92, 101, 114
    Dp-brane and string, 81, 182, 207, 236–242

Jacobi identity, 163, 279
Jet bundle, 246–248, 263
Join operation, 269–273
Jordan algebra, 67, 277, 284, 295–299, 303, 310, 315

Kac-Moody algebra, 19, 177–178
Kähler geometry, 183, 222–225, 238, 251, 271
Kaluza-Klein theory, 40–41, 194
Killing vectors, 21, 140–141, 177, 311
    Cosmological, 129, 328
    Killing tensor, 38–39
    Quantum gravity fluctuation, 21, 38–39, 43–45
    Riemann-Roch theorem, 167–168
    Rotating coordinate system, 144–146, 149
Kepler conjecture, 314
Knot theory, 14,
    Fields and horizon structure, 209–210, 222–225

Landau-Ginsburg potentials, 193, 206, 326

Langevin equation, 26, 96, 325
Lattice
    Compactified gauge, 160, 194–196, 228–230
    E_8 Lattice, 8, 19, 315, 321, 329
    Λ^{16} Lattice, 321
    Orthomodular, 8, 265, 267, 268–274
Length fluctuations, 4–12, 35, 43–44, 99–109, 131
LIGO (Light Interferometric Gravitational Observatory), 4, 17, 124–125
Lorentz algebra and boosts, 18, 65–75, 111–112, 114, 139, 143, 159, 161–164, 184, 189, 337
    Nonlinear Lorentz transformations, 149–154

M-theory, 135, 198–200, 234–242
Manifold
    Diffeomorphic, 34, 51, 169, 185–186, 220, 243, 248, 263, 286
    Fake, 169
Meet operation ∧, 269–273
Metric
    Black hole, 49, 238, 305
    Supergravity, 210–212, 238
    String, 201
    Weighted projective space, 177
Möbius function, 204, 230
Modularity, 188–190
Moduli space
    Conformal Killing equation, 167–168
    Donaldson polynomial, 160 170
    Holographic principle, 171–173, 184–188, 266–268
    Projective cones, 169, 173–174, 266–268
Moment of rotation, 345–352
Moose, 195

Neumann boundary condition, 237
Nonassociative algebra, 8, 274, 276–281, 91–294, 297, 299–303
    Co-associative algebra, 292–295
    Nonassociative quantum gravity, 308–310
Noncommutative geometry, 251–254, 308

Nonlocality, 10–11, 22, 28–29, 65–73
Nonlocalizability of momentum and energy
    Generalized uncertainty principle, 251–258, 339
Normed division algebra, 19, 67, 73, 276–278, 284, 291
NP complete, 76–81

Octahedron, 295–296, 314
Octonion
    24-cell, 19, 310–315, 328–329, 340
    Cayley form, 279–280
    Cayley numbers, 75, 273, 284, 290–292
    Cayley plane, 310, 315
    Cohomology of fermions, 299–301
    Exceptional groups
        E_6, 297–299
        E_8, 8, 19, 169, 234, 299, 310, 315, 321, 328–329
        $E_8 X E_8$, 310
        E_8 Lattice, 315, 321, 328–329
        F_4, 298–299, 310, 314–315
        G_2, 239–242, 297–298, 310, 329
        Fano plane, 277–278, 281, 291–292, 295–297
    Galois code GF(8), 317
    Gauge fields, 286, 295, 297–298, 302–304
    Hopf fibration, 239, 280–282, 284, 289–292, 303–304
    Mathieu group M_{24}, 320
    Nonassociative algebra, 8, 274, 276–281, 291–294, 297, 299–303
    Co-associative algebra, 292–295
    Octahedron, 295–296, 314
    Polyhedral representation, 295–297, 314–315, 328–329
    Quantum fluctuations, 236–242, 286–292
    Quantum gravity, 236–239, 286–292
    Quaternionic triplets, 276, 282–284, 291–294, 296, 299
    Supersymmetric, 301–303, 314–315, 321–322
One loop quantum gravity, 259–260

Open string theory, 11, 134, 142–149, 180, 234, 236–237
Operator
    B* algebra, 268–273
    Dirac, 231, 286, 300
    Nonassociative, 278–281, 290, 294, 297–299, 302
    Noncommutative geometric, 308
    Octonionic, 274, 276–281, 291–294, 297, 299–303
    Quantum gravity fluctuations, 238, 285–286
    Quaternionic, 66, 73, 183, 286–287, 299–301, 303, 308, 321
    Supersymmetric, 163–166, 182–183, 202–204, 301–303, 314–315, 321–322
Operator product expansion, 176
Orbifold, 194–195, 234, 237–238, 242
Orientifold, 194–195, 237–238

P complete, 76–81
Paradoxical equivalence, 341–348
Parity
    CP violations, 183, 194, 202, 204, 228
    Parity bit error, 316–317, 320
Partition functions, 33, 43, 122, 139–140, 182–183, 204, 220–222, 230, 259, 312, 326
Path integral
    ADM quantum gravity, 37–38, 287–289, 334
    Bohm's quantum mechanics, 31–32
    Euclidean, 167, 169, 182, 203, 221, 230
    Faddeev-Popov, 165
    Quaterionic quantum gravity, 287–289
    Superlagrangian, 207–209
    Topological, 165–166, 170, 221–224, 229–232
Peano's multilinear geometry, 268–273
Physical state (BRST quantization), 18, 52, 162–166, 183, 302
Planck scale
    Planck energy and mass, 3, 40–41, 50, 85, 121, 149, 217, 225, 235, 305, 322

Planck length, 1–8, 18, 41, 50, 58, 107, 132, 141, 149, 194, 211, 307, 314, 340, 352
Poincaré conjecture, 218–223
Poisson bracket, 249–251, 254, 320
  Bohm's quantum mechanics, 26–30,
  Gravitation, 35, 44, 251, 254, 320
Polyakov action, 169
Post-Newtonian relativity, 14–15, 88–92
  Post Newtonian quantum gravity, 15, 88–92
Projective space constructs
  Cones, 169, 173–174, 266–268
  Heavenly sphere, 65, 67–68, 72–75
  Null cones, 73
  Weighted projective space, 174–176
Propagator, 42, 148, 160, 276, 313
PSL(2, 7), 295–296, 310
  Automorphism of Fano plane, 310
PSL(2, C), 66–67, 189
PSL(2, Q), 66, 289
PSL(2, O), 66
PSO(3, 1), 67

Quantum cloning, 172
Quantum cohomology
  BRST, 18, 52, 162–166, 170, 183, 302
  Floer, 155–158
  Octonionic, 302–303
Quantum computer, 80–81, 139, 314, 317, 333–317
Quantum fluctuations
  Bohmian
    Mirror gauge field, 40–41, 193, 214, 242, 304
  Octonions, 236–242, 286–292
  Spacetime fluctuations, 4–12, 35, 43–44, 99–109, 131
  Squeezed states, 41, 49–51, 62–64
Quantum group, 208–210, 223–224
Quantum noise, 95–96, 109–110, 317
Quantum teleportation, 76–81, 314, 316, 322–323, 327, 340
Quantization
  B*, 258, 268–274, 276, 339
  BRST, 18, 52, 162–166, 183, 302
  Canonical, 35–39, 130, 333
Quantum gravity
  Detection, 7, 86–92, 101, 114
  Fluctuations, 7, 11, 36–49, 53–64, 99–110
  Octonions, 236–239, 286–292
  Quaternionic, 287–289
Quarks
  Co-associative representation, 295, 304
  U(5) and gravitation, 40
Quaternions
  Octonionic triplets, 276, 282–284, 291–294, 296, 299
  Qravitation, 287–289
Quintenssense, 41, 53, 127

Recursive set or code, 80, 329, 338, 355
Recursively enumerable, 333, 338
Reissnor-Nordstrom metric, 49, 53
Regge calculus, 352
Regge trajectory, 148
Regularization, 232, 259–260, 329
Renormalization of $\hbar$ and Planck length, 3, 6–8, 41, 50, 87, 117, 121, 125
Ricci curvature, 47, 157, 181, 211, 218–222
Ricci flow, 218–225
  Quantum groups, 223–224
Riemann curvature, 51, 102, 318, 320, 346
Riemann-Roch theorem, 141, 167, 182
Riemann sphere, 67–69, 189
Riemann zeta function, 230
Roots of Lie groups and classes, 231, 239–241, 310, 314–315, 329

S-duality, 205, 289
Sen connection, 42
SETI cosmic delayed choice experiment, 309
Skein relation, 209, 224
SL(2, 7), 295–296
  Coset SL(2, 7)/{4, 3, 2}, 297
SL(2, C), 66, 184, 187–189
SL(2, H) = SL(2, Q), 66, 281, 289, 292
SL(2, O), 66, 291
SL(2, R), 66, 176, 181

SL(2, Z), 190
SO(2, 1), 18, 66, 139, 161–163, 169–173, 198, 266
SO(3), 289, 303, 341–343
SO(3, 1), 34, 65–67, 169, 184, 196, 289, 303
SO(3, 2), 34, 134
SO(4), 47, 75, 89, 130, 147
SO(4, 1), 134, 196, 205, 273, 278, 280, 303
SO(5, 1), 66, 73–75, 205,
SO(6), 303
SO(7), 280, 282, 303
SO(8), 278–282, 284
SO(32), 234
SO(n), 180, 224,
Spin(4), 273, 321
Spin(5), 134, 294, 303
  Spin(2, 3), 278
Spin(6), 134, 294, 303
Spin(7), 239, 297
Spin(8), 239, 284, 295–296, 303, 317, 321
Spin(9), 310, 315
Spin(n), 284
Splitting of bundle, 170–173, 184, 205
Spurious state, 213
Squeezed state, 2, 21, 41, 48–51, 62–64, 151, 182, 197, 217–218, 243, 256–258, 301–302, 311
  Decoherence, 105–106, 114–116, 258
  Laser physics, 94–95, 105–06
  Quantum tunneling of universe from vacuum, 13, 124, 131–134, 258, 259–260
Standard model (GSW), 3, 193, 207, 213, 295, 304, 327
Stimulated emission of radiation, 14, 93–95
  Coherent states of strings and black holes, 148–149, 311–313
  Synchrotron radiation, 17, 117
String coupling, 194–200, 202, 206, 210–213
String length scale, 200–202, 205, 276, 308, 340
String and brane tension, 54, 145, 178–180, 200, 206

String types, 205
"Sum over all = 0" cosmology, 125–135
SU(2), 40, 75, 89, 176, 183–187, 205, 278, 286, 289, 295, 302–304, 327
SU(3), 238–242, 295, 298, 299, 304, 327
SU(3) × SU(2) × U(1), 295, 304, 327
SU(5), 40, 85,
SU(n), 41, 170, 193–196, 224, 232
Supersymmetry
  Graded algebra, 158–159, 162–163, 183, 207, 263–265, 302
  Holographic principle, 160–167
  N = 8 supersymmetry, 75, 274, 320–321
  Octonionic, 301–303, 314–315, 321–322
  Superstrings, 134–135, 178–180, 205–206, 233–242, 310, 315–317
  Twisted supersymmetry, 228–232
Symplectic geometry, 24–30, 249–250, 254–256

T-duality, 140, 202, 205–206, 234, 236–238
  Dp-branes, 178–180, 199, 205–206, 236–238
  Dyon, 196
Target map, 5, 169, 172, 201, 315
Teleparallel transport
Temperature, 12, 17, 50–51, 113–116, 326
  Black hole, 113, 139
  Hagedorn, 8, 140, 149, 207, 237, 290
  Rotating string, 143–147,
  Torsional, 41–48, 52
  Unruh, 11, 17, 41–48, 52, 110–114
Tetrad, 130, 153, 161–163, 168
Thermal noise, 95–96, 98,
Time operator, 146
Topological, 18, 52, 162–166, 183, 155–158, 302
  Torsional, 294
Torsion, 34–39, 42, 54,168, 201, 288–289
Transfinite numbers, 342
Tunnelling amplitude for a cosmology, 132–134, 227–262
Turing machine, 80, 336–338,
  Universal turing machine, 338, 358
Twisted supersymmetry, 228–232

Type I string, 178, 234–236, 237, 242
Type IIA string, 134–135, 198, 205–206, 234–236, 237, 242
Type IIB string, 134–135, 198, 205–206, 234–236, 237, 242

U(4), 73–74, 294–295, 303–305
U(5), 40–41, 73–74
U(n), 41, 73
Unitarity, 66, 148, 186, 193, 313, 336
Unitary equivalence, 63
Unitary transformation, 28, 63, 78–79, 105, 124, 174, 186, 256–258, 281, 313, 322, 336

Vacuum expectation, 127, 213, 228, 236
Venziano amplitudes, 149
Vertex operator, 148, 313
 Coherent states method, 148, 313
Vertical bundle, 52, 248, 251, 263

Virasoro algebra, 175–176
 Weighted projective space, 174–176, 208
Virtual gravitons, 15, 88

Weights of algebras, 178, 240–241, 295, 304
Weighted projective space: see Virasoro algebra Wheeler-DeWitt equation, 35, 43, 57, 286, 335, 354
Wilson lines and loops, 187, 213, 219, 222–225,
Witten index, 159, 203–204, 228–232
Wormhole, 6, 12–13, 49–64, 213, 219, 221–215, 244–245, 263, 266, 334–339
Wrapped Dp-brane, 199, 206, 213, 232–233, 237

Yukawa lagrangian, 202, 228

Zariski topology, 184–188